CHEMICAL PROCESS SYNTHESIS AND ENGINEERING DESIGN

Chemical Process Synthesis and Engineering Design

ANIL KUMAR
Indian Institute of Technology, Kanpur

McGRAW-Hill BOOK COMPANY
New Delhi New York St Louis San Francisco Auckland Bogotà Guatemala
Hamburg Lisbon London Madrid Mexico Montreal Panama Paris San Juan
São Paulo Singapore Sydney Tokyo Toronto

Editorial Manager: Suresh Gopal
Editorial Supervisor: Usha Tankha
Production Supervisor: M S Phogat

Library of Congress Cataloging in Publication Data

Kumar, Anil.
Chemical process synthesis and engineering design.
Includes bibliographical references and index.
1. Chemical plants--Design and construction.
I. Title.
TP 155.5.K78 1982 660.2'8 82-15178
ISBN 0-07-096470-X AACR2

Published by Tata McGraw-Hill Publishing Company Limited
12/4 Asaf Ali Road, New Delhi 110 002 and
printed at Raj Bandhu Industrial Co.,
C-61 Mayapuri Industrial Area, New Delhi-110064

To my father
who teaches me not to give up

To my mother
who gives the needed strength

The author gratefully acknowledges the financial support received from the Quality Improvement Programme at the Indian Institute of Technology, Kanpur for the preparation of the manuscript.

The author is also grateful to various publishers and societies for giving permission to use their copyright material.

Preface

A system approach to plant design has been adopted in this book which has immediate value as a textbook for graduate students of chemical design engineering, as well as a reference book for the practising engineer. An effort has been made to reduce the design into a stepwise procedure at the end of pertinent sections. Several problems have been given at the end of each chapter which are at times complementary to the text.

An attempt has been made to arrange the text in a sequence, which, in my opinion, a designer would normally require. The first step in chemical plant design consists of finding an optimal steady state flow sheet (Chapters 1 to 3). Actual plant operation, however, is dynamic in nature. As a result, at the design level, a process flow sheet must be synthesized (Chapter 4), safety factors must be considered (Chapter 5) and trouble-shooting analysis must be carried out (Chapter 6). At the operational level, after the plant has been commissioned, it must be monitored (Chapter 7) and maintained (Chapter 8). Since the decision to produce chemicals is always based on financial considerations, a book on plant design can never be complete without an adequate discussion on industrial economics (Chapters 9 and 10).

In 1973, I was requested by Prof. Arvind Kudchadkar (then our head of department and now with I.I.T. Bombay) to streamline the under-graduate design courses, but I required a very long discussion with Prof. M. M. Sharma, University Department of Chemical Technology, Bombay, before I could face the class with any confidence. I express my gratitude to my friend and colleague Prof. Santosh K. Gupta for providing the necessary motivation: I started work on this manuscript after he finished his "On Momentum Transfer Operations". He gave several excellent suggestions after reading the manuscript and these have been incorporated in the final version. It is impossible to forget the encouragement I received from Dr. D. N. Saraf who allowed me to experiment with design courses here in spite of disastrous initial results. Lastly I wish to thank Mr. D. S. Panesar for his efforts in converting rough and hand drawn drawings into presentable ink diagrams and Mr. U. S. Misra and T. R. Gupta for accurate typing.

To expect our children, Chetna and Bhawani, to sit quietly even for a moment is to expect too much. My wife, Renu, worked hard at keeping them away from me as I wrote this book. Coincidentally, she was expecting my second son when I started my work on this book and it was difficult to convince her that there was no correlation between the two.

ANIL KUMAR

Contents

1 Synthesis of Steady State Flow Sheet

1.1 INTRODUCTION

One needs to utilize tools of engineering to produce conditions different from the natural order. The necessity to produce a new condition is termed as the 'need' which ranges from creating a shopping centre to manufacturing consumer goods and chemicals.

Once a need (or goal) has been identified, different routes to satisfy it are examined. In engineering design, these constitute different alternatives of attaining the goal. As an example, if it is required to establish a passage across a river, one can either maintain a boat or ship service between the two sides or build a bridge across it. The necessity to move the traffic is the need (or goal) whereas establishing a boat service and a bridge are two alternatives of fulfilling it.

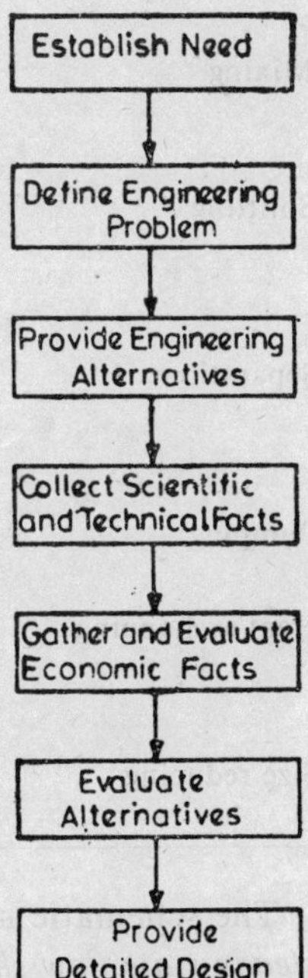

Fig. 1.1 Steps in complete engineering design.

Scientific, technical and economic facts are next collected on each feasible alternative. As an example, the information which has to be collected for building a bridge are the following: a survey of the area where the bridge is to be located (technical), the kind of bridge to be made (technical), estimation of the total load on it (technical), load bearing capacity of the material of construction (scientific) and the economy of bridge building (economic). A similar set of information is found for all alternatives and the superiority of one over the other is decided by economics. It may be appreciated that all these steps constitute the designing process and all are of equal importance. These have been shown as a flow diagram in Fig. 1.1.

In this book, the need is assumed to exist and the discussion is limited to the need of producing chemicals. To give an example in the chemical industry, the need to produce phenol in large quantities is considered. As it is not naturally available, it has to be synthesized and the different routes are shown in Fig. 1.2[1].

In Fig. 1.2, four routes have been shown, two of which are based on using benzene as the raw material while for the third and fourth, cumene and toluene are the starting points respectively. As the chemical reactions for each route are different, the processes conditions (i.e., pressure, temperature, etc. of reactors) are different; therefore the equipment which is utilized to attain these also differs. After the reactions have been carried out, the outgoing streams normally consist of unreacted raw materials and the by-products along with phenol. A separation of the mixture is required to obtain the pure product.

A chemical plant consists of an arrangement of equipment carrying out unit operations and reactors processing the raw materials to obtain the desired product. There are limited types of unit operations and their classification is given in Table 1.1[1]. There can be more than one arrangement to attain the same goal but there is only one which gives the best economy.

Table 1.1 Classification of Unit Operations

Unit operation	*Examples of Equipment*	*Symbols*
Mixing	A vessel, humidifier	
Splitting	Storage tank, T-joints for dividing streams	
Separation	Distillation columns, extractors, filter press, etc.	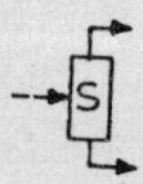
Pumping	Pumps, compressors etc.	
Heat exchanging	Condensers, exchangers, evaporators, boilers, furnaces etc.	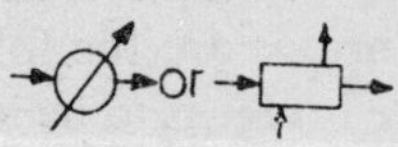
Size reducing	Grinders etc.	

The schematic representation of unit operations and reactors is called *steady state flow sheet*. A systematic procedure of its synthesis has been presented in this chapter.

There are several kinds of flow sheets besides the one mentioned above. One comes across them at various design levels as discussed in Sec. 1.2.

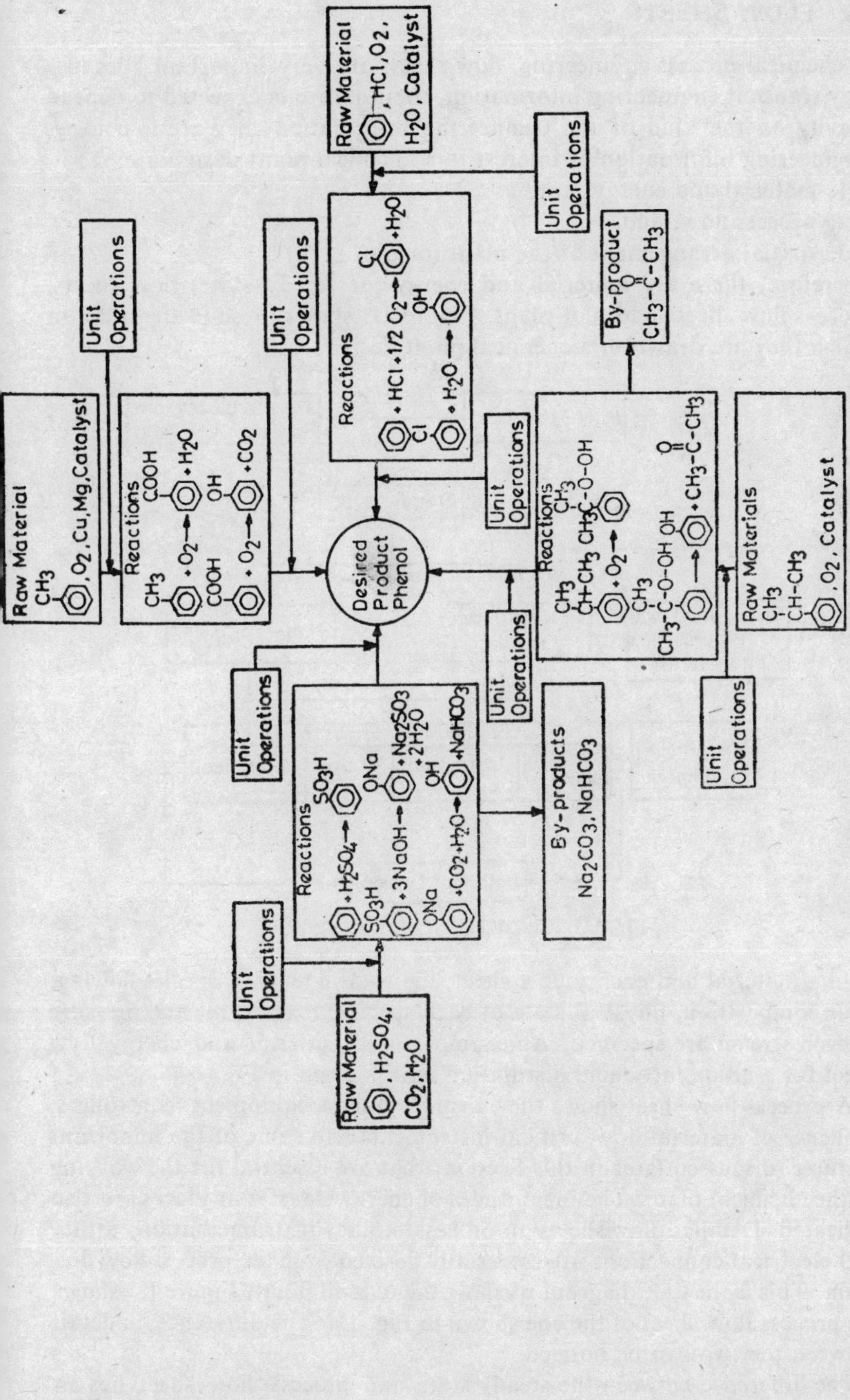

Fig. 1.2 Different process routes for production of phenol.

1.2 FLOW SHEETS

In chemical process engineering, flow sheets are very important. Because they transmit engineering information, their nature is expected to depend heavily on the kind of the engineering information they are to convey. Engineering information of interest in a chemical plant design are:

1. material and energy flow,
2. process flow, and
3. spatial arrangement of the plant (or plot plan).

Therefore, there are material and energy (or steady state) flow sheets, process flow sheets and plot plans. Figure 1.3 shows these in the order in which they are drawn in a chemical plant design.

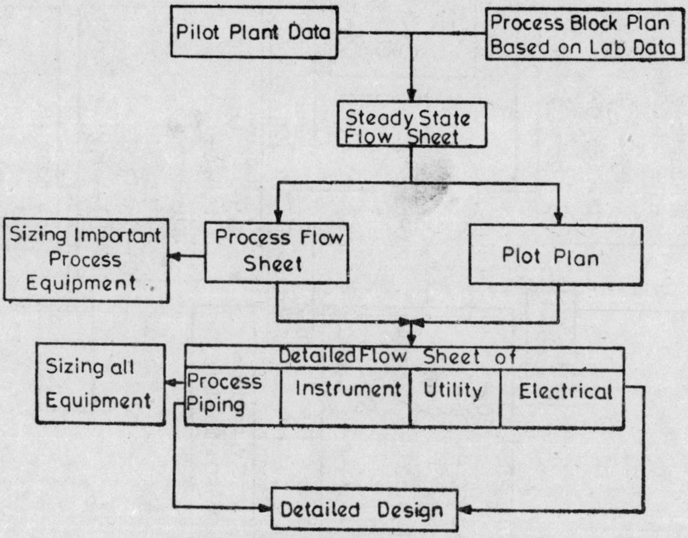

Fig. 1.3 Classification of flow sheets.

In a material and energy flow sheet, the total flow rate, species flowing, their composition, physical state of each species, temperatue and pressure of each stream are specified. An example of a material and energy flow sheet for a crude fatty acid distillation unit is given in Fig. 1.4.

A process flow sheet shows the essential process equipment, controllers, sequence of material flow, critical instruments and some of the important features (discussed later in this Section) that are essential for the working of the chemical plant. The magnitudes of energy transfer at places are also indicated. Detailed flow sheets on process piping, instrumentation, utility and electrical connections are essentially derived from the process flow diagram. This is the first diagram to show the overall plant. Figure 1.5 shows the process flow sheet of the one shown in Fig. 1.4. The difference in detail between the two can be noticed.

The difference between the steady state and process flow sheet lies in the fact that in drawing the latter, the dynamic nature of all process

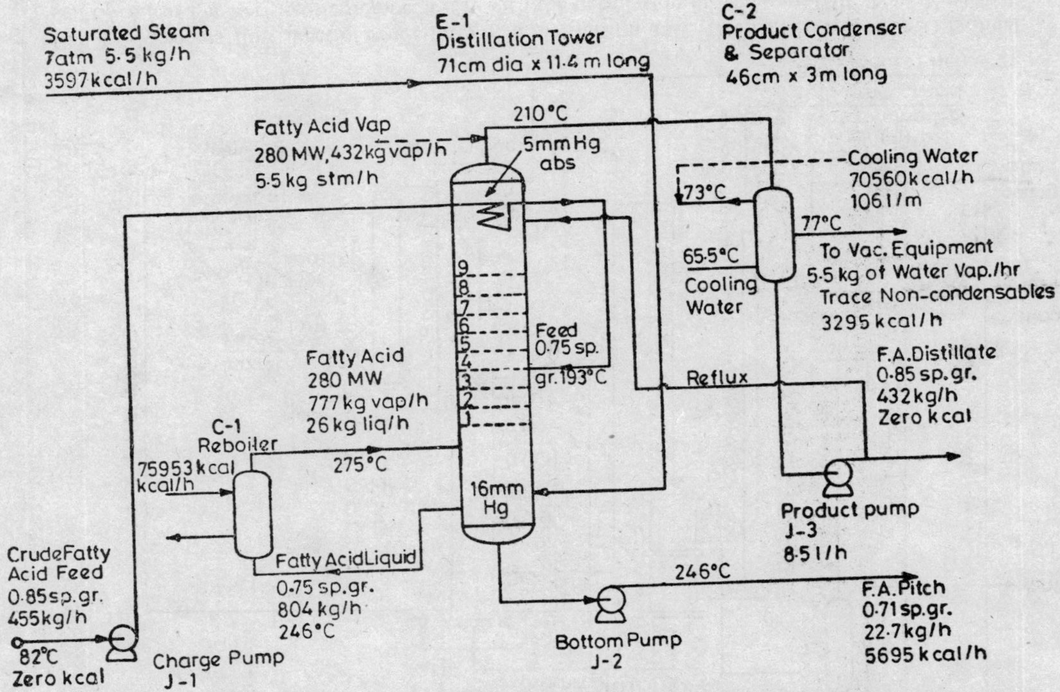

Fig. 1.4 Steady state flow sheet of a fatty acid distillation unit.[2] (Reprinted from J. P. O'Donnel, *Chemical Engineering*, September 1957, with the permission of McGraw-Hill, New York.)

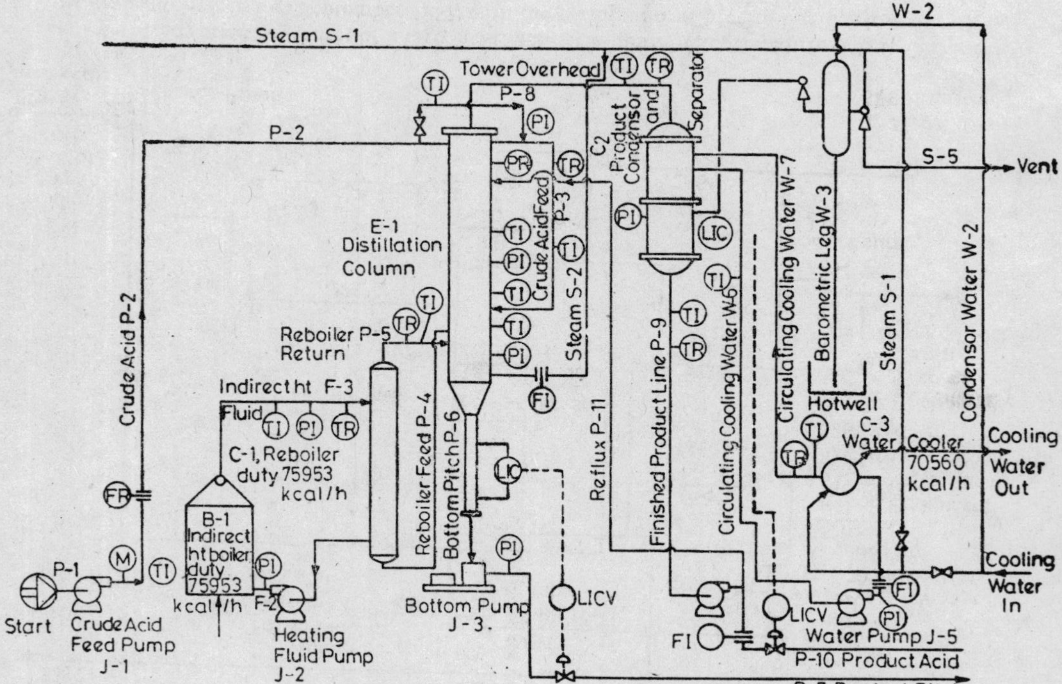

Fig. 1.5 Process flow sheet of crude fatty acid distillation unit.[2] (Reprinted from J. P. O'Donnel, *Chemical Engineering*, September 1957, with the permission of McGraw-Hill, New York.)

equipment must be considered. To illustrate this, a simpler steady state flow sheet for diethyl ether from ethanol is shown in Fig. 1.6. To run the

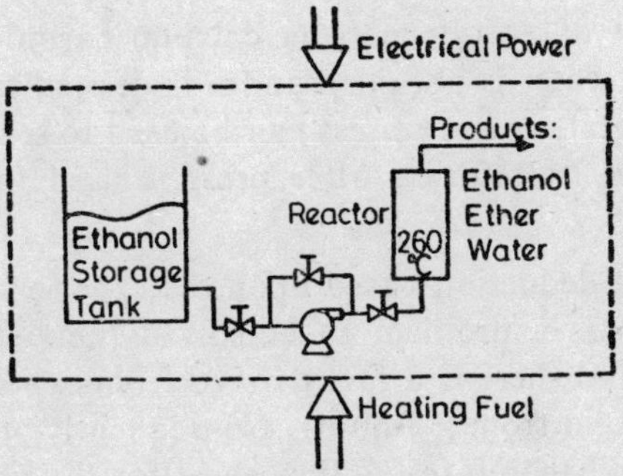

Fig. 1.6 A simplified flow sheet for diethyl ether.

plant, the availability of ethanol must be ensured. Since it is one of the by-products of sugar industries which is seasonal, to operate the reactor continuously, it is necessary to store a sufficient amount of alcohol. Also, electrical power is required for the pump and heating fuel for the reactor. Therefore, the availability of these must also be ensured. Not only this, fluctuations in the process variables are inevitable in actual practice which adversely affect the concentration of the product. To check these fluctuations, instruments for the control and measurement of these process variables are provided. Details of these are included in the process flow sheet, this way making it a very important document.

In this chapter the problem of synthesizing a steady state flow sheet has been discussed and is stated formally in Sec. 1.3.

1.3 THE PROBLEM OF STEADY STATE FLOW SHEETING

In a chemical plant, the raw material is the starting point whereas the product is the goal. The structural arrangement of equipment constitutes the steady state flow sheet. In general there can be more than one flow sheet to attain the goal (Fig. 1.2) and the task of a chemical engineer is to create one which is the best.

The strategy of obtaining the optimal flow sheet is to generate a first guess which is subsequently improved. This is done by employing the Theorem Proving Principle (TP). The methods of Propositional Logic to help in understanding TP are explained in Sec. 1.4.

1.4 SCOPE OF PROPOSITIONAL LOGIC IN PROCESS SYNTHESIS

Symbolic logic considers languages whose purpose is to symbolize 'reasoning'. Since there is a reason to have a specific structure of a chemical plant, one can employ the method of propositional logic. As an example, the following argument is considered:

"In the production of acetylene from calcium carbide, a slurry of calcium hydroxide in water is obtained and it is desired to separate calcium hydroxide from water. A filter press can be used to separate solid suspensions from a liquid. Therefore a filter press is used to separate calcium hydroxide." (1.4.1)

Above, there is a deduction based upon the experience of working with a filter press. There is a problem to be solved (i.e., separation of lime from the slurry) and there is a fact derived from experience (i.e., a filter press separates a solid from a liquid). Consequently, the use of a filter press is deduced and therefore the flow sheet for the production of acetylene must include it. This conclusion may appear trivial, but it illustrates that the flow sheet is derived through reasoning and consequently can be expressed through principles of propositional logic. In the following, the latter is described in brief.

1.5 PROPOSITIONAL LOGIC[3, 4]

Propositional logic deals with statements which can be either true or false. No other quantification is permitted. As an example, the statement "distillation columns separate miscible liquids exploiting the difference in their volatilities" is a scientific fact which is always true (with the exception of azeotropes). "Pump increases the pressure of fluid" or "heat exchangers are used for transfer of thermal energy" are some of the facts which are true. These can be expressed through symbols as follows:

P_1 = pump increases the pressure of fluid (1.5.1a)

H = heat exchangers are used for transfer of heat energy (1.5.1b)

P_1 and H are called propositions. These symbols can be used interchangeably with these statements. Propositions can express only simple statements as in Eq. 1.5.1; however for representation of more complex ideas as in eqn. 1.4.1, logical connectives are used. Table 1.2 gives four logical connectives: negation (not), conjunction (and), disjunction (or) and implication (if . . . then).

Table 1.2 Logical Connectives

Symbol	*Name*
$-$	Negation (not)
$\cdot$	Conjunction (and)
$+$	Disjunction (or)
$\rightarrow$	Implication (if . . . then)

As an example, to express symbolically the statement: "A chemical engineer can operate a pump as well as a filter press", propositions P and F are defined as:

$$P = \text{a chemical engineer can operate a pump} \tag{1.5.2a}$$

$$F = \text{a chemical engineer can operate a filter press} \tag{1.5.2b}$$

As the original statement amounts to saying that a chemical engineer can operate a pump *and* a filter press, it can be expressed as a conjunction of P and F or $(P \cdot F)$. If a chemical engineer cannot operate a pump, it can be symbolized as $\bar{P}$ and if he cannot operate the pump but can operate the filter press, it is symbolized as $(\bar{P} \cdot F)$. If it is desired to express symbolically "the pressure of a fluid can be increased either by centrifugal pump or by a reciprocating pump", propositions R and C are defined as:

$$R = \text{pressure can be increased by a reciprocating pump} \tag{1.5.3a}$$

$$C = \text{pressure can be increased by a centrifugal pump} \tag{1.5.3b}$$

The above statement would be $(R + C)$. Implication is a very important logical connective and is used where the occurrence of a given statement is linked with the occurrence of some other. As an example of compound statement, the following is considered: "if a liquid consists of a solid in the form of suspension, by passing through a filter press the solid can be separated from the liquid". The propositions are defined as below:

$$A = \text{liquid consists of solid in the form of suspension} \tag{1.5.4a}$$

$$B = \text{one passed the liquid through a filter press} \tag{1.5.4b}$$

$$C = \text{solid is separated from the liquid} \tag{1.5.4c}$$

The compound statement is reinterpreted as "*if* the liquid consists of solid as suspension *and* is passed through a filter press, *then*, solid will be separated from the liquid", and the propositional representation would therefore be $[(A \cdot B) \rightarrow C]$.

As another example of implication, a chemical reaction between hydrogen and chlorine is considered.

$$H_2 + Cl_2 = 2HCl \tag{1.5.5}$$

This reaction is known to occur spontaneously when hydrogen and chlorine are mixed in the presence of sunlight at room temperature with the liberation of heat. To express this reaction symbolically, there is only one information sufficient for coding, i.e., the reaction can go in the forward direction only in the presence of the reactants. In view of this, Eq. 1.5.5 can be re-expressed as "if hydrogen *and* chlorine both are present, *then* the product hydrochloric acid will be formed". Consequently following propositions are defined:

$$H_2 = \text{hydrogen is present} \tag{1.5.6a}$$

$$Cl_2 = \text{chlorine is present} \tag{1.5.6b}$$

$$HCl = \text{hydrochloric acid is formed} \tag{1.5.6c}$$

The chemical reaction of Eq. 1.5.5 is written as

$$(H_2 \cdot Cl_2) \rightarrow HCl \qquad (1.5.7)$$

Simple propositions, like P, H, F, R, C, etc., are called '*atoms*', and compound propositions like (P·F), (R + C) and $[(H_2 \cdot Cl_2) \rightarrow HCl]$ are called formula, semantic equations or just equations. Strictly speaking an 'atom' is also an equation, but in normal usage, the two are distinguished. If there is an equation, G, then its negation, $\bar{G}$, is also an equation. Also, if there are two equations G and H, the result of any combination of these two, i.e. (G·H), (G + H) or (G → H), is also an equation.

The truth value of an atom refers to whether it is true (T) or false (F). If the truth values of atoms constituting a given formula, G, are specified, it is of great practical as well as theoretical significance in deducing the truth value of G. Some of basic laws of propositional logic are stated below:

1. $\bar{A}$ is true when A is false and $\bar{A}$ is false when A is true.
2. (A·B) is true only when both A and B are true; otherwise (A·B) is false.
3. (A + B) is true if any one of A and B is true. It is false only when both A and B are false.
4. (A → B) is false only when A is true and B is false.

Table 1.3 Basic Laws of Propositional Logic

A	B	$\bar{A}$	A · B	A + B	A → B
T	T	F	T	T	T
T	F	F	F	T	F
F	T	T	F	T	T
F	F	T	F	F	T

These are also expressed in Table 1.3. Using these laws it is possible to derive the truth value of any equation by preparing a truth table. Two equations are equal only when their truth tables are identical. As an example, two functions G and H are considered which are defined as

$$G = A \rightarrow B \qquad (1.5.8a)$$

$$H = \bar{A} + B \qquad (1.5.8b)$$

Table 1.4 Truth Values of A → B and $\bar{A}$ + B

A	B	$\bar{A}$	$\bar{A}$ + B	A → B (from Table 1.3)
T	T	F	T	T
T	F	F	F	F
F	T	T	T	T
F	F	T	T	T

Their truth values are given in Table 1.4 and it is seen that they are same for all choices of A and B.

$$\therefore \quad G = H \tag{1.5.9}$$

By equality, it is meant that whenever G occurs, it can be replaced by H. There are some equations which are true under all possible interpretations of their atoms and they are called valid equations denoted by ■. As opposed to this, there are some equations which are always false and are called inconsistent equations denoted by □. Table 1.5 gives some basic relations, which can be proved by preparing truth tables. They are used so extensively in propositional logic that they are sometimes called laws.

Table 1.5 Basic Relations in Propositional Logic

1. $A \rightarrow B = \bar{A} + B$
2. Commutative law
   $A \cdot B = B \cdot A$, $A + B = B + A$
3. Associative law
   $(A \cdot B) \cdot C = A \cdot (B \cdot C)$, $(A + B) + C = A + (B + C)$
4. Distributive law
   $A + (B \cdot C) = (A + B) \cdot (A + C)$
   $A \cdot (B + C) = A \cdot B + A \cdot C$
5. De Morgan's law
   $(\overline{A + B}) = \bar{A} \cdot \bar{B}$, $(\overline{A \cdot B}) = \bar{A} + \bar{B}$
6. Others
   (a) $A \cdot \bar{A} = \square$; $(\bar{\bar{A}}) = A$; $A + \bar{A} = \blacksquare$
   (b) $A \cdot \square = \square$; $A + \blacksquare = \blacksquare$
   (c) $A \cdot \blacksquare = A$; $A + \square = A$

Using the laws of Table 1.5, it is possible to simplify an equation. As an example, the following equation is considered:

$$G = (A \rightarrow B) \cdot A \tag{1.5.10}$$

The simplification is achieved within the parenthesis first. Use of law 1 in Table 1.5 gives

$$G = (\bar{A} + B) \cdot A \tag{1.5.11}$$

Applying the distributive law 4, G becomes

$$G = (A \cdot \bar{A}) + (A \cdot B) \tag{1.5.12}$$

Noting law 6(a) in Table 1.5,

$$G = \square + (A \cdot B) = (A \cdot B) + \square \tag{1.5.13}$$

and finally according to law 6(c),

$$G = (A \cdot B) \tag{1.5.14}$$

The most simplified form of an equation occurs when atoms occurring are joined by conjunction or disjunction type of connectives and with the help of Table 1.5, it is possible to reduce a given equation to this. This is particularly necessary when one wishes to find the logical consequence discussed in Sec. 1.6.

1.6 DEDUCTION THEOREM OF PROPOSITIONAL LOGIC

It has already been stated that the desired products can be manufactured if one has suitable reactors, equipment for unit operations and necessary raw materials. If one defines R, U, RM and P as

$$\mathrm{R} = \text{one has suitable reactors} \tag{1.6.1a}$$

$$\mathrm{U} = \text{one has equipment to carry out unit operations} \tag{1.6.1b}$$

$$\mathrm{RM} = \text{one has necessary raw materials} \tag{1.6.1c}$$

$$\mathrm{P} = \text{one manufactures the desired products} \tag{1.6.1d}$$

then P would be a logical consequence of R, U and RM if and only if

$$(\mathrm{R} \cdot \mathrm{U} \cdot \mathrm{RM}) \rightarrow \mathrm{P} \tag{1.6.2}$$

R, U and RM above need not be atoms but can also be equations themselves. It may be recalled that a steady state flow sheet is nothing but R·U. With the knowledge of RM and P in Eq. 1.6.2, it is possible to find the flow sheet using the deduction theorem stated below.

The *Deduction Theorem* of propositional logic states that if G is a logical consequence of statements $F_1, F_2, \ldots, F_n$, then the $[(F_1 \cdot F_2 \cdot F_3 \cdot \ldots \ldots \cdot F_n) \rightarrow G]$ is valid, i.e., it is true always. This is equivalent to saying

$$(F_1 \cdot F_2 \cdot \ldots\ldots \cdot F_n) \cdot \bar{G} \text{ is inconsistent} \tag{1.6.3}$$

which can be easily proved as follows.

Since $[(F_1 \cdot F_2 \cdot \ldots \cdot F_n) \rightarrow G]$ is always true, $[\overline{(F_1 \cdot F_2 \cdot \ldots \cdot F_n) \rightarrow G}]$ will always be false (or inconsistent). But

$$X = [\overline{(F_1 \cdot F_2 \cdot \ldots \cdot F_n) \rightarrow G}] = [\overline{\overline{(F_1 \cdot F_2 \cdot \ldots \cdot F_n)} + G}] \tag{1.6.4}$$

which follows from law 1 of Table 1.5. Using De Morgan's law,

$$X = \overline{\overline{(F_1 \cdot F_2 \cdot \ldots \cdot F_n)}} \cdot \bar{G}$$

$$= (F_1 \cdot F_2 \cdot \ldots \cdot F_n) \cdot \bar{G} \tag{1.6.5}$$

i.e., X is to be proven inconsistent. This is a very important relation and is the *basis of resolution based synthesis procedure* of flow sheeting discussed next.

1.7 RESOLUTION-BASED SYNTHESIS PROCEDURE[3,5]

This particular procedure of flow sheet development can best be illustrated through the following example. One has MgO, H_2, O_2 and C as raw materials and can perform the following chemical reactions

$$MgO + H_2 = Mg + H_2O \tag{1.7.1a}$$

$$C + O_2 = CO_2 \tag{1.7.1b}$$

$$CO_2 + H_2O = H_2CO_3 \tag{1.7.1c}$$

These are written in propositional logic as equations

$$F_1 = (MgO \cdot H_2) \rightarrow (Mg \cdot H_2O) \tag{1.7.2a}$$

$$F_2 = (C \cdot O_2) \rightarrow CO_2 \tag{1.7.2b}$$

and $F_3 = (CO_2 \cdot H_2O) \rightarrow H_2CO_3$ (1.7.2c)

F_1 can be simplified using law 1 of Table 1.5 to

$$F_1 = \overline{(MgO \cdot H_2)} + (Mg \cdot H_2O) \tag{1.7.3}$$

Using associative law (3 of Table 1.5),

$$F_1 = [\overline{(MgO \cdot H_2)} + Mg] \cdot [\overline{(MgO \cdot H_2)} + H_2O] \tag{1.7.4}$$

Similarly F_2 and F_3 can be rewritten as

$$F_2 = \overline{(C \cdot O_2)} + CO_2 \tag{1.7.5}$$

$$F_3 = \overline{(CO_2 \cdot H_2O)} + H_2CO_3 \tag{1.7.6}$$

The available raw materials are also represented as facts F_4, F_5, F_6 and F_7.

$$F_4 = MgO \tag{1.7.7a}$$

$$F_5 = H_2 \tag{1.7.7b}$$

$$F_6 = O_2 \tag{1.7.7c}$$

$$F_7 = C \tag{1.7.7d}$$

From the deduction theorem given in Eq. 1.6.5, one proves X defined by

$$X = (F_1 \cdot F_2 \cdot F_3 \cdot F_4 \cdot F_5 \cdot F_6 \cdot F_7) \cdot \overline{H_2CO_3} \tag{1.7.8}$$

as inconsistent. It is only then can one conclude that the product carbonic acid can be produced through these reactions and raw materials. As shown now, the flow sheet is generated in the process of doing so.

Substituting for F_1 to F_7 in Eq. 1.7.8, X becomes

$$X = [\overline{(MgO \cdot H_2)} + Mg] \cdot [\overline{(MgO \cdot H_2)} + H_2O] \cdot [\overline{(C \cdot O_2)} + CO_2] \cdot [\overline{(CO_2 \cdot H_2O)} + H_2CO_3] \cdot MgO \cdot H_2 \cdot O_2 \cdot C \cdot \overline{H_2CO_3} \tag{1.7.9}$$

In Eq. 1.7.9 F_3 is combined with $\overline{H_2CO_3}$ first using law 4 of Table 1.5

$$\begin{aligned} F_3 \cdot \overline{H_2CO_3} &= [\overline{(CO_2 \cdot H_2O)} + H_2CO_3] \cdot \overline{H_2CO_3} \\ &= [\overline{(CO \cdot H_2O)} \cdot \overline{H_2CO_3}] + [H_2CO_3 \cdot \overline{H_2CO_3}] \end{aligned} \tag{1.7.10}$$

As $H_2CO_3 \cdot \overline{H_2CO_3}$ is inconsistent,

$$\begin{aligned} F_3 \cdot \overline{H_2CO_3} &= [\overline{(CO_2 \cdot H_2O)} \cdot \overline{H_2CO_3}] + \square \\ &= \overline{(CO_2 \cdot H_2O)} \cdot \overline{H_2CO_3} \end{aligned} \tag{1.7.11}$$

Substituting this in Eq. 1.7.9,

$$X = X_1 \cdot \overline{(CO_2 \cdot H_2O)} \tag{1.7.12}$$

where X_1 is given by

$$X_1 = [\overline{(MgO \cdot H_2)} + Mg] \cdot [\overline{(MgO \cdot H_2)} + H_2O] \cdot [\overline{(C \cdot O_2)} + CO_2] \cdot MgO \cdot H_2 \cdot O_2 \cdot C \cdot \overline{H_2CO_3} \tag{1.7.13}$$

On rearrangement Eq. 1.7.12 becomes

$$X = X_1 \cdot \overline{(CO_2 + H_2O)} \tag{1.7.14}$$

and using law 4 of Table 1.5, the following is obtained

$$X = (X_1 \cdot \overline{CO_2}) + (X_1 \cdot \overline{H_2O}) \tag{1.7.15}$$

Therefore to prove X inconsistent, it is necessary to prove $(X_1 \cdot \overline{CO_2})$ as well as $(X_1 \cdot \overline{H_2O})$ inconsistent. $(X_1 \cdot \overline{CO_2})$ is considered first in the following:

$$(X_1 \cdot \overline{CO_2}) = X_2 \cdot [\overline{(C \cdot O_2)} + CO_2] \cdot \overline{CO_2} \cdot C \cdot O_2 \tag{1.7.16}$$

where

$$X_2 = [\overline{MgO \cdot H_2)} + Mg] \cdot [\overline{(MgO \cdot H_2)} + H_2O] \cdot MgO \cdot H_2 \cdot \overline{H_2CO_3} \tag{1.7.17}$$

Equation 1.7.16 is expanded using law 4 of Table 1.5,

$$X_1 \cdot \overline{CO_2} = X_2 \cdot \{[\overline{(C \cdot O_2)} \cdot (CO_2 \cdot C \cdot O_2)] + [CO_2 \cdot \overline{CO_2} \cdot C \cdot O_2]\} \tag{1.7.18}$$

The first square parenthesis is inconsistent because $\overline{(C \cdot O_2)} \cdot (C \cdot O_2)$ occurs and the second one is inconsistent because $CO_2 \cdot \overline{CO_2}$ occurs.

$$\therefore \quad X_1 \cdot \overline{CO_2} = \Box \tag{1.7.19}$$

Similarly

$$X_1 \cdot \overline{H_2O} = X_3 \cdot [\overline{(MgO \cdot H_2)} + H_2O] \cdot \overline{H_2O} \cdot MgO \cdot H_2 \tag{1.7.20}$$

where

$$X_3 = [\overline{(MgO \cdot H_2)} + Mg] \cdot [\overline{(C \cdot O_2)} + CO_2] \cdot O_2 \cdot C \cdot \overline{H_2CO_3} \tag{1.7.21}$$

On expanding Eq. 1.7.20 becomes

$$X_1 \cdot \overline{H_2O} = X_3 \cdot \{[\overline{(MgO \cdot H_2)} \cdot \overline{H_2O} \cdot MgO \cdot H_2] + [H_2O \cdot \overline{H_2O} \cdot MgO \cdot H_2]\} \tag{1.7.22}$$

Similarly the first square bracket is inconsistent because of the occurrence of $\overline{(MgO \cdot H_2)} \cdot (MgO \cdot H_2)$ and the second one is inconsistent because of $H_2O \cdot \overline{H_2O}$.

$$\therefore \quad X_1 \cdot \overline{H_2O} = \Box \tag{1.7.23}$$

Consequently,

$$X = (X_1 \cdot \overline{CO_2}) + (X_1 \cdot H_2O) = \Box \tag{1.7.24}$$

In other words H_2CO_3 can be prepared from the fact list consisting of reactions F_1 to F_3 and raw materials F_4 to F_7. The resolution steps given in Eqs. 1.7.9 to 1.7.24 have been pictorially represented in Fig. 1.7.

A careful examination of Fig. 1.7 suggests the following systematic procedure, which becomes the basis of the resolution based synthesis procedure discussed in Sec. 1.8. The goal is compared with the raw materials in the fact list. When it is not found as one of them, the right

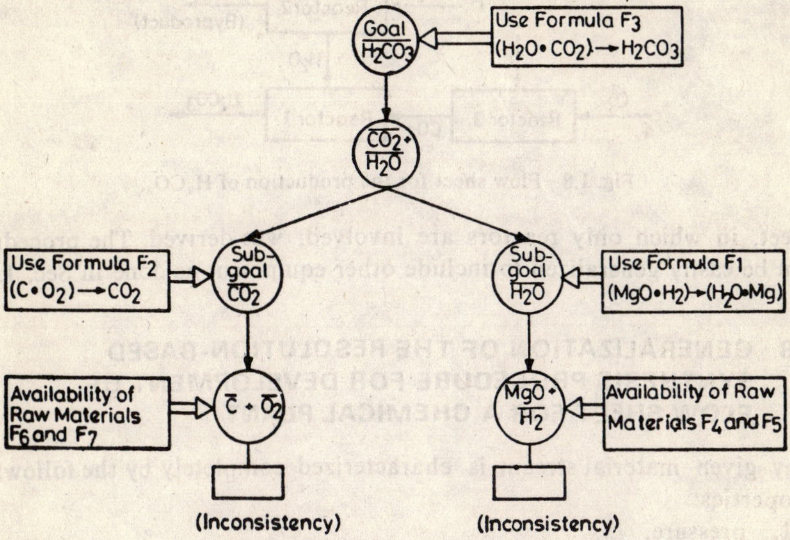

Fig. 1.7 A pictorial representation of resolution steps to show the inconsistency of $(F_1 \cdot F_2 \ldots F_7) \cdot \overline{G}$.

hand side of all reactions producing it is scanned. In the solved example above F_3 contains the goal H_2CO_3 in the right hand side. The result of the conjunction of $\overline{H_2CO_3}$ with F_3 is $(\overline{CO_2} + \overline{H_2O})$. The mathematical step of $(F_3 \cdot \overline{G})$ physically amounts to generating G from CO_2 and H_2O through the reaction F_3. Now the availability of CO_2 and H_2O is checked from the fact list and it is found that they are not available as raw materials; however CO_2 appears in the right hand side of F_2 whereas H_2O appears in the right hand side of F_1. The union of $(F_1 \cdot F_2)$ with the subgoal $(\overline{CO_2} + \overline{H_2O})$ amounts to generating CO_2 from reaction F_2 and generating H_2O from reaction F_1. As raw materials required for these reactions are available, one stops here and is also confirmed through the propositional logic by the appearance of inconsistency.

In a particular problem, if the goal list G consists of more than one chemical to be manufactured, one starts similarly from the top of this list. After attaining it completely, the next one in G is considered and the process is repeated till the entire list of G is covered.

Synthesis of a flow sheet is now a simple matter and in Fig. 1.8, the flow sheet for the example of manufacturing H_2CO_3 is shown. The scanning is started from the end of the flow sheet; reaction F_3, for which H_2O and CO_2 are required, is carried out in reactor 1. As each of these chemicals is not available, reactions F_2 and F_1 are carried out in reactors 2 and 3. One stops at this because the raw materials for them are available (formula F_4 to F_7). In the example considered, a special case of the flow

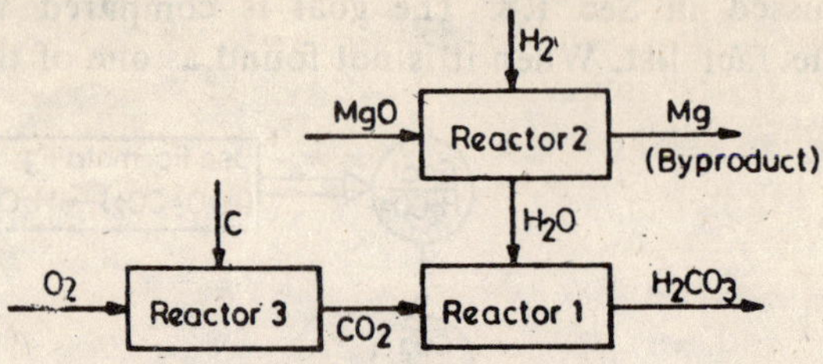

Fig. 1.8 Flow sheet for the production of H_2CO_3.

sheet, in which only reactors are involved, was derived. The procedure can be easily generalized to include other equipment as done in Sec. 1.8.

1.8 GENERALIZATION OF THE RESOLUTION-BASED SYNTHESIS PROCEDURE FOR DEVELOPMENT OF FLOW SHEET FOR A CHEMICAL PLANT

Any given material stream is characterized completely by the following properties:

1. pressure,
2. temperature (or enthalpy),
3. total flow rate,
4. number of phases,
5. composition of each phase, and
6. species.

Two streams are equal only when all these properties are equal which means that liquid benzene at 60°C and one atmosphere is not equivalent to a stream of liquid benzene at 30°C at 2 atmosphere, even though the species of both these streams is the same. However, they can be made equal if the former stream is cooled to 30°C and its pressure increased to two atmosphere for which a cooler and a pump are required.

One considers two streams s_1 and s_2 having different compositions with all other properties being the same. Stream s_1 is assumed to be a mixture of liquid benzene (x_B) and toluene (x_T) and s_2 is pure toluene. Streams s_1 and s_2 can be equated in two ways. The first one (easiest way) is to mix another stream (s_3) of pure benzene with s_2 to obtain the same composition as in s_1 and the second (a more difficult) method is to use a separator to fractionate s_1. The first method hinges on the availability of stream s_3 and the shortcoming of the second method is that a completely pure component can never be obtained by distillation even though the purity can be improved by increasing the number of equilibrium plates. In view of this, to preserve a pure stream (i.e. pure toluene in s_2), distillation of s_1 to match the stream may be preferred in spite of higher cost of separation.

For convenience, the total fact list is broken into raw material and reaction list. The goal list may have more than one product and the first item (say G_1) in it is identified and the raw material list is scanned for its availability. If there is one or more than one stream containing G_1, these streams are adjusted and matched using mixer, separator, pump and heat exchanger. If the item is not available in the raw material list, the right hand side of the reaction list is scanned and the reaction which gives it as a reaction product is identified. If reactants are available in the raw material list, G_1 of the goal list has been successfully attained. If it is not, one rescans the reaction list for suitable reactions that would produce these reactants, and so on. At every stage, streams are matched using a combination of mixers, separators, pumps and heat exchangers. This stepwise procedure is shown in Fig. 1.9.

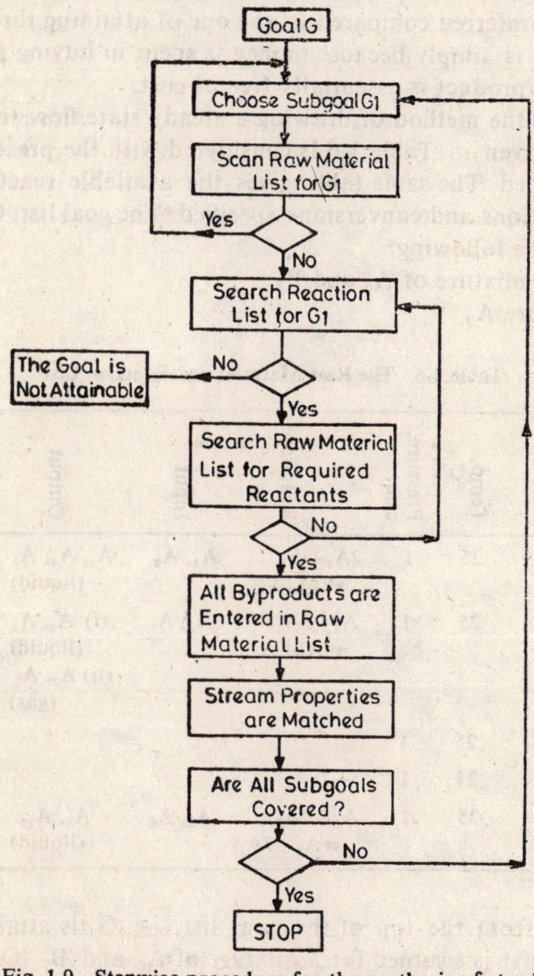

Fig. 1.9 Stepwise procedure for the synthesis of steady state flow sheet for a chemical plant

There are a few special situations of drawing of flow sheet which need to be considered now. The first one is when there is more than one reaction route to make the desired product. For example, in Fig. 1.2, there are four sets of reactions that would produce phenol. For each of these reaction routes, there are different flow sheets, and the best can be found only through economic analysis.

The second situation arises when the chemical required to attain a given subgoal is present in more than one stream of the raw material list. To decide to use one, or the other, or both of them with some weighting ratio, the total cost of obtaining the desired composition has to be minimized.

Lastly the situation, when a by-product stream is generated in the synthesis process, is considered. If it is possible to attain the goal (or subgoal) through further processing of this particular stream, then this alternative is preferred compared to the one of attaining through the raw material. This is simply because money is spent in buying raw materials whereas the byproduct is essentially free of cost.

To illustrate the method of drawing a steady state flow sheet, the raw material list given in Table 1.6 is considered with the pressure and temperature specified. The same table gives the available reaction list with reaction conditions and conversions specified. The goal list, G, is assumed to consist of the following:

1. G_1 = A mixture of A_6 and B_7
2. G_2 = Pure A_9

Table 1.6 The Raw Material and Reaction List

Raw material	Species (liquid)	Temp, °C	Pressure, atm.	Reactions	Input	Output	Temp, °C	Pressure, atm.
s_1	A_5, B_4	25	1	$2A_1+A_2 \rightleftharpoons 2A_3+A_4$	A_1, A_2	A_1, A_3, A_4 (liquid)	100	2
s_2	A_2, B_3	25	1	$A_3+2A_5 \rightleftharpoons A_6+A_7$	A_3, A_5	(i) A_6, A_3 (liquid) (ii) A_7, A_5 (gas)	80	1
s_3	A_1, B_1, B_2	25	1					
s_4	A_8, B_9	25	1					
s_5	A_{11}, B_7	25	1	$A_3+2A_8 \rightleftharpoons A_9+7A_{10}$	A_3, A_8	A_9, A_{10} (liquid)	50	1

One starts from the top of the goal list, i.e. G_1 is attained first. The raw material list is scanned for a mixture of A_6 and B_7. Only stream s_5 is found to consist of B_7 in mixture with A_{11}. Therefore B_7 is separated from this. As A_6 is not available in the raw material list, it must be synthesized through a suitable reaction in the reaction list. On separation of B_7 from stream s_5, stream s_{15} consisting of pure A_{11} is generated, as shown in Fig. 1.10. This is added into the revised ıaw material list RM1.

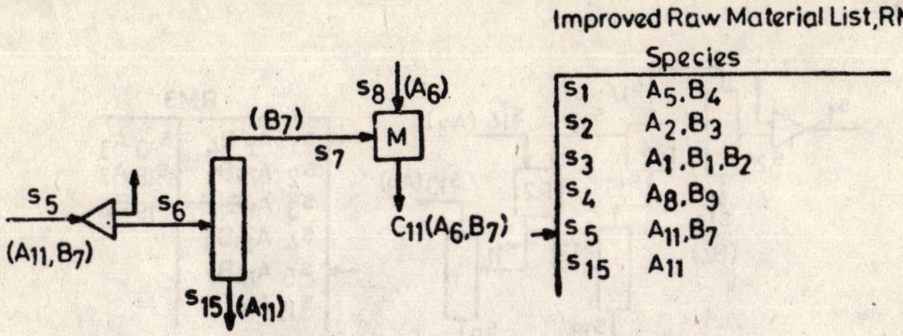

Fig. 1.10 First step for attaining G_1

In the second step, the current raw material list, RM1, is scanned for the component A_6. As it is not found to be available, the reaction list is scanned for A_6 in the right hand side of reactions and it is found that reaction 2 must be carried out for obtaining it. There are two outputs of this reaction, and the one containing A_6 is a mixture of A_6 and A_3. Consequently, it is separated and compared with s_8 for the equality of pressure and temperature. As pressures are the same, but temperatures are different (80° versus 25°), a cooler is placed as in Fig. 1.11, which represents the second step of the steady state flow sheeting. s_{10} and s_{12} are by-products and are added into the raw material list RM1 to give RM2 shown in Fig. 1.11.

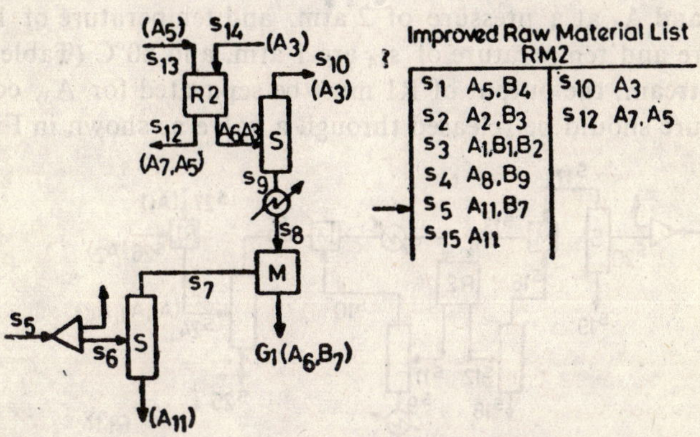

Fig. 1.11 Second step for attaining G_1

In the third step, a mixture of A_5 and A_3 is obtained to run the reaction 2 of Table 1.6. Availability of A_5 is examined first in the current raw material list RM2. A_5 is found to be available in s_1 as well as s_{12}. As s_{12} is the by-product, it should be used first. As its utilization amounts to recycling of the unreacted A_5, this would not fulfil the total demand of A_5 and s_1 should also be used along with it. Consequently A_5 is separated from s_1. As the pressure and temperature of this separated stream is the same as that required for R2, it is fed directly to the reactor as shown in

Fig. 1.12. Stream s_{12} has been utilized fully and therefore it is removed from RM3.

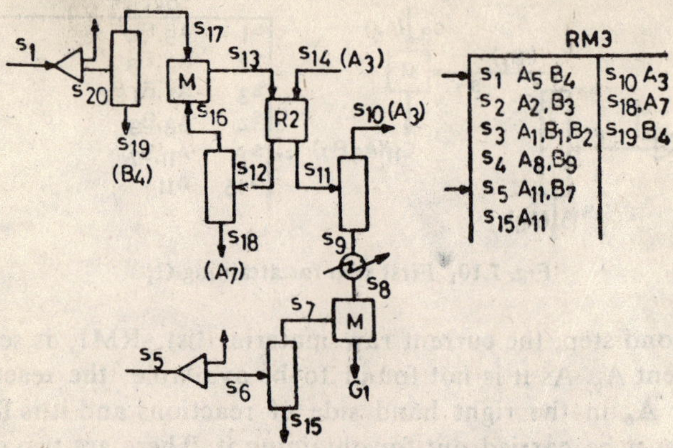

Fig. 1.12 Third step for attaining G_1

In the fourth step, to obtain pure A_3, RM3 is scanned. s_{10} is found to contain A_3 but as this is the unreacted part of what is fed in R2, it is necessary to seek an additional source for A_3. As no other stream contains the desired compound, the right hand side of reactions of the list must be checked. A_3 is found to be a product of reaction 1, and is carried out in reactor R1. According to Table 1.6, the output stream of R1 is a mixture of A_1, A_3 and A_4 at a pressure of 2 atm. and temperature of 100°C. As the pressure and temperature of s_{14} are 1 atm. and 80°C (Table 1.6), and is a pure stream, the output of R1 must be separated for A_3, cooled and then pressure should be released through a valve as shown in Fig. 1.13.

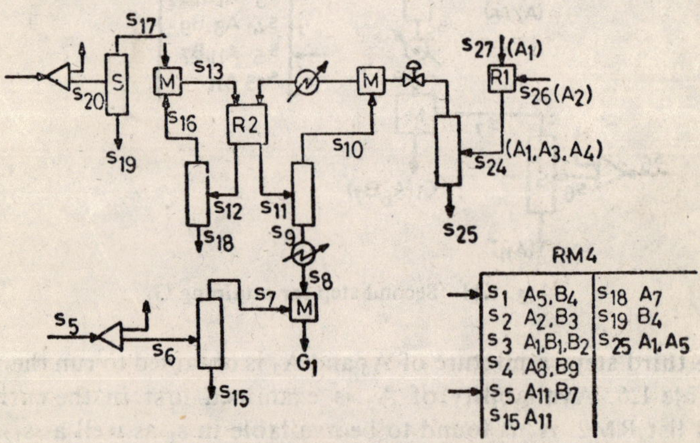

Fig. 1.13 Fourth step for attaining G_1

To run R1, it is important that A_1 and A_2 be made available first. RM4 is scanned for this, and it is found that A_1 is available in s_3 as a mixture with A_1, B_1 and B_2 whereas A_2 is available in s_2 as a mixture of

A_2 and B_3. A_1 and A_2 are consequently separated from s_2 and s_3 through suitable separators and then pressures and temperatures are adjusted using pumps and heat exchangers respectively. One stops at this point because G_1 has now been obtained through available raw materials and by-product streams and the resultant flow sheet for attaining the subgoal G_1 is shown in Fig. 1.14.

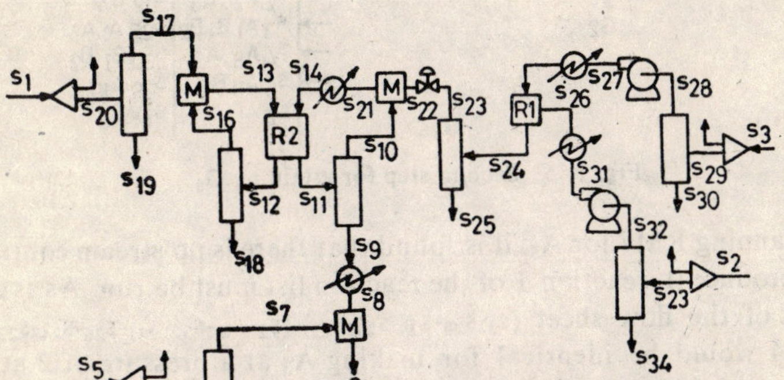

RM5

s_1 A_5, B_4	s_{18} A_7
s_2 A_2, B_3	s_{19} B_4
s_3 A_1, B_1, B_2	s_{25} A_1, A_5
s_4 A_8, B_9	s_{30} B_1, B_2
s_5 A_{11}, B_7	s_{34} B_3
s_{15} A_{11}	

Fig. 1.14 Complete flow sheet for attaining G_1

After having attained G_1, one now goes on to G_2. The current raw material list RM5 is checked for A_9, but as it is not present with any of available streams, the reaction list is scanned. The relevant reaction which produces A_9 is reaction 3 in Table 1.6 and is carried out in reactor R_3. The product of R3 according to Table 1.6 is a mixture of A_8, A_9 and A_{10} at a pressure of 1 atm. and temperature of 50°C. Therefore G_2 is connected through a separator to R3 as shown in Fig. 1.15.

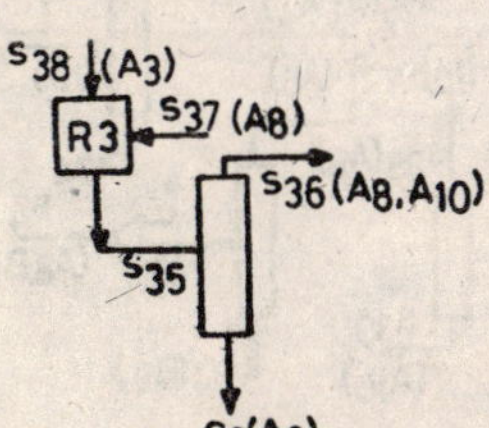

RM6

s_1 A_3, B_4	s_{18} A_7
s_2 A_2, B_3	s_{19} B_4
s_3 A_1, B_1, B_2	s_{25} A_1, A_5
s_4 A_8, B_9	s_{30} B_1, B_2
s_5 A_{11}, B_7	s_{34} B_3
s_{15} A_{11}	s_{36} A_8, A_{10}

Fig. 1.15 First step for attaining G_2

To produce A_8, one can use s_4 as well as s_{36}. As s_{36} consists of only unreacted A_8, and is a recycle stream, s_4 must also be used along with it as shown in Fig. 1.16. In the raw material list RM7, consequently, s_{36} is removed and s_{39} and s_{41} are added.

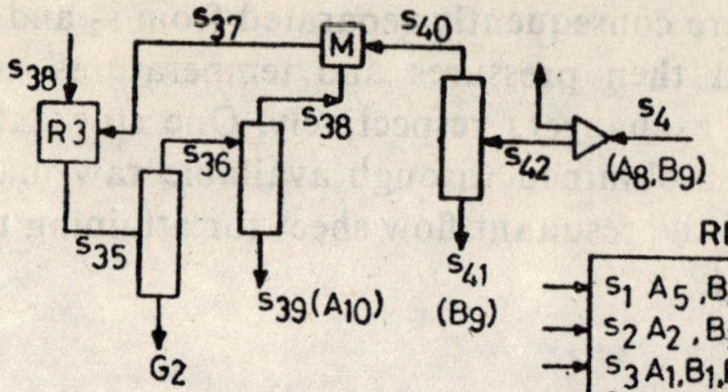

RM7

s_1 A_5, B_4	s_{18} A_7
s_2 A_2, B_3	s_{19} B_4
s_3 A_1, B_1, B_2	s_{25} A_1, A_5
s_4 A_8, A_9	s_{30} B_1, B_2
s_5 A_{11}, B_7	s_{39} A_{10}
s_{15} A_{11}	s_{41} B_9

Fig. 1.16 Second step for attaining G_2

On scanning RM7 for A_3, it is found that there is no stream containing it and to produce it, reaction 1 of the reaction list must be run. As is evident, the part of the flow sheet (s_3-s_{29}-s_{28}-s_{27}-s_{25})-(s_2-s_{33}-s_{32}-s_{31}-s_{26}-s_{24}-s_{23}-s_{12}) of Fig. 1.14 would be identical for making A_3 at a pressure of 2 atm. and temperature of 100°C. One has two choices now. The first is to increase the capacity of R1 and produce more to meet the entire demand of A_3 or have a separate reactor R1 with its accessories for attaining G_2. In the latter case, flow sheets for producing G_1 and G_2 are completely independent of each other and the whole chemical plant producing G consists of two small plants running unrelated to each other. Normally this is not a good strategy because it is cheaper to plan one bigger reactor compared to two small, independent ones.

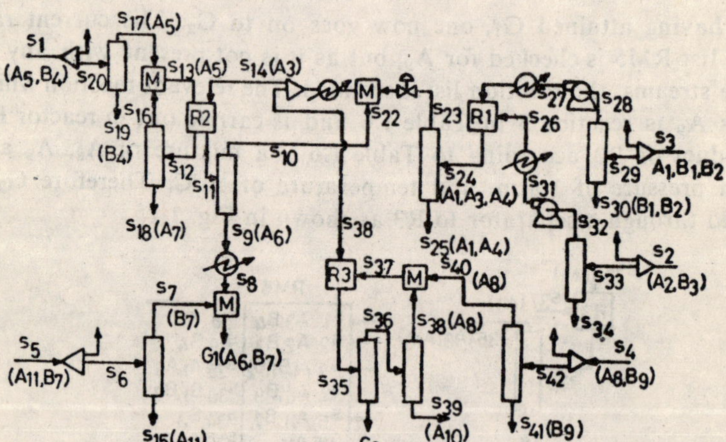

RM8

s_1 A_5, B_4	s_{18} A_7
s_2 A_2, B_3	s_{19} B_4
s_3 A_1, B_1, B_2	s_{25} A_1, A_5
s_4 A_8, B_9	s_{30} B_1, B_2
s_5 A_{11}, B_7	s_{34} B_3
s_{15} A_{11}	s_{39} A_{10}, A_{11}, B_1

Fig. 1.17 Final flow sheet for the process

It is now decided to increase the size of R2 and produce enough A_3 for attaining both G_1 and G_2. As streams s_{23}, s_{22}, s_{21} and s_{14} all contain pure streams of A_3 differing only in temperature and pressure, the stream after reactor R3 is chosen for which a minimum adjustment is required to be made to match it with s_{38} of Fig. 1.16. As the pressure and temperature of s_{38} is 1 atm. and 100°C respectively (Table 1.6), which is the same as that of s_{14}, s_{14} is divided using a splitter. The final flow sheet is shown in Fig. 1.17 which manufactures both G_1 and G_2 and the overlapping region is the section of the flow sheet producing A_3. RM8, which is also given in the figure, specifies all the byproducts formed in the process of achieving the goal.

1.9 CONCLUSIONS

Flow sheeting is a synthesis problem, i.e. one would like to find out the structure of a system that would produce the desired product. In literature the approach has been to decompose the synthesis problem into smaller tasks and one continues the decomposition till one can solve subtasks thus generated. The method of tearing is based upon:[5, 8–10]

1. general problem solving approach (G.P.S.), and
2. theorem proving approach (T.P.).

Both these methods differ significantly in their solution procedure. The G.P.S. method starts from the raw material end and generates all plausible paths which lead to the final product. As opposed to this in the T.P. method, one starts from the final product end and generates a path upto raw materials.

The theorem proving method of flow sheeting discussed in this section has the advantage of being conceptually simpler: one can draw the flow sheet without any aid of computer whereas this is not possible in G.P.S. method. However, people have programmed the T.P. method in LISP and solved the flow sheet of Fig. 1.17 by computer.[5]

As explained in this chapter, there is more than one flow sheet possible for manufacturing a specified final product from given raw materials. It is not an easy task to find the optimal one because it involves an optimization with respect to the structure of the chemical plant. There have been a few attempts to express structure in numerical mathematics, in literature.[8] One can express the flow sheet in semantic mathematics very conveniently but then it is not amenable to usual numerical optimization procedures known to chemical engineers. In view of these difficulties, the following procedure is suggested for finding *an* economical flow sheet.[9]

In a chemical plant, normally there are two major tasks besides chemical reaction, namely heat exchanging and separation of mixtures. These tasks are individually optimized with respect to the network structure and in this way, the flow sheet generated in this chapter is

improved[9]. In the next chapter optimization of the heat exchanger network is discussed.

REFERENCES

1. Rudd, D. F., Powers, G. J. and Siirola, J. J., *Process Synthesis*, Prentice-Hall, N.Y., 1973.
2. O'Donnel, J. P., *Chem. Engg.*, **64**, 249, Sept. 1957.
3. Chang, C. L. and Lee, R. C., *Symbolic Logic and Mechanical Theorem Proving*, Academic Press, N.Y., 1973.
4. Robbin, J. W., *Mathematical Logic, A First Course*, 1st edn., W. A. Benjamin, N.Y., 1969.
5. Mahalec, V. and Motard, R. L., *Computer and Chem. Eng.*, **1**, 57, 1977.
6. Rudd, D. F., *A.I.Ch.E.J.*, **14**, 343, 1968.
7. Masso, A. H. and Rudd D. H., *A.I.Ch.E.J.*, **15**, 10, 1969.
8. Hendry, J. E., Rudd, D. F. and Seader, J. D., *A.I.Ch.E.J.*, **19**, 1, 1973.
9. Hlavaček, V., *Computers and Chem. Eng.*, **1**, 77, 1977.
10. Powers, G. J. and Rudd, D. F., *Basic Question of Design Theory*, Spiller, W. R. (ed.), Elsevier, N.Y., 1974.
11. Perry, R. H. (ed.), *Chemical Engineers' Handbook*, 5th edn., McGraw-Hill, N.Y., 1974.

PROBLEMS

1.1 Prove the following semantic relations through truth tables as well as through simplification[8]:

(a) $(P \rightarrow Q) \cdot (P \rightarrow R) = P \rightarrow (Q \cdot R)$

(b) $(P \rightarrow Q) \rightarrow (P \cdot Q) = (\overline{P} \rightarrow Q) \cdot (Q \rightarrow P)$

(c) $P \cdot Q \cdot (\overline{P} + \overline{Q}) = \overline{P} \cdot \overline{Q} \cdot (P + Q)$

(d) $P + (P \rightarrow (P \cdot Q)) = \overline{P} + \overline{Q} + (P \cdot Q)$

(e) $P \rightarrow Q = (\overline{Q} \rightarrow \overline{P})$

1.2 Simplify following relations[3]:

(a) $(P \cdot (Q \rightarrow R)) \rightarrow P$

(b) $P \rightarrow ((Q \cdot R) \rightarrow S)$

(c) $(\overline{P \rightarrow Q}) + (P + Q)$

(d) $(\overline{P \cdot Q}) \cdot (P + Q)$

(e) $((P \rightarrow Q) \cdot \overline{Q}) \rightarrow \overline{P}$

(f) $(P \rightarrow S) \cdot (S \rightarrow U) \cdot P$

(g) $(P + \overline{Q}) \rightarrow R$

1.3 In Eq. 1.5.7, semantic equation of a chemical reaction has been given. On a similar line, suggest semantic equations for:
(a) mixing,
(b) splitting,
(c) separating of streams,
(d) pumping, and
(e) heat exchanging.

1.4 In the representation of a chemical reaction, only the property of the existence of raw materials has been used. Modify this representation to include the fact that one should also maintain the reaction temperature and pressure.

1.5 Consider the example of Sec. 1.7. In the production of H_2CO_3, different process conditions required for running reactors 1, 2 and 3 have been ignored. If these are run at different temperatures intuitively one feels that suitable exchangers should be placed as a part of the flow sheet. Use results of P. 1.3 and modify the analysis of this section to show mathematically the position of the exchanger.

1.6 Assume that the formation of H_2CO_3 in reactor 1 occurs at high pressure. Show mathematically the position of the pump.

1.7 Assume that the reaction for the formation of H_2CO_3 above does not go to complete conversion. How do you build this fact in the semantic representation of the chemical reaction. If the goal is defined to be a pure stream of H_2CO_3, show mathematically the position of the separator.

1.8 In the process of micro-encapsulation of dyes, the dye solution is dispersed as small globules in a polymer solution. On drying and pasting these granules on a paper, one can obtain a non-smearing carbon paper. One defines the goal G as these non-smearing carbon papers. The raw material list consists of:
(a) solid polymer,
(b) a suitable solvent for the polymer,
(c) a suitable solvent for the dye immiscible in the polymer solvent,
(d) the dye, and
(e) glue to stick granules on paper (water soluble).
Draw the flow sheet for this process. It may be noted that one should expand the list of unit operations and equipment given in Table 1.1.

1.9 Examine reference 11 carefully and suggest most suitable equipment for each unit operation in P. 1.8.

1.10 A typical Indian meal consists of chapatis, rice, dal, one vegetable and curd. Design a kitchen which can feed five hundred people at a time if the cooking is mechanized.
(a) Define your goal clearly and the raw material list very carefully.
(b) Synthesize a steady state flow sheet.

1.11 Suggest equipment for each unit operation involved in P. 1.10. Also suggest a rough layout of the kitchen.

1.12 In the 'deep breeder system' of poultry farming, chickens are not allowed to get out of the cage during their entire life span. The problems that one faces in such farming are the following:
(a) Maintaining correct humidity and temperature.
(b) Treatment of chicken droppings.
(c) Keeping them free from diseases.
The goal of poultry farming is to obtain meat and eggs. State your raw material

list clearly and using the method outlined in this chapter, synthesize the process flow sheet carefully.

1.13 For each of the process steps in P. 1.12, suggest a suitable method of carrying it out. Give a proper layout of the farm. Specify a suitable design of the cage to contain the bird so that the droppings do not stick to its bottom. Also eggs must be classified into small and big ones. How are you going to do it in reality.

1.14 In a spaceship where an astronaut has to live for a long stretch of time, design a self-contained system with a human as a part of it. While doing it, it may be remembered that the total space available in the spaceship is limited and it is not possible to store enough water necessary for the whole duration. In view of this, the raw material list consists of

(a) packed food for the entire duration, and
(b) two days' supply of water.

First find out how much water one requires in a day. Then list the daily activity of a person. This constitutes the goal list. Next, give the process route of achieving it keeping in mind that everything has to be recycled. Find out how best you can recycle human excreta.

1.15 Consider an acetaldehyde plant starting from ethanol as the raw material. Prepare the reaction list. If the goal consists of pure acetaldehyde, suggest a process route using the method outlined in this chapter.

1.16 Do the same as in P. 1.15 for an ether plant. Write clearly the raw material, reaction and goal list.

1.17 Methanol is a very important raw material for several chemical industries and it is decided to manufacture it in large quantities. Examine the literature for different raw materials which would form methanol. These would constitute different process routes for attaining the goal. Draw the flow sheet for each route.

2
Optimization of Flow Sheet with respect to Heat Exchanger Network

2.1 INTRODUCTION

In a chemical plant, the following major tasks are involved:

1. chemical reaction,
2. change in pressure,
3. change in temperature, and
4. change in concentration.

As pointed out in Chapter 1, a steady state flow sheet shows the arrangement of equipment to carry these out and to find an optimal flow sheet, one optimizes each of these individually with respect to the plant arrangement. The sequencing of reactors would consist of choosing between stirred tank, batch, semibatch, or flow reactors or a combination of these. This is discussed at great length in any standard book on kinetics[1] and is not dealt with here. Sequencing to achieve the pressure change is not important as it is normally carried out in a single pump (or compressor).

In this chapter, the optimization of the steady state flow sheet with respect to the task of achieving temperature change is considered. In a chemical plant, some streams are required to be heated and some, to be cooled. There will be saving of considerable energy if streams, to be cooled, are used to heat up streams, to be heated. Such a strategy will give rise to the formation of a heat exchanger network. It is shown that the total cost of the network depends very strongly upon the order of sequencing and subsequently a systematic procedure has been given to optimize this network with respect to its structure. Appendix 2.1[2] gives the design procedure of a single heat exchanger and shows that a countercurrent exchanger is better for a given task. In view of this, the discussion in this chapter is limited to countercurrent exchangers.

2.2 NETWORK OF HEAT EXCHANGERS[3–5]

One assumes there is a hot stream, S_{h1}, in a chemical plant with an inlet temperature of 400°C and heat capacity flow rate ($=MC_p$) of 100 units to be cooled to 200°C. In the same chemical plant, there are two more cold streams, S_{c1} and S_{c2}, with their inlet temperatures as 100° and 200°C to be heated to 150° and 250°C respectively. Heat capacity flow rates of S_{c1} and S_{c2} are assumed to be 200 units each.

One alternative to achieve this task is to use cooling water for S_{h1} and high pressure steam to heat S_{c1} and S_{c2} to get temperatures at the desired level. A more economic alternative is to utilize S_{h1} to heat S_{c1} and S_{c2} to the desired temperatures. As can be seen, the heat to be taken out from S_{h1} to bring its temperature to 200°C is the same as required by S_{c1} and S_{c2}.

In Fig. 2.1, the second alternative is considered. S_{c1} and S_{c2} can be made to come into contact with S_{h1} in two ways: In Fig. 2.1a, S_{c2} makes

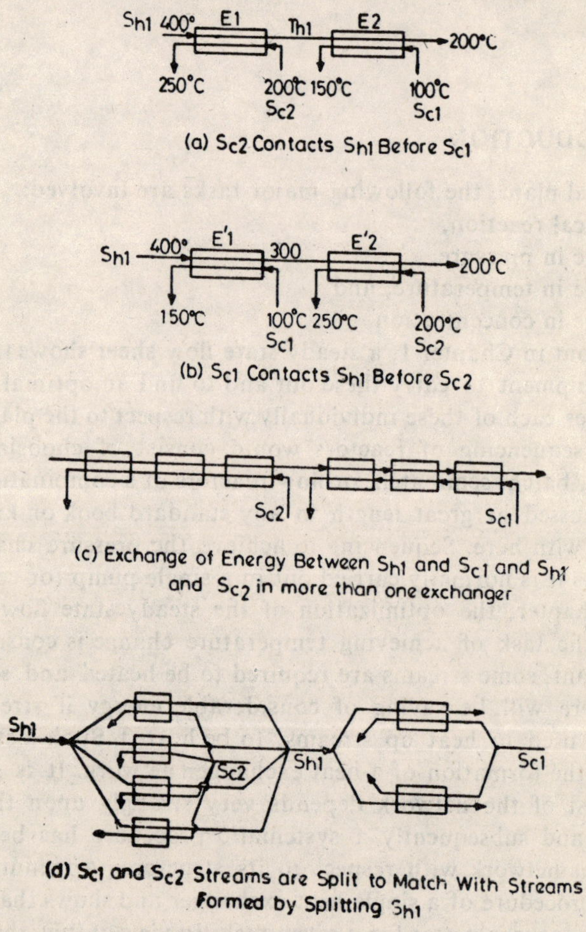

Fig. 2.1 Possible configurations of exchangers for one hot stream and two cold streams

contact first and in Fig. 2.1b, S_{c1} makes contact first. Other plausible alternatives can be generated either by contacting S_{h1} and S_{c1} (or S_{c2}) in more than one exchanger or split the streams to contact in separate exchangers. It will be shown later that these two classes of alternatives are exactly equivalent to those shown in Figs. 2.1a and b which are analysed in the following.

Heat exchangers are priced according to the surface area required. Therefore minimization of total surface area, A_T, is indicative of the minimum cost network. A_T of the network of Fig. 2.1a is calculated assuming the heat transfer coefficient, U, as 150 units. The intermediate hot stream temperature, T_{h1}, is found by

$$T_{h1} = 400 - \frac{200\,(250 - 200)}{200} = 300°C \tag{2.2.1a}$$

Therefore

$$(\Delta t_{lm})_{E1} = \frac{(400 - 250) - (300 - 200)}{\ln \dfrac{400 - 250}{300 - 200}} = 123.5°C \tag{2.2.1b}$$

and

$$(\Delta t_{lm})_{E2} = \frac{(300 - 150) - (200 - 100)}{\ln \dfrac{300 - 150}{100 - 100}} = 123.5°C \tag{2.2.1c}$$

Consequently

$$A_{E1} = A_{E2} = \frac{100 \times 100}{150 \times 123.5} = 0.540 \tag{2.2.2a}$$

and

$$A_T = A_{E1} + A_{E2} = 1.08 \tag{2.2.2b}$$

However, for the arrangement of Fig. 2.1 (b)

$$(\Delta t_{lm})_{E'1} = \frac{(400 - 150) - (300 - 100)}{\ln \dfrac{400 - 150}{300 - 100}} = 223°C \tag{2.2.3a}$$

$$(\Delta t_{lm})_{E'2} = 0 \tag{2.2.3b}$$

Therefore

$$A_{E'1} = 0.30 \tag{2.2.4a}$$

$$A_{E'2} = \infty \tag{2.2.4b}$$

and

$$A'_T = A_{E'1} + A_{E'2} = \infty \tag{2.2.4c}$$

$A_{E'2}$ is infinite because T_{c2} is the same as the T_h^* which is in reality never permitted in any finite heat exchanger. T_h^* is always greater than the inlet cooling temperature. The temperature difference, ΔT_m $(=T_h^* - T_{c2})$ for a given heat exchanger is a problem of optimization which has been solved by Kern[a] and has been shown as 10° to 12°C ($\simeq$ 20°F). In view of this, the scheme of Fig. 2.1b should be modified and two more alternatives are generated and shown in Fig. 2.2a and b. The first one shown in Fig. 2.2a occurs when it is decided to exchange heat upto the point till the temperature of the hot stream, S_h, falls to 210°C. Then S_{h1} is cooled to 200°C in a water cooler. Also S_{c2} will not reach the desired outlet temperature of 250°C, for which a hot utility (for example, steam) is used. The second method shown in Fig. 2.2b would be to cool S_{c2} first to 190°C and then

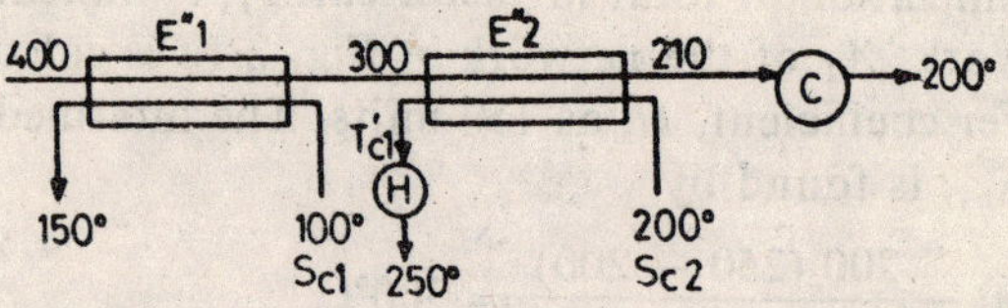

(a) Sh1 Cooled upto 210°C only

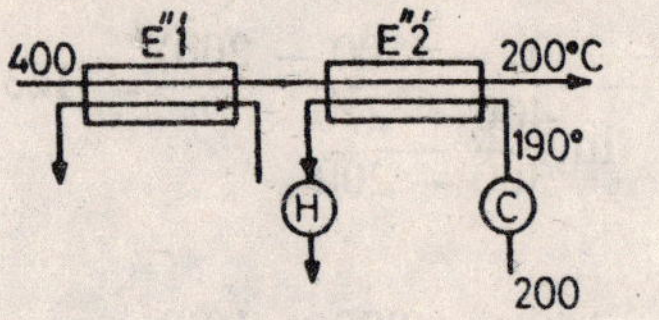

(b) Sc2 First Cooled to 190° before Exchanging with Sh1

Fig. 2.2 Modification of the scheme of Fig. 2.1b in view of $T_h^* \neq T_{c2}$

heat to the temperature of 250°C after it emerges from the exchanger $E2'$. Networks of Figs. 2.2a and b are analysed now with the assumption that the heating utility is steam at 300°C and the cooling utility is water at 30°C.

Since the areas of exchangers $E''1$ and $E'''1$ are going to be unaffected by changes in the latter exchanger,

$$A_{E''1} = A_{E'''1} = A_{E'1} = 0.30 \tag{2.2.5}$$

For $E''2$,

$$(\Delta t_{lm})_{E''2} = \frac{(300 - 245 - (210 - 200)}{\ln \frac{55}{10}} = 26.39 \tag{2.2.6a}$$

and

$$A_{E''2} = \frac{(100 \times 90)}{150 \times 26.39} = 2.27 \tag{2.2.6b}$$

Heaters and coolers required in Fig. 2.3 are going to be of small capacity and hence their areas are neglected. Therefore the total area, A_T'', for Fig. 2.2a is

$$A_T'' \simeq 0.30 + 2.27 = 2.57 \tag{2.2.7}$$

which is finite but is larger than that for the arrangement of Fig. 2.1a. Similarly the total area A_T''' for the arrangement in Fig. 2.2b is calculated and is found to be 2.69 units, which is also larger than A_T.

From the simple example considered above, it is shown that the order of contacting hot and cold streams is extremely important. It is found to affect not only the operating cost (the added cost of utility of Figs. 2.2a and b) but also the fixed cost of the total heat exchanger network. This particular problem was first solved by Rudd and Masso[3] and was formally stated as follows.

There are M hot streams S_{hi} to be cooled from T_{hi} to T_{hi}^* where $i = 1, 2, \ldots, M$, and there are N cold streams S_{cj} to be heated from T_{cj} to T_{cj}^* where $j = 1, 2, \ldots, N$ where M and N need not be equal. Cold and hot utilities are also available so that they can be used to obtain the desired temperatures in case heat exchanging with existing streams is not sufficient. All heat exchangers are assumed to be countercurrent shell and tube type and no phase change in either hot or cold streams is assumed to occur. The total cost of the exchanger network is desired to be minimized with respect to the choice of the network structure.

2.3 SOME NECESSARY CONDITIONS FOR THE EXISTENCE OF AN OPTIMAL HEAT EXCHANGER NETWORK[4]

Finding the best network is not a regular optimization problem because the representation of the network structure mathematically is a hard task. One procedure would be to formulate all possible arrangements, and find out the best configuration by brute force for the task stated above. Not all of these configurations may be feasible, and the necessary conditions discussed below place a limit on the choice of the network.

2.4 MAXIMUM HEAT TRANSFER IN A SINGLE EXCHANGER (RULE 1)

For a single heat exchanger, if T_c is greater than T_h^*, the direction of heat transfer is reversed and the log mean temperature difference becomes undefined. Consequently T_c can at best be equal to T_h^*, for which Δt_{lm} is zero. For such an exchange, infinite surface area is required and therefore any configuration involving an exchanger with $T_{ci} = T_{hj}^*$ is regarded as a nonfeasible one. Indeed on this basis, the network of Fig. 2.1b was neglected and in its place networks of Fig. 2.2 were considered. Maximum heat transfer can be calculated assuming that heat exchange can occur between hot and cold fluids as long as the temperature T_h^* drops to within $(T_c + \Delta T_m)$ where ΔT_m is 10°–12°C (~ 20°F)[2].

2.5 HOT AND COLD UTILITIES (RULE 2)

One considers the situation when the hot utility is hotter than the hottest input temperature of hot streams and the cold utility is colder than the coldest input temperature of cold streams. If cold and hot streams are made to exchange energies, the output temperature of the former can never be greater than the hottest input temperature of the latter. Therefore hot utility is used only when the required output temperature of any cold stream is more than the input value of the hottest stream. As a result, if there is a network of exchangers, the exchangers with hot utility (heaters) fall only in the end of the network as shown in Fig. 2.3. For similar reasons, exchangers with cold utility (coolers) are placed only in the end of the network. This constraint arises because of the need to maximize the heat recovery.

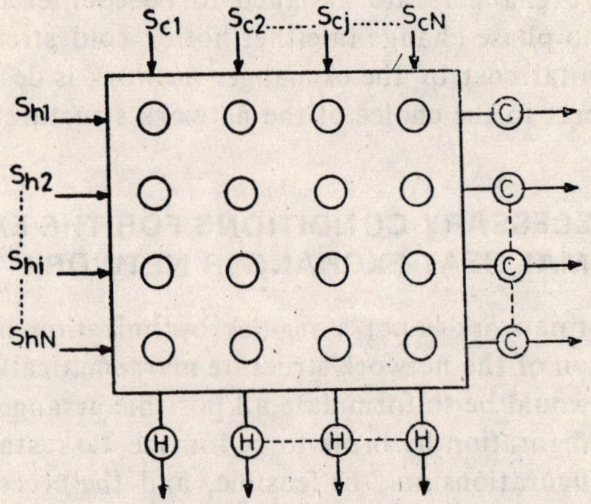

Fig. 2.3 Division of a heat exchanger network into internal and peripheral regions. (Reprinted from Ref. 4 with the permission of Pergamon Press, Oxford.)

Looking at the network shown in Fig. 2.3, it is seen that the entire network can be divided into two sections:

(a) the internal region where S_{hi} and S_{cj} (for all possible values of i and j) are matched; and

(b) the peripheral region where auxiliary equipment like heaters and coolers are placed.

This kind of partitioning, however, runs into difficulty if the hot and cold utility temperatures are not the highest and lowest. In such circumstances, heaters and coolers appear as a part of the internal structure. If these cold or hot utilities are treated as one of the exchanging streams, the structure of the network remains unchanged from Fig. 2.3.

Therefore the second necessary condition for the optimality can be stated as follows. If the hot and cold utilities are the hottest and coldest, heaters and coolers become the part of the peripheral structure of the network. Otherwise utilities are treated as one of the streams, and this constitutes rule 2 of feasible networks.

2.6 CONDITION OF OPTIMALITY FOR THE MINIMUM AREA NETWORK

The division of networks according to rule 2 makes it possible to optimize the internal and peripheral regions separately with respect to the choice of structure. After optimizing each of these regions individually, they can be put together to obtain the overall optimal structure. Since the peripheral structure as seen from Fig. 2.3, involves one exchanger (a heater or a cooler) at a time and is not a network, its optimization, therefore, is an easy task and not considered here. In view of this, the optimality of the internal structure is analysed in Sec. 2.6.1.

2.6.1

For the interior region, the structure giving minimum exchanger area is analysed first. The network structure is decided by the order of matching of S_{hi} with S_{cj} and is expressed mathematically as

$$\text{Network} = (S_{hi}, S_{cj})_n \quad i = 1, 2, \ldots, M \quad j = 1, 2, \ldots, N \quad n = 1, 2, \ldots \tag{2.6.1.1}$$

where (S_{hi}, S_{cj}) means that S_{hi} is matched with S_{cj} in nth exchanger and the problem consists of minimization of total surface area, A_T, with respect to (S_{hi}, S_{cj}) for all choices of i and j. As an explicit function cannot be formulated relating A_T with (S_{hi}, S_{cj}) as the variable, the commonly known methods of optimization cannot be applied. However, it is evident that if there is an optimum structure denoted by $(S_{hi}, S_{cj})_n^*$ for all possible values of i, j and n, for which A_T is minimum $(=A_T^*)$, then any permutation of hot and cold streams will increase A_T from A_T^* or at best, will maintain it on the same level.

Utilizing this property, conditions of pairing (or matching) hot and cold stream can be found for the most efficient heat transfer. In Sec. 2.7, three very special situations in energy transfer have been analyzed and later generalized for any exchanger network.

2.7 THREE SPECIAL SITUATIONS IN ENERGY TRANSFER

2.7.1 Order of Matching Hot and Cold Streams for Equal Heat Exchanged

In this section a special situation is considered in which number of hot and cold streams are equal, i.e.,

$$M = N = m \text{ (say)} \tag{2.7.1.1}$$

Also,

$$W_{h1} = W_{h2} = \ldots = W_{hm} = W_h \text{ (say)} \tag{2.7.1.2a}$$

and

$$W_{c1} = W_{c2} = \ldots = W_{cm} = W_c \text{ (say)} \tag{2.7.1.2b}$$

It is further assumed that a given hot stream exchanges energy once only with another cold stream and the heat duty of all exchangers is equal, i.e.,

$$Q_1 = Q_2 = \ldots = Q_m = \left(\frac{Q_T}{m}\right) = Q \text{ (say)} \tag{2.7.1.3}$$

where Q_T is the total heat exchanged for the entire network. Before optimizing the network, Eq. A2.1.8 for a single exchanger is rewritten as

$$UA = \frac{W_c W_h}{W_c - W_h} \ln \frac{T_h - T_c^*}{T_h^* - T_c} = \frac{W_c W_h}{W_h - W_c} \ln \frac{T_h^* - T_c}{T_h - T_c^*} \tag{2.7.1.4}$$

Eliminating T_h^* and T_c^* with the help of Eq. A2.1.3a one gets

$$A = \frac{Q}{U(\alpha - \beta)} \ln \frac{(T_h - T_c - \beta)}{(T_h - T_c - \alpha)} \tag{2.7.1.5}$$

where

$$\beta = Q/W_h \tag{2.7.1.6}$$

and

$$\alpha = Q/W_c \tag{2.7.1.7}$$

For the network considered in this section, α and β are the same for all exchangers, but α need not be equal to β because W_c is not necessarily equal to W_h. If hot stream, S_{hi} is matched with cold stream, S_{cj} then the area $A_{(ij)}$ for this exchanger is given by

$$A_{(ij)} = \frac{Q}{U(\alpha - \beta)} \ln \left(\frac{T_{hi} - T_{cj} - \beta}{T_{hi} - T_{cj} - \alpha}\right) \tag{2.7.1.8}$$

On summation of Eq. 2.7.1.8 for the whole structure, the total area of the network, A_T, can be found; but to do so, the matching order must be known.

To find the optimal order of matching, an optimal sequence is assumed to exist in which S_{hi} is matched with S_{ci} and S_{hj} with S_{cj}. In this optimal network, the permutation of streams of exchangers i and j, is made, i.e. matching is changed from (S_{hi}, S_{ci}) and (S_{hj}, S_{cj}) to (S_{hi}, S_{cj}) and (S_{hj}, S_{ci}). On this permutation the remaining exchangers are unaffected and the change in the total area, ΔA_T, of the network because of this is given by

$$\Delta A_T = \frac{Q}{U(\alpha - \beta)} \left[\left\{\ln \frac{T_{hi} - T_{cj} - \beta}{T_{hi} - T_{cj} - \alpha} + \ln \frac{T_{hj} - T_{ci} - \beta}{T_{hj} - T_{ci} - \alpha}\right\} - \left\{\ln\left(\frac{T_{hi} - T_{ci} - \beta}{T_{hi} - T_{ci} - \alpha}\right) + \ln\left(\frac{T_{hj} - T_{cj} - \beta}{T_{hj} - T_{cj} - \alpha}\right)\right\}\right] \tag{2.7.1.9}$$

or

$$\Delta A_T = \frac{Q}{U(\alpha-\beta)} \ln\left[\frac{(T_{hi}-T_{cj}-\beta)(T_{hj}-T_{ci}-\beta)(T_{hi}-T_{ci}-\alpha)(T_{hj}-T_{cj}-\alpha)}{(T_{hi}-T_{ci}-\beta)(T_{hj}-T_{cj}-\beta)(T_{hi}-T_{cj}-\alpha)(T_{hj}-T_{ci}-\alpha)}\right]$$

$$\equiv \frac{Q}{U(\alpha-\beta)} \ln X \tag{2.7.1.10}$$

Simplification of terms in parentheses can be carried out by noting

$$\frac{T_{hi}-T_{cj}-\beta}{T_{hi}-T_{ci}-\beta} = 1 + \frac{T_{ci}-T_{cj}}{(T_{hi}-T_{ci}-\beta)} \tag{2.7.1.11a}$$

$$\frac{T_{hj}-T_{ci}-\beta}{T_{hj}-T_{cj}-\beta} = 1 - \frac{T_{ci}-T_{cj}}{(T_{hj}-T_{cj}-\beta)} \tag{2.7.1.11b}$$

$$\frac{T_{hi}-T_{ci}-\alpha}{T_{hi}-T_{cj}-\alpha} = 1 - \frac{T_{ci}-T_{cj}}{T_{hi}-T_{cj}-\alpha} \tag{2.7.1.11c}$$

and

$$\frac{T_{hj}-T_{cj}-\alpha}{T_{hj}-T_{ci}-\alpha} = 1 + \frac{T_{ci}-T_{cj}}{T_{hj}-T_{ci}-\alpha} \tag{2.7.1.11d}$$

Also,

$$\frac{(T_{hi}-T_{cj}-\beta)(T_{hj}-T_{ci}-\beta)}{(T_{hi}-T_{ci}-\beta)(T_{hj}-T_{cj}-\beta)}$$

$$= \left\{1 + \frac{T_{ci}-T_{cj}}{(T_{hi}-T_{ci}-\beta)}\right\}\left\{1 - \frac{T_{ci}-T_{cj}}{T_{hj}-T_{cj}-\beta}\right\}$$

$$= 1 - \frac{(T_{ci}-T_{cj})(T_{hi}-T_{hj})}{(T_{hi}-T_{ci}-\beta)(T_{hj}-T_{cj}-\beta)} \tag{2.7.1.12}$$

Similarly

$$\frac{(T_{hi}-T_{ci}-\alpha)(T_{hj}-T_{cj}-\alpha)}{(T_{hi}-T_{cj}-\alpha)(T_{hj}-T_{ci}-\alpha)}$$

$$= \left\{1 - \frac{T_{ci}-T_{cj}}{T_{hi}-T_{cj}-\alpha}\right\}\left\{1 + \frac{T_{ci}-T_{cj}}{T_{hj}-T_{ci}-\alpha}\right\}$$

$$= 1 + \frac{(T_{ci}-T_{cj})(T_{hi}-T_{hj})}{(T_{hj}-T_{ci}-\alpha)(T_{hi}-T_{cj}-\alpha)} \tag{2.7.1.13}$$

Therefore

$$X = \left\{1 - \frac{(T_{ci}-T_{cj})(T_{hi}-T_{hj})}{(T_{hi}-T_{ci}-\beta)(T_{hj}-T_{cj}-\beta)}\right\}\times$$

$$\left\{1 + \frac{(T_{ci}-T_{cj})(T_{hi}-T_{hj})}{(T_{hj}-T_{ci}-\alpha)(T_{hi}-T_{cj}-\alpha)}\right\}$$

$$= 1 + \frac{(T_{ci}-T_{cj})(T_{hi}-T_{hj})\{(T_{hi}-T_{ci}-\alpha)+(T_{hj}-T_{cj}-\beta)\}}{(T_{hi}-T_{ci}-\beta)(T_{hj}-T_{cj}-\beta)(T_{hi}-T_{cj}-\alpha)(T_{hj}-T_{ci}-\alpha)}$$

$$\tag{2.7.1.14}$$

If (S_{hi}, S_{ci}) and (S_{hj}, S_{cj}) is optimal, A_T will either increase or at best remain unchanged by the permutation, i.e.

$$\Delta A_T \geqslant 0 \tag{2.7.1.15}$$

Substituting this in Eq. 2.7.1.9, one gets

$$X \geqslant 1 \tag{2.7.1.16}$$

or

$$\frac{(T_{ci}-T_{cj})(T_{hi}-T_{hj})\{(T_{hi}-T_{ci}-\alpha)+(T_{hj}-T_{cj}-\beta)\}}{(T_{hi}-T_{ci}-\beta)(T_{hj}-T_{cj}-\beta)(T_{hi}-T_{cj}-\alpha)(T_{hj}-T_{ci}-\alpha)} \geqslant 0 \tag{2.7.1.17}$$

As heat transfer occurs from the hot to cold fluid, $(T_{hi}-T_{ci}-\alpha)$, $(T_{hj}-T_{cj}-\beta)$ each is positive (Eqs. 2.7.1.5 and 2.7.1.6).

Therefore from Eq. (2.7.1.17), it follows

$$(T_{ci}-T_{cj})(T_{hi}-T_{hj}) \geqslant 0 \tag{2.7.1.18}$$

i.e.,

$$T_{hi} \geqslant T_{hj} \tag{2.7.1.19a}$$

and

$$T_{ci} \geqslant T_{cj} \tag{2.7.1.19b}$$

Therefore the following relation is obtained for all i and j

$$T_{h1} \geqslant T_{h2} \geqslant T_{h3} \ldots \geqslant T_{hm} \tag{2.7.1.20a}$$

and

$$T_{c1} \geqslant T_{c2} \geqslant T_{c3} \ldots \geqslant T_{cm} \tag{2.7.1.20b}$$

Consequently the optimal structure is (S_{h1}, S_{c1}), $(S_{h2}, S_{c2}) \ldots (S_{hm}, S_{cm})$ under the assumption stated in Eqs. 2.7.1.1, 2.7.1.2 and 2.7.1.3. Supposing there are two hot streams of input temperatures of 400°C and 300°C of each having heat capacity flow rates of 100 units and two cold streams of input temperatures 100°C and 200°C of each having heat capacity flow rates of 200 units. According to the result derived in this section, they should be arranged as

$$S_{h1} \text{ at } T_{h1} = 400°\text{C}, \quad S_{c1} \text{ at } T_{c1} = 200°\text{C} \tag{2.7.1.21a}$$

$$S_{h2} \text{ at } T_{h2} = 300°\text{C}, \quad S_{c2} \text{ at } T_{c2} = 100°\text{C} \tag{2.7.1.21b}$$

and matching must be done as (S_{h1}, S_{c1}) and (S_{h2}, S_{c2}).

The result derived in this section can be easily applied to the problem of matching one hot stream and two cold streams discussed in Fig. 2.1 as follows. If exchangers E_1 and E_2 are disconnected by breaking the connecting stream S_{h1}, the matching problem is exactly given by Eq. 2.7.1.21. Thus, the results of Eq. 2.7.1.20 can be applied even when the assumption of Eq. 2.7.1.1 is not strictly valid in a given problem.

2.7.2 One Hot Stream Exchanging with One Cold Stream in One Exchanger versus Several Exchangers

One considers the situation when one hot stream, S_{h1}, is exchanging energy with one cold stream, S_{c1}. Heat duty for this process is known because inlet and outlet temperatures of both these streams are fixed at T_{h1}, T_{c1} and T^*_{h1}, T^*_{c1}. Two basic questions now need to be answered for a task of this nature. The first one is whether the energy exchange should occur in one or several series exchangers. If it is decided that S_{c1} matches with S_{h1} in m exchangers one would like to find the solution of the second question: what should be order of matching and the value of m.

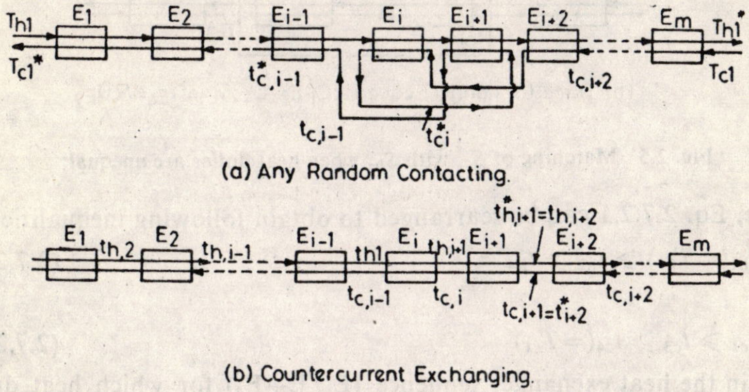

Fig. 2.4 **Matching of one hot stream in *m* exchanger in series with one cold stream**

Figure 2.4 shows a sequence of m heat exchangers for the exchange of energy between S_{h1} and S_{c1}. In Fig. 2.4a, a random matching is shown where the outgoing cold stream of E_{i+2} is sent to E_i with hot stream passing sequentially. In Fig. 2.4b, heat exchanging is strictly countercurrent. Inlet and outlet temperatures have the same subscript as that of the exchanger. For the moment the heat duty of all exchangers are assumed to be the same. If the hot and the cold streams entering in a given exchanger are treated as distinct, then the assumptions of Sec. 2.7.1 are strictly met: there are m hot streams and m cold streams which are exchanging once only with each other. Therefore according to Eq. 2.7.1.20, matching must be done in order of decreasing input temperatures and the countercurrent matching shown in Fig. 2.4b is the best. Now it can be proved that the countercurrent matching is the best *even when* heat duties of exchangers are not equal as follows.

A sequence of four heat exchangers represented in Fig. 2.5 are matched countercurrently. When their heat duties are equal, the sequencing is given by Eq. 2.7.1.20 as

$$t_{h1}(=T_{h1}) \geqslant t_{h2} \geqslant t_{h3} \geqslant t_{h4} \tag{2.7.2.1a}$$

and

$$t_{c1} \geqslant t_{c2} \geqslant t_{c3} \geqslant t_{c4}\ (=T_{c1}) \tag{2.7.2.1b}$$

Exchangers E_2 and E_3 can always be combined into a single one E_2', because this is a physical process. As inlet and outlet temperatures do not

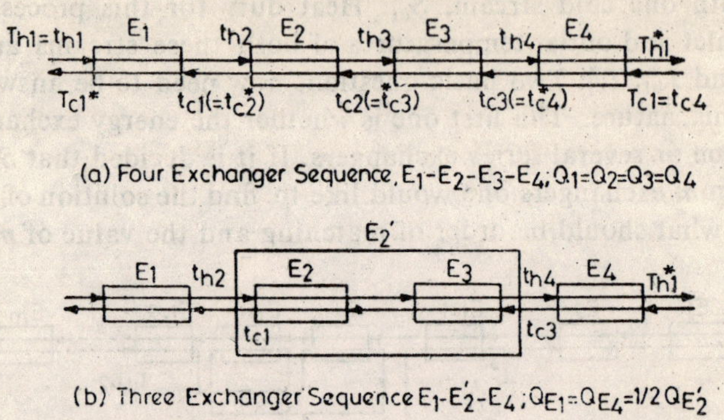

Fig. 2.5 Matching of S_{c_1} with S_{h_1} when heat duties are unequal

change, Eq. 2.7.2.1 can be rearranged to obtain following inequalities

$$t_{h1}(=T_{h1}) \geqslant t_{h2} \geqslant t_{h4} \tag{2.7.2.2a}$$

and

$$t_{c1} \geqslant t_{c3} \geqslant t_{c4}(=T_{c1}) \tag{2.7.2.2b}$$

Thus, in the heat exchanger sequence (E_1, E_2', E_4) for which heat duties are not equal, matching is once again in order of decreasing input temperature of the hot stream.

Therefore the countercurrent configuration with matching in order of decreasing temperature is optimal even when heat duty of exchangers are not equal. Now to find the value of m the heat duty of exchanger E_i is assumed as Q_i such that

$$Q_T = \sum_{i=1}^{m} Q_i \tag{2.7.2.3}$$

where Q_T is the total energy exchanged. Intermediate hot temperatures can be found in terms of the heat exchanged and T_{h1}, i.e.,

$$t_{h2} = T_{h1} - \frac{Q_1}{W_{h1}} \tag{2.7.2.4a}$$

$$t_{h3} = T_{h1} - \left(\frac{Q_1}{W_{h1}} + \frac{Q_2}{W_{h1}}\right) \tag{2.7.2.4b}$$

or in general

$$t_{hi} = T_{h1} - \sum_{j=1}^{i-1} \left(\frac{Q_j}{W_{h1}}\right) \tag{2.7.2.4c}$$

Similarly for intermediate cold temperatures

$$t_{c(m-1)} = T_{c1} + \frac{Q_m}{W_{c1}} \tag{2.7.2.5a}$$

$$t_{c(m-2)} = T_{c1} + \left(\frac{Q_m}{W_{c1}} + \frac{Q_{m-1}}{W_{c1}}\right) \tag{2.7.2.5b}$$

or in general

$$t_{ci} = \left(T_{c1} + \sum_{j=i+1}^{m} \frac{Q_j}{W_{c1}} \right) \tag{2.7.2.5c}$$

Using Eq. 2.7.1.4, the surface area A_{Ei} of the heat exchanger E_i can be found to be

$$A_{Ei} = \frac{W_{c1}W_{h1}}{U(W_{h1} - W_{c1})} \ln \frac{t_{hi+1} - t_{ci}}{t_{hi} - t_{ci-1}}$$

$$= \frac{W_{c1}W_{h1}}{U(W_{h1} - W_{c1})} \ln \frac{T_{h1} - T_{c1} - \sum_{j=1}^{i}\left(\frac{Q_j}{W_{h1}}\right) - \sum_{j=i+1}^{m}\left(\frac{Q_j}{W_{c1}}\right)}{T_{h1} - T_{c1} - \sum_{j=1}^{i-1}\left(\frac{Q_j}{W_{h1}}\right) - \sum_{j=i}^{m}\left(\frac{Q_j}{W_{c1}}\right)} \tag{2.7.2.6}$$

The above equation (Eq. 2.7.2.6) is obtained by eliminating intermediate temperatures using Eq. 2.7.2.4 and 2.7.2.5. The total heat exchanger area, A_T, is found by the summation of the areas of all exchangers and is given by

$$A_T = \frac{W_{c1}W_{h1}}{U(W_{h1} - W_{c1})} \sum_{i=1}^{m} \ln \frac{T_{h1} - T_{c1} - \sum_{j=1}^{i}\left(\frac{Q_j}{W_{h1}}\right) - \sum_{j=i+1}^{m}\left(\frac{Q_j}{W_{c1}}\right)}{T_{h1} - T_{c1} - \sum_{j=1}^{i-1} \frac{Q_j}{W_{h1}} - \sum_{j=i}^{m} \frac{Q_j}{W_c}}$$

$$= \frac{W_{c1}W_{h1}}{U(W_{h1} - W_{c1})} \sum_{i=1}^{m} \ln X_i \tag{2.7.2.7}$$

It is further noted that

$$(X_iX_{i+1}) = \left\{ \frac{T_{h1} - T_{c1} - \sum_{j=1}^{i} \frac{Q_j}{W_{h1}} - \sum_{j=i+1}^{m} \frac{Q_j}{W_{c1}}}{T_{h1} - T_{c1} - \sum_{j=1}^{i-1} \frac{Q_j}{W_{h1}} - \sum_{j=i}^{m} \frac{Q_j}{W_{c1}}} \right\} \cdot \left\{ \frac{T_{h1} - T_{c1} - \sum_{j=1}^{i+1} \frac{Q_j}{W_h} - \sum_{j=i+2}^{m}\left(\frac{Q_j}{W_{c1}}\right)}{T_{h1} - T_{c1} - \sum_{j=1}^{i} \frac{Q_j}{W_{h1}} - \sum_{j=i+1}^{m}\left(\frac{Q_j}{W_{c1}}\right)} \right\}$$

$$= \frac{T_{h1} - T_{c1} - \sum_{j=1}^{i+1} \frac{Q_j}{W_h} - \sum_{j=i+2}^{m}\left(\frac{Q_j}{W_{c1}}\right)}{T_{h1} - T_{c1} - \sum_{j=1}^{i-1} \frac{Q_j}{W_{h1}} - \sum_{j=i}^{m}\left(\frac{Q_j}{W_{c1}}\right)} \tag{2.7.2.8}$$

Consequently

$$\prod_{i=1}^{m} X_i = X_1X_2 \ldots X_m = \frac{T_{h1} - T_{c1} - \sum_{j=1}^{m} \frac{Q_j}{W_{h1}}}{T_{h1} - T_{c1} - \sum_{j=1}^{m} \frac{Q_j}{W_{c1}}} \tag{2.7.2.9}$$

Therefore,

$$A_T = \frac{W_{c1}W_{h1}}{U(W_{h1} - W_{c1})} \sum_{i=1}^{m} \ln X_i = \frac{W_{h1}W_{c1}}{U(W_{c1} - W_{h1})} \ln \left(\prod_{i=1}^{m} X_i \right) \tag{2.7.2.10}$$

or

$$A_T = \frac{W_{c1} W_{h1}}{U(W_{h1} - W_{c1})} \ln \frac{(T_{h1} - T_{c1} - Q_T/W_{h1})}{(T_{h1} - T_{c1} - Q_T/W_{c1})} \quad (2.7.2.11)$$

which is the same as that of a single exchanger in which hot and cold streams enter at temperatures T_{h1} and T_{c1} respectively and exchange an amount of heat, Q_T.

It is thus seen that for a task of a hot stream to be cooled from T_h to T_h^* by a cold stream to be heated from T_c to T_c^*, it is of no consequence whether the exchanging is done in one or several exchangers.

2.7.3 One Hot Stream Contacting One Cold Stream in Several Parallel Heat Exchangers

One assumes that the hot stream is split in m heat exchangers to contact similarly split cold stream as shown in Fig. 2.6. It is desired to find out if stream splitting is a better strategy compared to the energy exchange in one exchanger.

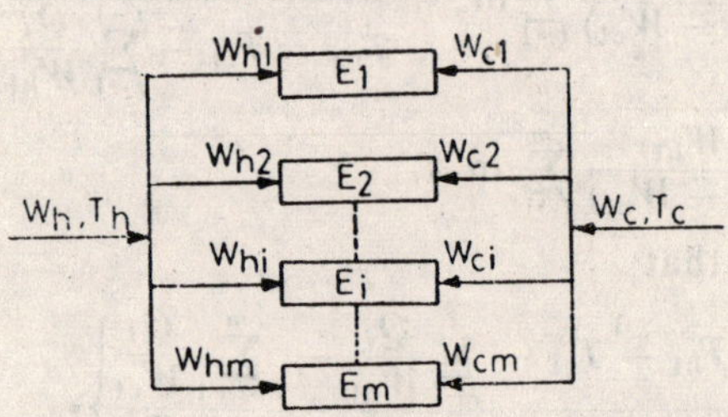

Fig. 2.6 Exchange of energy in m parallel exchangers

If Q_i is the heat transferred in exchanger E_i, its surface area is given by Eq. 2.7.1.4.

$$A_i = \frac{W_{ci} W_{hi}}{U(W_{hi} - W_{ci})} \ln \frac{T_h - T_c - Q_i/W_{hi}}{T_h - T_c - Q_i/W_{ci}} \quad (2.7.3.1)$$

Therefore the total area A_T is found by summing over i

$$A_T = \sum A_i = \sum_{i=1}^{m} \frac{W_{ci} W_{hi}}{U(W_{hi} - W_{ci})} \ln \frac{T_h - T_c - Q_i/W_{hi}}{T_h - T_c - Q_i/W_{ci}} \quad (2.7.3.2)$$

where

$$W_h = \sum_{i=1}^{m} W_{hi}, \quad (2.7.3.3a)$$

$$W_c = \sum_{i=1}^{m} W_{ci}, \quad (2.7.3.3b)$$

and

$$Q_T = \sum_{i=1}^{m} Q_i \tag{2.7.3.3c}$$

To minimize A_T with respect to the choice of Q_i, W_{ci} and W_{hi} with constraints expressed by Eq. 2.7.3.3 three arbitrary variables λ_1, λ_2 and λ_3, called Lagrange multipliers, are defined. A_T^* is then written as

$$A_T^* = \sum \frac{1}{U\left(\dfrac{1}{W_{ci}} - \dfrac{1}{W_{hi}}\right)} \ln \frac{T_h - T_c - Q_i/W_{hi}}{T_h - T_c - Q_i/W_{ci}} + \lambda_1(W_h - \Sigma W_{hi})$$
$$+ \lambda_2(W_c - \Sigma W_{ci}) + \lambda_3(Q_T - \Sigma Q_i) \tag{2.7.3.4}$$

Instead of minimizing Eq. 2.7.3.2 with its constraints, Eq. 2.7.3.4 is minimized without constraints. In this equation, variables are now W_{ci}, W_{hi}, Q_{hi}, λ_1, λ_2 and λ_3 with respect to which A_T^* has to be optimized. To minimize A_T^* with respect to λ_1, one sets $\partial A_T^*/\partial\lambda_1$ equal to zero, and the equation thus generated is the same as Eq. 2.7.3.3a. Similar results are found for λ_2 and λ_3.

To minimize A_T^* with respect to W_{hi}, W_{ci} and Q_i one sets $\partial A_T^*/\partial W_{hi}$, $\partial A_T^*/\partial W_{ci}$ and $\partial A_T^*/\partial Q_i$ equal to zero as follows

$$\frac{\partial A_T^*}{\partial W_{hi}} = \frac{1}{U} \frac{-\dfrac{1}{W_{hi}^2}}{\left(\dfrac{1}{W_{ci}} - \dfrac{1}{W_{hi}}\right)^2} \ln \frac{T_h - T_c - \dfrac{Q_i}{W_{hi}}}{T_h - T_c - \dfrac{Q_i}{W_{ci}}}$$
$$+ \frac{1}{U\left(\dfrac{1}{W_{ci}} - \dfrac{1}{W_{hi}}\right)} \frac{\dfrac{Q_i}{W_{hi}^2}}{\left(T_h - T_c - \dfrac{Q_i}{W_{hi}}\right)} - \lambda_1 = 0 \tag{2.7.3.5}$$

On rearranging after the use of Eq. 2.7.3.1,

$$-\frac{\dfrac{A_i}{W_{hi}}}{\left(\dfrac{1}{W_{ci}} - \dfrac{1}{W_{hi}}\right)} + \frac{\dfrac{Q_i}{W_{hi}}}{U\left(\dfrac{1}{W_{ci}} - \dfrac{1}{W_{hi}}\right)\left(T_h - T_c - \dfrac{Q}{W_{hi}}\right)} - \lambda_1 W_{hi} = 0 \tag{2.7.3.6}$$

Summing over i and using Eq. 2.7.3.3a, λ_1 is found to be

$$\lambda_1 W_h = \lambda_1 \sum_{i=1}^{m} W_{hi} = \sum_{i=1}^{m} \frac{\dfrac{Q_i}{W_{hi}}}{U\left(\dfrac{1}{W_{ci}} - \dfrac{1}{W_{hi}}\right)\left(T_h - T_c - \dfrac{Q_i}{W_{hi}}\right)}$$
$$- \sum_{i=1}^{m} \frac{\dfrac{A_i}{W_{hi}}}{\left(\dfrac{1}{W_{ci}} - \dfrac{1}{W_{hi}}\right)} \tag{2.7.3.7}$$

Eliminating λ_1 between Eqs. 2.7.3.6 and 2.7.3.7

$$-\frac{\dfrac{A_i}{W_{hi}}}{\left(\dfrac{1}{W_{ci}}-\dfrac{1}{W_{hi}}\right)}+\frac{\dfrac{Q_i}{W_{hi}}}{U\left(\dfrac{1}{W_{ci}}-\dfrac{1}{W_{hi}}\right)\left(T_h-T_c-\dfrac{Q_i}{W_{hi}}\right)}$$

$$=\frac{W_{hi}}{W_h}\left\{\sum_{j=1}^{m}\frac{\dfrac{Q_j}{W_{hj}}}{U\left(\dfrac{1}{W_{cj}}-\dfrac{1}{W_{hj}}\right)-\left(T_h-T_c-\dfrac{Q_j}{W_{hj}}\right)}-\sum_{j=1}^{m}\frac{\dfrac{A_j}{W_{hj}}}{\left(\dfrac{1}{W_{cj}}-\dfrac{1}{W_{hj}}\right)}\right\} \tag{2.7.3.8}$$

One similarly sets $\dfrac{\partial A_T^*}{\partial W_{ci}}$ and $\dfrac{\partial A_T^*}{\partial Q_i}$ equal to zero and eliminates λ_2 and λ_3 using Eqs. 2.7.3.3b and 2.7.3.3c to obtain

$$\frac{\dfrac{A_i}{W_{ci}}}{\left(\dfrac{1}{W_{ci}}-\dfrac{1}{W_{hi}}\right)}-\frac{\dfrac{Q_i}{W_{ci}}}{U\left(\dfrac{1}{W_{ci}}-\dfrac{1}{W_{hi}}\right)\left(T_h-T_c-\dfrac{Q_i}{W_{ci}}\right)}$$

$$=\frac{W_{ci}}{W_c}\left\{\sum_{j=1}^{m}\frac{\dfrac{A_j}{W_{cj}}}{\left(\dfrac{1}{W_{cj}}-\dfrac{1}{W_{hj}}\right)}-\sum_{j=1}^{m}\frac{\dfrac{Q_j}{W_{cj}}}{\left(\dfrac{1}{W_{ci}}-\dfrac{1}{W_{hj}}\right)\left(T_h-T_c-\dfrac{Q_j}{W_{cj}}\right)}\right\} \tag{2.7.3.9}$$

and

$$\frac{\dfrac{Q_i}{W_{ci}}}{U\left(\dfrac{1}{W_{ci}}-\dfrac{1}{W_{hi}}\right)\left(T_h-T_c-\dfrac{Q_i}{W_{ci}}\right)}-\frac{\dfrac{Q_i}{W_{hi}}}{U\left(\dfrac{1}{W_{ci}}-\dfrac{1}{W_{hi}}\right)\left(T_h-T_c-\dfrac{Q_i}{W_{hi}}\right)}$$

$$=\frac{Q_i}{Q_T}\left\{\sum_{j=1}^{m}\frac{\dfrac{Q_j}{W_{cj}}}{U\left(\dfrac{1}{W_{cj}}-\dfrac{1}{W_{hj}}\right)\left(T_h-T_c-\dfrac{Q_j}{W_{cj}}\right)}\right.$$

$$\left.-\sum_{j=1}^{m}\frac{\dfrac{Q_j}{W_{hj}}}{U\left(\dfrac{1}{W_{cj}}-\dfrac{1}{W_{hj}}\right)\left(T_h-T_c-\dfrac{Q_j}{W_{hj}}\right)}\right\} \tag{2.7.3.10}$$

Eqs. 2.7.3.8 and 2.7.3.9 are added to obtain

$$A_i+\frac{1}{U\left(\dfrac{1}{W_{ci}}-\dfrac{1}{W_{hi}}\right)}\left\{\frac{\dfrac{Q_i}{W_{hi}}}{T_h-T_c-\dfrac{Q_i}{W_{hi}}}-\frac{\dfrac{Q_i}{W_{ci}}}{T_h-T_c-\dfrac{Q_i}{W_{ci}}}\right\}$$

$$= \frac{W_{cl}}{W_c}\sum_{j=1}^{m}\frac{\dfrac{A_j}{W_{cj}}}{\dfrac{1}{W_{cj}}-\dfrac{1}{W_{hj}}} - \frac{W_{hl}}{W_h}\sum_{j=1}^{m}\frac{\dfrac{A_j}{W_{hj}}}{\dfrac{1}{W_{cj}}-\dfrac{1}{W_{hj}}}$$

$$+ \frac{W_{hl}}{W_h}\sum_{j=1}^{m}\frac{\dfrac{Q_j}{W_{hj}}}{\left(\dfrac{1}{W_{cj}}-\dfrac{1}{W_{hj}}\right)U\left(T_h - T_c - \dfrac{Q_i}{W_{hj}}\right)}$$

$$- \frac{W_{cl}}{W_c}\sum_{j=1}^{m}\frac{\dfrac{Q_j}{W_{cj}}}{U\left(\dfrac{1}{W_{cj}}-\dfrac{1}{W_{hj}}\right)\left(T_h - T_c - \dfrac{Q_j}{W_{cj}}\right)} \tag{2.7.3.11}$$

In Eq. 2.7.3.11 the expression in the parantheses of the left hand side is eliminated using Eq. 2.7.3.10 to obtain

$$\left\{A_l - \frac{W_{cj}}{W_c}\sum_{j=1}^{m}\frac{\dfrac{A_j}{W_{cj}}}{\left(\dfrac{1}{W_{cj}}-\dfrac{1}{W_{hj}}\right)} + \frac{W_{hl}}{W_h}\sum_{j=1}^{m}\frac{\dfrac{A_j}{W_{hj}}}{\left(\dfrac{1}{W_{cj}}-\dfrac{1}{W_{hj}}\right)}\right\}$$

$$-\left\{\left(\frac{W_{hl}}{W_h}-\frac{Q_l}{Q_T}\right)\sum_{j=1}^{m}\frac{\dfrac{Q_j}{W_{hj}}}{U\left(\dfrac{1}{W_{cj}}-\dfrac{1}{W_{hj}}\right)\left(T_h - T_c - \dfrac{Q_j}{W_{hj}}\right)}\right\}$$

$$-\left\{\left(\frac{W_{cl}}{W_c}-\frac{Q_l}{Q_T}\right)\sum_{j=1}^{m}\frac{\dfrac{Q_j}{W_{cj}}}{U\left(\dfrac{1}{W_{cj}}-\dfrac{1}{W_{hj}}\right)\left(T_h - T_c - \dfrac{Q_j}{W_{cj}}\right)}\right\} \doteq 0 \tag{2.7.3.12}$$

One (trivial) solution of this equation is $A_l = Q_l = W_{cl} = W_{hl} = 0$. Another (non-trivial) solution is given by

$$\frac{W_{hl}}{W_h} = \frac{Q_l}{Q_T} \tag{2.7.3.13a}$$

$$\frac{W_{cl}}{W_c} = \frac{Q_l}{Q_T} \tag{2.7.3.13b}$$

and

$$\frac{A_l}{A_T} = \frac{W_{cl}}{W_c} \tag{2.7.3.13c}$$

which can be checked by substituting these in Eq. 2.7.3.12. With the help of these relations, A_T^* is found to be

$$A_T^* = \sum_{i=1}^{m}\frac{Q_l}{\left(\dfrac{Q_l}{W_{cl}}\right)-\left(\dfrac{Q_l}{W_{hl}}\right)}\ \ln\frac{T_h - T_c - \dfrac{Q_T}{W_h}}{T_h - T_c - \left(\dfrac{Q_T}{W_c}\right)}$$

$$= \frac{\Sigma Q_i}{\dfrac{Q_T}{W_c} - \dfrac{Q_T}{W_h}} \ln \frac{T_h - T_c - \dfrac{Q_T}{W_h}}{T_h - T_c - \dfrac{Q_T}{W_c}}$$

$$= \frac{W_c W_h}{W_h - W_c} \ln \frac{T_h - T_c - Q_T/W_h}{T_h - T_c - Q_T/W_c} \quad (2.7.3.14)$$

i.e., the minimum area of the ensemble of exchangers is the same as that of the one transferring the total heat Q_T.

In Sec. 2.8, a heat content diagram is discussed which is a convenient method of representing a general heat exchanger network problem.

2.8 HEAT CONTENT DIAGRAM REPRESENTATION OF THE NETWORK PROBLEM

A heat content diagram is a plot of stream temperatures and heat capacity flow rates and is a means by which the total heat exchanged and the input and output temperatures of hot and cold streams are shown. As an example if there is a stream to be cooled from 400° to 200°C with heat capacity flow rate of 100 units, this appears as a rectangle in the heat content diagram of Fig. 2.7a. The area of this rectangle is equal to the total heat exchanged. The horizontal relative position of this rectangle is unimportant as long as the area of the block remains unchanged which

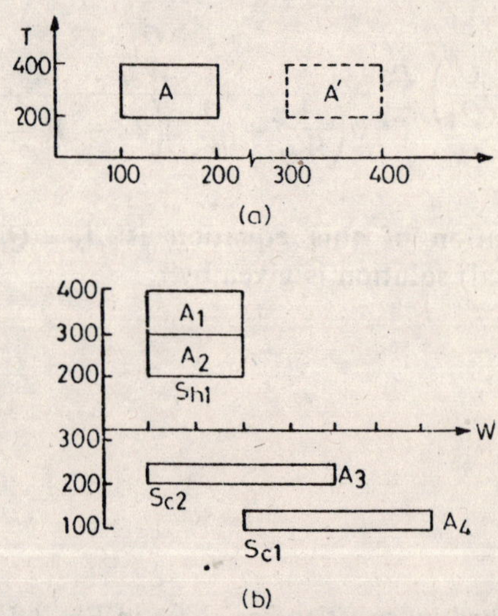

Fig. 2.7 Heat content diagram of one hot and two cold stream: (a) effect of lateral shift, (b) total diagram according to the convention

means that the lateral shift of the block A in Fig. 2.7a to A′ does not change the representation.

While showing several hot and cold streams, the following convention is observed to avoid overlapping of rectangles. Hot streams are shown above the horizontal line, and cold streams below it. All hot streams are shown in order of decreasing *input* temperatures and all cold streams, in order of their decreasing *output* temperatures. As an example, the heat content diagram of the problem of Fig. 2.1 is drawn in Fig. 2.7b. In this, the rectangle corresponding to S_{c2} is drawn first.

Horizontal splitting of a rectangular block in a heat content diagram indicates that total heat transfer occurs in two exchangers in series. Thus division of the block S_{h1} into A_1 and A_2 in Fig. 2.7b stands for the fact that the total heat exchange occurs in two exchangers E1 and E2. If A_1 is exchanged by S_{c1}, then A_1 is set equal to A_4 and A_2 equal to A_3 (Fig. 2.8a). Similarly if A_1 is exchanged by S_{c2}, A_2 is set equal to A_3 and A_2 equal to A_4 (Fig. 2.8b). The two resulting network structures are the same as those shown in Fig. 2.1a and b.

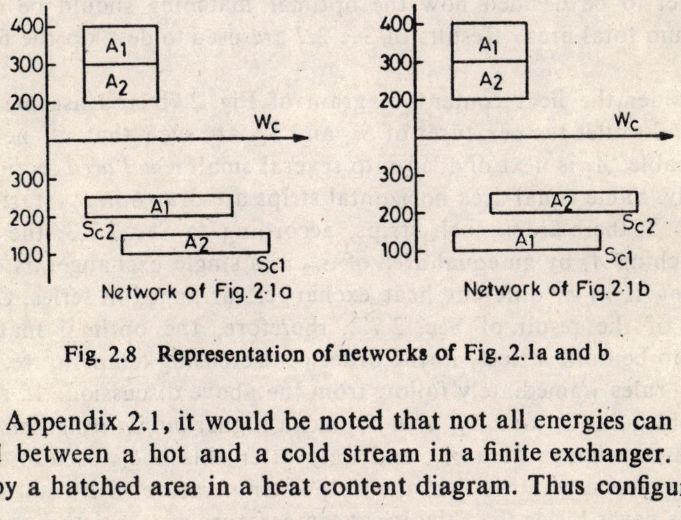

Fig. 2.8 Representation of networks of Fig. 2.1a and b

From Appendix 2.1, it would be noted that not all energies can be exchanged between a hot and a cold stream in a finite exchanger. This is shown by a hatched area in a heat content diagram. Thus configurations

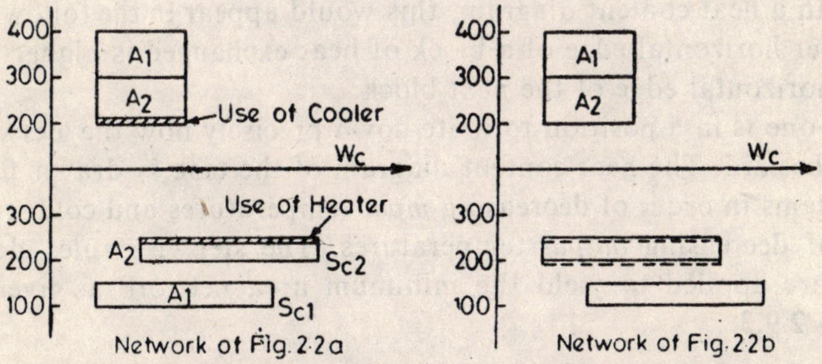

Fig. 2.9 Heat content diagram of structures shown in Fig. 2.2

of Fig. 2.2a and b are depicted by heat content diagrams of Fig. 2.9a and b. The hatched areas are exchanged with hot or cold utilities: the lower

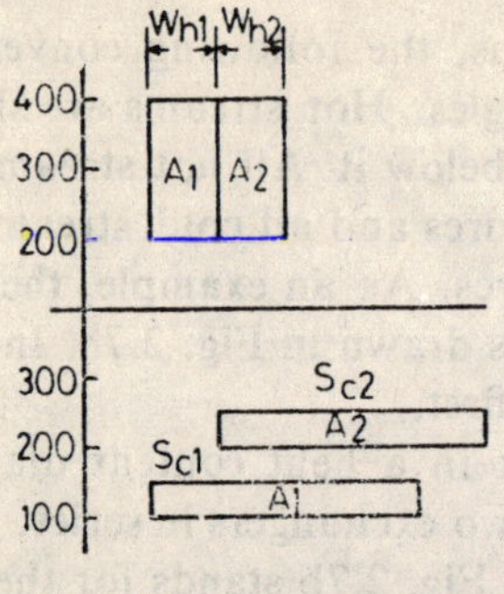

Fig. 2.10 Representation of stream splitting on heat content diagram

hatched area indicates the use of a cooler and the upper one, the use of a heater.

Finally splitting of streams are represented by vertical lines on the heat content diagram as done in Fig. 2.10. In the figure, the S_{h1} block has been split into two streams with heat capacity flow rates, W_{h1} and W_{h2}. Each of these streams are cooled from 400°C to 200°C using S_{c1} and S_{c2} as shown.

2.9 MATCHING OF HEAT CONTENT DIAGRAM FOR MINIMUM NETWORK AREA

It is yet to be decided how the optimal matching should be done for minimum total area. Results of Sec 2.7 are used to develop the following rules.

Consider the heat content diagram of Fig. 2.7b and assume that the inlet and outlet temperatures of S_{c1} and S_{c2} are such that all heat is exchangeable. A_1 is next divided into several small *equal-area* sections horizontally. Same equal-area horizontal strips are drawn in S_{c2} starting from the top. If there are m such strips, according to Sec. 2.7.3, the problem of matching A_1 by an equal area of S_{c2} in a single exchanger is identical to doing it in m different heat exchangers arranged in series. On application of the result of Sec. 2.7.2, therefore, the optimal matching is found to be done strictly in the order of decreasing temperature.

Two rules immediately follow from the above discussion. If there are two cold fluids S_{c1} and S_{c2}, and S_{c2} has a higher *output temperature*, then S_{c2} must exchange area A_1, and only after this is S_{c1} used to exchange A_2. The second result is that the output temperature of ith hot (or cold) block is never lower than the input temperature of $(i + 1)$th hot (or cold) block. In a heat content diagram, this would appear in the following way: the lower horizontal edge of a block of heat exchanged is higher than the upper horizontal edge of the next block.

Now one is in a position to write down precisely how the network can be synthesized. The heat content diagram of the task is drawn first with hot streams in order of decreasing *input* temperatures and cold streams in order of decreasing *output* temperatures. The stepwise rules developed below are applied to yield the minimum area network as given in Sec. 2.9.1 to 2.9.3.

Step 1 One finds out if the hot utility temperature is higher than the highest input temperature of hot streams. If it is not, it should be regarded as one of the hot streams and should be shown as a point corresponding to its temperature on the heat content diagram. If it is, it is ignored in the diagram. Similarly, the cold utility is shown only when it is not the coldest. When the hot (or cold) utility is shown in the heat content diagram, it means that it is treated as one of the exchanging streams. The matching in order of decreasing temperature is done exactly as for other streams in the network.

Step 2 The peripheral network of Fig. 2.3 can be easily found as follows. If there is a cold stream, S_{ck}, whose input temperature, T_{ck}, is such that $(T_{ck} + \Delta T_m) > T^*_{hk}$ (the output temperature of S_{hk}), the hot block is deleted below $(T_{ck} + \Delta T_m)$, and the remainder heat is exchanged by a cooler. Similarly, if there is a hot stream, S_{hk}, such that $(T_{hk} - \Delta T_m) < T^*_{ck}$ (the output temperature of S_{ck}) then the region above $(T_{hk} - \Delta T_m)$ is deleted from this cold block to be exchanged in a heater. This step fixes the number of auxiliary heaters and coolers in the peripheral network.

Step 3 According to the discussion at the beginning of this section, the matching of areas is done as follows. The natural divisions of heat content diagrams are drawn by extending all horizontal edges of different hot and cold blocks, and while matching, heat is exchanged in the amounts given by the area of these so generated. Hot streams with the *hottest input* temperature is matched with the cold stream with the *highest output* temperature. The remainder hot blocks are matched in the order of their decreasing input temperatures with cold blocks with decreasing output temperatures. In this allocation, the original difference between two streams are ignored and two hot (or cold) blocks with same input (or output) temperatures are exchanged simultaneously even though representing two distinct fluids. When heat exchange occurs between two (or several) hot blocks having the same input and output temperature with a single cold block, the cold block is split vertically. As has been pointed out earlier, this amounts to transfer of heat in parallel exchangers.

The above matching principle is best illustrated through an example[3,4]. Figure 2.11 shows the heat content diagram for two hot streams S_{h1} and S_{h2} with input temperatures of 400°C and 320°C and output temperatures of 250°C and 100°C respectively. Simultaneously there is one cold stream S_{c1} to be heated from 120°C to 410°C and it is assumed that the hot and cold utilities have the highest and the lowest temperatures. Therefore according to step 1 (Sec. 2.9.1), these are not shown in Fig. 2.11a. In the second step (Sec. 2.9.2) the unexchangeable areas are deleted. These are hatched and consist of the region S_{h2} below 130° $(=T_{ck} + \Delta T_m)$ and region S_{c1} above 390° $(=T_{hk} - \Delta T_m)$. In the third step (Sec. 2.9.3), all horizontal lines are extended as shown in Fig. 2.11b, which leads to four blocks numbered 1, 2, 3 and 4 in hot streams. These are natural divisions and heat should be exchanged in this quantity in this order, i.e., area 1 should be exchanged first, areas 2 and 3 simultaneously and then area 4. As there is only one cold stream, there is one natural division in it and therefore one simply divides it into four sections equal to areas 1, 2, 3 and 4 now.

According to the matching rule, section M is cut equal to area 1. As areas 2 and 3 have to be exchanged simultaneously, section N is cut equal to the sum of areas 2 and 3. As 2 and 3 correspond to two distinct hot

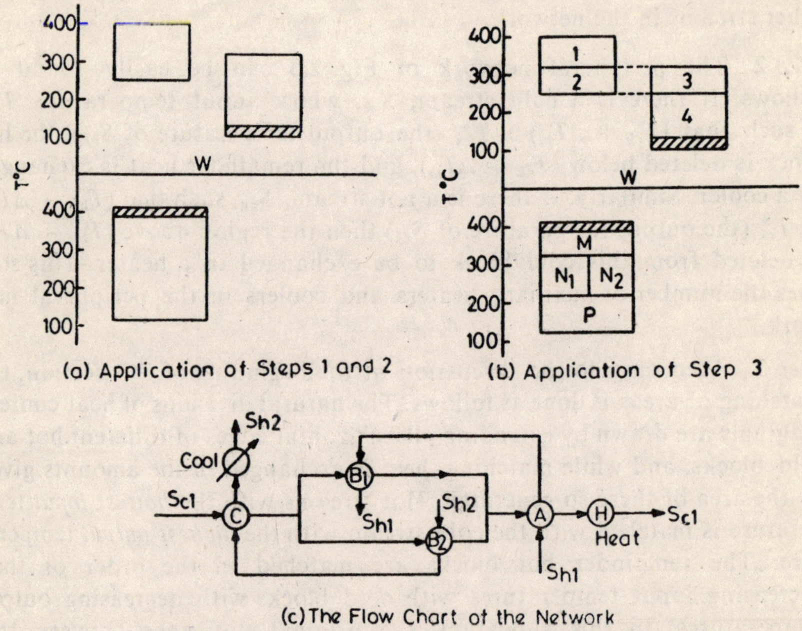

Fig. 2.11 Matching of two hot streams and one cold stream and flow sheet of the minimum area network. (Reprinted from Ref. 5 with the permission of American Institute of Chemical Engineers, New York.)

streams S_{h1} and S_{h2}, the section N is divided vertically into N_1 and N_2 equal to areas 2 and 3 respectively. The remainder of section P is matched with area 4 in the end. Drawing of the flow sheet based upon the heat content diagram of Fig. 2.11b is a simple matter and is given in Fig. 2.11c. When one arbitrarily follows stream S_{c1} it is found that it exchanges heat in exchanger C equal to the area of section P; then it is split into two streams of heat capacity flow rates W_{c1} and W_{c2} (Fig. 2.11b) to exchange heat equal to areas of section N_1 and N_2 in exchangers B_1 and B_2. Then W_{c1} and W_{c2} are once again mixed for heat transfer in exchanger. Finally to bring the cold stream to the desired temperature, one uses a heater. S_{h1} exchanges heat in exchangers A and B_1. S_{h2} exchanges heat in exchangers B_2 and C and then a cooler as shown.

To take care of a matching situation when the hot utility has a temperature lower than the hottest input temperature of hot streams, one considers the problem solved by Ponton and Donaldson[6]. There are three hot streams and one cold stream whose input and output temperatures are given in Table 2.1. Hot utility is steam at 235.56°C (456°F) which is lower than the highest input temperature of the hot streams

Table 2.1[5,6] **The Problem of Three Hot and One Cold Streams Exchanging Heat where Utility does not have the Highest Temperature**

Stream	Heat capacity flow rate kcal/°C (Btu/°F)	Input temperature °C (°F)	Output temperature °C (°F)
S_{c1}	31,746.03(70,000)	−3.89(25)	215.56(420)
S_{h1}	9,070.29(20,000)	260 (500)	43.33(110)
S_{h2}	22,675.74(50,000)	221.11(430)	110 (230)
S_{h3}	13,605.44(30,000)	204.44(400)	43.33(110)
Steam (hot utility)	—	235.56(456)	—

(Reprinted from Reference 5 with the permission of American Institute of Chemical Engineers, New York.)

(i.e. S_{h1}). The heat content diagram of this matching problem is given in Fig. 2.12 in which a dot has been shown corresponding to steam temperature.

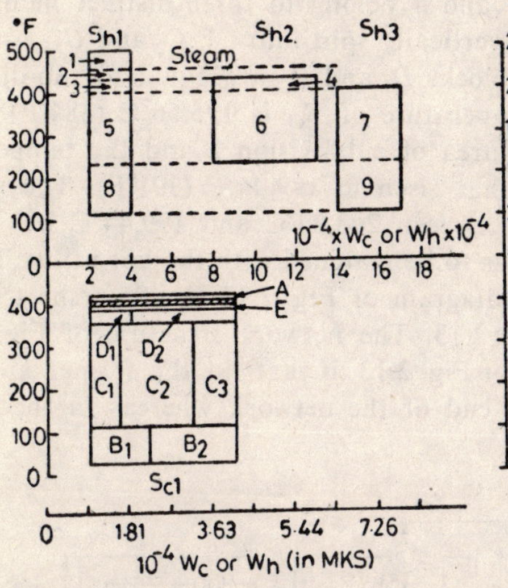

Fig. 2.12 Matching of three hot streams with one cold stream with utility temperature less than the hottest stream temperature. (Reprinted from Reference 5 with the permission of American Institute of Chemical Engineers, New York.)

In application it is found that in the 1st step, there exists no unexchangeable area. As the steam temperature is not the highest, it is regarded as one of the hot streams and, therefore, one not only extends horizontal edges of blocks for S_{h1}, S_{h2} and S_{h3} but also draws a horizontal line through the point corresponding to the steam utility. This leads to five divisions of S_{h1} block, two divisions of S_{h2} block and two divisions of

S_{h3} block. As the matching must be done in order of decreasing input temperatures of these blocks, steam is used after the block 1 of S_{h1} has been exchanged. However, procedure presented in the beginning of this section does not fix the total amount of heat exchanged between S_{c1} and the hot utility. To take care of this situation, one starts matching from subsection 1 of S_{h1} and after finding the presence of the hot utility, stops at that and begins the matching from bottom upwards, i.e. 8 and 9 is matched simultaneously, then 5, 6 and 7 and so on. After all areas are matched, the area of S_{c1} left unmatched is exchanged with the hot utility. This is demonstrated on this example as follows.

Subsection A between 215.56°C (420°F) and 208.33°C (407°F) equal to area of block 1 of S_{h1} is cut first. Next divisions 8 of S_{h1} and 9 of S_{h3} are matched simultaneously with S_{c1}. This necessitates stream splitting of S_{c1} such that the areas of sub-blocks B_1 and B_2 are equal to those of sub-divisions 8 and 9. It is because these subdivisions belong to distinct fluid streams. The cold stream emerges out of these at 43.33°C (110°F). Next sub-divisions 5, 6 and 7 are matched simultaneously and after the exchange of the heat, the temperature of S_{c1} rises to 178.89°C (354°F). As subdivisions 5, 6 and 7 belong to three distinct fluids S_{h1}, S_{h2} and S_{h3}, block C of S_{c1} is vertically split into C_1, C_2 and C_3 as shown in the figure. Similarly sub-blocks D_1 and D_2 are equal to sub-divisions 3 and 4 and the output temperature of S_{c1} is 195.56°C (384°F). Block E has been cut equal to the area of subdivision 2 and the temperature of the cold stream by now has risen to 199.44°C (391°F). The remaining unmatched area of S_{c1} between 208.33°C and 199.44°C has been hatched indicating that this has to be matched with the hot utility, steam. Based on the heat content diagram of Fig. 2.12 the flow sheet can be drawn and is shown in Fig. 2.13. The network structure of Fig. 2.11 must be contrasted with that of Fig. 2.13 in that, in the former auxiliary equipment fall only in the end of the network whereas in the latter, it is a part of the network.

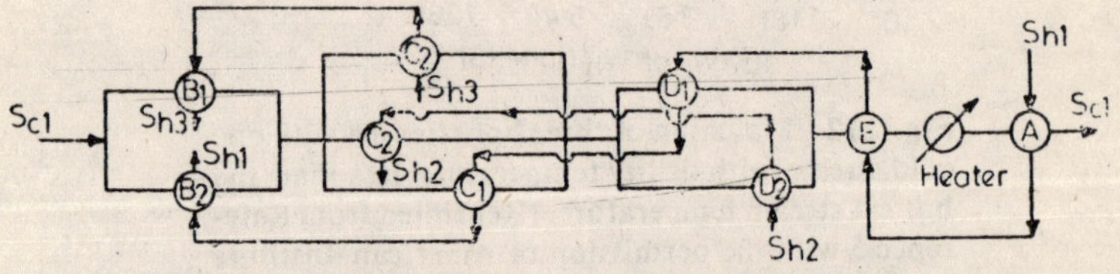

Fig. 2.13 The flow sheet corresponding to Fig. 2.12

The method given in this section gives minimum area network which need not be a minimum cost network. Indeed, to find the optimal solution, adjustments to this 'near-optimal' flow sheet are made till one obtains the actual minimum by trial-and-error. In Sec. 2.10, a trial-and-error procedure has been given to obtain the desired result.

2.10 RULES OF ADJUSTMENT OF THE MINIMUM AREA HEAT EXCHANGER NETWORK TO FIND THE OPTIMAL SOLUTION

The method suggested in the previous section gives a minimum area network. However, minimum area network is not the same as minimum total cost network. The cost of a single exchanger, C_i, is correlated by

$$C_i = aA_i^b \tag{2.10.1}$$

where A_i is the exchanger surface area and a and b are two constants. If b is constant for all exchangers, the total fixed cost, F_c, of the network would be

$$F_c = \sum_i aA_i^b \leqslant a(\sum_i A_i)^b \tag{2.10.2}$$

i.e., the minimum surface area network gives the upper limit of F_c. However, one wishes to minimize the total cost, C_T, which is the sum of depreciated fixed cost and utility requirements

$$C_T = \delta(\Sigma aA_i^b) + C_{hu}S_{hu} + C_{cu}S_{cu} \tag{2.10.3}$$

where C_{hu} is the cost of hot utility per unit mass, C_{cu} is the cost of cold utility and δ is the rate of depreciation (or the annual rate of return). In view of Eqs. 2.10.2 and 2.10.3, the minimum area network is far from an optimal one.

Mathematical rules which give minimum C_T are difficult to derive and there are rules-of-thumb which have been found to work (sometimes called heuristic rules) for obtaining minimum cost network. These heuristic rules for heat exchanger network have been developed by Nishida et al[5] and are given below:

Rule 1 Areas of different blocks are considered. The smallest block is deleted by expanding the neighbouring block (or blocks). This is repeatedly applied till the total cost of the network can no longer be reduced further.

Rule 2 If a network consists of a sub network in which a hot (or cold) stream contacts a cold (or hot) stream which it had contacted already at least once, either of these two matches are deleted.

Rule 3 One replaces stream splitting by stream unsplitting. In the unsplitted network, hot and cold streams are matched in decreasing order of their arithmetic averages of input and output temperatures.

Rule 1 is applied first. To the best network generated after applying it, rule 2 is applied. If there is more than one network generated on application of rule 2, they are compared for the minimum total cost. To the network so generated, rule 3 is applied. There can be several points in the network where rule 3 may be applied and this is done one at a time to find the minimum.

To demonstrate the application of the heuristic rules in obtaining the minimum cost network, the problem solved by Nishida et al[6] is considered. There are two hot streams and two cold streams. Their input, output temperatures and heat capacity flow rates are given in Table 2.2.

Table 2.2[5] The Problem of Two Hot and Two Cold Streams Exchanging Heat. The Hot Utility has the Hottest Temperature and the Cold Utility has the Lowest Temperature

Stream	Heat capacity flow rate kcal/°C (Btu/°F)	Input temperature °C (°F)	Output temperature °C (°F)
S_{c1}	6553.3 (14,450.1)	60 (140)	160 (320)
S_{c2}	5229.0 (11,530.1)	115.6 (240)	260 (500)
S_{h1}	7559.5 (16,668.8)	160 (320)	93.3 (200)
S_{h2}	9070.3 (20,000)	248.9 (480)	137.8 (280)

Hot utility:

Steam at 282.2°C (540°F)

Latent heat = 364.85 kcal/kg (656.6 Btu/lb)

Cold utility:

Cooling water at 37.8°C (100°F)

Maximum water output temperature = 82.2°C (180°C)

Approach temperature:

Heat exchanger = 11.1°C (20°F)

Steam heater = 13.9°C (25°F)

Water cooler = 11.1°C (20°F)

Overall heat transfer coefficients:

Heat exchanger = 732 kcal/h.m². °C (150 Btu/h. ft². °F)

Steam heater = 976 ,, (200 ,,)

Water cooler = 732 ,, (150 ,,)

Cost parameters:

$a = 350$, $b = 0.6$

Annual rate of return, $\delta = 0.1$

Cooling water = 2.27×10^{-5} cost unit/kg (5×10^{-5} cost/lb)

Steam = 4.54×10^{-4} -do- (1×10^{-3} -do-)

(Reprinted from Reference 5 with the permission of American Institute of Chemical Engineers, New York.)

For the problem of Table 2.2, the heat content diagram is drawn first as in Fig. 2.14. Natural divisions in S_{h2} and S_{h1} are 1, 2, 3 and 4 and in S_{c2} and S_{c1} are A, B, C and D. Because S_{c2} has the outlet temperature of 260°C (500°F), which is more than the inlet temperature of S_{h2}, the upper region of the block S_{c2} from 238°C (460°F) has to be reached with the help of a hot utility. As the area of A is less than that of section 1, it is divided into two subsections 1′, and 1″. The height of 1′ is found by heat balance and occurs at a temperature of 204.1°C (399.4°F). Now B and C sub-blocks have to be exchanged simultaneously. As the area of the sub-division 1″ is less than the combined area of sub-blocks B and C, B and

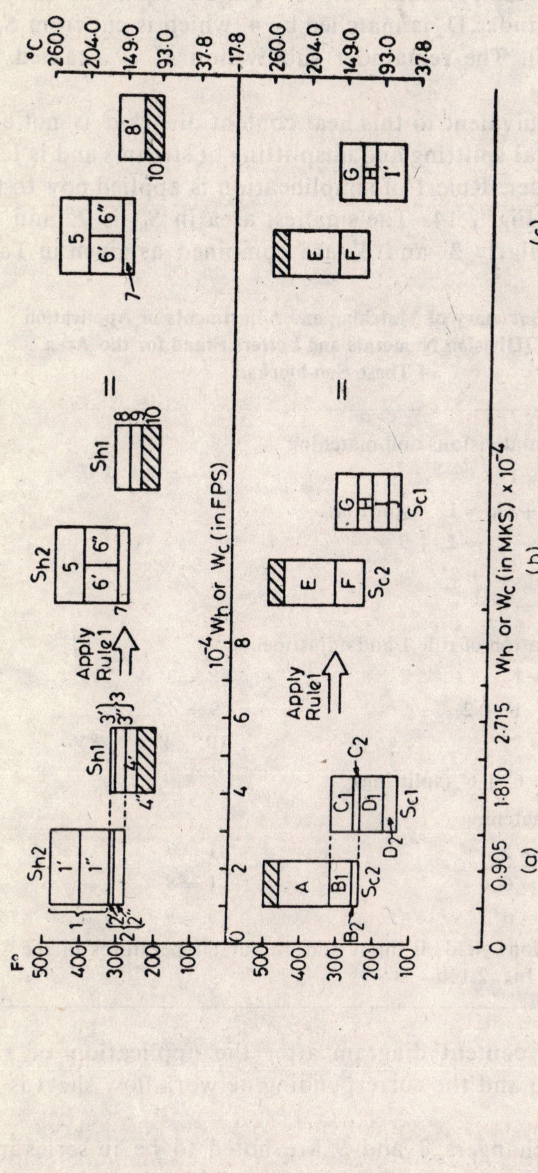

Fig. 2.14 Heat content diagram of the problem given in Table 2.2. Simplification after rule 1 is applied

C are divided into B_1 and C_1 and B_2 and C_2 by one horizontal line, such that the area of subdivision 1″ equals the combined area of B_1 and C_1 and this occurs at temperature 126.1°C (259°F). After this, B_2 and C_2 must be exchanged with 2 and 3 simultaneously. As the sum of areas 2 and 3 is larger, these are together divided into 2′, 3′ and 2″, 3″ at temperature 152.9°C (307.3°F) to exchange B_2 and C_2 simultaneously. D_1 is next cut at temperature 78.9°C (174°F) equal to the area of (2″ + 3″). The remainder D_2 is matched by 4′ which is cut from S_{h1} block at 121.4°C (250.5°F). The remainder subdivision 4″ is attained by cold utility.

The flow sheet equivalent to this heat content diagram is not a simple one, requiring several splitting and unsplitting of streams and is left as an exercise to the reader. Rule 1 of simplification is applied now to the heat content diagram of Fig. 2.14. The smallest area in S_{h1} is 2′ and is combined with 1″. Similarly 3′ and 3″ are combined as given in Table 2.3.

Table 2.3 Summary of Matching and Adjustments in Application of Rule 1. (Division Numerals and Letters Stand for the Area of These Sub-blocks.)

(1) Original divisions and matching

$A = 1'$

$B_1 + C_1 = 1''$ (splitting)

$B_2 + C_2 = 2' + 3'$

$D_1 = 2'' + 3''$

$D_2 = 4'$

(2) Application of rule 1 and adjustments

$5 = 1'$	$8 = 3' + 3''$
$6 = 1'' + 2'$	$9 = 4'$
$7 = 2''$	$10 = 4''$

$6 = 6' + 6''$ (splitting)

(3) New matching

$E = 5$	$H = 7$
$F = 6'$	$I = 8$
$G = 6''$	$J = 9$

These relations yield all input and output temperatures of the network of Fig. 2.14b.

The resultant heat content diagram after the application of rule 1 is shown in Fig. 2.14b and the corresponding network flow sheet is given in Fig. 2.15a.

In Fig. 2.15a exchangers 8 and 9 are noted to be in series involving S_{c1} and S_{h1}. According to Sec. 2.7.2, the entire heat can be equivalently transferred in a single heat exchanger without affecting the optimality of the problem. In view of this, divisions 8 and 9 in the heat content diagram have been combined in Fig. 2.14c and the first trial flow sheet of the network is given in Fig. 2.16b. All input and output temperatures of exchangers have been calculated from the knowledge of heat duties (Table 2.3). In Table 2.4, log mean temperatures for all exchangers have

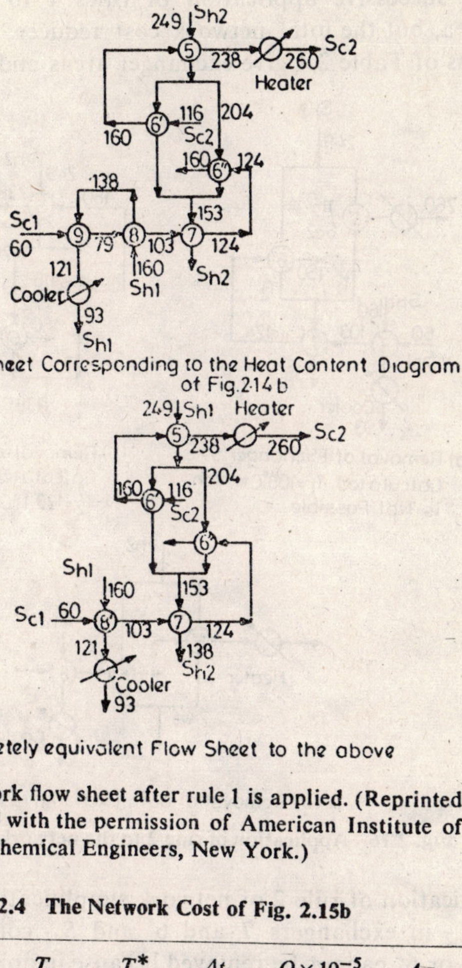

Fig. 2.15 Network flow sheet after rule 1 is applied. (Reprinted from Reference 5 with the permission of American Institute of Chemical Engineers, New York.)

Table 2.4 The Network Cost of Fig. 2.15b

Exchanger Number	T_h, °C	T_h^*, °C	T_c, °C	T_c^*, °C	Δt_{lm}, °C	$Q \times 10^{-5}$, kcal/h	A_i, m²	$\left(\frac{C_i}{a}\right) = A_i^{0.6}$
5	248.9	204.4	160.0	237.8	24.05	4.032	22.85	6.54
6′	204.4	152.5	115.6	160.0	40.54	2.331	7.90	3.46
6″	204.4	152.5	124.4	160.0	35.63	2.331	9.03	3.74
7	152.5	137.8	103.3	124.4	31.61	1.386	5.82	2.88
8′	160.0	121.4	60.0	103.3	59.03	2.923	6.74	3.14
Cooler	121.4	93.3	37.8	82.2	46.83	2.117	6.13	2.97
Heater	282.2	282.2	237.8	260.0	32.07	1.167	3.72	2.20

$A_T = 62.19$, $\left(\frac{F_c}{a}\right) = 24.93$.

been given and their surface areas are found from the knowledge of the overall heat transfer coefficients, U. As pointed out earlier, the network resulting from the heat content diagram of Fig. 2.14a, will have a minimum total area, but it need not be minimal cost network. Consequently, a successive application of rules 1 to 3 increases the total network area, but the total network cost reduces. To show this the last two columns of Table 2.4 give exchanger areas and their costs.

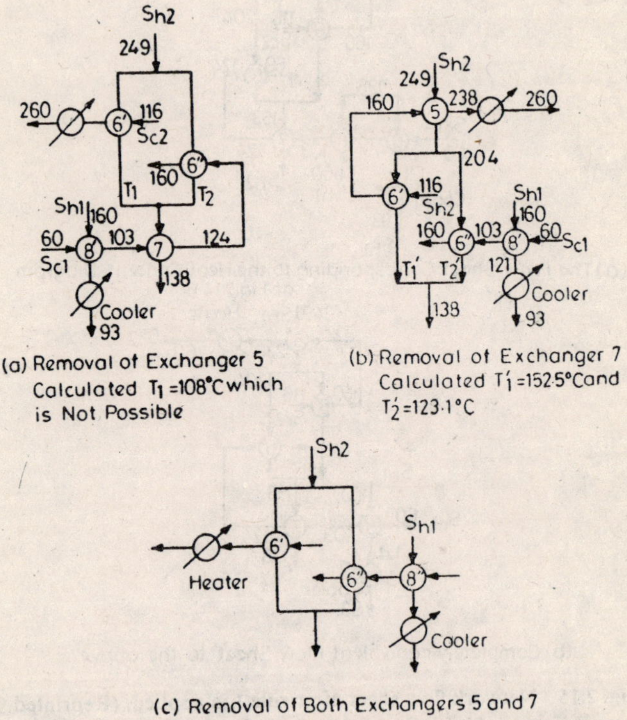

Fig. 2.16 Application of rule 2 to the network of Fig. 2.15b

For application of rule 2 of network simplification, it is noted that S_{c1} contacts S_{h2} in exchangers 7 and 6″ and S_{c2} contacts S_{h2} in 6′ and 5. Further, 6′ or 6″ cannot be removed because in doing so, one also removes the stream splitting. Therefore as suggested in rule 2, exchangers 5 and 7 are removed one by one as in Figs. 2.16a and b. In Fig. 2.16c both are removed. One calculates T_1, T_1' and T_2' by heat balance as 107.8°C (226°F), 152°C (306.5°F) and 123.1°C (253°F) respectively. As exchanger 6′ is impossible to attain, the network of Fig. 2.16a is rendered impossible. On the same ground, the network of Fig. 2.16c is also rejected. The only feasible network is that given in Fig. 2.16b, and in Table 2.5, calculations for the exchanger area and cost are given. The total network area is found to increase, but the network fixed cost, C_F, has reduced further. It may be noted that utility requirements remain unchanged in the simplification resulting from rule 2.

Table 2.5 The Network Cost of Fig. 2.16b

Exchanger Number	T_h, °C	T_h^*, °C	T_c, °C	T_c^*, °C	Δt_{lm}, °C	$Q\times10^{-5}$, kcal/h	A_i, m^2	$\left(\frac{C_i}{a}\right)=A_i^{0.6}$
5	248.9	204.4	160.0	237.8	24.05	4.032	22.85	6.54
6′	204.4	152.5	115.6	160.0	40.54	2.356	7.90	3.46
6″	204.4	123.1	103.3	160.0	30.44	3.715	16.63	5.40
8′	160.0	121.4	60.0	103.3	59.03	2.923	6.69	3.13
Cooler	121.4	93.3	37.8	82.2	46.83	2.117	6.13	2.97
Heater	282.2	282.2	237.8	260.0	32.07	1.167	3.72	2.20

$$A_T = 63.92 \text{ m}^2, \quad \left(\frac{F_c}{a}\right) = 23.70$$

Now rule 3 is applied to Fig. 2.16b, according to which, stream splitting is replaced by stream-unsplitting. As the average of input and output temperature of S_{c2} is higher, it is to be contacted first as shown in Fig. 2.17a and exchangers 5 and 6′ are combined into one, 5′. The temperature T_1' is found from energy balance and it is 177.8°C (352°F). Exchanger areas and the cost are then calculated and the results are given in Table 2.6. Even though the total area, A_T, has increased, the total fixed cost, F_c, has gone down further. Once again the total utility requirement has remained the same. Thus the network of Fig. 2.17 is better than ones drawn earlier in Figs. 2.15 and 2.16 and is chosen to carry out the task of Table 2.2.

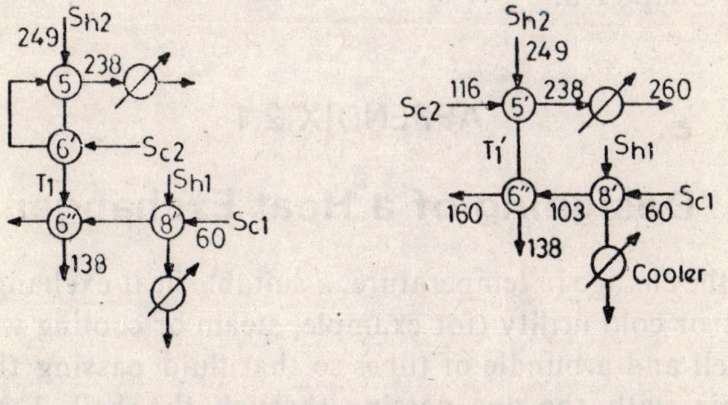

Fig. 2.17 Application of rule 3 to the network of Fig. 2.16b. (a) Removal of stream splitting; (b) Combine 5 and 6′ into one single exchanger 5′. According to Section 2.7.3 T_1' is calculated to be 177.8°C. (Reprinted from Reference 5 with the permission of American Institute of Chemical Engineers, New York.)

Table 2.6 The Network Cost of Fig. 2.17b

Exchanger number	T_h, °C	T_h^*, °C	T_c, °C	T_c^*, °C	Δt_{lm} °C	$Q \times 10^{-5}$ kcal/h	A_i, m²	$\left(\frac{C_i}{a}\right) = A_i^{0.60}$
5′	248.9	177.8	115.6	237.8	29.67	6.451	29.73	7.65
6″	177.8	137.8	103.3	160.0	25.22	3.629	19.97	6.03
8′	160.0	121.4	60.0	103.3	59.03	2.923	6.78	3.15
Cooler	121.4	93.3	37.8	82.2	46.83	2.117	6.13	2.97
Heater	282.2	282.2	237.8	260.0	32.07	1.167	3.72	2.20

$A_T = 66.33 \text{ m}^2, \quad \left(\frac{F_c}{a}\right) = 22.00$

2.11 CONCLUSION

The problem of heat exchanger network consists of finding an optimal arrangement of exchangers for a given heat exchanging task. Some mathematical constraints have been developed in this chapter which give the minimum area network. Since this is not the same as the minimum cost network, heuristic rules by Nishida et al have been given which leads to a near optimal arrangement of exchangers (if not the optimal one). These rules have been applied to some practical problems as an example and is shown to improve the network.

In literature, there have been several attempts to solve this problem using various techniques[5,7], and the approach discussed here was specifically chosen because of its amenability to hand calculations. Within reasonable time, the optimal (or near-optimal) network can be obtained using the algorithm presented in this chapter without the use of a computer. One goes on to the problem of sequencing of distillation columns discussed in Chapter 3.

APPENDIX 2.1

Designing of a Heat Exchanger

To achieve the change in temperature, a suitable heat exchanger is used by using a hot or cold utility (for example, steam or cooling water). It consists of a shell and a bundle of tubes so that fluid passing through tubes does not mix with the one passing through the shell. The exchange of energy occurs across tube walls and in the analysis presented below, the phase change is assumed not to occur.

Heat exchangers are normally classified according to the pattern of fluid flow in the tubes relative to the flow in the shell. Common flow patterns found in commercial exchangers are:

1. countercurrent, when they flow in opposite directions;
2. cocurrent, i.e. fluids in the shell and in the tube have the same flow direction; and
3. multipass cross flow.

These flow patterns are shown schematically in Fig. 2.18. Multipass cross flow is essentially countercurrent and can be treated exactly as in 1. Consequently, only cocurrent and countercurrent flow patterns are considered in the following analysis[a].

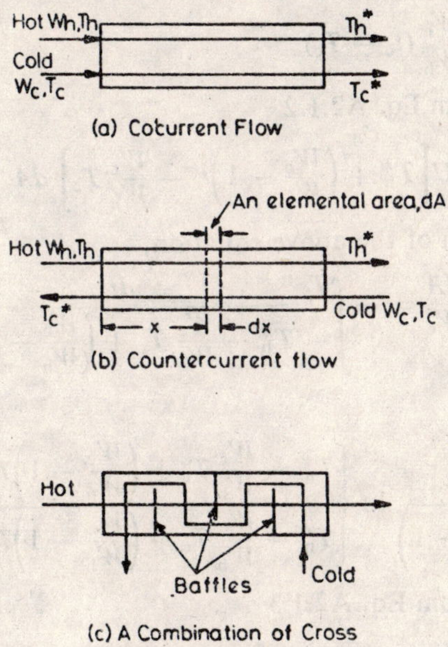

Fig. 2.18 Flow patterns of fluids in a heat exchanger

Countercurrent flow is considered first in which a hot stream at T_h with the heat capacity flow rate W_h (heat capacity flow rate, $W = mc_p$ where m is the mass flow rate and c_p is the average heat capacity of the stream) is exchanging energy with a cold stream at temperature T_c with heat capacity flow rate, W_c. Exit temperatures of hot and cold streams are T_h^* and T_c^* respectively. An energy balance on the elemental area of the exchanger shown in Fig. 2.18b is made. The amount of energy transferred, dQ, is given by

$$dQ = UdA(t_h - t_c) \tag{A2.1.1}$$

where U is the overall heat transfer coefficient, assumed to be constant over the entire heat exchanger. The t_h and t_c are temperatures of hot and cold streams respectively within the exchanger at a distance x from the beginning. If dt_c is the temperature change of the cold stream over the differential surface area, dA, then

$$dQ = -W_c\,dt_c = U(t_h - t_c)\,dA. \qquad \text{(A2.1.2)}$$

A total energy and energy balance on length x of the exchanger are given by

$$Q = W_h(T_h - T^*) = W_c(T_c^* - T_c) \qquad \text{(A2.1.3a)}$$

and

$$W_h(T_h - t_h) = W_c(T_c^* - T_c) \qquad \text{(A2.1.3b)}$$

where Q is the total energy transferred. From these equations, one has

$$t_h = T_h^* + \frac{W_c}{W_h}(t_c - T_c) \qquad \text{(A2.1.4)}$$

Eliminating t_h from Eq. A2.1.2

$$-W_c\,dt_c = U\left[T_h^* + \left(\frac{W_c}{W_h} - 1\right)t_c - \frac{W_c}{W_h}T_c\right]dA \qquad \text{(A2.1.5)}$$

and on integration of the above equation,

$$\int_0^A \frac{U\,dA}{W_c} = \frac{UA}{W_c} = -\int_{T_c^*}^{T_c} \frac{dt_c}{T_h^* - \dfrac{W_c}{W_h}T_c + \left(\dfrac{W_c}{W_h} - 1\right)t_c} \qquad \text{(A2.1.6)}$$

or

$$\frac{UA}{W_c} = \frac{1}{\left(\dfrac{W_c}{W_h} - 1\right)} \ln\left\{\frac{T_h^* - \dfrac{W_c}{W_h}T_c + \left(\dfrac{W_c}{W_h} - 1\right)T_c^*}{T_h^* - \dfrac{W_c}{W_h}T_c + \left(\dfrac{W_c}{W_h} - 1\right)T_c}\right\} \qquad \text{(A2.1.7)}$$

Substituting T_h from Eq. A2.1.3

$$UA = \frac{W_c W_h}{W_c - W_h}\ln\frac{T_h - T_c^*}{T_h^* - T_c}$$

$$= \frac{1}{\dfrac{1}{W_h} - \dfrac{1}{W_c}}\ln\frac{T_h - T_c^*}{T_h^* - T_c}$$

$$= \frac{Q}{\dfrac{Q}{W_h} - \dfrac{Q}{W_c}}\ln\frac{T_h - T_c^*}{T_h^* - T_c} \qquad \text{(A2.1.8)}$$

Eliminating Q/W_h and Q/W_c with the help of Eq. A2.1.3a, the equation above simplifies to

$$UA = \frac{Q}{(T_h - T_h^*) - (T_c^* - T_c)}\ln\frac{T_h - T_c^*}{T_h^* - T_c} \qquad \text{(A2.1.9)}$$

or

$$Q = UA\cdot\frac{\Delta_1 - \Delta_2}{\ln\left(\dfrac{\Delta_1}{\Delta_2}\right)} = UA\,\Delta t_{lm} \qquad \text{(A2.1.10)}$$

where

$$\Delta_1 = T_h - T_c^* \tag{A2.1.11a}$$

and

$$\Delta_2 = T_h^* - T_c \tag{A2.1.11b}$$

A similar derivation for cocurrent flow pattern of Fig. 2.18a can be carried to obtain

$$Q = UA\,\frac{\Delta_1' - \Delta_2'}{\ln \dfrac{\Delta_1'}{\Delta_2'}} = UA\,\Delta t_{lm} \tag{A2.1.12}$$

where

$$\Delta_1' = T_h - T_c \tag{A2.1.13a}$$

and

$$\Delta_2' = T_h^* - T_c^* \tag{A2.1.13b}$$

The effect of cocurrent and countercurrent flow patterns on energy transfer can be easily found by calculating Δt_{lm}, the log mean temperature difference. Supposing a hot stream at 400°C is desired to be cooled to 200°C. If cold water is available at temperature of 25°C to be heated to a temperature of 100°C, Δt_{lm} for countercurrent exchanger is

$$(\Delta t_{lm})_{counter} = \frac{(400 - 100) - (200 - 25)}{\ln \dfrac{300}{175}} = 232°\text{C} \tag{A2.1.14}$$

However, if these fluids are contacted cocurrent-wise, $(\Delta t_{lm})_{co}$ becomes 208°C, which is lower than that for the countercurrent flow. Thus if U is the same for both these exchangers, a higher surface area is required for the desired heat transfer for cocurrent heat exchanger. The constancy of U between the two exchangers is not a serious limitation in this analysis because heat transfer coefficient is a strong function of the Reynold's number. Therefore under normal circumstances, for a desired change in temperature a countercurrent exchanger is preferred.

REFERENCES

1. Levenspiel, O., *Chemical Reaction Engineering*, 2nd edn., Wiley, New York, 1972.
2. Kern, D. Q., *Process Heat Transfer*, 1st edn., McGraw-Hill, Kogakusha, Tokyo, 1950.
3. Masso, A. H. and Rudd, D. H., *A.I.Ch.E.J.*, **15**, 10, 1969.
4. Nishida N., Kobayashi, S. and Ichikawa, A., *Chem. Engg. Sci.*, **26**, 1841, 1971.
5. Nishida, N., Liu, Y. A. and Lapidus, L., *A.I.Ch.E.J.*, **23**, 77, 1977.

6. Ponton, J. W. and Donaldson, R. A. B., *Chem. Engg. Sci.*, **29**, 2375, 1974.
7. Linnhoff, B. and Flower, J. R., *A.I.Ch.E.J.*, **24**, I & II, 633 and 642, 1978.

PROBLEMS

2.1 Read reference 2 and find out correlations for estimating U for liquid-liquid system. Carry out detailed design of exchangers of Fig. 2.2a if the flow rate of the hot fluid is 100,000 kg/h. Repeat your calculations for the flow rate of 1000 kg/h, 100 kg/h and 10 kg/h. You will find that the efficiency of the heat exchanger system reduces with the reduction of the flow rate.

2.2 What happens to your calculations above if the hot stream condenses at 300°C with $\Delta H_c = 100$ kcal/kg. Would you like to condense all hot stream in the first exchanger or in the second exchanger completely. (Remember T_h^* will not be 200°C here).

2.3 Supposing there are one saturated hot stream (pure component) and two cold streams S_{c1} and S_{c2} at 100°C and 200°C with their heat capacity flow rates as 100 each. Hot stream condenses at 400°C with the latent heat of 100 kcal/kg and flow rate of 100 kg/h.

(a) Does it matter if S_{c1} exchanges first or S_{c2} exchanges first, if S_{c1} is to be heated to 150°C and S_{c2} to 250°C?

(b) Supposing S_{c1} is to be heated to 180°C and S_{c2} to 260°C, is it economical to heat S_{c1} completely, exchange as much of heat as available with S_{c2} and then use a hot utility? Or is it the other-way-round?

(c) Compare scheme (b) with the following. S_{c1} and S_{c2} are heated to 100° and 250°C respectively whereas the remainder temperature difference is attained in auxiliary exchangers.

2.4 In Sec. 2.7.1, the rule of order of matching hot and cold streams was derived for liquid and non-condensing vapours. Derive a similar rule for condensing vapours.

2.5 Is the rule of Sec. 2.7.2 valid for condensing vapours ? Prove your claim.

2.6 Is the rule of Sec. 2.7.3 valid for condensing vapours ? Prove your claim.

2.7 Supposing the hot stream is a mixture of condensing vapours, what happens to the rule developed in P. 2.4 to 2.6.

2.8 Consider the flow sheet of Fig. 1.17. Formulate the total heat exchanging problem and optimize the flow sheet with respect to it. Find out if you require any hot and/or cold utility and what should be their temperatures.

2.9 Consider the problem of matching three hot streams and one cold stream of Table 2.1. Suppose there are two hot utilities, one at 235.6°C (456°F) and the other at 148.9°C (300°F). How do you do the matching now ? Can it be done using the algorithm of Sec. 2.9 ?

2.10 Solve the following problem of matching for minimum area network[5]:

	Heat capacity flow rate	Input temp., °C	Output temp., °C
S_{c1}	21,600	100	400
S_{c2}	24,500	150	360
S_{c3}	24,700	200	400
S_{h1}	31,500	480	250
S_{h2}	25,200	400	150

Draw the flow sheet.

2.11 Solve the problem of matching of three hot and three cold streams for minimum area network[5]:

	W_c	Input temp., °C	Output temp., °C
S_{c1}	16,000	100	430
S_{c2}	32,760	180	350
S_{c3}	26,350	200	400
S_{h1}	28,000	440	150
S_{h2}	23,800	520	300
S_{h3}	33,600	390	150

Draw the flow sheet.

2.12 Solve the problem of matching of five hot and five cold streams[5]:

	W_c	Input temp., °C	Output temp., °C
S_{c1}	14,450	140	320
S_{c2}	11,530	240	431
S_{c3}	16,000	100	430
S_{c4}	32,760	180	350
S_{c5}	26,350	200	400
S_{h1}	16,670	320	200
S_{h2}	20,000	480	280
S_{h3}	28,000	440	150
S_{h4}	23,800	520	300
S_{h5}	33,600	390	150

Draw the flow sheet.

2.13 Find the minimum cost network for P. 2.10. Do the same for P. 2.11. Use cost data given in this chapter.

2.14 Optimize the flow sheets of P. 1.15, 1.16 and 1.17 with respect to heat exchanging.

2.15 Find out the requirements of P. 1.14. Do you feel one should bother optimizing it? It may be remembered that at several occasions one does not bother doing it because the total saving is a small fraction of the total expenditure involved.

3

Optimization of Steady State Flow Sheet with respect to Adjustment of Concentration

3.1 INTRODUCTION

It was explained in Chapter 1 that a stream is completely characterized by specification of species, concentration, temperature, number of phases and pressure. Since a chemical plant consists of equipment to achieve changes in these, the best flow sheet is the one in which each of these tasks is carried out at the optimal level. In Chapter 2, the task of heat exchanging for the whole plant was optimized, and in this chapter, the optimality of the task of concentration adjustment is discussed.

Concentration adjustment can be done either by mixing or by separation[1–3]. Mixing, being a cheaper operation, is always preferred, but not all concentration levels can be attained through it and in view of this, at times, a combination of mixing and separation is used. In this chapter, the optimization of mixing operation is discussed first followed by that of the combination of mixing and separation. In the end, the optimization of the task of separation is elaborated. A major portion of this chapter is devoted to the latter because of its industrial importance.

3.2 ADJUSTMENT OF CONCENTRATION BY MIXING

Mixers for solids and liquids have been described in Perry's Handbook[2] and a suitable unit can be found depending upon the nature of the material to be handled. Invariably they have much lower fixed cost and do not require sophisticated controls for their operation as separation equipment do.

The problem of mixing is introduced as follows:[3] It is desired to blend N number of species, denoted by S_i, each having cost c_i' per unit weight. After blending them, M products are obtained. If x_{ij} *weight units of* S_i are mixed in product P_j, the constraint on the availability of raw material is given by

$$x_{i1} + x_{i2} + x_{i3} + \ldots + x_{iM} \leqslant b_i \tag{3.2.1}$$

where b_i is the total amount of the species S_i available. Similar constraint equations are written for all the species as follows:

$$x_{11} + x_{12} + x_{13} + \ldots + x_{1M} \leqslant b_1 \tag{3.2.2a}$$

$$x_{21} + x_{22} + x_{23} + \ldots + x_{2M} \leqslant b_2 \tag{3.2.2b}$$

$$\vdots \qquad \vdots \qquad \vdots \qquad \vdots \quad \vdots$$

$$x_{N1} + x_{N2} + x_{N3} + \ldots + x_{NM} \leqslant b_N \tag{3.2.2c}$$

The total profit, C_P, of selling M products can then be written as

$$C_P = c_1'S_1 + c_2'S_2 + \ldots + c_N'S_N \tag{3.2.3}$$

where S_i is the total weight of the species S_i used in forming the blend, which is given by

$$S_i = \sum_{j=1}^{M} x_{ij} \quad i = 1, 2, \ldots, N \tag{3.2.4}$$

The problem of blending consists of deciding upon the composition of each product subject to the constraint that the total profit, C_P, is maximum. It may be appreciated that the problem posed here is not a hypothetical one, but has great significance. Rudd et al.[4] point out that in refineries several products are obtained by distillation of crude petroleum and thermal cracking of heavies but consumer materials like gasoline, diesel, kerosene, etc. are obtained by blending these. As the demand and price of these consumer products are subject to variation due to fluctuating market forces, Eqs. 3.2.2 to 3.2.4 need to be adjusted from day-to-day and consequently the optimal blending has to be evaluated everyday.

The optimization of linear function with respect to linear constraints is called *linear programming* and the problem given in Eqs. 3.2.2 to 3.2.4 belongs to this class. The solution procedure commonly used for solving these equations is known as the *simplex method* and its algorithm is described below. Equation 3.2.2 is first rewritten in terms of the total weight of the product P_j and weight fractions, a_{ij} of component S_i in it as

$$\text{Maximize } (c_1P_1 + c_2P_2 + \ldots c_MP_M) \tag{3.2.5a}$$

$$a_{11}P_1 + a_{12}P_2 + \ldots + a_{1M}P_M \leqslant b_1 \tag{3.2.5b}$$

$$a_{21}P_1 + a_{22}P_2 + \ldots + a_{2M}P_M \leqslant b_2 \tag{3.2.5c}$$

$$\vdots \qquad \vdots \qquad \vdots \quad \vdots$$

$$a_{N1}P_1 + a_{N2}P_2 + \ldots + a_{NM}P_M \leqslant b_N \tag{3.2.5d}$$

Coefficients a_{ij} above are known and variables P_1 to P_M are to be solved with the constraint that they cannot be negative.

$$P_i \geqslant 0, \quad i = 1, \ldots, M \tag{3.2.6}$$

c_1 to c_M are sale price of P_1 to P_M. The number of equations, N, is evidently less than the number of unknowns P_1 to P_M (i.e., M unknowns). In the simplex method, the inequalities of Eq. 3.2.5 are converted to equalities by floating additional $(M - N)$ variables, P_{M+1} to P_{M+N} as

$$\begin{aligned}
a_{11}P_1 &+ a_{12}P_2 + \ldots + a_{1M}P_M + P_{M+1} = b_1 \\
a_{21}P_1 &+ a_{22}P_2 + \ldots + a_{2M}P_M + P_{M+2} = b_2 \\
&\vdots \\
a_{N1}P_1 &+ a_{N2}P_2 + \ldots + a_{NM}P_M + P_{N+M} = b_N
\end{aligned} \tag{3.2.7}$$

Now the total number of unknowns are $(N + M)$ and to solve these sets of equations, M variables must be specified arbitrarily to find the remaining unknowns. The algorithm specifying M variables which leads to the optimal solution constitutes the simplex method and is described below:

Equation 3.2.7 is written in the form of a matrix as shown in Table 3.1. The left of the matrix shows variables that are to be solved through Eq. 3.2.7 and their unit costs. In the first trial, variables P_1 to P_M are assumed to be zero and P_{M+1} to P_{M+N} are solved. As these are fictitious variables, their unit weight cost are taken as zero. Naturally the first trial is far from the best one. The first guess of the solution is improved by removing one variable at a time from the left of the main matrix and introducing one of the chosen variables in its place.

3.2.1 The Choice of the Entering Variables

The choice of the entering variable is decided by calculating a variable Δ_j given by

$$\Delta_j = c_j - \sum_{i=1}^{N} a_{ij}c_i \,; \quad j = 1, 2, \ldots M. \tag{3.2.8}$$

where c_i is read from the left of the matrix of Table 3.1 and a_{ij} is found from the main matrix. c_j is the cost values of $P_j (j = 1 \text{ to } M)$ which also appears at the bottom of the main matrix in the table.

One finds the subscript of the maximum Δ and the entering variable has the same subscript. For example, if Δ_3 is maximum, then P_3 should replace one of the 'calculated variables' (P_{M+1} to P_{M+N}). The choice of the leaving variable is made as follows.

3.2.2 The Choice of Leaving Variables

One assumes the subscript of the entering variable as e. If the values of the calculated variables are p_i, a quantity, q_i, is calculated which is

Table 3.1 Matrix Representation of the Linear Programming Problem

Matrix 1		Matrix 2 ($i\downarrow$ $j\longrightarrow$)										Vector 3
c_i	P_i	P_1	P_2	P_3		P_M	P_{M+1}	P_{M+2}		P_{M+N}	B	$q_E p_i/a_{ie}$
0	P_{N+1}	a_{11}	a_{12}	a_{13}	...	a_{1M}	1	0	...	0	b_1	(p_1/a_{1e})
0	P_{N+2}	a_{21}	a_{22}	a_{23}	...	a_{2M}	0	1	...	0	b_2	(p_2/a_{2e})
⋮	⋮	⋮	⋮	⋮	...	⋮	⋮	⋮	...	⋮	⋮	⋮
0	P_{N+M}	a_{N1}	a_{N2}	a_{N3}	...	a_{NM}	0	0	...	1	b_N	(p_N/a_{Ne})

In the first trial, one sets $P_1 = P_2 = \ldots = P_M = 0$. These are called 'chosen variables'.

c_j (cost values of P_1 to P_M) $\quad c_1 \quad c_2 \quad c_3 \quad \ldots \quad c_M$

Calculated values of P_i, $p_i = \ldots$ $\quad b_1 \quad b_2 \quad b_3 \quad \ldots \quad b_N$ (first trial values)

Calculations of $\Delta_j = \left(c_j - \sum_{i=1}^{N} c_i a_j\right)$; $j = 1$ to M $\quad \Delta_1 \quad \Delta_2 \quad \underset{\uparrow\, e = 3}{\Delta_3} \quad \Delta_4 \quad \ldots \quad \Delta_M$

defined as

$$q_i = (p_i/a_{ie}) \tag{3.2.9}$$

and is shown on the right of the main matrix for all values of $i = 1$ to N. One finds the subscript of the minimum q_i and the variables leaving the main matrix would be P_i and in its place, P_e would appear. This is carried out till the optimal solution is found.

3.2.3 The Test of Optimality

If all Δ's are greater than zero, one has a minimum C_T and if all Δ's are less than or equal to zero, one has a maximum C_T. If any one of Δ's does not satisfy the criteria of optimality, the procedure to remove and add one variable is continued.

To illustrate the simplex method of solution, the following three dimensional problem is solved

$$\text{Maximize } (4x_1 + 3x_2 + x_3) \tag{3.2.10a}$$

$$2x_1 + 4x_2 + x_3 \leqslant 20 \tag{3.2.10b}$$

$$3x_1 + 2x_2 + 2x_3 \leqslant 25 \tag{3.2.10c}$$

$$x_1 + 3x_2 + x_3 \leqslant 23 \tag{3.2.10d}$$

$$x_1, x_2, x_3 \geqslant 0 \tag{3.2.10e}$$

The inequalities above are converted to equalities by floating dummy variables x_4, x_5 and x_6 such that

$$\text{Maximize } (4x_1 + 3x_2 + x_3) \tag{3.2.11a}$$

$$2x_1 + 4x_2 + x_3 + x_4 = 20 \tag{3.2.11b}$$

$$3x_1 + 2x_2 + 2x_3 + x_5 = 25 \tag{3.2.11c}$$

$$x_1 + 3x_2 + x_3 + x_6 = 23 \tag{3.2.11d}$$

$$x_1 \text{ to } x_6 \geqslant 0 \tag{3.2.11e}$$

In the first trial x_1, x_2 and x_3 are set equal to zero as given in Table 3.2. In matrix 1, the calculated variables along with their costs are shown. The Δ_j are then calculated using Eq. 3.2.8. To Find Δ_1, one multiplies entries under column c_i in matrix 1 with entries under x_1 in matrix 2, and the sum of these is subtracted from c_1. Because all entries under c_1 of matrix 1 are zero, in the first iteration $\Delta_j = c_j$ for all j as given in Table 3.2a. Since Δ_1 is the highest, e = 1 and vector 3 is generated by dividing the entries of column B with entries of column x_1 of matrix 2. As q_2 corresponding to x_5 of matrix 1 is the smallest, in the second trial, x_1 replaces x_5.

For the second trial, x_1 is eliminated from Eqs. 3.2.11a and 3.2.11c with the help of Eq. 3.2.11b and one rearranges these equations such that the

Table 3.2 Solution of Eq. 3.2.10 using Simplex Method

(a) *First Trial*

Calculated variables are x_4, x_5 and x_6.

Matrix 1		Matrix 2						Vector 3	
c_i	x_i	x_1	x_2	x_3	x_4	x_5	x_6	B	q_i
0	x_4	2	4	1	1	0	0	20	10
0	x_5	3	2	2	0	1	0	25	8.33←
0	x_6	1	3	1	0	0	1	23	23
									x_5 leaves

$i, j =$	1	2	3	4	5	6
$c_j =$	4	3	1	0	0	0
$p_i =$	0	0	0	20	25	23
$\Delta_j =$	4	3	1	0	0	0
	↑e = 1					

(b) *Second Trial*

(i) x_1 is entered and x_5 is removed. Calculated variables are x_4, x_1 and x_6.
(ii) Matrix 2 is rearranged to obtain a unit diagonal matrix in x_4, x_1 and x_6.

Matrix 1		Matrix 2						Vector 3	
c_i	x_i	x_1	x_2	x_3	x_4	x_5	x_6	B	q_i
0	x_4	0	2.66	−0.33	1	−0.66	0	3.33	1.25←
4	x_1	1	0.66	0.66	0	0.33	0	8.33	12.62
0	x_6	0	2.33	0.33	0	−0.33	1	14.67	6.30
									x_4 leaves

$i, j =$	1	2	3	4	5	6
$c_j =$	4	3	1	0	0	0
$p_i =$	8.33	0	0	3.33	0	14.67
$\Delta_j =$	0	0.36	−1.64	0	−1.32	0
		↑ e = 2	x_2 enters			

(c) *Third Trial*

(i) x_2 is entered in place of x_4. Calculated variables are now x_2, x_1 and x_6.
(ii) Matrix 2 is rearranged to obtain a unit diagonal matrix in x_2, x_1 and x_6.

c_i	x_i	x_1	x_2	x_3	x_4	x_5	x_6	B
3	x_2	0	1	−0.13	0.38	−0.24	0	1.25
4	x_1	1	0	0.76	−0.25	0.5	0	7.49
0	x_6	0	0	0.63	0.89	0.91	1	11.76

$i, j =$	1	2	3	4	5	6
$c_j =$	4	3	1	0	0	0
$p_i =$	7.49	1.25	0	0	0	11.76
$\Delta_j =$	0	0	−1.65	−0.14	−1.25	0

The solution is optimal because $\Delta_j \leqslant 0$ for all j.

coefficients of x_4, x_1 and x_6 are 1. On such rearrangement,

$$2.66x_2 - 0.34x_3 + x_4 - 0.67x_5 = 3.33 \quad (3.2.12a)$$
$$x_1 + 0.67x_2 + 0.67x_3 + 0.33x_5 = 8.33 \quad (3.2.12b)$$
$$2.33x_2 + 0.33x_3 - 0.33x_5 + x_6 = 14.67 \quad (3.2.12c)$$

In Table 3.2(b), entries under x_i are x_4, x_1 and x_6 and under c_i, 0, 4 and 0 as their costs. Δ_j is calculated similarly by multiplying the column of c_i of matrix 1 with the column of x_j in matrix 2 and subtracting the sum from c_j. As Δ_2 is found to have the highest value, the entering variable is x_2. To find the leaving variable, vector 3 is generated by dividing entries of column B by entries of column x_2 in matrix 2. The minimum q_i is found to be 1.25 corresponding to x_4 of matrix 1. Therefore x_4 leaves and the calculated variables are now x_2, x_1 and x_6.

One similarly rearranges Eq. 3.2.12 as done earlier to find the new set of equations as

$$x_2 - 0.13x_3 + 0.38x_4 - 0.25x_5 = 1.25 \quad (3.2.13a)$$
$$x_1 + 0.76x_3 - 0.25x_4 + 0.5x_5 = 7.49 \quad (3.2.13b)$$
$$0.63x_3 - 0.89x_4 + 0.91x_5 + x_6 = 11.76 \quad (3.2.13c)$$

Δ_j are similarly calculated and they are all found to be either zero or negative, which means that the optimal solution is found.

3.3 THE COMBINATION OF SEPARATION AND MIXING FOR MINIMUM SEPARATION LOAD[3,5]

It has already been mentioned that any level of concentration change cannot be attained through mixing. In such cases, one is left with no choice but to use suitable separators along with mixers. However, the use of separators is kept to a minimum because of the latter's higher costs. To find the optimum combination of these, linear programming discussed in Sec. 3.2 is used.

To demonstrate the use of linear programming, the example given in Table 3.3 is considered. There are two streams, s_1 and s_2, having three components, A, B, and C, out of which a fixed mixture of A and B with

Table 3.3 The Problem of Two Raw Material Streams and a Single Goal

Component	C_A	C_B	C_C	Total quantity	CGS_A	CGS_B	CGS_C	$\lvert C_{AS} - C_{AG}\rvert$	$\lvert C_{BS} - C_{BG}\rvert$	$\lvert C_{CS} - C_{CG}\rvert$
Stream, s_1	0.50	0.30	0.20	500.00	0.40	2.67	0.00	0.30	0.50	0.20
Stream, s_2	0.05	0.40	0.55	600.00	4.00	2.00	0.00	0.15	0.40	0.55
Goal, G	0.20	0.80	0.00	150.00	—	—	—			

$Q_{S1} = 1.0$, $\quad Q_{S2} = 1.10$

no C is desired. Three possible arrangements can be used to obtain the desired product, G; in Fig. 3.1a, only s_1 is used, in Fig. 3.1b only s_2 is

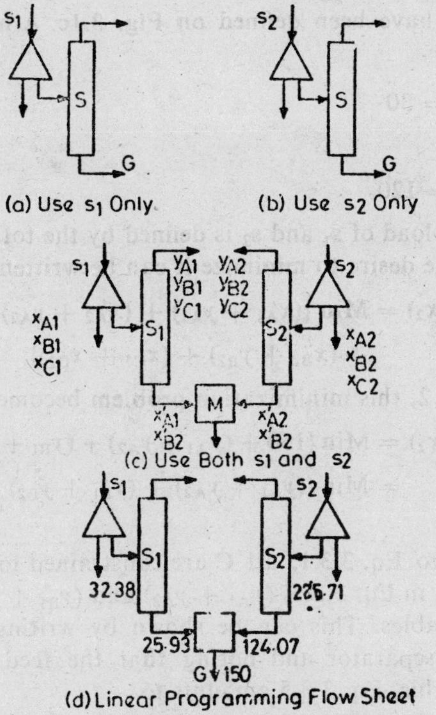

Fig. 3.1 Possible separation flow sheets for the problem of Table 3.3. (Reprinted from reference 5 with the permission of Pergamon Press, Oxford.)

used, whereas in Fig. 3.1c both s_1 and s_2 are used and the final goal is obtained by mixing. Naturally, the last flow sheet is better as it tends to minimize the total amount of material to be processed in separators as shown below.

One assumes that x_1 weight of stream s_1 and x_2 weight of stream s_2 are used to attain the goal G. x_1 consists of x_{A1} of A, x_{B1} of B and x_{C1} of C. Similarly x_2 consists of x_{A2} of A, x_{B2} of B and x_C of C. Since G does not have any C, the mass balance equations on separators s_1 and s_2 for the component are

$$x_{C1} = y_{C1} \tag{3.3.1a}$$

$$x_{C2} = y_{C2} \tag{3.3.1b}$$

Similarly for components A and B are

$$x_{A1} = x'_{A1} + y_{A1} \tag{3.3.2a}$$

$$x_{A2} = x'_{A2} + y_{A2} \tag{3.3.2b}$$

$$x_{B1} = x'_{B1} + y_{B1} \tag{3.3.2c}$$

and

$$x_{B2} = x'_{B2} + y_{B2} \tag{3.3.2d}$$

These variables have been defined on Fig. 3.1c. A mass balance on the mixer gives

$$x'_{A1} + x'_{A2} = 30 \tag{3.3.3a}$$

and

$$x'_{B1} + x'_{B2} = 120 \tag{3.3.3b}$$

If the separation load of s_1 and s_2 is defined by the total amount of material processed, the desire to minimize it can be written as

$$\text{Min } (x_1 + x_2) = \text{Min } \{(x'_{A1} + y_{A1}) + (x'_{A2} + y_{A2}) + (x'_{B1} + y_{B1}) + (x'_{B2} + y_{B2}) + (x_{C1} + x_{C2})\} \tag{3.3.4}$$

On using Eq. 3.3.2, this minimization problem becomes

$$\begin{aligned}\text{Min } (x_1 + x_2) &= \text{Min } \{150 + (y_{A1} + y_{A2}) + (y_{B1} + y_{B2}) + (y_{C1} + y_{C2})\} \\ &= \text{Min } \{(y_{A1} + y_{A2}) + (y_{B1} + y_{B2}) + (y_{C1} + y_{C2})\}\end{aligned} \tag{3.3.5}$$

Since according to Eq. 3.3.1, all C are constrained to appear at the top of the separators, in Eq. 3.3.5, $(y_{A1} + y_{A2})$ and $(y_{B1} + y_{B2})$ are treated as independent variables. This can be shown by writing all mass balance relations on the separator and noting that the feed compositions are fixed. In view of this, Eq. 3.3.5 amounts to

$$\begin{aligned}\text{Min } (x_1 + x_2) &= \text{Min } \{(y_{A1} + y_{A2}) + (y_{B1} + y_{B2})\} \\ &\quad \text{Min } (y_{A1} + y_{A2}) + \text{Min } (y_{B1} + y_{B2})\end{aligned} \tag{3.3.6}$$

It is further noted that

$$(y_{A1} + y_{A2}) \leqslant (280 - 30) \tag{3.3.7}$$

and

$$(y_{B1} + y_{B2}) \leqslant (390 - 120) \tag{3.3.8}$$

Examination of Eq. 3.3.6 to 3.3.8 reveals that the minimization of separation load results in the following two linear programming problems

$$(y_{A1} + y_{A2}) \leqslant 250 \tag{3.3.9a}$$

$$(y_{A1} + y_{A2}) \geqslant 0 \tag{3.3.9b}$$

$$\text{Minimize } (y_{A1} + y_{A2}) \tag{3.3.9c}$$

and

$$(y_{B1} + y_{B2}) \leqslant 270 \tag{3.3.10a}$$

$$(y_{B1} + y_{B2}) \geqslant 0 \tag{3.3.10b}$$

$$\text{Minimize } (y_{B1} + y_{B2}) \tag{3.3.10c}$$

The optimal solution of Eq. 3.3.9 obviously is

$$(y_{A1} + y_{A2}) = 0 \tag{3.3.11}$$

and as y_{A1} and y_{A2} are independent variables, the above condition implies that

$$y_{A1} = y_{A2} = 0 \tag{3.3.12}$$

One similarly concludes that

$$y_{B1} = y_{B2} = 0 \tag{3.3.13}$$

Consequently

$$\frac{x'_{A1}}{x'_{B1}} = \frac{x_{A1}}{x_{B1}} = \frac{0.5}{0.3} = \frac{5}{3} \tag{3.3.14a}$$

and

$$\frac{x'_{A2}}{x'_{B2}} = \frac{x_{A2}}{x_{B2}} = \frac{0.05}{0.4} = \frac{1}{8} \tag{3.3.14b}$$

On solving mass balance equations for the mixer, one gets

$$x'_{A1} = 16.22 \tag{3.3.15a}$$

$$x'_{B1} = 9.71 \tag{3.3.15b}$$

$$x'_{A2} = 13.78 \tag{3.3.15c}$$

and

$$x'_{B2} = 110.29 \tag{3.3.15d}$$

x_1 and x_2 are

$$x_1 = 32.38 \tag{3.3.16a}$$

and

$$x_2 = 275.71 \tag{3.3.16b}$$

The final flow sheet resulting from the linear programming is shown in Fig. 3.1d and is better than the ones shown in Fig. 3.1a and b where streams s_1 and s_2 alone are used. One can thus always generate the separation flow sheet for any concentration adjustment problem.

Motard et al.[5] have suggested an easy but empirical method of drawing a separation flow sheet without solving the corresponding linear programming problem. It is worth mentioning here because the result is very close to the one obtained by the rigorous procedure of optimization. According to the method suggested, a quantity CGS_j for component is calculated which is defined by

$$CGS_j = \left(\frac{C_{jG}}{C_{js}}\right) \tag{3.3.17}$$

where C_{jG} is the concentration of component j in the goal stream G and C_{js} is the concentration of j in stream s. The required amount of component j is obtained from the stream which has minimum CGS_j. Also the order in which they are utilized is as follows. One finds the absolute concentration difference for species j between the goal G and the stream s and calculates a quantity Q_s defined by

$$Q_s = \sum_j |C_{js} - C_{jG}| \tag{3.3.18}$$

Calling of streams is done in order of increasing Q_s. To demonstrate the working of the method, CGS_j and Q_s for species A, B and C are calculated for the example of Table 3.3. Q_{s1} is more than Q_{s2} and therefore stream s_1 is ordered first. As in stream s_1, CGS_A is the lowest, the entire 20 weight units of A required for G is ordered from this stream. Since

$$0.5x_1 = 30 \tag{3.3.19}$$

x_1 is found to be

$$x_1 = 60 \tag{3.3.20}$$

But with 60 weight units of s_1, 0.3×60 or 18 weight units of B is obtained which is also allowed to come along with stream s_1 (Fig. 3.1c). The remainder of B required for attaining the goal, G, i.e. 68 weight units, is attained through stream s_2. Thus stream s_4 is pure B and the mass balance yields x_2 as

$$0.4x_2 = 102 \quad \text{(a)}$$

or

$$x_2 = 255 \quad \text{(b)} \tag{3.3.21}$$

The resultant flow sheet, according to this procedure, is given in Fig. 3.2 which should be compared with Fig. 3.1d. It is noted that the basic structure of these two is the same and also stream quantities are very close to each other.

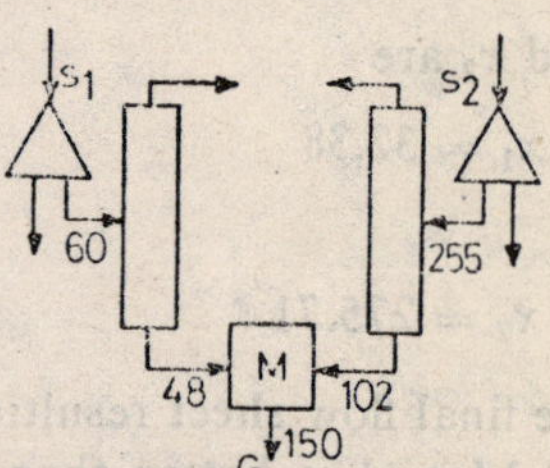

Fig. 3.2 Separation flow sheet from the empirical method

The problem solved by linear programming above is an easy one, but for a more difficult problem, the empirical method outlined proves to be very handy. Table 3.4 gives a more difficult situation where four streams

Table 3.4 The Problem of Two Raw Materials and Four Products

Stream	Components			
	x_A	x_B	C_A	C_B
s_1	4	6	0.40	0.60
s_2	6	4	0.60	0.40
G_1	2	1	0.67	0.33
G_2	3	7	0.30	0.70
G_3	3	0	1.00	0.00
G_4	2	2	0.50	0.50

(subgoals G_1, G_2, G_3 and G_4) are obtained from two streams s_1 and s_2. In Table 3.5, CGS and Q_s values have been calculated. From Q_s values

Table 3.5 CGS and Q_s Values for the Problem of Table 3.4

	G_1		G_2		G_3		G_4	
	s_1	s_2	s_1	s_2	s_1	s_2	s_1	s_2
CGS_A	1.68	1.12	0.75	0.50	2.5	1.67	1.25	0.83
CGS_B	0.55	0.83	1.17	1.75	0	0	0.83	1.25
$\lvert C_A-C_G \rvert$	0.27	0.07	0.10	0.30	0.60	0.40	0.10	0.10
$\lvert C_B-C_G \rvert$	0.27	0.07	0.10	0.30	0.60	0.40	0.10	0.10
Q_s	0.54	0.14	0.20	0.60	1.20	0.80	0.20	0.20

one finds that for attainment of G_1, stream s_1 is used and then s_2, if required. As CGS_B for s_1 is lower, all the requirement of B of G_1 is met by this stream. This means that 1.67 weight units of s_1 (consisting of 0.67A and 1B) is ordered. The remainder of A comes from stream s_2 meeting all its requirement of A. Similarly for G_2, stream s_2 is used first meeting all its requirement of A. But, before s_2 is used, the byproduct stream of the separator S_1 is completely consumed. For G_3, Q_s and CGS values indicate that all requirement of A of G_3 is met by stream s_1. G_4 is formed by mixing the remainder of streams as shown in Fig. 3.3.

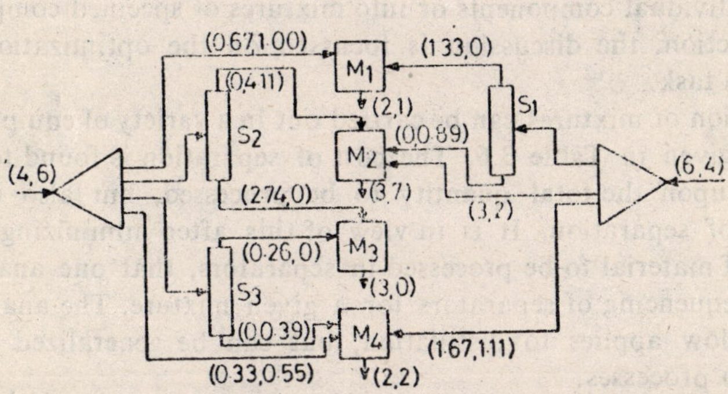

Fig. 3.3 Separation flow sheet for the problem of Table 3.4

The method suggested by Motard et al.[6] to arrive at the flow sheet unfortunately depends upon the order in which the goal list is attained. As an example, the problem discussed above is reconsidered; the mixtures are ordered in the sequence of G_2, G_1, G_3 and G_4. Even though the Q_s and CGS values remain unchanged, the separation flow sheet drawn in Fig. 3.4 is found to be completely altered.

Lastly the flow sheets of Figs. 3.3 and 3.4 are compared with the strategy of separating s_1 and s_2 into pure components A and B and mixing them next into products G_1 to G_4. The separation load then will be 20 weight units which is more than those of Fig. 3.3 (=9.72) and

Fig. 3.4 (=10.50).

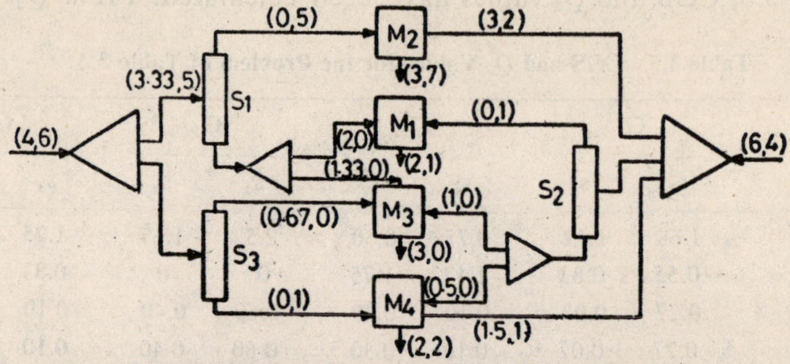

Fig. 3.4 The effect of order of subgoal attainment upon the network structure

In this section, the procedure to minimize the separation load for a given separation task has been given. Through the linear programming method, an optimal flow sheet consisting of mixers and separators was derived. Subsequently the empirical method suggested by Motard et al.[8] was discussed for a quick but an approximate separation of flow sheet.

3.4 THE TASK OF SEPARATION

In earlier sections, it was assumed that a mixture can always be separated into its individual components or into mixtures of specified compositions. In this section, the discussion is focussed on the optimization of the separation task.

Separation of mixtures can be carried out in a variety of equipment and they are given in Table 3.6. The cost of separation is found to depend not only upon the total quantity to be processed, but also upon the sequence of separation. It is in view of this after minimizing the total amount of material to be processed in separators, that one analyzes the optimal sequencing of separators for a given mixture. The analysis presented below applies to distillation, but can be generalized for other separation processes.

Consider a mixture of benzene and toluene to be separated into its pure components. Single distillation column can be used obtaining benzene as an overhead and toluene as the bottom product. Since only one column is required, the question of sequencing does not arise. If a three component mixture consisting of benzene (B), toluene (T) and *o*-xylene (X), is required to be separated into its pure components, two distillation columns will be required for the task. Pure benzene can be obtained as overhead and toluene and *o*-xylene at the bottom of the first column (Fig. 3.5a) and in the second column, the binary mixture is distilled into pure toluene and *o*-xylene. The second possible sequence would be to obtain a mixture of benzene and toluene as an overhead in the first column (Fig. 3.5b) and to distill this mixture into pure components in

Table 3.6[1] Classes of Separation Processes

S.No.	Name	Feed	Separating Agent	Products	Principle of Separation
	Equilibrium Separation Processes				
1.	Evaporation	Liquid	Heat	Liquid + vapour	Difference in volatility
2.	Flash expansion	Liquid	Pressure reduction	Liquid + vapour	-do-
3.	Distillation	Liquid and/or vapour	Heat	Liquid + vapour	-do-
4.	Stripping	Liquid	Noncondensable gas	Liquid + vapour	-do-
5.	Absorption	Gas	Nonvolatile liquid	Liquid + vapour	Preferential solubility
6.	Extraction	Liquid	Immiscible liquid	Two liquids	Different solubilities of different species in the two liquid phases
7.	Crystallization	Liquid	Cooling or heat causing simultaneous evaporation	Liquid and solids	Difference in freezing tendencies
8.	Adsorption	Gas or liquid	Solid adsorbent	Fluid and solid	Difference in adsorption potential
9.	Ion exchange	Liquid	Solid resin	Liquid + solid resin	Law of mass action applied to available anions or cations
10.	Drying of solids	Moist solid	Heat	Dry solid + humid vapour	Evaporation of water
11.	Leaching or washing	Solids	Solvent	Liquid + solid	Preferential solubility
12.	Clathration	Liquid	Clathrating molecule and cooling	Liquid + solid	Preferential participation in crystal structure
13.	Osmosis	Salt solution	More concentrated salt solution; membrane	Two liquids	Tendency to achieve uniform osmotic pressure removes water from more dilute solution

(*Contd.*)

Table 3.6 *(Contd.)*

S.No.	Name	Feed	Separating Agent	Product	Principle of Separation
14.	Bubble fractionation; foam fractionation	Liquid	Rising air bubbles; sometimes also complexing surfactants	Two liquids	Tendency of surfactant molecules to accumulate at the gas-liquid interface and rise with air bubble
15.	Flotation	Mixed powdered solids	Added surfactants; rising bubbles	Two solids	Tendency of surfactants to adsorb preferentially on one solid species
16.	Magnetic separation	Mixed powdered solids	Magnetic field	Two solids	Attraction of materials in magnetic field
17.	Paper chromatography	Liquid	Capillary; paper or gel phase	Regions of moistened paper	Preferential solubilities and adsorption potentials in two phases
18.	Freeze drying	Frozen water containing solid	Heat	Dry solid and water vapour	Sublimation of water
19.	Desublimation	Vapour	Cooling	Solid and vapours	Preferential condensation, preferential participation in crystal structure
20.	Gel filtration	Liquid	Solid gel (e.g., cross-linked dextran)	Gel phase and liquid	Difference in molecular size and hence inability to penetrate swollen gel matrix
21.	Dual-temperature exchange reactions	Fluid	Heating and cooling	Two fluids	Difference in reaction equilibrium constant at two different temperatures
22.	Zone melting	Solid	Heat	Solids of non-uniform composition	Same as crystallization
Mechanical Separation Processes					
1.	Filtration	Liquid + solid	Pressure reduction; filter medium	Liquid + solid	Size of solid greater than pore size of filter medium
2.	Mesh demister	Gas + solid or liquid	Pressure reduction; wire mesh	Gas + solid or liquid	-do-

3.	Settling	Liquid + solid or another immiscible liquid	Gravity	Liquid + solid or another immiscible liquid	Density difference
4.	Centrifuge (sedimentation type)	-do-	Centrifugal force	-do-	-do-
5.	Centrifuge (filtration type)	Liquid + solid	Centrifugal force	Liquid + solid	Size of solid greater than pore size of filter medium
6.	Cyclone	Gas + liquid or liquid	Flow (inertia)	Gas + solid	Density difference
7.	Electrostatic precipitation	Gas + fine	Electric field	Gas + fine solid	Charge on fine solid particles

(Reprinted from C. J. King, *Separation Processes*, 1971, with the permission of McGraw-Hill Book Company, New York.)

the second column. The arrangement to split the ternary mixture (Fig. 3.5c) into two binary mixtures in the first column and subsequent distill-

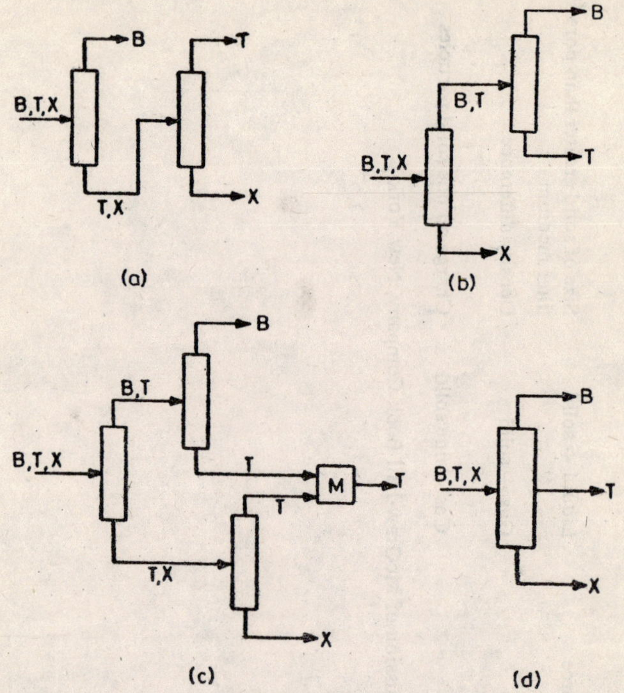

Fig. 3.5 Possible sequences of distillation columns. Sequence (c) is nonoptimal whereas (d) is impossible

ation into pure components in another two columns requires three columns and is worse than the earlier two. It may be mentioned here that for the task of obtaining pure components one cannot employ single distillation column (Fig. 3.5d) because of the following reason. When a multicomponent distillation is carried out, heavier components tend to concentrate in lower plates whereas lighter components accumulate in the top of the column. For any equilibrium stage falling between the top and bottom, an intermediate concentration falling between these extreme values is obtained and this can never be a pure component. Thus for obtaining pure components, it is necessary to use at least two distillation columns for ternary mixture, three columns for a quaternary mixture, and products should be withdrawn at the top and the bottom of every column.

The example above can now be generalized for an R-component mixture to be separated into its pure components. First to obtain pure components, it is necessary that only two streams are withdrawn per distillation column as overhead and bottom products. For an R-component mixture, at least $(R - 1)$ number of columns are required.

Thus, sequencing is a very important parameter which affects not only the fixed cost but the total operating cost of the column network also. Consequently, a given separation task must be optimized with respect to it. However, before this is taken up, it is instructive to find the total number of possible sequences for an R-component mixture. In the first column, j components are assumed to be collected at the overhead and the remaining $(R-j)$ components at the bottom. A quantity S_R is defined as equal to the number of different column sequences for an R-component mixture. If S_j is the number of column sequences for separation of j-component mixture, and S_{R-j} the number of column sequences for $(R-j)$ component mixture, then S_R is equal to the product of S_j and S_{R-j}. As j can take any value from 1 to $(R-1)$, S_R is given by summing $(S_j S_{R-j})$ for all values of j as

$$S_R = \sum_{j=1}^{R-1} S_{R-j} S_j \tag{3.4.1}$$

S_1 is equal to one and S_2 is also found to be 1 by substituting $R=2$ in the above equation. S_3 is generated from values of S_2 and S_1, S_4 generated from S_3, S_2 and S_1 etc., and results are summarized in Table 3.7. Having 10 or 11 components in a given mixture is not a very unrealistic system, and as the number of possible sequences are very large, a brute force technique for optimization with respect to the network sequence becomes an impossible task.

Table 3.7[1] Number of Column Sequences for Different Values of R

R	S_R
2	1
3	2
4	5
5	14
6	42
7	132
8	429
9	1,430
10	4,862
11	16,796

(Reprinted from C. J. King, *Separation Processes*, 1971, with the permission of McGraw-Hill Book Company, New York.)

In Appendix 3.1, designing of a multicomponent distillation is reviewed[1,6] and in Sec. 3.5, the importance of sequencing is established.

3.5 IMPORTANCE OF DISTILLATION COLUMN SEQUENCING

Supposing there is a mixture containing more than two components and it is desired to separate them into pure components. As has been pointed out in Sec. 3.4, $(R - 1)$ columns are required for an R-component mixture and the number of possible configurations are given by Table 3.7. In this section, all these configurations are examined to find that they are not identical but distinguish themselves in terms of total column volume and operating costs.

To demonstrate this, the separation of a ternary mixture of propane, *iso*-butane and *n*-butane into (almost) pure components is considered. Figure 3.6 shows the two possible configurations and the specified

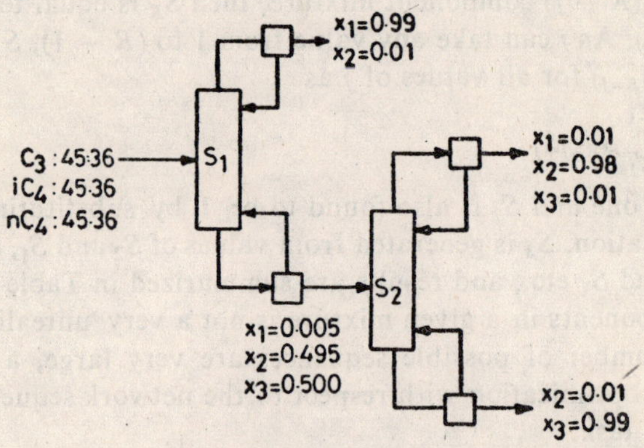

(a) Configuration I: Key Components for S_1 are C_3 and iC_4.

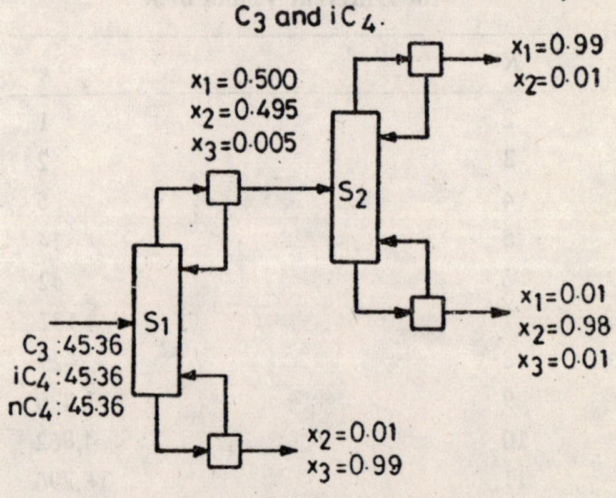

(b) Configuration II: Key Components for S_1 are i-C_4 and n-C_4.

Fig. 3.6 Separation of propane (C_3), *iso*-butane (iC_4) and *n*-butane (nC_4)

splitting on key comopnents. It is assumed that all distillation columns are at a uniform temperature of 85°C (185°F). This is not necessary and is made only to simplify calculations. Partition coefficients, K, for these compounds at this temperature are found to be 3.25, 1.70 and 1.30[6]. It is also assumed that all feeds are saturated liquids. The stepwise calculations of Table A3.1 are performed for both these configurations as follows.

Configuration I of Fig. 3.6

STEPS A, B AND C

Column C1

Key Components: C_3 and iC_4

$$\alpha_{C_3-iC_4} = \frac{3.25}{1.70} = 1.91$$

$$N_{m1} = \frac{2 \ln 99}{\ln \alpha_{C_3-iC_4}} = \frac{2 \times \ln 99}{\ln 1.91} = 14.22$$

Column C2

$$\alpha_{iC_4-nC_4} = 1.31$$

$$N_{m2} = \frac{\ln (99 \times 98)}{\ln 1.31} = 34.03$$

STEP D

Column C1

$$\alpha_1 = \frac{K_{C_3}}{K_{nC_4}} = 2.5$$

$$\alpha_2 = 1.31$$

$$\alpha_3 = 1.00$$

For saturated liquid feed

$$\frac{2.5}{2.5 - \theta_1} + \frac{1.31}{1.31 - \theta_1} + \frac{1}{1 - \theta_1} = 0$$

$$\theta_1 = 1.829$$

Column C2

$$\frac{1.31 \times 0.495}{1.31 - \theta_2} + \frac{0.50}{1 - \theta_2} = 0$$

$$\theta_2 = 1.133$$

STEP E

$$(1 + R_{m1}) = \frac{2.5 \times 0.99}{2.5 - 1.829} + \frac{1.31 \times 0.01}{1.31 - 1.829}$$

$$= 3.66$$

or $R_{m1} = 2.66$

$$(1 + R_{m2}) = \frac{1.31 \times 0.98}{1.31 - 1.133} + \frac{0.01}{1 - 1.133}$$

$$= 7.25$$

Configuration II of Fig. 3.6

Column C1

Key Components: iC_4 and nC_4

$$\alpha_{iC_4-nC_4} = 1.31$$

$$N_{m1} = \frac{2 \ln 99}{\ln 1.31} = 34.03$$

Column C2

$$\alpha_{C_3-iC_4} = 1.91$$

$$N_{m2} = \frac{\ln (99 \times 98)}{\ln 1.91} = 14.22$$

Column C1

$$\alpha_1 = 2.50$$

$$\alpha_2 = 1.31$$

$$\alpha_3 = 1.00$$

For saturated liquid feed

$$\frac{2.5}{2.5 - \theta_1} + \frac{1.31}{1.31 - \theta_1} + \frac{1}{1 - \theta_1} = 0$$

$$\theta_1 = 1.117$$

Column C2

$$\frac{2.5 \times 0.50}{2.5 - \theta_2} + \frac{0.495 \times 1.31}{1.31 - \theta_2} = 0$$

$$\theta_2 = 1.713$$

$$(R_{m1} + 1) = \frac{2.5 \times 0.5}{2.5 - 1.117} + \frac{1.31 \times 0.495}{1.31 - 1.117}$$

$$= 4.26$$

or $R_{m1} = 3.26$

$$(R_{m2} + 1) = \frac{0.98 \times 2.5}{2.5 - 1.713}$$

$$= 3.113$$

or $R_{m2} = 6.25$

$R_{m2} = 2.113$

STEP F

$R_1 = 1.30\ R_{m1} = 3.46$

$R_2 = 1.30\ R_{m2} = 8.13$

$$\frac{R_1 - R_{m1}}{R_1 + 1} = \frac{0.3 \times 2.66}{4.66} = 0.178$$

$$\frac{R_2 - R_{m2}}{R_2 + 1} = \frac{0.3 \times 6.25}{9.13} = 0.205$$

From Fig. 3.13, one obtains

$$\frac{N_1 - N_{m1}}{N_1 + 1} \equiv g_1 = 0.44$$

$$\frac{N_2 - N_{m2}}{N_2 + D_1} \equiv g_2 = 0.43$$

$$\therefore\ N_1 = \frac{N_{m1} + g_1}{1 - g_1} = \frac{14.66}{0.56} = 26.18$$

$$N_2 = \frac{N_{m2} + g_2}{1 - g_2} = \frac{34.46}{0.57} = 60.46$$

$R_1 = 1.30 \cdot R_{m1} = 4.24$

$R_2 = 1.30 \cdot R_{m2} = 2.75$

$$\frac{R_1 - R_{m1}}{R_1 + 1} = \frac{0.3 \times 3.26}{5.24} = 0.187$$

$$\frac{R_2 - R_{m2}}{R_2 + 1} = \frac{0.3 \times 2.113}{3.75} = 0.169$$

From Fig. 3.13, one obtains

$g_1 = 0.44$

$g_2 = 0.45$

$$N_1 = \frac{N_{m1} + g_1}{1 - g_1} = \frac{34.47}{0.56} = 61.55$$

$$N_2 = \frac{N_{m2} + g_2}{1 - g_2} = \frac{14.67}{0.55} = 26.67$$

STEP G

$D_1 = 45.36$ kg. mol $(=100$ lb-mol)

$L_1 = R_1D_1 = 156.94;\ V_1 = 202.30$

$L_1' = 293.02,\ V_1' = 202.30$

$D_2 = 45.36$ kg-mol (100 lb-mol)

$L_2 = R_2D_2 = 368.78,\ V_2 = 414.14$

$L_2' = 459.50,\ V_2' = 414.14$

$D_1 = 90.72$ kg-mol/h

$(= 200$ lb-mol/h)

$L_1 = R_1D_1 = 384.55,\ V_1 = 475.37$

$L_1' = 520.73,\ V_1' = 475.37$

$D_2 = 45.36$ kg-mol (100 lb-mol)

$L_2 = R_2D_2 = 124.74,\ V_2 = 170.10$

$L_2' = 215.46,\ V_2' = 170.10$

STEP H

$\rho_v = 0.0024 \times 10^3\ kg/m^3\ (=0.15\ lb/ft^3)$

$\rho_l = 0.72 \times 10^3\ kg/m^3\ (0.45\ lb/ft^3)$

$$F_{lv_1} = \frac{293.02}{202.30}\sqrt{\frac{0{,}0024}{0.72}} = 0.0837$$

$F_{lv_2} = 0.0641$

Plate spacing is taken as 60.96 cm (or 2 ft)

$C_{sb_1} \simeq C_{sb_2} = 0.34$

$\therefore\ U_{n1} \simeq U_{n2} = 1.80$ m/sec $(=5.89$ ft/s)

$\rho_v = 2.4\ kg/m^3$

$\rho_l = 720\ kg/m^3$

$F_{lv_1} = 0.0632$

$F_{lv_2} = 0.0731$

Plate spacing is taken as 60.96 cm (or 2 ft)

$C_{sb_1} \simeq C_{sb_2} = 0.34$

$\therefore\ U_{n1} \simeq U_{n2} = 1.80$ m/sec (5.89 ft/s)

STEP I

Molecular weight of the vapour at the bottom of C1 as well as C2 is 58 kg/mol.

Vapour flow rates at the bottom are

$\therefore \quad W_1 = 11{,}733.40$ kg/h

$W_2 = 24{,}020.12$ kg/h

$$\therefore \quad A_1 = \frac{11733.40}{3600 \times 0.0024 \times 1.80 \times 10^3}$$

$$= 0.754 \text{ m}^2$$

$$A_2 = \frac{24020.12}{3600 \times 0.0024 \times 10^3 \times 1.80}$$

$$= 1.544 \text{ m}^2$$

$$\therefore \quad \text{Vol}_1 = (26.18 \times 0.6096) \times 0.754\text{m}^3$$

$$= 12.03 \text{ m}^3$$

$$\text{Vol}_2 = (60.46 \times 0.6096) \times 1.544$$

$$= 56.91 \text{ m}^3$$

$$V_T = 68.94 \text{ m}^3$$

Molecular weight of the vapour at the bottom of C1 and C2 is 58 kg/kg-mol.

Vapour flow rates at the bottom are

$W_1 = 27571.46$ kg/h

$W_2 = 9865.80$ kg/h

$$A_1 = \frac{27571.46}{3600 \times 0.0024 \times 10^3 \times 1.80}$$

$$= 1.773 \text{ m}^2$$

$$A_2 = \frac{9865.80}{3600 \times 0.0024 \times 10^3 \times 1.80}$$

$$= 0.634 \text{ m}^2$$

$$\text{Vol}_1 = (61.55 \times 0.6096) \times 1.773$$

$$= 66.52 \text{ m}^3$$

$$\text{Vol}_2 = (26.67 \times 0.6096) \times 0.634$$

$$= 10.31 \text{ m}^3$$

$$V_T = 76.83 \text{ m}^3$$

TOTAL HEAT INPUT TO REBOILERS

$\Delta H_{C_3} = 2561.4$ kcal/mol

$\Delta H_{iC_4} = 3619.3$ kcal/mol

$\Delta H_{nC_4} = 4009.1$ kcal/mol

(Data at 10 atm. 85°C)

Composition of the vapour at the reboiler

$y_1 \cong 0$

$$y_2 = \frac{\alpha_2 x_2}{\alpha_1 x_2 + \alpha_2 x_2 + \alpha_3 x_3}$$

$$= 0.565$$

$y_3 = 0.435$

$$\therefore \quad \dot{Q}_1 = (V_1'0.565) \times 3619.3 + (V_1'0.435) \times 4009.1$$

$$= (2044.9 + 1743.9) \times 202.3$$

$$= 7.66 \times 10^5 \text{ kcal/h}$$

$$\dot{Q}_2 = 4009.1 \times 414.14 \text{ kcal/h}$$

$$= 16.60 \times 10^5 \text{ kcal/h}$$

$$\therefore \quad \dot{Q}_T = \dot{Q}_1 + \dot{Q}_2$$

$$= 24.26 \times 10^5 \text{ kcal/h.}$$

$$\dot{Q}_1 = 475.37 \times \Delta H_{nC_4}$$

$$= 19.06 \times 10^5 \text{ kcal/h}$$

$$\dot{Q}_2 = 170.1 \times 3619.3$$

$$= 6.16 \times 10^5 \text{ kcal/h}$$

$$\dot{Q}_T = 25.22 \times 10^5 \text{ kcal/h}$$

The network cost can be found as follows. The cost, C_i, of a distillation column, i, can be expressed as

$$C_i = a(\mathrm{Vol}_i)^b \simeq a(\mathrm{Vol}_i) \tag{3.5.1}$$

where a and b are constants, Vol_i is the volume of the column and b is approximated as equal to 1 in the above representation. The total fixed cost C_{FT} of the distillation column network therefore is given by

$$C_{FT} = a \sum_i (\mathrm{Vol}_i) \tag{3.5.2}$$

If the annual rate of return is δ and is assumed that columns last forever, the total operating cost, OC_T of the network is given by (Chapter 10)

$$OC_T = a\delta \sum_i (\mathrm{Vol})_i + c \sum \dot{Q}_i \tag{3.5.3}$$

where $\dot{Q}_i$ is the heat supplied in the reboiler of the ith column and c is the cost of supplying unit heat.

The basis of comparison of the two configurations of Fig. 3.6 is to find OC_T for both these, but for this example, it is really not necessary because both $\sum_i (\mathrm{Vol}_i)$ and $\sum_i \dot{Q}_i$ for configuration I is lower. Thus configuration I is better than configuration II for all choices of a, δ and c for the separation of the mixture of propane, *iso*-butane and *n*-butane into its pure components. This example demonstrates that the order of sequencing affects not only the total fixed cost but the total operating cost of the network also.

For a three component mixture, there are only two possible networks and, consequently, it was easy to find the best. However, for higher order mixtures this is not an easy task. Also unlike the design of heat exchanger network dealt in Chapter 2, there is no theoretical lead for deciding on the network structure. In Sec. 3.6, a few heuristic rules have been derived and it is shown how these are used.

3.6 HEURISTIC RULE OF REMOVING COMPONENTS ONE-BY-ONE (RULE 1)[1,10]

One considers an R component liquid mixture to be separated into its pure components. In Fig. 3.7, only two columns of the separator network is considered in which three mixtures are produced. The first mixture consists of 1 to L components, the second $(L + 1)$ to K components and the third $(K + 1)$ to R components. In the first sequence, the first separator, D_L, splits the mixture into two: mixture A, consisting of 1 to L components and mixture B, consisting of $(L + 1)$ to R components. Then the separator, D_K, splits mixture B into another two: mixture C consisting of $(L + 1)$ to K components and mixture E consisting of $(K + 1)$ to R components. In the second sequence, separator D_K splits the original mixture into mixture E and a combination of A and C. Then the separator D'_L splits the latter into A and C. It is desired to find out which of them is a better arrangement.

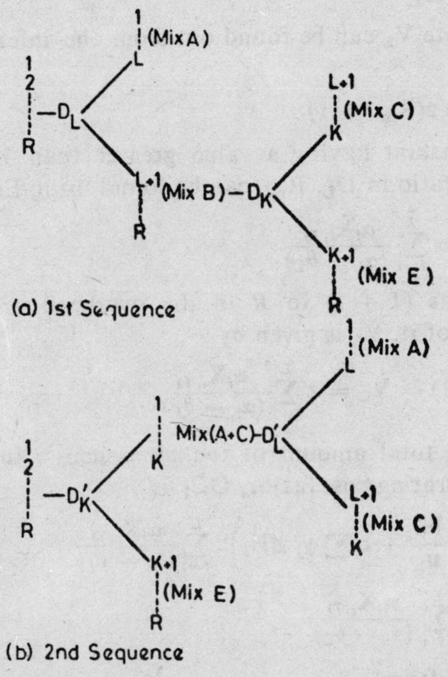

Fig. 3.7 Splitting of an *R*-component mixture into three mixtures A, B and C. (Reprinted from reference 10 with the permission of the American Institute of Chemical Engineers, New York.)

To be able to do so, the total operating cost function, OC_T, for both these sequences is written and compared. All feeds to columns are assumed to be at bubble point and column D_L is considered first. Its volume, Vol_L, is given by

$$Vol_L = N_L S_L \tag{3.6.1}$$

where N_L is the number of stages and S_L is the cross-sectional area of D_L. If u_L is the flooding velocity under operating conditions of D_L,

$$S_L = (V_L/u_L) \tag{3.6.2}$$

where V_L is the volumetric flow rate of vapours in the stripping section of D_L. The heat required $\dot{Q}_L$, in the reboiler of D_L is estimated as

$$\dot{Q}_L = (\sum_i y_i \Delta H_i) V_L \tag{3.6.3}$$

where y_i is the mole fraction of the component i in the vapour leaving the reboiler and ΔH_i is its latent heat. The operating cost OC_L thus becomes (Eq. 3.5.3)

$$OC_L = \left(a\delta \frac{N_L}{u_L} + c \sum_i y_i \Delta H_i\right) V_L \tag{3.6.4}$$

Vapour flow rate V_L can be found out from the information on reflux ratio, R_L, as

$$(R_L + 1) = z(R_{mL} + 1) \tag{3.6.5}$$

where z is a constant having a value greater than 1 and R_{mL} is the minimum reflux ratio in D_L. R_{mL} can be found using Eq. A3.1.38a as

$$(R_{mL} + 1) = \sum_{i=1}^{L} \frac{\alpha_i X_{i,D}}{(\alpha_i - \theta_L)} \tag{3.6.6}$$

where components $(L + 1)$ to R in the overhead are assumed to be absent. In terms of it, V_L is given by

$$Dz(R_{mL} + 1) = V_L = z \sum_{i=1}^{L} \frac{\alpha_i X_{i,D}}{(\alpha_i - \theta_L)} \tag{3.6.7}$$

where $X_{i,D}$ is the total amount of the component i in the overhead of D_L. Thus the operating cost factor, OC_L is

$$OC_L = \left(a\delta \frac{N_L}{u_L} + c \sum_i y_i \Delta H_i\right) z \sum_{i=1}^{L} \frac{\alpha_i X_{i,D}}{(\alpha_i - \theta_L)}$$

$$= A_L \sum_{i=1}^{L} \frac{\alpha_i X_{i,D}}{(\alpha_i - \theta_L)} \tag{3.6.8}$$

where θ_L is found from

$$\sum_{i=1}^{R} \frac{\alpha_i X_{i,F}}{(\alpha_i - \theta_L)} = 0 \tag{3.6.9}$$

$X_{i,F}$ above is the total amount of the component i in the feed. The total operating cost OC_{T1} of the 1st sequence of Fig. 3.7 can be written as

$$OC_{T1} = A_L \sum_{i=1}^{L} \frac{\alpha_i X_{i,D}}{(\alpha_i - \theta_L)} + A_K \sum_{i=L+1}^{K} \frac{\alpha_i X_{i,D}}{(\alpha_i - \theta_K)} \tag{3.6.10}$$

where A_L and A_K are constants for columns D_L and D_K. Similarly, the total operating cost of the second sequence, OC_{T2}, is found to be

$$OC_{T2} = A'_K \sum_{i=1}^{K} \frac{\alpha_i X_{i,D}}{(\alpha_i - \theta'_L)} + A'_L \sum_{i=1}^{L} \frac{\alpha_i X_{i,D}}{(\alpha_i - \theta'_K)} \tag{3.6.11}$$

The difference in cost between the two, ΔC, is obtained by subtracting Eq. 3.6.11 from 3.6.10

$$\Delta C = (OC_{T1} - OC_{T2}) = \left\{A_L \sum_{i=1}^{L} \frac{\alpha_i X_{i,D}}{(\alpha_i - \theta_L)} + A_K \sum_{i=L+1}^{K} \frac{\alpha_i X_{i,D}}{(\alpha_i - \theta_K)}\right\}$$

$$- \left\{A'_K \sum_{i=1}^{K} \frac{\alpha_i X_{i,D}}{(\alpha_i - \theta'_K)} + A'_L \sum_{i=1}^{L} \frac{\alpha_i X_{i,D}}{(\alpha_i - \theta'_L)}\right\} \tag{3.6.12}$$

Now A_L and A'_L are compared. In the example solved in Sec. 3.5, the flooding velocity u_L was seen to be unaffected for a vide range of vapour and liquid flow rates. As key components of D_L and D'_L and their splitting (here, specified to be almost total) are the same, the number of stages required in D_L and D'_L are expected to be the same (example of Sec. 3.5). Therefore if the components of the mixtures have almost the same latent heats, ΔH_i, the following approximate relations hold good

$$A_L \simeq A'_L \tag{3.6.13}$$

and

$$A_K \simeq A'_K \tag{3.6.14}$$

Also if the amounts of all components in the mixture and the difference in successive specific volatilities are of the same order, Eq. 3.6.9 can be approximated by the following two terms[10]

$$\frac{\alpha_{K+1}X_{K+1,F}}{\alpha_{K+1}-\theta_K}+\frac{\alpha_K X_{K,F}}{\alpha_K-\theta_K}=0 \tag{3.6.15a}$$

and

$$\frac{\alpha_{L+1}X_{L+1,F}}{\alpha_{L+1}-\theta_L}+\frac{\alpha_L X_{L,F}}{\alpha_L-\theta_L}=0 \tag{3.6.15b}$$

The two term approximation arises because terms in Eq. A3.1.41 separated far away from the key components contribute negligibly to the value of θ calculated. From these equations, the roots θ_K and θ_L can be conveniently found with very small error. Similar equations can be written, for θ'_K and θ'_L as

$$\frac{\alpha_{K+1}X_{K+1,F}}{\alpha_{K+1}-\theta'_K}+\frac{\alpha_K X_{K,F}}{\alpha_K-\theta'_K}=0 \tag{3.6.16a}$$

$$\frac{\alpha_{L+1}X_{L+1,F}}{\alpha_{L+1}-\theta'_L}+\frac{\alpha_L X_{L,F}}{\alpha_L-\theta'_L}=0 \tag{3.6.16b}$$

On comparison of Eqs. 3.6.15 and 3.6.16,

$$\theta_K = \theta'_K \tag{3.6.17a}$$

and

$$\theta_K = \theta'_L \tag{3.6.17b}$$

and substituting Eqs. 3.6.13, 3.6.14 and 3.6.17 in Eq. 3.6.12, one has

$$\Delta C = A_L\left\{\sum_{i=1}^{L}\frac{\alpha_i X_{i,D}}{(\alpha_i-\theta_L)}-\sum_{i=1}^{L}\frac{\alpha_i X_{i,D}}{\alpha_i-\theta_L}\right\} + A_K\left\{\sum_{i=L+1}^{K}\frac{\alpha_i X_{i,D}}{(\alpha_i-\theta_K)}-\sum_{i=1}^{K}\frac{\alpha_i X_{i,D}}{(\alpha_i-\theta_K)}\right\} \tag{3.6.18}$$

or

$$\Delta C = -A_K\sum_{i=1}^{L}\frac{\alpha_i X_{i,D}}{(\alpha_i-\theta_K)} \tag{3.6.19}$$

As A_K and $\sum_{i=1}^{L}\left\{\frac{\alpha_i X_{i,D}}{\alpha_i-\theta_K}\right\}$ (terms belonging to components lighter than light key) both are positive quantities, ΔC is negative, which means that the first sequence of Fig. 3.7 has a lower cost. Because K and L are two arbitrary numbers such that $K > L$, the best separation procedure is a sequential removal of lighter components as shown in Fig. 3.8. It may

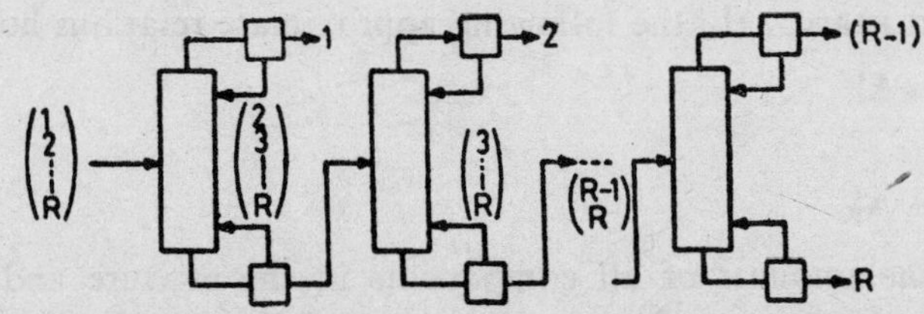

Fig. 3.8 Optimal separation sequence for a multicomponent mixture

be understood that the optimality of this sequence derived in this section hinges upon the two term approximation of Eqs. 3.6.15 and 3.6.16. When this breaks down, the sequential separation is not the best and suitable changes in the network structure should be made. In the example given in Sec. 3.5, moles of C_3, and iC_4 and nC_4 were of the same order, and so were their relative volatilities. Consequently, the sequential separation was found to be a better strategy. The following heuristic rules (rules 2 and 3) arise because the two term approximation breaks down.

3.7 HEURISTIC RULE OF REMOVING COMPONENTS IN LARGE EXCESS EARLY (RULE 2)[10]

One considers the situation when components S (S > K) is present in large excess. In this case, the two term approximation is modified by including the terms for the component S. Thus θ_K, θ_L, θ'_K and θ'_L are determined by following equations:

$$\frac{\alpha_{L+1}X_{L+1,F}}{\alpha_{L+1}-\theta_L}+\frac{\alpha_L X_{L,F}}{\alpha_L-\theta_L}+\frac{\alpha_S X_{S,F}}{\alpha_S-\theta_L}=0, \tag{3.7.1a}$$

$$\frac{\alpha_{K+1}X_{K+1,F}}{\alpha_{K+1}-\theta_K}+\frac{\alpha_K X_{K,F}}{\alpha_K-\theta_K}+\frac{\alpha_S X_{S,F}}{\alpha_S-\theta_K}=0, \tag{3.7.1b}$$

$$\frac{\alpha_{K+1}X_{K+1,F}}{\alpha_{K+1}-\theta'_K}+\frac{\alpha_K X_{K,F}}{\alpha_K-\theta'_K}+\frac{\alpha_S X_{S,F}}{\alpha_S-\theta'_K}=0, \tag{3.7.1c}$$

and

$$\frac{\alpha_{L+1}X_{L+1,F}}{\alpha_{L+1}-\theta'_L}+\frac{\alpha_L X_{L,F}}{\alpha_L-\theta'_L}=0. \tag{3.7.1d}$$

The term corresponding to the component S appears in equations above because of the large quantity of S. Because A_L, A_K, A'_L and A'_K are not affected by θ_L, θ_K, θ'_L and θ'_K, relations given in Eqs. 3.6.13 and 3.6.14 remain unaltered. Comparison of Eq. 3.7.1a, b, c and d gives

$$\theta_K=\theta'_K \tag{3.7.2}$$

but θ_L and θ'_L are different, the difference being dependent upon the contribution of $\alpha_S X_{S,F}/(\alpha_S - \theta'_K)$ to the value of θ'_K. The more removed the component S is from key components, the smaller shall be its effect. Equation 3.6.18 is rewritten as

$$\Delta C = A_L\left(\sum_{j=1}^{L} \frac{\alpha_j X_{j,F}}{\alpha_j - \theta_L} - \sum_{j=1}^{L} \frac{\alpha_j X_{j,F}}{\alpha_j - \theta'_L}\right) - A_K \sum_{j=1}^{L} \frac{\alpha_j X_{j,F}}{\alpha_j - \theta_K} \tag{3.7.3}$$

which could be greater than zero. This implies that the sequence $D'_K - D'_L$ could become economical compared to the $D_L - D_K$ sequence depending upon the value of $X_{S,F}$.

The effect of $X_{S,F}$ on the sequence of separation is examined now. If X_S compares with quantities of other components, the arrangement given in Fig. 3.8 is symbolically represented in Fig. 3.9a. In this figure D_S stands

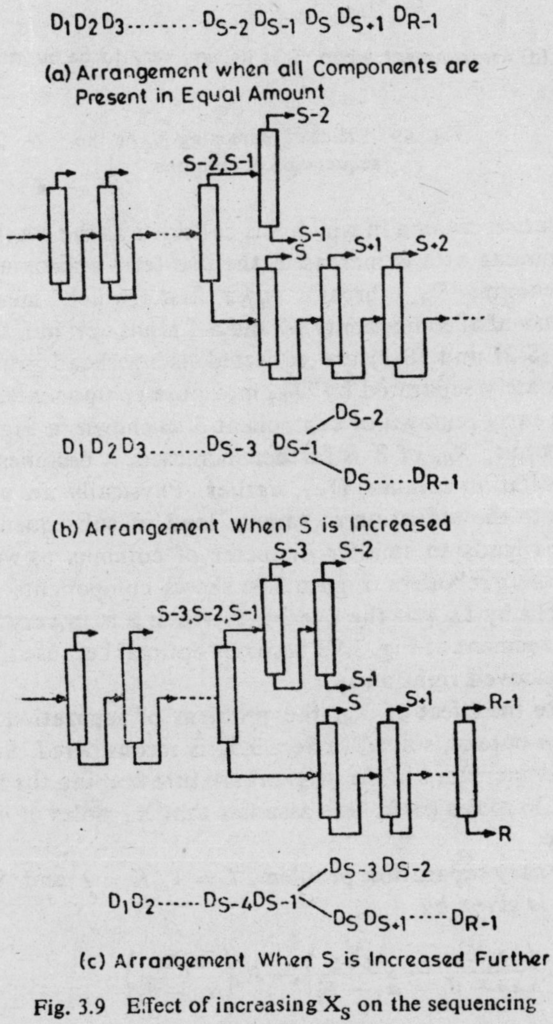

Fig. 3.9 Effect of increasing X_S on the sequencing of columns (*contd.*)

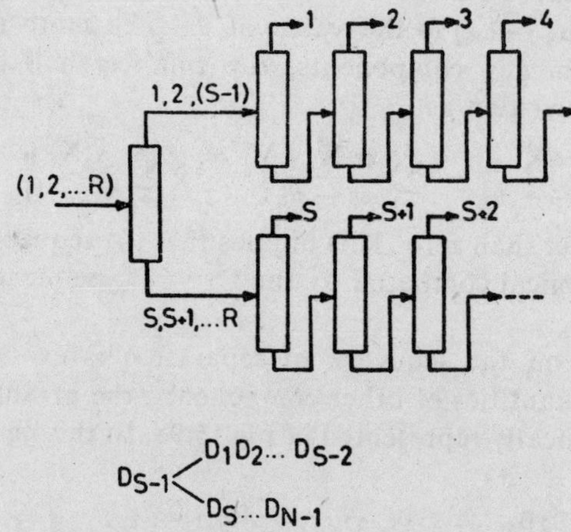

(d) Arrangement when S is in very very large quantity

Fig. 3.9 Effect of increasing X_S on the sequencing of columns

for the distillation column in which S is collected as the overhead product. When the amount of S is increased, the two term approximation for the distillation column D_{S-1} breaks down first, which means that the successive removal of components S-2 and S-1 is not optimal. Consequently, components (S-2) and (S-1) are collected as overhead product of D_{S-1}, which is ultimately separated by D_{S-2} into pure components. This scheme facilitates an early removal of component S as shown in Fig. 3.9b.

As the amount, X_S, of S is further increased, it becomes necessary to place the distillation column, D_{S-1}, earlier. Physically an early removal of S amounts to the reduction in vapour load of subsequent distillation columns. This leads to smaller diameter of columns as well as smaller heat load of their reboilers. Figure 3.9c shows components S-3, S-2 and (S-1) being split by D_{S-1} at the overhead. When S is in very large proportion, the arrangement of Fig. 3.9d becomes optimal because the component needs to be removed right away.

To illustrate the effect of X_S, the problem of separation of propane, *i*-butane and *n*-butane, solved in Sec. 3.5, is reconsidered. Supposing the moles of nC_4 is increased in the original mixture keeping the amount of C_3 and iC_4 at 45.36 moles each. One assumes that X_3 moles of nC_4 is present in the mixture.

For this ternary separation problem, $L = 1$, $K = 2$ and $S = 3$ in Eq. 3.7.3 and ΔC is given by

$$\Delta C = A_1\left(\frac{\alpha_1 X_1}{\alpha_1 - \theta_1} - \frac{\alpha_1 X_1}{\alpha_1 - \theta_1'}\right) - A_2\left(\frac{\alpha_1 X_1}{\alpha_1 - \theta_2}\right) \qquad (3.7.4)$$

where X_1 is moles of propane and *iso*-butane which is equal to 45.36 moles.

θ_1, θ_1' and θ_2 are given by

$$\frac{45.36\alpha_1}{\alpha_1 - \theta_1} + \frac{45.36\alpha_2}{\alpha_2 - \theta_1} + \frac{\alpha_3 X_3}{\alpha_3 - \theta_1} = 0 \tag{3.7.5a}$$

$$\frac{45.36\alpha_2}{\alpha_2 - \theta_2} + \frac{\alpha_3 X_3}{\alpha_3 - \theta_2} = 0 \tag{3.7.5b}$$

and

$$\frac{45.36\alpha_1}{\alpha_1 - \theta_1'} + \frac{45.36\alpha_2}{\alpha_2 - \theta_1'} = 0 \tag{3.7.5c}$$

For different values of X_3, θ_1, θ_1' and θ_2 have been calculated from Eq. 3.7.5 and results are given in Table 3.8 for $\alpha_1 = 2.50$, $\alpha_2 = 1.31$ and $\alpha_3 = 1$. These relative volatility values are same as those used in Sec. 3.5.

In calculation of A_1 and A_2, the constant, a, is assumed to be 36,000, the rate of return, δ, as 10% and $(c\Sigma y_i \, \Delta H_i)$ a constant value of 25 units. As A_1 and A_2 involve actual number of stages N_1 and N_2, they are going to vary as the reflux ratio is changed. However, they are not expected to be a strong function of it and A_1 and A_2 in Eq. 3.7.4 are taken to be 68 and 125 units for reasons explained earlier. ΔC has been calculated in Table 3.8 and is found to be negative first and then changes its sign as X_3 is increased beyond 226.8 moles. Therefore for a mixture containing 45.36 moles of propane, 45.36 moles of *i*-butane and 317.52 moles of *n*-butane, the distillation column sequence of Fig. 3.6b (i.e. configuration II) is better than that of Fig. 3.6a. This can also be checked from detailed calculations done in Sec. 3.5, where the two term approximation has not been made.

3.8 HEURISTIC RULE OF DIFFICULT SEPARATION (RULE 3[1])

The situation of two successive components K and K + 1, having their relative volatilities very close to each other is considered now. The number of stages required for separation of these would be very large (Eq. A3.1.12). Also when α_K and α_{K+1} are very close, the two term approximation for calculation of θ (i.e. Eq. 3.6.15) breaks down once again and the successive removal of lighter components need not be an optimal strategy. The general rule for separation of components K and (K + 1) is to postpone this separation to the last because of the following reason[10]. As the cost of a distillation column is proportional to its volume, for a tall column one would like to minimize its diameter to reduce the total cost. This can be done by reducing the total vapour flow rate through it and is achieved by removing non-key components earlier and postponing the separation of these components to the last.

To demonstrate the effect of relative volatilities on the sequence of separation, an equimolar mixture of *iso*-butane, *n*-butane and *i*-pentane is considered. The degree of splitting and quantities have been shown in

Table 3.8 Calculations of Eqs. 3.8.4 and 3.8.5 for ΔC ($\alpha_1 = 2.5$, $\alpha_2 = 1.31$, $\alpha_3 = 1.00$, $A_1 = 68$ and $A_2 = 125$)

X_2	θ_1	θ_1'	θ_2	$(2.50 - \theta_1)$	$(2.50 - \theta_1')$	$(2.50 - \theta_2)$	$\frac{125}{(2.50 - \theta_2)}$	$\frac{68}{(2.50 - \theta_1)}$	$68\left(\frac{1}{2.5 - \theta_1} - \frac{1}{2.5 - \theta_1'}\right)$	$\frac{\Delta C}{\alpha_1 X_1}$
45.36	1.83	1.73	1.13	0.67	0.77	1.37	91.3	100.5	12.5	−78.8
90.72	1.91	1.73	1.18	0.59	0.77	1.32	94.7	115.2	17.2	−77.5
226.80	2.10	1.73	1.19	0.40	0.77	1.31	95.4	170.0	82.0	−13.4
317.52	2.17	1.73	1.26	0.33	0.77	1.24	100.9	206.0	118.0	+17.1
453.60	2.24	1.73	1.28	0.26	0.77	1.22	102.5	262.0	174.0	+71.5

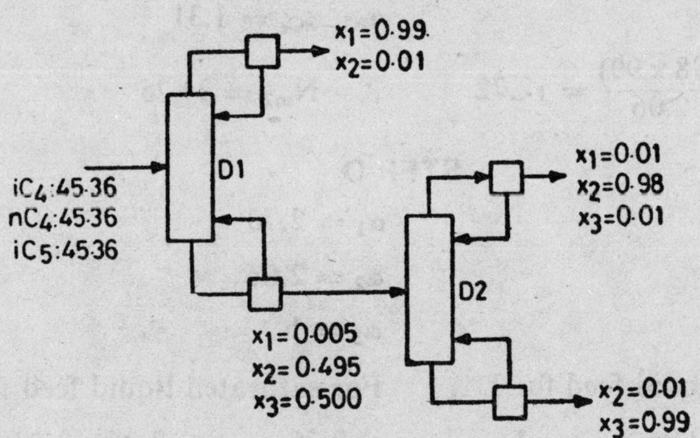

(a) Sequence I For the Separation of i-C_4, n-C_4 and i-C_5. Key Components for the First Column are iC_4 and nC_4.

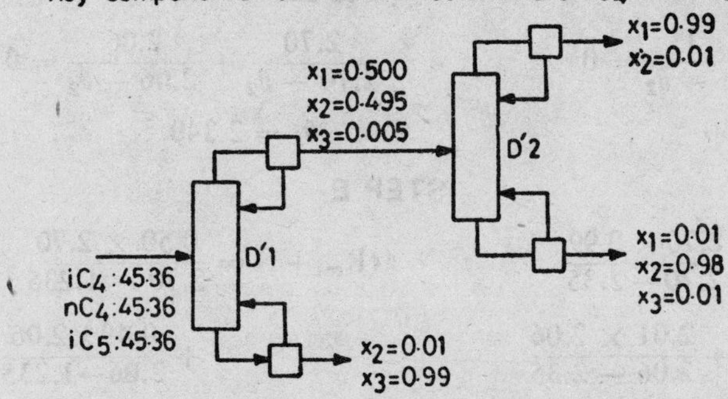

$K_{iC_4} = 1.70$, $K_{nC_4} = 1.30$, $K_{iC_5} = 0.63$

(b) Sequence II Key Components for the First Column are nC_4 and iC_5

Fig. 3.10 Separation sequences of a mixture of *iso*-butane (iC_4), *n*-butane (nC_4) and *iso*-pentane (iC_5)

Fig. 3.10. Once again to simplify calculations, all distillation columns are assumed to be at a uniform temperature of 85°C (185°F). In the following, two sequences I and II are evaluated for the optimum.

STEPS A, B, C

Sequence I of Fig. 3.10

Key components are iC_4 and nC_4 for D1.

$$\alpha_{iC_4-nC_4} = \frac{1.70}{1.30} = 1.31$$

$$N_{m1} = \frac{2 \ln 99}{\ln 1.31} = 34.26$$

Sequence II of Fig. 3.10

Key components are nC_4 and iC_5 for D1.

$$\alpha_{nC_4-iC_5} = 2.06$$

$$N_{m1} = 12.72$$

$\alpha_{nC_4-iC_5} = 2.06$

$\therefore \quad N_{m2} = \dfrac{\ln(98 \times 99)}{\ln 2.06} = 12.72$

$\alpha_{iC_4-nC_4} = 1.31$

$\therefore \quad N_{m2} = 34.26$

STEP D

$\alpha_1 = 2.70$

$\alpha_2 = 2.06$

$\alpha_3 = 1$

For saturated liquid feed for D1,

$$\frac{2.70}{2.70-\theta_1} + \frac{2.06}{2.06-\theta_1} + \frac{1}{1-\theta_1} = 0$$

$\theta_1 = 2.350$

For D2

$$\frac{2.06}{2.06-\theta_2} + \frac{1}{1-\theta_2} = 0$$

or $\theta_2 = 1.350$

$\alpha_1 = 2.70$

$\alpha_2 = 2.06$

$\alpha_3 = 1$

For saturated liquid feed for D1,

$$\frac{2.70}{2.70-\theta_1} + \frac{2.06}{2.06-\theta_1} + \frac{1}{1-\theta_1} = 0$$

or $\theta_1 = 1.235$

For D2

$$\frac{2.70}{2.70-\theta_2} + \frac{2.06}{2.06-\theta_2} = 0$$

or $\theta_2 = 2.340$

STEP E

$$(R_{m1}+1) = \frac{2.70 \times 0.99}{2.70-2.35} + \frac{2.01 \times 2.06}{2.06-2.35} = 7.566$$

or $R_{m1} = 6.566$

$$(R_{m2}+1) = \frac{0.98 \times 2.06}{2.06-1.35} + \frac{0.01}{1-1.35} = 2.82$$

$\therefore \quad R_{m2} = 1.82$

$$(R_{m1}+1) = \frac{0.50 \times 2.70}{2.70-1.235} + \frac{0.49 \times 2.06}{2.06-1.235} - \frac{0.01}{0.235} = 2.14$$

$\therefore \quad R_{m1} = 1.14$

$$(R_{m2}+1) = \frac{0.99 \times 2.70}{2.70-2.340} - \frac{0.01 \times 2.06}{0.28}$$

$\therefore \quad R_{m2} = 5.29$

STEP F

$R_1 = 1.30 \cdot R_{m1} = 8.54$

$R_2 = 1.30 \cdot R_{m2} = 2.37$

$$\frac{R_1 - R_{m1}}{R_1 + 1} + \frac{0.3 \times 6.566}{9.54} = 0.206$$

$$\frac{R_2 - R_{m2}}{R_2 + 1} = \frac{0.3 \times 1.82}{3.37} = 0.162$$

From Fig. 3.13

$g_1 = 0.42$

$R_1 = 1.30 \cdot R_{m1} = 1.48$

$R_2 = 1.30 \cdot R_{m2} = 6.88$

$$\frac{R_1 - R_{m1}}{R_1 + 1} = \frac{0.30 \times 1.14}{2.48} = 0.138$$

$$\frac{R_2 - R_{m2}}{R_2 + 1} = \frac{0.30 \times 5.29}{7.88} = 0.201$$

From Fig. 3.13

$g_1 = 0.48$

$g_2 = 0.44$

Therefore

$$N_1 = \frac{N_{m1} + g_1}{(1 - g_1)} = \frac{34.68}{0.58} = 59.79$$

$$N_2 = \frac{N_{m2} + g_2}{(1 - g_2)} = \frac{13.16}{0.54} = 24.37$$

$g_2 = 0.42$

Therefore

$$N_1 = \frac{13.20}{0.52} = 25.38$$

$$N_2 = \frac{34.68}{0.58} = 59.79$$

STEP G

$D_1 = 45.36$ (=100 lb-mol)

$L_1 = R_1D_1 = 387.37$, $V_1 = 432.73$

$L_1' = 523.45$, $V_1' = 432.73$

$D_2 = 45.36$ kg-mol (=100 lb-mol)

$L_2 = R_2D_2 = 107.50$, $V_2 = 152.86$

$L_2' = L_2 + F = 198.22$, $V_2' = 155.86$

$D_1 = 90.72$ kg-mol (=200 lb-mol)

$L_1 = R_1D_1 = 134.27$, $V_1 = 224.99$

$L_1' = 270.35$, $V_1' = 224.99$

$D_2 = 45.36$ kg-mol (=100 lb-mol)

$L_2 = 312.08$, $V_2 = 357.44$

$L_2' = 402.8$, $V_2' = 357.44$

STEP H

$U_{n1} \simeq U_{n2} = 1.80$ m/s (=5.89 ft/s)

Tray spacing as 60.96 cm (2 ft)

$U_{n1} \simeq U_{n2} = 1.80$ m/s (=5.89 ft/s)

Tray spacing as 60.96 cm (2 ft)

STEP I

Mole fraction of nC_4, y_2, in the vapour leaving the reboiler of D1 is given by

$$y_2 = \frac{2.06 \times 0.495}{2.06 \times 0.495 + 0.500 \times 1}$$

$$= 0.671$$

and $y_3 = 0.329$

$\therefore$ Average molecular weight

$$M_{w,a} = 0.67 \times 58 + 0.33 \times 72$$

$$= 62.68 \text{ kg/mol}$$

Therefore

$$A_1 = \frac{432.73 \times 62.68}{3600 \times 0.0024 \times 10^3 \times 1.80}$$

$$= 1.74 \text{ m}^2$$

$$A_2 = \frac{152.86 \times 72}{3600 \times 2.4 \times 1.80}$$

$$= 0.71 \text{ m}^2$$

Vapour flow rates at the reboilers

$$W_1 = 224.99 \times 72 \text{ kg/h}$$

$$= 16199.28 \text{ kg/h}$$

$$W_2 = 357.44 \times 58$$

$$= 20731.52 \text{ kg/h}$$

Therefore

$$A_1 = \frac{16199.28}{3600 \times 2.4 \times 1.8}$$

$$= 1.04 \text{ m}^2$$

$$A_2 = \frac{20731.52}{3600 \times 2.4 \times 1.8}$$

$$= 1.33 \text{ m}^2$$

$Vol_1 = (59.75 \times 0.6096) \times 1.74\ m^3$
$= 63.38\ m^3$
$Vol_2 = (24.39 \times 0.6096) \times 0.71$
$= 10.56\ m^3$
$\therefore$ Total vol $V_T = 73.94\ m^3$

$Vol_1 = (25.38 \times 0.6096) \times 1.04$
$= 16.09$
$Vol_2 = (59.79 \times 0.6096) \times 1.33$
$= 48.48\ m^3$
$\therefore V_T = 64.57\ m^3$

TOTAL REBOILER DUTY

$\varDelta H_{iC_4} = 3619.3$ kcal/mol
$\varDelta H_{nC_4} = 4009.1$ kcal/mol
$\varDelta H_{iC_5} = 4788.6$ kcal/mol

$\dot{Q}_1 = 432.73(0.671 \times 4009.1 + 4788.6 \times 0.329)$
$= 18{,}45{,}835.5$ kcal/h
$\dot{Q}_2 = 152.86 \times 4788.6$ kcal/h
$= 7{,}31{,}985.4$ kcal/h
$\therefore Q_T = 25{,}77{,}820.9$ kcal/h

$\dot{Q}_1 = 224.99 \times 4788.6$ kcal/h
$= 10{,}77{,}387.1$ kcal/h
$\dot{Q}_2 = 357.44 \times 4009.1$ kcal/h
$= 14{,}33{,}012.7$ kcal/h
$\dot{Q}_T = 25{,}10{,}399.8$ kcal/h

Sequence II is found to be better because isomers are difficult to split and should be separated last.

3.9 PROCEDURE OF OPTIMAL SEQUENCING[11]

In Secs. 3.6 to 3.8, three heuristic rules of optimal separation were discussed. These rules at times become conflicting and they fail to lead to an optimal separation sequence for a given mixture. As an example, a mixture consisting of five components (C_3, iC_4, nC_4, iC_5 and nC_5 hydrocarbons) is considered with its composition given in Table 3.9. This is

Table 3.9 Separation of a Mixture of Propane (C_3), *Iso*-butane (iC_4), Normal butane (nC_4), *Iso*-pentane (iC_5) and Normal Pentane (nC_5) by Distillation

Component	Quantity (moles)	K at 85°C
C_3	2.27	3.25
i-C_4	6.80	1.70
n-C_4	11.34	1.30
i-C_5	9.07	0.63
n-C_5	15.87	0.54

to be separated into its individual constituents. If the heuristic rule of removing components one by one (rule 1) is used, propane should be separated first to be followed by *iso*-butane and so on. If the heuristic rule of removing components in excess first (rule 2) is used, nC_5 should be separated in the first column. Lastly if the rule of difficult separation is used iC_4 and nC_4 and iC_5 and nC_5 should be separated in individual column when other components are not present. Thus three heuristic rules lead to three different flow sheets, and as a result, they cannot be used as more than mere guidelines. The best separation flow sheet can always be generated by optimizing it with the help of the Principle of Dynamic Programming and the method is demonstrated by solving the problem of Table 3.10 as follows.

Table 3.10 Possible Mixtures Generated in Single Splitting of Five Component Separation

Four component mixtures	*Further possible splittings*
(i) C_3 i-C_4 n-C_4 i-C_5	(i) C_3/i-C_4 n-C_4 i-C_5, C_3 i-C_4/n-C_4 i-C_5, C_3 i-C_4 n-C_4/i-C_5
(ii) i-C_4 n-C_4 i-C_5 n-C_5	(ii) i-C_4/n-C_4 i-C_5 n-C_5, i-C_4 n-C_4/i-C_5 n-C_5, i-C_4 n-C_4 i-C_5/n-C_5
Three components	
(iii) C_3 i-C_4 n-C_4	(iii) C_3/i-C_4 n-C_4, C_3 i-C_4/n-C_4
(iv) i-C_4 n-C_4 i-C_5	(iv) i-C_4/n-C_4 i-C_5, i-C_4 n-C_4/i-C_5
(v) n-C_4 i-C_5 n-C_5	(v) n-C_4/i-C_5 n-C_5, n-C_4 i-C_5/n-C_5
Two components	
(vi) C_3 i-C_4	(vi) C_3/i-C_4
(vii) i-C_4 n-C_4	(vii) i-C_4/n-C_4
(viii) n-C_4 i-C_5	(viii) n-C_4/i-C_5
(ix) i-C_5 n-C_5	(ix) i-C_5/n-C_5

The five component mixture C_3, i-C_4, n-C_4, i-C_5 and n-C_5 can be split in four ways:

C_3/i-C_4 nC_4 i-C_5 n-C_5 (3.9.1a)

C_3 i-C_4/n-C_4 i-C_5 n-C_5 (3.9.1b)

C_3 i-C_4 n-C_4/i-C_4 n-C_5 (3.9.1c)

and

C_3 i-C_4 n-C_4 i-C_4/n-C_5 (3.9.1d)

The dash above specifies key components and it is seen that in a single column, a five component mixture can give rise to two possible four-component mixtures, three three-component mixtures and four two-component mixtures. These are given in Table 3.10 and also against each mixture; their further possible splittings have been specified. It may be noted that the next splitting does not generate any new groups of mixtures from what is given in the first column of the table.

In Table 3.11, volumes of columns are calculated for different mixtures given in Table 3.10 using the method given in Table A 3.1 for multicomponent distillation. In this calculation the actual reflux R is taken as 1.3 R_m and the whole column is assumed to be at a uniform temperature of 85°C (185°F). The vapour density is assumed to be 2.4 kg/m^3, the maximum allowable vapour velocity is taken as 1.80 m/sec and all feeds are assumed to be at their bubbles points. Minimum number of plates N_m are calculated for 99% splitting of key components; however, while calculating R_m, the residual amount of the separated key component in the other stream was assumed to be zero (i.e., perfect split). These are not very serious assumptions and the results can be easily improved if so desired.

There is no optimization with respect to choice of key components in two component systems; however, it is so in three component mixtures. Table 3.12 gives all calculations and optimal splitting is found when the total volume of the column network is minimum. For the calculation of the optimum, the cost of energy has been neglected, but if it is desired they can be easily included. The optimum split has been marked with an asterisk in Table 3.12. The optimal split for all three component mixtures is found first. One then goes on to a four component mixture and it may be noted that, when distilled, it gives either two or three component mixtures depending upon the choice of the key components. For example, the distillation of (C_3 i-C_4 n-C_4/i-C_5) generates (C_3 i-C_4 n-C_4) and for optimization of split point of four component mixture, the optimal splitting in subsequent columns is assumed. This fact is represented by marking the three component mixture by an asterisk in the 4th, 5th and 6th row of Table 3.12. This way, the information is built up to find the optimal splitting of the five component mixture and the corresponding optimal network is given in Fig. 3.11. Optimally, the five component

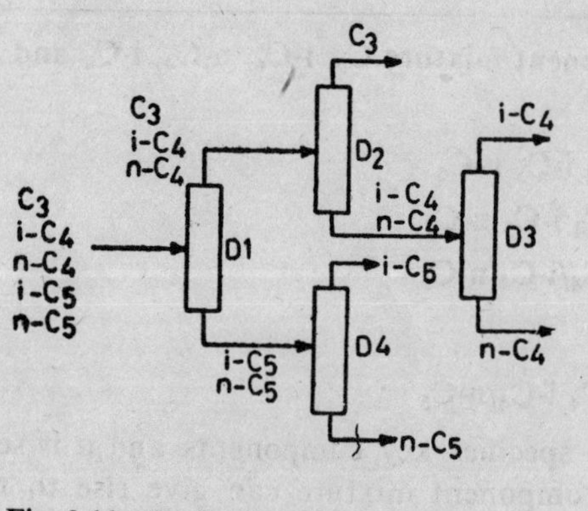

Fig. 3.11 Optimal sequence for the mixture given in Table 3.9

mixture is split at n-C_4 in distillation column D1. The distillate (C_3 i-C_4

Table 3.11 Calculation of Volumes of Distillation Columns for Splits as Given in Table 3.9

	N_m	R_m	R	N	D (kg-mol)	V (kg-mol)	$A_i \times 10^{-1}$ (m²)	Vol$_i$ (m³)
1. C_3/iC_4	14.20	4.43	5.75	25.2	2.27	15.31	0.58	0.88
2. iC_4/nC_4	34.03	8.63	11.22	57.5	6.80	83.14	3.10	10.88
3. nC_4/iC_5	12.69	1.69	2.20	27.1	11.34	36.29	1.68	2.78
4. iC_5/nC_5	58.54	16.20	21.07	98.3	9.07	200.21	9.27	55.56
5. $C_3/iC_4\ nC_4$	14.20	8.55	11.12	24.3	2.27	27.49	1.02	1.52
6. $C_3\ iC_4/nC_4$	34.03	7.34	9.54	57.5	9.07	95.62	3.57	12.50
7. $iC_4/nC_4\ iC_5$	34.03	9.16	11.91	57.5	6.80	87.84	3.62	12.71
8. $iC_4\ nC_4/iC_5$	12.69	1.23	1.60	27.6	18.14	47.17	2.18	3.68
9. $nC_4/iC_5\ nC_5$	12.69	2.57	3.34	23.5	11.34	49.21	2.28	3.26
10. $nC_4\ iC_5/nC_5$	58.54	7.05	9.16	98.3	20.41	207.38	9.61	57.54
11. $C_3/iC_4\ nC_4\ iC_5$	14.20	8.10	10.53	25.2	2.27	26.17	1.06	1.62
12. $C_3\ iC_4/nC_4\ iC_5$	34.03	6.86	8.92	59.5	9.07	89.99	3.72	13.48
13. $C_3\ iC_4\ nC_4/iC_5$	12.69	1.33	1.73	24.0	20.41	55.72	2.58	3.87
14. $iC_4/nC_4\ iC_5\ nC_5$	34.03	9.92	12.90	58.5	6.80	94.57	4.12	14.67
15. $iC_4\ nC_4/iC_5\ nC_5$	12.69	1.66	2.16	23.5	18.14	57.33	2.66	3.80
16. $iC_4\ nC_4\ iC_5/nC_5$	58.54	7.45	9.69	101.5	27.22	290.93	13.46	83.30
17. $C_3/iC_4\ nC_4\ iC_5\ nC_5$	12.20	9.91	12.88	25.7	2.27	31.51	1.34	2.09
18. $C_3\ iC_4/nC_4\ iC_5\ nC_5$	34.03	7.34	9.54	59.5	9.07	95.62	4.15	15.08
19. $C_3\ iC_4\ nC_4/iC_5\ nC_5$	12.69	1.48	1.92	25.0	20.41	59.60	2.76	4.23
20. $C_3\ iC_4\ nC_4\ iC_5/nC_5$	58.54	5.89	7.66	103.5	29.48	255.33	11.83	74.60

$R = 1.3\ R_m$; plate spacing = 0.6096 m (=2 ft)
Vapour density = 2.4 kg/m³, maximum vapour velocity = 1.80 m/s.

Table 3.12 Calculation of the Optimal Network

1.	$(C_3/iC_4\ nC_4)^* = 1.52 + (iC_4\ nC_4)$ $= 1.52 + 10.88 = 12.40$	$(C_3\ iC_4/nC_4) = 12.50 + (C_3\ iC_4)$ $= 12.50 + 0.88 = 13.38$
2.	$(iC_4/nC_4\ iC_5) = 12.71 + (nC_4\ iC_5)$ $= 12.71 + 2.78 = 15.49$	$(iC_4\ nC_4/iC_5)^* = 3.68 + (iC_4\ nC_4)$ $= 3.68 + 10.88 = 14.56$
3.	$(nC_4/iC_5\ nC_5)^* = 3.26 + (iC_5\ nC_3)$ $= 3.26 + 55.56 = 58.82$	$(nC_4\ iC_5/nC_5) = 57.54 + (nC_4\ iC_5)$ $= 57.54 + 2.78 = 60.32$
4.	$(C_3/iC_4\ nC_4\ iC_5)^* = 1.62 + (iC_4\ nC_4\ iC_5)^*$ $= 1.62 + 14.56 = 16.18$ $(C_3\ iC_4\ nC_4/iC_5) = 3.77 + (C_3\ iC_4\ nC_4)^*$ $= 3.87 + 12.40 = 16.27$	$(C_3\ iC_4/nC_4\ iC_5) = 13.48 + (C_3\ iC_4) + (nC_4\ iC_5)$ $= 13.48 + 0.88 + 2.78$ $= 17.14$
5.	$(iC_4/nC_4\ iC_5\ nC_5) = 14.67 + (nC_4\ iC_5\ nC_5)^*$ $= 14.67 + 58.82 = 73.49$ $(iC_4\ nC_4\ iC_5/nC_5) = 83.30 + (iC_4\ nC_4\ iC_5)^*$ $= 83.30 + 14.56 = 97.86$	$(iC_4\ nC_4/iC_5\ nC_5)^* = 3.80 + (iC_4\ nC_4) + (iC_5\ nC_5)$ $= 3.80 + 10.88 + 55.56$ $= 70.24$
6.	$(C_3/iC_4\ nC_4\ iC_5\ nC_5) = 2.09 + (iC_4\ nC_4\ iC_5\ nC_5)^*$ $= 2.09 + 70.24 = 72.33$ $(C_3\ iC_4/nC_4\ iC_5\ nC_5) = 15.08 + (C_3\ iC_4) + (nC_4\ iC_5\ nC_5)^*$ $= 15.08 + 0.88 - 58.82 = 74.78$ $(C_3\ iC_4\ nC_4/iC_5\ nC_5)^* = 4.23 + (C_3\ iC_4\ nC_4)^* + (iC_5\ nC_5)$ $= 4.23 + 12.40 + 55.56 = 72.19$ $(C_3\ iC_4\ nC_4\ iC_5/nC_5) = 74.60 + (C_3\ iC_4\ nC_4\ iC_5)^*$ $= 74.60 + 16.18 = 90.78$	

*The optimum split.

n-C_4) is next separated by removing C_3 in D_2 and then i-C_4 and n-C_4 in D_3. The bottom product of D1 is separated into pure components in column D4.

3.10 CONCLUSION

Use of the separation process to achieve concentration change is an expensive process and in view of this, mixing operation is utilized whenever possible to minimize separation load. This is a linear programming problem and in general can be solved using the algorithm presented in this chapter. Some quick empirical methods have also been suggested which leads to almost similar results. After minimizing the total separation load, multicomponent mixtures need to be distilled. It was shown that the total column network cost depends upon the order of distillation.

A few heuristic rules were later developed for an optimal network. However, they serve as mere guidelines while finding the best network design. In the end, a procedure based upon dynamic programming was suggested for finding an optimum network for a given mixture. A careful examination of the flow sheet derived (Fig. 3.15) for the problem of separation of five components reveals that a combination of heuristic rules 1 and 3 decides the optimality.

APPENDIX 3.1

Design of a Distillation Column for Multicomponent Systems[1,6]

The most rigorous method of designing distillation columns consists of carrying out plate-to-plate calculations, keeping track of variation in plate temperature, relative volatilities and flow rates of liquid and vapour. In the normal course of designing, this detailed calculation is not necessary and decent results can be obtained using the following short-cut method.

In the short-cut method given here, a minimum number of trays, N_m, is calculated for total reflux condition and then the minimum reflux, R_m, for the desired splitting of the product. From the knowledge of these, Gilliland correlation is used to find the number of actual theoretical plates required for any reflux rate more than the minimum reflux rate.

A3.1.1 Calculation of Minimum Number of Trays, N_m, for Total Reflux Condition

Figure 3.12 shows the schematic diagram of a distillation column. Feed F is introduced on the (M + 1)th plate and distillate D and the bottom product B are withdrawn as shown in the figure. The reboiler is counted as one plate because the leaving vapour, V_1, is in equilibrium with the liquid, L_2. As the exchanger at the top condenses the entire vapour, it is not counted as a theoretical plate. Plates are numbered from the bottom to the top. Heat duty of thc reboiler and the condenser are Q_B and Q_C respectively. The condition of the total reflux arises when F, B and D are all equal to zero; liquid is vaporized at the bottom and whatever vapour (V_N) leaves the top is completely condensed and fed back to the column.

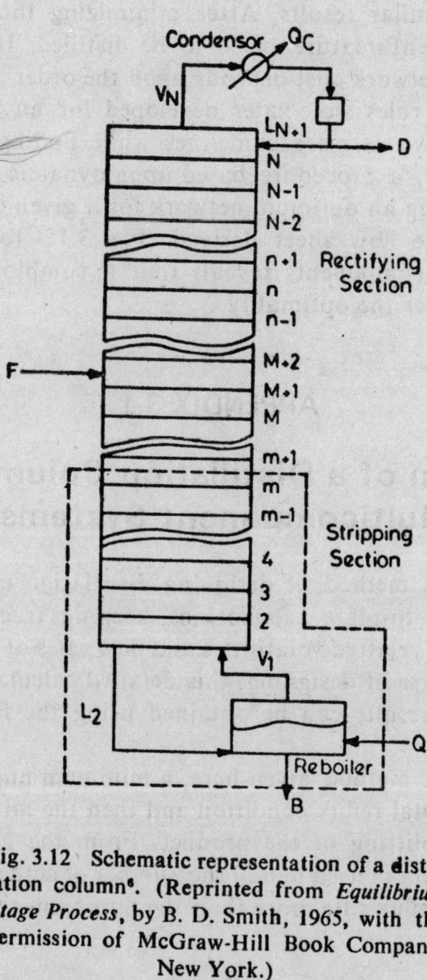

Fig. 3.12 Schematic representation of a distillation column[6]. (Reprinted from *Equilibrium Stage Process*, by B. D. Smith, 1965, with the permission of McGraw-Hill Book Company, New York.)

The basic assumption of the derivation presented here for the case of total reflux is that the total liquid and vapour flow rates in stripping and rectifying sections (Fig. 3.12) remain unchanged from plate to plate, i.e.,

$$L_2 = L_3 = \ldots = L_M = L' \quad \text{(say)} \quad \text{(a)}$$

$$V_1 = V_2 = \ldots = V_M = V' \quad \text{(say)} \quad \text{(b)}$$

$$L_{M+2} = L_n = \ldots = L_{N+1} = L \quad \text{(say)} \quad \text{(c)}$$

and

$$V_{M+2} = \ldots = V_N = V \quad \text{(say)} \quad \text{(d)} \qquad \text{(A3.1.1)}$$

When D and B are not zero, L and L′ and V and V′ are related through the quality of the feed. However, for the condition of total reflux, these would be equal because of the absence of feed to the column.

If the mole fraction of component i on plate m is denoted as $x_{i,m}$ in the liquid phase and $y_{i,m}$ in the vapour phase, a mass balance on the dotted box of Fig 3.12 gives

$$VY_{i,m} = Lx_{i,m+1} \qquad \text{(A3.1.2)}$$

As V and L are equal, one has

$$y_{i,m} = x_{i,m+1} \qquad \text{(A3.1.3)}$$

If relative volatility, α_i, of the component i with respect to a reference component, r, is defined as

$$\alpha_i = \frac{(y_i/x_i)}{(y_r/x_r)} \qquad \text{(A3.1.4)}$$

for plate 1, one has

$$\frac{y_{i,1}}{y_{r,1}} = \alpha_i\left(\frac{x_{i,1}}{x_{r,1}}\right) \qquad \text{(A3.1.5)}$$

Using relation A3.1.3,

$$\left(\frac{y_{i,1}}{y_{r,1}}\right) = \left(\frac{x_{i,2}}{x_{r,2}}\right) \qquad \text{(A3.1.6)}$$

and Eq. A3.1.5 is written as

$$\frac{x_{i,2}}{x_{r,2}} = \alpha_i\left(\frac{x_{i,1}}{x_{r,1}}\right) \qquad \text{(A.3.1.7)}$$

For plate 2, similar results are found

$$\frac{y_{i,2}}{y_{r,2}} = \alpha_i\left(\frac{x_{i,2}}{y_{r,2}}\right) = \alpha_i^2\left(\frac{x_{i,1}}{x_{r,1}}\right) \qquad \text{(A3.1.8)}$$

On using Eq. A3.1.3, one has

$$\frac{y_{i,2}}{y_{r,2}} = \frac{x_{i,3}}{x_{r,3}} \qquad \text{(A3.1.9)}$$

which simplifies to

$$\left(\frac{x_{i,3}}{x_{r,3}}\right) = \alpha_i^2\left(\frac{x_{i,1}}{x_{r,1}}\right) \qquad \text{(A3.1.10)}$$

Equation A3.1.10 relates the liquid composition of the third plate to the liquid composition of the first one. The composition of the *i*th component on the *N*th plate can be similarly related to that at the bottom as

$$\left(\frac{y_{i,\mathrm{N}}}{y_{r,\mathrm{N}}}\right)=\left(\frac{x_{i,\mathrm{D}}}{x_{r,\mathrm{D}}}\right)=(\alpha_i^{\mathrm{N}})\left(\frac{x_{i,1}}{x_{r,1}}\right) \qquad \text{(A3.1.11)}$$

where $x_{i,\mathrm{D}}$ and $x_{r,\mathrm{D}}$ are composition in the distillate.
Thus

$$N=(\equiv N_{\mathrm{m}})=\frac{\ln\left[\left(\frac{x_{i,\mathrm{D}}}{x_{r,\mathrm{D}}}\right)\Big/\left(\frac{x_{i,1}}{x_{r,1}}\right)\right]}{\ln \alpha_i} \qquad \text{(A3.1.12)}$$

The value of N calculated from the above equation represents the minimum number of equilibrium plates, N_{m}, for a given separation between the distillate and the bottom product.

A3.1.2 Underwood's Equation for Minimum Reflux, R_{m}

One defines the minimum reflux ratio for a given splitting of products below which, even when decreased by an infinitesimal quantity, an infinite number of equilibrium plates are required for the task.

In any finite distillation column, there is a change in concentration from plate-to-plate. For an infinite column for a given separation, there will be a plate, above which the composition stops changing. This is called the occurrence of pinch point. Some important points concerning R_{m} are emphasized now: the first is that F, B and D have values equal to what is in the column to be designed. The second point concerning R_{m} is that it corresponds to the reflux ratio which produces the pinch point in *both* rectifying as well as stripping section. Lastly in multicomponent distillation, the two components for which separation is specified are called key components, and R_{m} is calculated for the desired separation of key components. The distillation column operating at R_{m} is not infinite, but is large enough to be regarded as 'near infinite'. In the derivation of $\mathrm{R_m}$ below, the molar flow rates of vapour and liquid are assumed to be constant from plate-to-plate.

For a given plate with vapour and liquid in equilibrium, the vapour composition, y_i, of the component i, is written in terms of the relative volatility α_i (Eq. A3.1.4)

$$\mathrm{y_i}=\alpha_i x_i\left(\frac{y_{\mathrm{r}}}{x_{\mathrm{r}}}\right); \quad i=1, 2, \ldots, \mathrm{C} \qquad \text{(A3.1.13)}$$

Adding these for all components, it becomes

$$1=\sum_i y_i=\left(\frac{y_{\mathrm{r}}}{x_{\mathrm{r}}}\right)\sum_i \alpha_i x_i \qquad \text{(A3.1.14)}$$

Eliminating (y_r/x_r) between Eqs. A3.1.13 and A3.1.14

$$y_i = \frac{\alpha_i x_i}{\sum_{i=1}^{C} \alpha_i x_i} \tag{A3.1.15}$$

The mass balance equation of the nth plate of the rectifying section is given by

$$V y_{i,n} = L x_{i,n+1} + D x_{i,D} \tag{A3.1.16}$$

where V and L are vapour and liquid flow rates in the section. The reflux ratio, R, is defined as

$$R = (L/D) \tag{A3.1.17}$$

and in terms of it, Eq. A3.1.16 is rewritten as

$$y_{i,n} = \frac{R}{(R+1)} x_{i,n+1} + \frac{x_{i,D}}{(R+1)} \tag{A3.1.18}$$

Eliminating $y_{i,n}$ from above with the help of Eq. A3.1.15,

$$\frac{\alpha_i x_{i,n}}{\Sigma \alpha_i x_{i,n}} = \frac{R}{R+1} x_{i,n+1} + \frac{x_{i,D}}{(R+1)} \tag{A3.1.19}$$

If the pinch point is assumed to occur in the rectifying section at the nth plate, it means

$$x_{i,n} = x_{i,n+1} = x_{i,z} \quad \text{(say)} \tag{A3.1.20}$$

and Eq. A3.1.19, consequently becomes

$$\left(\frac{\alpha_i}{\Sigma \alpha_i x_{i,z}} - \frac{R_m}{R_m+1}\right) x_{i,z} = \frac{x_{i,D}}{(R_m+1)} \tag{A3.1.21}$$

In the above equation, R has been replaced by R_m because the use of Eq. A3.1.20 amounts to the occurrence of the minimum reflux. A quantity ϕ is defined as

$$\phi = \frac{R_m}{R_m+1} \sum_{i=1}^{C} \alpha_i x_{i,z} \tag{A3.1.22}$$

in terms of which Eq. A3.1.21 becomes

$$\frac{R_m}{(R_m+1)} \frac{(\alpha_i - \phi)}{\phi} x_{i,z} = \frac{x_{i,D}}{(R_m+1)} \tag{A3.1.23}$$

or

$$x_{i,z} = \frac{\phi}{R_m} \frac{x_{i,D}}{(\alpha_i - \phi)} \tag{A3.1.24}$$

R_m is found by substituting this in Eq. A3.1.22 as

$$\phi = \frac{R_m}{(R_m+1)} \frac{\phi}{R_m} \sum_{i=1}^{C} \frac{\alpha_i x_{i,D}}{(\alpha_i - \phi)} \tag{A3.1.25}$$

or

$$(R_m+1) = \sum_{i=1}^{C} \frac{\alpha_i x_{i,D}}{(\alpha_i - \phi)} \tag{A3.1.26}$$

As the minimum reflux ratio, R_m, is the one for which the pinch point occurs on the stripping section also, the latter must be analyzed simultaneously. The mass balance equation for the *m*th plate in the stripping section is

$$y_{i,m} = \frac{L'}{V'} x_{i,m+1} - \frac{B}{V'} x_{i,B} \tag{A3.1.27}$$

or

$$y_{i,m} = \frac{R'+1}{R'} x_{i,m+1} - \frac{x_{i,B}}{R'} \tag{A3.1.28}$$

where

$$R' = (V'/B) \tag{A3.1.29}$$

Eliminating $y_{i,m}$ from the equilibrium relation (Eq. A3.1.13),

$$\frac{\alpha_i x_{i,m}}{\sum\limits_{i=1} \alpha_i x_{m,m}} = \frac{R'+1}{R} x_{i,m+1} - \frac{x_{i,B}}{R'} \tag{A3.1.30}$$

If the pinch point occurs at the *m*th plate such that

$$x_{i,m} = x_{i,m+1} = x'_{i,z} \text{ (say)} \tag{A3.1.31}$$

Eq. A3.1.30, after replacing R' by R'_m, reduces to

$$\left(-\frac{\alpha_i}{\sum\limits_i \alpha_i x'_{i,z}} + \frac{R'_m+1}{R'_m}\right) x'_{i,z} = \frac{x_{i,B}}{R'_m} \tag{A3.1.32}$$

A parameter ψ for the stripping section is similarly defined by

$$\psi = \frac{R'_m+1}{R'_m} \sum_{i=1}^{C} \alpha_i x'_{i,z}, \tag{A3.1.33}$$

in terms of which Eq. A3.1.32 is written as

$$-\frac{R'_m+1}{R'_m} \frac{\alpha_i - \psi}{\psi} x'_{i,z} = \frac{x_{i,B}}{R'} \tag{A3.1.34}$$

or

$$x'_{i,z} = \frac{-\psi(x_{i,B})}{(R'_m+1)(\alpha_i - \psi)} \tag{A3.1.35}$$

R'_m is obtained by substituting this in Eq. A3.1.33,

$$\psi = -\frac{\psi}{R'_m} \sum_{i=1}^{C} \frac{\alpha_i x_{i,B}}{(\alpha_i - \psi)} \tag{A3.1.36}$$

or

$$-R'_m = \sum_{i=1}^{C} \frac{\alpha_i x_{i,B}}{(\alpha_i - \psi)} \tag{A3.1.37}$$

If F, B, R_m and their compositions are specified, R'_m can be found through mass balance. In view of this, for a given R_m, the roots of ϕ are found from Eq. A3.1.26. If the pinch point occurs simultaneously in the stripping section, the same values of ϕ should also satisfy Eq. A3.1.37. If the common roots of ϕ and ψ (Eq. A3.1.26 and A3.1.25) is denoted by θ, the latter will satisfy both these equations, i.e.,

$$\sum_{i=1}^{C} \frac{\alpha_i x_{i,D}}{(\alpha_i - \theta)} = (R_m + 1) \tag{A3.1.38a}$$

and

$$\sum_{i=1}^{C} \frac{\alpha_i x_{i,B}}{(\alpha_i - \theta)} = -R'_m \tag{A3.1.38b}$$

Multiplying Eq. A3.1.38a by D and A3.1.38b by B and adding these, one obtains

$$D(R_m + 1) - BR'_m = V - V' = \sum_{i=1}^{C} \frac{\alpha_i[Dx_{i,D} + Bx_{i,B}]}{(\alpha_i - \theta)} \tag{A3.1.39}$$

or

$$\frac{V - V'}{F} = \sum \frac{\alpha_i[Dx_{i,D} + Bx_{i,B}]/F}{(\alpha_i - \theta)} = \sum_{i=1}^{C} \frac{\alpha_i x_{i,F}}{(\alpha_i - \theta)} \tag{A3.1.40}$$

If q denotes the quality of the feed, defined as the ratio of the mole of saturated liquids present per mole of the feed ($q = 1$ for saturated liquid feed and $q = 0$ for saturated vapour feed), then

$$(1 - q) = \frac{V - V'}{F} = \sum_{i=1}^{C} \frac{\alpha_i x_{i,F}}{(\alpha_i - \theta)} \tag{A3.1.41}$$

The procedure of finding the minimum reflux can now be written as follows: From the quality of the feed, the values of θ which satisfy Eq. A3.1.41 are found. If the light and heavy key components have relative volatilities α_l and α_h respectively, the root of θ falling between these two limits (i.e., $\alpha_l < \theta < \alpha_h$) is chosen and substituted in Eq. A3.1.38a to find the minimum reflux ratio, R_m.

A3.1.3 Designing a Multicomponent Distillation Column for a Specified Splitting

N_m and R_m are the two parameters for the desired separation which completely characterizes the distillation column. Thus, if a reflux ratio, R, is fixed which is greater than R_m, the specified key component separation can be obtained in finite number of stages, N, which is uniquely determined. The relation between N and R is expressed by Gilliland correlation given in Fig. 3.13. This is a plot of $\left(\frac{N - N_m}{N + 1}\right)$ versus $\left(\frac{R - R_m}{R + 1}\right)$. For a given R, $\frac{R - R_m}{(R + 1)}$ is calculated with the knowledge of R_m, $\frac{N - N_m}{(N + 1)}$ is then found from the plot and from it, N is calculated from the knowledge of N_m.

Actual R is normally found through cost optimization of a given column. As R is increased, fewer and fewer equilibrium stages are

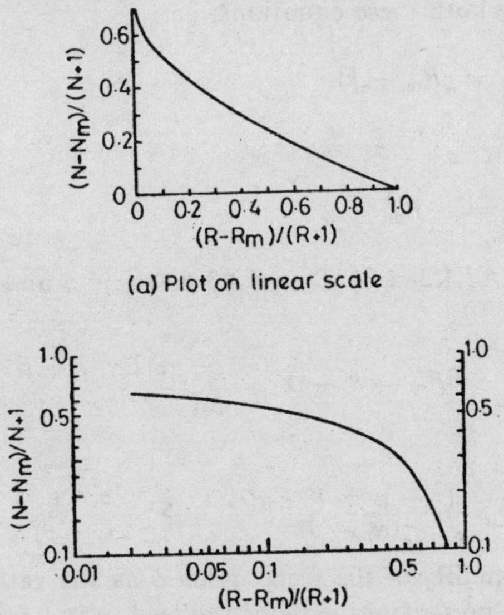

(a) Plot on linear scale

(b) Plot on log-log scale

Fig. 3.13 Gilliland correlation between $(N - N_m)/(N + 1)$ and $(R - R_m)/(R + 1)$. Plots (a) and (b) are identical[7] (Reprinted from E. R. Gilliland, *Ind. Eng. Chem.*, **32**, 1220, 1940 with the permission of American Chemical Society, Washington.)

required or a smaller column for the specified separation. This implies a smaller fixed cost. But larger R also means a larger operating cost. Between these opposing factors, there is a minimum in the total cost at R_{opt} as shown in Fig. 3.14. To calculate R_{opt} is a tedious process and is normally not worthwhile to carry out for a given separation task. Experience has shown that R_{opt} normally lies between 1.2 to 1.5 times R_m[6].

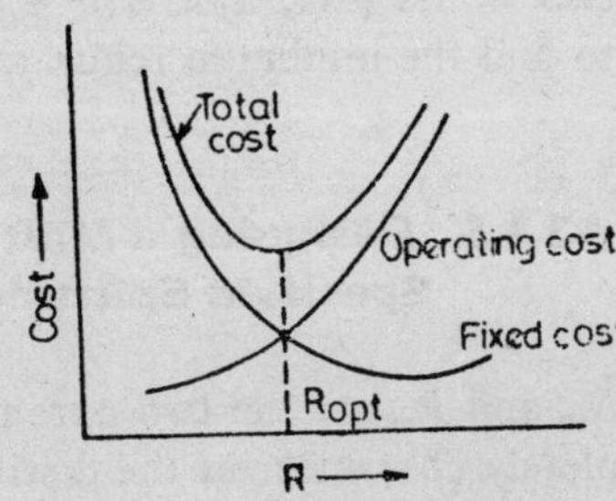

Fig. 3.14 Calculation of optimum reflux ratio

After calculating N for R_{opt}, the diameter of the column is calculated. This is specified by the vapour flooding velocity, U_n, which is estimated using Fig. 3.15. The liquid-vapour flow parameter, F_{lv}, is defined as

$$F_{lv} = \frac{\text{maximum liquid flow rate}}{\text{maximum vapour flow rate}}\sqrt{\frac{\rho_v}{\rho_l}} \equiv \frac{w}{W}\sqrt{\frac{\rho_v}{\rho_l}} \qquad \text{(A3.1.42)}$$

where ρ_v and ρ_l are vapour and liquid densities respectively. The capacity parameter, C_{sb}, defined by

$$C_{sb} = U_n\sqrt{\rho_v/(\rho_l - \rho_v)} \qquad \text{(A3.1.43)}$$

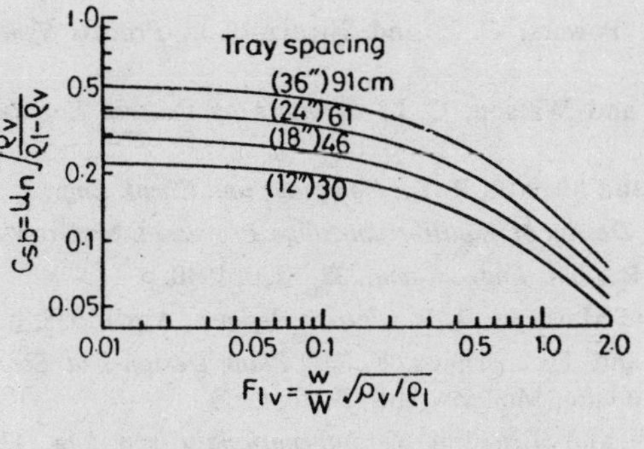

Fig. 3.15 Correlation of flooding on bubble cap trays. (Reprinted from ref. 8 with the permission of Gulf Publishing Co., Texas.)

fixes the flooding velocity, U_n. From the knowledge of the vapour flow rate and U_n, the cross-sectional area of the column, is calculated. The stepwise procedure of calculation is given in Table A3.1.

Table A3.1 Stepwise Procedure of Designing a Multicomponent Distillation Column

Step A	Define key components. Let the heavy and light key components have relative volatilities α_h and α_l.
Step B	Calculate relative volatility, α_{lh}, of the light key with respect to the heavy key as $$\alpha_{lh} = \frac{\alpha_l}{\alpha_h}$$
Step C	Use Eq. A3.1.12 to determine N_m using the specified splitting of key components and α_{lh}.
Step D	Use Eq. A3.1.41 to determine θ. Let the root of θ lying between α_l and α_h be θ^*. $$\alpha_l < \theta^* < \alpha_h$$
Step E	Substitute θ^* calculated in step D in Eq. A3.1.38 to calculate R_m.
Step F	Take $R = 1.3R_m$ and using the Gilliland correlation, determine N.
Step G	Calculate L, V, L' and V' using information on D, F and the feed condition $$L = RD,\ V = L + D$$ $$L' = L + qF,\ V' = V + (1 - q)F.$$
Step H	Calculate F_{lv} using Eq. A3.1.43. Take vapour and liquid flow rates in this equation at the same point. Find C_{sb} using Fig. 3.15 from which calculate U_n.
Step I	Calculate cross-sectional area of the column from the information on maximum vapour flow rate. Find its cost using information on the volume of the column.

REFERENCES

1. King, C. J., *Separation Processes*, McGraw-Hill, New York, 1971.
2. Perry, R. H. (ed.), *Chemical Engineers Handbook*, 5th edn., McGraw-Hill, N.Y., 1974.
3. Rudd, D. F., Powers, G. J. and Siirola, J. J., *Process Synthesis*, Prentice-Hall, N.J., 1973.
4. Rudd, D. F. and Watson, C. J., *Strategy of Process Engineering*, Wiley, N.Y., 1972.
5. Mahalec, V. and Motard, R. L., *Computer and Chem. Eng.*, **1**, 57, 1977.
6. Smith, B. D., *Design of Equilibrium Stage Processes*, McGraw-Hill, N.Y., 1965.
7. Gilliland, E. R., *Ind. Engg. Chem.*, **32**, 1220, 1940.
8. Fair, J. R. and Mathews, R. L., *Petrol. Refiner.*, April 1958, p. 153.
9. Peter, M. S. and Timmerhaus, K. D., *Plant Design and Economics for Chemical Engineers*, 2nd edn., McGraw-Hill, N.Y., 1968.
10. Nishimura, N. and Hirazumi, Y., *International Chem. Eng.*, **11**, 188, 1971.
11. Hendry, J. E. and Hughes, R. R., *C.E.P.*, **68**(6), 71, 1972.

PROBLEMS

3.1 Supposing there is a mixture of sand and sugar of practically the same particle size. Suggest ways, as many as possible, of separating them when sugar is the desired product. Do the same for the sand and wax mixture.

3.2 Crude petroleum oil is transported in ships from the Middle East to different countries. There have been several accidents in the past leading to spillage of large amounts of oil in the sea. This has been leading to major water pollution. As a chemical engineer give different alternatives of transportation and evaluate them.

3.3 Styrene is formed by pyrolysis of ethyl benzene. Since the reaction does not go to completion, the stream from the reactor is a binary mixture of the two and it is decided to separate them by distillation. Find out at what pressure and temperature the column should be operated when it is known that styrene polymerizes spontaneously beyond the temperature of 30°C.

3.4 Supposing there is a mixture of ethane and propane and it is decided to separate them by distillation. One can either operate at a very low temperature and 1 atm. or at ambient temperature but at high pressure. Model this problem and find the optimal pressure and temperature by minimizing the total cost of separation. Make suitable approximations.

3.5 Solve the problem of Table 3.4 by linear programming:

3.6 Find the optimal solution for following five streams to be divided into 8 goals:

	S_1	S_2	S_3	S_4	S_5	G_1	G_2	G_3	G_4	G_5	G_6	G_7	G_8
A	5	8	3	4	3	3	3	5	3	3	2	2	2
B	3	1	3	4	1	0	3	3	1	2	1	2	0
C	2	1	4	2	6	3	3	2	1	2	2	0	2
Total	10	10	10	10	10	6	9	10	5	7	5	4	4

Also solve this problem using the heuristic rule given in this chapter. Find the distillation load and compare with the solution obtained from linear programming.

3.7 The criterion used in sequencing the distillation column was the relative volatility and the components were arranged in the order of their decreasing value. What property would you use in sequencing multicomponent liquid-liquid extraction. Attempt this problem after reading references 1 and 6 for the design procedure.

3.8 Supposing there is a mixture of three components A, B and C. Components A and B have the same relative volatility which means that components A and B would appear together in distillation. To separate A from B, supposing one can use liquid-liquid extraction. Can one now have two strategies given in Fig. P 3.7.

Find out which is better by calculating the total volume.

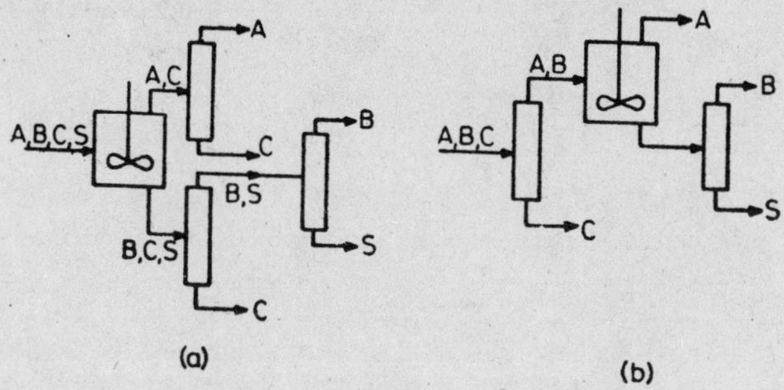

Fig. P 3.7

Properties:

K^* for extraction	K for distillation
$K^*_A = 0$	$K_A = 2$
$K^*_B = \infty$	$K_B = 2$
$K^*_C = 2$	$K_S = 1.5$
	$K_C = 1.1$

$A = 100, B = 100, C = 100, S = 100.$

3.9 Develop similar heuristic rules as of Sec. 3.7 for liquid-liquid extraction. Are they the same as those for distillation? If not, how do they differ?

3.10 Suggest a numerical procedure for finding the optimal separating sequence in multicomponent liquid-liquid extraction.

3.11 There are instances where one initially carries out separation by distillation, but later switches to other means of separation (for example, extraction). Suggest a systematic numerical procedure to find the optimal separation network. After you have worked out this problem, examine reference 11 where a complex system has been solved.

3.12 In Table 3.6, there are 22 methods given under the equilibrium separation processes. Methods which are restricted to binary mixture only would not be dependent on the order of separation. Examine the table in this light and find those methods in which sequencing is important.

3.13 Is the order of separation important in mechanical separation processes? Examine processes one by one under this heading in Table 3.6 and prove your claim.

3.14 When an ethanol stream emerges from the reactor, it consists of hydrogen, water and acetaldehyde. Find out the optimal separation network. You can easily generate the result of this problem to any vapour mixture containing one or more than one noncondensable vapour.

3.15 When the boiling points of the components differ significantly, the temperature of separation becomes extremely important. This is so in the light of P. 3.14. As an example, consider a diethyl ether plant. The product stream consists of four components: ethylene, diethyl ether, water and ethanol. Obtain (or estimate) necessary properties from reference 2 and find the temperatures of different separators.

3.16 Consider a six component mixture to be separated by distillation. Optimize the separation network.

Components	K	x
A	2.1	0.2
B	1.8	0.1
C	1.6	0.2
D	1.3	0.3
E	0.9	0.1
F	0.5	0.1

4

Synthesis of Process Flow Sheet

4.1 INTRODUCTION

In Chapter 1, the procedure of synthesizing a steady state flow sheet was given for specified products starting from given raw materials. For a given task, more than one flow sheet is possible; however, only one of them is optimal, and Chapters 2 and 3 outline the procedure for obtaining it. After the plant structure has been established, the energy and mass balances are performed to obtain different process conditions while designing a new plant and only after carrying these out, individual equipment is designed.

In this procedure of plant design, a steady state operation of individual equipment is assumed, i.e., each equipment operates exactly as required. This is never possible in actual practice and at best they can be operated around the steady state values. To demonstrate this, the following example of a polymerization unit is considered.

A polymerization reaction of styrene represented by

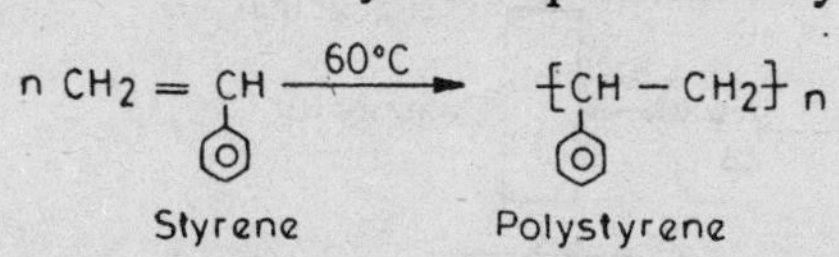

(4.1.1)

is normally carried out batchwise because at high conversions the reaction mass becomes very viscous. This makes the steady flow of material very difficult. Consequently, one of the strategies adopted is to carry out the reaction to low conversions batchwise and then to dump in an evaporator surge tank (EST). Since the final product is required in the form of polymer flakes without the unreacted monomer, the product collected in EST is first sent through an evaporator and then a flaker. Evidently to obtain the final product continuously the flaker and the evaporator should run continuously, and by adjusting the volume of this tank and maintaining a proper inventory, the operation of the evaporator can be delinked from that of the reactor. The steady state flow sheet for this process is given in Fig. 4.1 in which the output of the

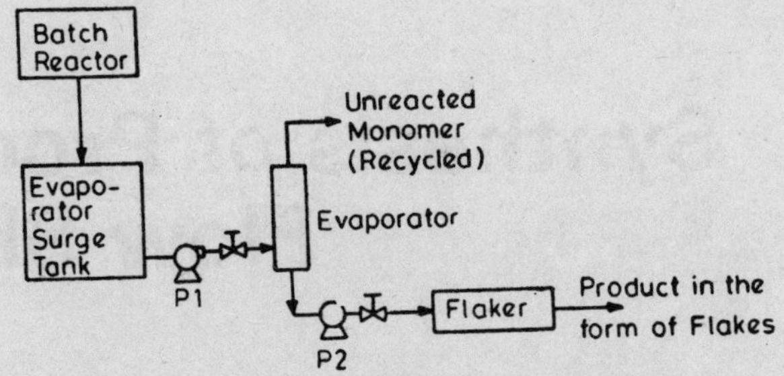

Fig. 4.1 Flow sheet for styrene polymerization plant

evaporator is directly sent to the flaker. The process as a whole is a semi-batch type because the product output is continuous even though some of its components in it are operated batchwise.

Molten polymeric liquids have very high viscosities[1] and a gravity feed system can never be used. Consequently pump Pl is placed between EST and the evaporator and pump P2 between the evaporator and the flaker. In the flow sheet, there are two pumps in series which must have identical flow rates always. This is impossible to maintain for the entire time domain because any perturbation in electrical power input to any of the motors, or any fall in their efficiency of working would cause an imbalance. As a result, the pump P2 can run 'dry' when it is pumping at higher rates and therefore to delink these, a flaker surge tank (FST) is placed between the evaporator and P2 as shown in Fig. 4.2.

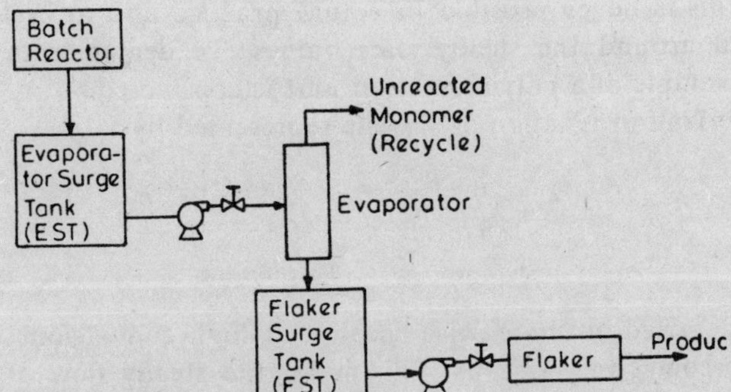

Fig. 4.2 Process flow sheet for the styrene polymerization of Fig. 4.1.[2] (Reprinted from Davis, J. J. and Kermode, R. I., I & EC, *Proc. Des. Dev.*, 14, 459, 1975, with the permission of American Chemical Society, Washington.)

Even the flow sheet of Fig. 4.2 cannot be guaranteed to work in all circumstances. Different operational routes that the plant can take are shown in Fig. 4.3. To understand how this has been constructed, it is

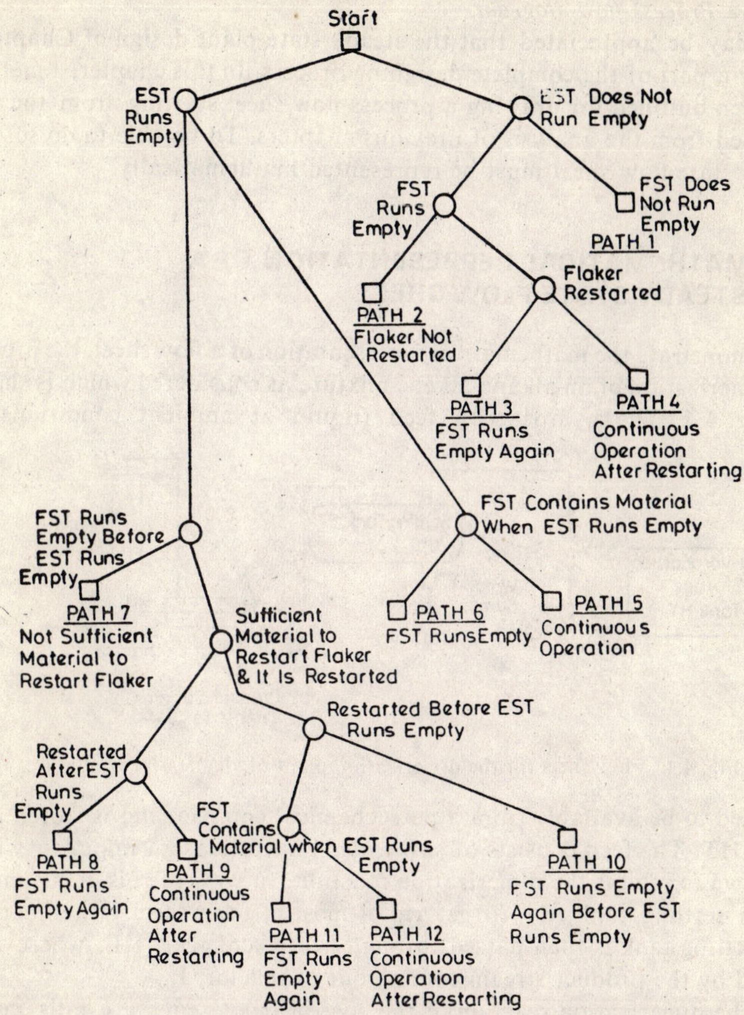

Fig. 4.3 Different events in the polymerization plant of Fig. 4.2. (Reprinted from Davis, J. J. and Kermode, R. I., I & EC, *Proc. Des. Dev.*, 14, 459, 1975, with the permission of American Chemical Society, Washington.)

noted that the flaker stops functioning when FST runs dry. When EST and FST have material, the operation is continuous (path 1). If EST runs empty and FST has material, the operation is still continuous but it is so only for a while (path 5). If both EST and FST run empty, the flaker has to be shut down temporarily and it can be put to operation only after sufficient material has collected in FST (path 4).

The purpose of this example is to demonstrate that the operation of a chemical plant is *dynamic in nature*. In view of this, in a plant design, provisions must be built in it not only to prevent failures but also to restore it to working condition in case it occurs. Such a flow sheet is called a *Process flow diagram*.

It may be appreciated that the steady state plant design of Chapter 1 is only a part of the complete designing process. In this chapter, a method has been outlined for drawing a process flow sheet starting from the one obtained from the analysis of previous chapters. To be able to do so, the steady state flow sheet must be represented mathematically.

4.2 MATHEMATICAL REPRESENTATION OF A STEADY STATE FLOW SHEET

To demonstrate the mathematical representation of a flow sheet, the process for dimerization of an alkane/alkene mixture[8] is considered which is shown in Fig. 4.4. The hydrocarbon feed (liquid at ambient conditions) is

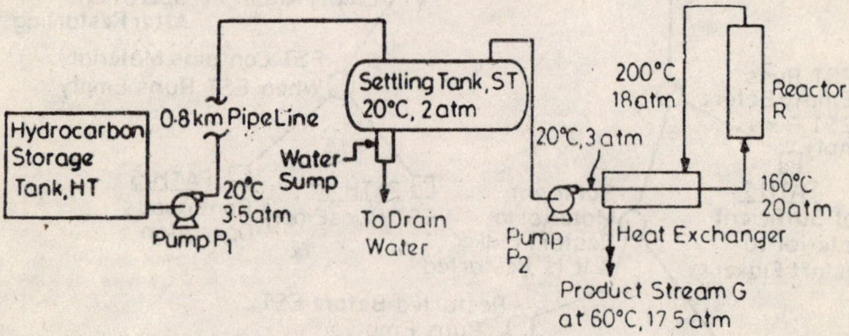

Fig. 4.4 Flow sheet for the dimerization plant of alkane/alkene mixture

assumed to be available from a petrochemical complex and is stored in a tank, HT. The feed consists of suspended moisture as an impurity which is known to inhibit the dimerization reaction. In view of this, it is pumped into a settling tank, ST to get rid of most of the water. The effluent of the settling tank is then passed through heat exchanger, HE, where it is heated by the product stream coming out of reactor, R.

In the dimerization reaction of the hydrocarbon, pressure falls in the reactor because there is contraction in volume, whereas the temperature rises because it is an exothermic reaction. Because the dimerization reaction occurs through radical mechanism and the reaction intermediates can polymerize, there is also a tendency of polymer formation. To suppress the unwanted polymerization in reactor, R, the reaction should be carried out under extremely controlled conditions, and it is in this context the residence time, τ_R, and the inlet feed temperature, T_{Ri}, of the reactor are extremely important parameters.

To represent the steady state flow sheet of Fig. 4.4 mathematically, the method of propositional logic developed in Chapter 1 is used. It is observed that the dimerization plant consists of hydrocarbon tank HT, pump P_1, 1/2-mile-long pipe line PL, settling tank ST, pump P_2, heat exchanger HE and tubular reactor R in this sequence. Because the attainment of the final product, G, is the result of the arrangement of these equipment, this fact can be written as

$$(HT \cdot P_1 \cdot PL \cdot ST \cdot P_2 \cdot HE \cdot R) \rightarrow G \tag{4.2.1}$$

where HT to G are semantic variables and are defined as

$$\left.\begin{aligned} HT &= \text{there is a hydrocarbon storage tank,} \\ P_1 &= \text{there is a pump,} \\ PL &= \text{there is 1/2-mile-long pipe line,} \\ ST &= \text{there is a settling tank,} \\ P_2 &= \text{there is a pump,} \\ HE &= \text{there is a heat exchanger,} \\ R &= \text{there is a reactor, and} \\ G &= \text{the goal is attained.} \end{aligned}\right\} \tag{4.2.2}$$

The semantic equation of the flow sheet is therefore given by

$$(HT \cdot P_1 \cdot PL \cdot ST \cdot P_2 \cdot HE \cdot R) \tag{4.2.3}$$

with the restriction that the order of variables in the above equation is maintained and the commutative rule is assumed to be invalid. Since a process flow sheet is meant to accommodate the time-dependent behaviour of all this equipment, the equipment behaviour is written in propositional logic as given in Sec. 4.3.

4.3 GENERAL SEMANTIC EQUATION OF EQUIPMENT

The approach taken in the studies of unit operations for describing the working of a given equipment is to express its characteristics either by a differential equation or numerical results. In contrast to this, its semantic equation describes its working in words and both these representations are completely equivalent.

Since a flow sheet represents the structure of a chemical plant and can be written only in the form of a semantic function (Eq. 4.2.3), it is necessary to write the working of an equipment in similar form. To demonstrate the procedure a heat exchanger is considered as an example. Figure 4.5 shows W_1, W_2, W_3 and W_4 as flow rates and T_1, T_2, T_3 and T_4

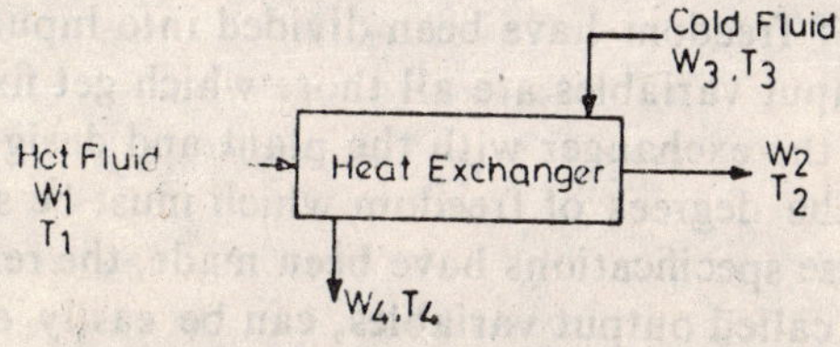

Fig. 4.5 Variables involved in heat exchanger calculations

Table 4.1 Variables and Design Equations for a Heat Exchanger

List of variables:

(1) Type of heat exchanger, K

(2) Four temperatures, T_1, T_2, T_3, T_4

(3) Four flow rates, W_1, W_2, W_3, W_4

(4) Total heat exchanged, Q

(5) Total heat transfer coefficient, U

(6) Total exchanger area, A

(7) Log mean temperature difference, $(\Delta T)_{lm}$

Design relation for steady state operation:

(a) Mass balance

$$W_1 = W_2$$
$$W_3 = W_4$$

(b) Log mean temperature difference

$$(\Delta T)_{lm} = \frac{(T_1 - T_4) - (T_2 - T_3)}{\ln\left(\frac{T_1 - T_4}{T_2 - T_3}\right)}$$

(c) $Q = W_1 C_P(T_1 - T_2) = W_2 C_P(T_4 - T_3) = UA(\Delta T)_{lm}$

(d) $U = f(K, W_1, W_3, T_1, T_2, T_3, T_4)$

as temperatures of different streams. Table 4.1 gives a list of all variables and design relations between them[4]. There are thirteen variables and seven equations which have been written assuming steady state. These can be modified accordingly for unsteady state region if so desired. However, this does not alter the number of these relations because it merely adds the time domain to the analysis. *Degrees of freedom* of a given equipment is defined as the difference between the total number of variables and the total number of relations, and for the case of heat exchanger, it is (13 — 7) or 6. It is indeed the number of variables which must be specified either by the designer or some imposed requirement to solve the seven design equations of Table 4.1 uniquely. The imposed requirement depends largely upon the situation; for example, for the design of a new heat exchanger, area is to be calculated for given process conditions and for an existing exchanger (of specified surface area), the exit temperature must be controlled. In either of these cases four out of the six degrees of freedom are committed by virtue of the fact that the exchanger is connected to the remainder of the system. This situation is depicted in Fig. 4.6, in which the total degrees of freedom have been divided into input variables and design variables. Input variables are all those which get fixed because of the connectivity of the exchanger with the plant and design variables are the remainder of the degrees of freedom which must be specified by the designer. When these specifications have been made, the remainder of unknowns, which are called output variables, can be easily calculated. For example in Fig. 4.6a the exchanger area A is the output variable and in

Fig. 4.6b and c, the changes in utility flow rates W_3 is the output variable for maintaining the specified output temperature T_2 of the process stream.

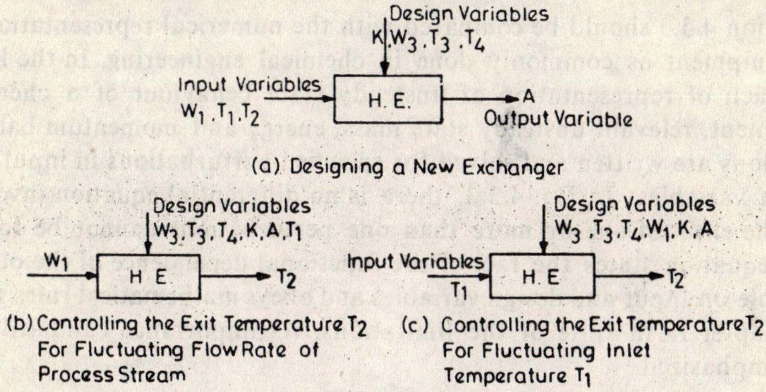

Fig. 4.6 Division of degrees of freedom into input and design variables for a heat exchanger

The discussion above is true not only for heat exchangers but for other equipment also. Every equipment has some degrees of freedom, out of which some are lost due to its connectivity and these are called input variables. The fact that the output variables of an equipment are exactly determined when its input and design variables are known, is written as

$$\left.\begin{array}{l}(\mathrm{IV}\cdot\mathrm{DV})\rightarrow\mathrm{OV}\\ \text{where}\\ \mathrm{IV}=\text{input variables of the equipment are known}\\ \mathrm{DV}=\text{design variables of the equipment are known}\\ \text{and}\\ \mathrm{OV}=\text{output variables of the equipment are known}\end{array}\right\}\quad(4.3.1)$$

The above equation can be written in terms of fluctuations of an equipment operating around steady state as follows. One defines

$$\left.\begin{array}{l}\mathrm{I}=\text{there is no deviation in input variables from their steady state values}\\ \mathrm{D}=\text{there is no deviation in design variables from their steady state values}\\ \mathrm{O}=\text{There is no deviation in output variables}\end{array}\right\}\quad(4.3.2)$$

If there is any deviation either in input variables or in design variables, it is bound to produce deviation in output variable. In this statement the knowledge of whether there is a perturbation in one or all of input or design variables is unimportant. This can be written as

$$(\bar{\mathrm{I}}+\bar{\mathrm{D}})\rightarrow\mathrm{O}\qquad(4.3.3)$$

Conversely, if there is no deviation in output variable it would imply that there is no deviation in input variables (I) and in design variable (D). Therefore

$$(I \cdot D) \rightarrow 0 \tag{4.3.4}$$

Equation 4.3.3 should be compared with the numerical representation of an equipment as commonly done in chemical engineering. In the latter approach of representation of unsteady state behaviour of a chemical equipment, relevant unsteady state mass, energy and momentum balance equations are written and solved for specified perturbations in input and design variables. In Eq. 4.3.3, there is no differential equation involved and the effect of one or more than one perturbations cannot be found. This equation states the fact of the functional dependence of the output variable on input and design variables and obeys mathematical rules given in Chapter 1. In spite of the limitations, its importance can hardly be overemphasized.

4.4 DEFINITION OF GOAL AND SEMANTIC EQUATIONS OF EQUIPMENT OF FIG. 4.4

Since it is desired that the operation of the plant is maintained at the optimal level of steady state, definition of the goal G can be written as

G = the plant is operated at the optimum level at the steady state (4.4.1)

Any deviation from the steady state operation of the plant is understood to cause a fall in the concentration of the dimer in the product stream as well as the total through-put. In view of the fact that any deviation from optimal value is undesirable, the goal, G, is equated to

$$G = Q_P \tag{4.4.2}$$

where

Q_P = there is no deviation in total through-put of the product from its steady state value (4.4.3)

Having defined the goal mathematically the unsteady state behaviour of each equipment is written. The flow sheet of Fig. 4.4 consists of a reactor, heat exchanger, settling tank, two pumps and a storage tank and one considers the reactor first. From basic studies of the reaction kinetics, the reactor is known to give the desired level of conversion to dimer provided the reactor pressure, temperature, residence time and the temperature of inlet streams are at the desired level. Further, for a given volume of the reactor, the inlet flow rate must be controlled to maintain a constant residence time. Thus the unsteady state operation of the reactor can be written as

$$(P_R \cdot T_R \cdot T_{Ri} \cdot Q_{Ri} \cdot C_{WR}) \rightarrow Q_P \tag{4.4.4}$$

where

C_{WR} = there is no deviation in water concentration in the inlet stream from its steady state value

P_R = there is no deviation in the reactor pressure from its steady state value

T_R = there is no deviation in the reactor temperature from its steady state value

T_{Ri} = there is no deviation in the inlet reactor stream temperature from its steady state value

Q_{Ri} = there is no deviation in the inlet reactor stream flow rate from its steady state value (4.4.5)

Symbol Q_P in Eq. 4.4.4 has the same meaning as in Eq. 4.4.3. The semantic equation for a heat exchanger is written with the help of Fig. 4.6b and c as

$$(Q_{EI} \cdot T_{EI} \cdot Q_H \cdot T_{H1} \cdot T_{H2}) \rightarrow T_{EO} \tag{4.4.6}$$

where

Q_{Ei} = there is no deviation in the exchanger inlet flow rate from its steady state value

T_{Ei} = there is no deviation in the exchanger's inlet stream temperature from its steady state value

Q_H = there is no deviation in the flow rate of the heating fluid from its steady state value

T_{H1} = there is no deviation in the inlet temperature of the heating fluid from its steady state value

T_{H2} = there is no deviation in the outlet temperature of the heating fluid from its steady state value

and

T_{Eo} = there is no deviation in the output temperature of the cold (process) stream from its steady state value (4.4.7)

A pump delivers fluid material not only when the choice of the pump has been made correctly but there is enough material to be pumped. The choice of a pump is decided by its ΔP-Q characteristic where ΔP is the head developed and Q is the delivered flow rate. However, once the choice of a pump is made, the pump works satisfactorily as long as sufficient material is available for pumping, i.e.,

$$M_P \rightarrow Q_P \tag{4.4.8}$$

where

M_P = there is sufficient material to deliver

Q_P = the pump is delivering without any deviation from its steady state flow rate (4.4.9)

Above, the suction pressure is assumed constant and its variation can be easily included if so desired.

The semantic equation for the settling tank, ST, is written next. A settling tank is used for separating suspended solid or liquid in another fluid medium and to achieve this, sufficient residence time is provided for the suspended medium to settle. This is done by controlling the inlet and outlet flow rates and maintaining a proper level in it. Therefore, the settling tank can be represented as

$$(Q_{STi} \cdot Q_{STO} \cdot L_{ST}) \rightarrow (C_{WST} \cdot Q_{STO}) \tag{4.4.10}$$

where

$$\left.\begin{aligned} Q_{STi} &= \text{there is no deviation in the inlet flow rate from its steady state value} \\ Q_{STO} &= \text{there is no deviation in the outlet flow from its steady state value} \\ L_{ST} &= \text{there is no deviation in the liquid level from its steady state value} \\ &\text{and} \\ C_{WST} &= \text{there is no deviation in suspended water concentration in the outlet stream of the settling tank from its steady state value} \end{aligned}\right\} \tag{4.4.11}$$

Equation 4.4.10 can be simplified by expanding the implication and then opening the bracket which leads to

$$\begin{aligned} (Q_{STi} \cdot Q_{STO} \cdot L_{ST}) \rightarrow (C_{WST} \cdot Q_{STO}) &= (\overline{Q_{STi} \cdot Q_{STO} \cdot L_{ST}}) + (C_{WST} \cdot Q_{STO}) \\ &= (\bar{Q}_{STi} + \bar{Q}_{STO} + \bar{L}_{ST}) + (C_{WST} \cdot Q_{STO}) \\ &= \bar{Q}_{STi} + \bar{L}_{ST} + [\bar{Q}_{STO} + (C_{WST} \cdot Q_{STO})] \\ &= \bar{Q}_{STi} + \bar{L}_{ST} + [(Q_{STO} + \bar{Q}_{STO}) \cdot (C_{WST} + Q_{STO})] \\ &= \bar{Q}_{STi} + \bar{L}_{ST} + [(C_{WST} + \bar{Q}_{STO}) \cdot \blacksquare] \\ &= (\overline{Q_{STi}\, L_{ST}\, Q_{STO}}) + C_{WST} \end{aligned} \tag{4.4.12}$$

Thus the settling tank behaviour is given by

$$(Q_{STi} \cdot L_{ST} \cdot Q_{STO}) \rightarrow C_{WST} \tag{4.4.13}$$

Also its level will rise or fall when the rate of withdrawal is more or less than the input. Thus

$$(Q_{STi} \cdot Q_{STO}) \rightarrow L_{ST} \tag{4.4.14}$$

which stands for the fact that there is no deviation in the level of the settling tank provided input and output flow rates are maintained at the steady state level. Since the behaviour of the settling tank is given by the combination of both these facts,

$$\begin{aligned} &[(Q_{STi} \cdot Q_{STO}) \rightarrow L_{ST}] \cdot [(Q_{STi} \cdot Q_{STO} \cdot L_{ST}) \rightarrow C_{WST}] \\ &= (Q_{STi} \cdot Q_{STO}) \rightarrow C_{WST} \end{aligned} \tag{4.4.15}$$

A storage tank has the function of storing sufficient fluid so that the entire or a section of the chemical plant can be operated continuously. Its semantic equation is written from the fact that there is going to be material in it as long as input and output flow rates are regulated. Or,

$$(Q_{\mathrm{HTi}} \cdot Q_{\mathrm{HTO}}) \rightarrow M_{\mathrm{HT}} \tag{4.4.16}$$

where

$$\left.\begin{aligned} M_{\mathrm{HT}} &= \text{material in the tank does not fall below a minimum level} \\ Q_{\mathrm{HTi}} &= \text{there is an input of the material} \\ Q_{\mathrm{HTO}} &= \text{material is withdrawn at a steady rate} \end{aligned}\right\} \tag{4.4.17}$$

Lastly, the operation of a pipe line is considered. It is noted that it will deliver material at its outlet equal to its input provided it does not leak or does not develop a crack. This is written as

$$(Q_{\mathrm{PLi}} \cdot LK_{\mathrm{PL}} \cdot B_{\mathrm{PL}}) \rightarrow Q_{\mathrm{PLO}} \tag{4.4.18}$$

where

$$\left.\begin{aligned} LK_{\mathrm{PL}} &= \text{there is no leak at flanges} \\ B_{\mathrm{PL}} &= \text{there is no crack anywhere in the pipeline} \\ Q_{\mathrm{PLO}} &= \text{there is no deviation in the output flow rate from the steady state value} \\ &\text{and} \\ Q_{\mathrm{PLi}} &= \text{there is no deviation in the input flow rate from the steady state value} \end{aligned}\right\} \tag{4.4.19}$$

To show that the plant will function under all fluctuations, the conjunction of the negation of the goal with the ensemble of equipment must be proved as inconsistent. This is a mathematical requirement. Before doing this, the nature of these fluctuations must be examined further.

4.5 RELATION BETWEEN FLUCTUATIONS OF DIFFERENT EQUIPMENT

In the flow sheet of Fig. 4.4, the hypothetical case when the half-mile pipeline develops a crack is analyzed. When this occurs, the delivered flow rate to the settling tank is reduced. Because of the imbalance in the input-output flow rates, the level in it begins to fall. The pump P_2 delivers the desired flow rate till the level has completely fallen and when this occurs, the plant stops operating. Before the plant-shutdown occurs, the concentration of water in the exit stream will be higher than the desired value because adequate separation does not occur due to the reduced residence time. The higher water concentration is felt at the reactor only after this disturbance in the concentration reaches there. However, at a given time, the magnitude of the perturbation in water concentration at the exit stream of the settling tank and that at the reactor are completely unrelated. This is true not only for concentration but for other kinds of perturbations and the time lag arises because it takes a finite time for it to travel between two points separated by a finite distance.

The situation when all possible deviations are allowed (i.e., while moving along with the process stream from the storage tank all the way upto the reactor) $\overline{Q}_{HTO}$, $\overline{Q}_{STi}$, $\overline{C}_{WST}$, $\overline{L}_{ST}$, $\overline{Q}_{P2O}$ and $\overline{T}_{EO}$ are all true. Because these are output variables of different equipment, it is reasonable to expect them not to be correlated to each other. This is true only when these deviations are measured at a single instant of time. *Two deviations at a given time are equal only when they occur at the same location or when they are made equal by means of some external mechanisms called controllers.*

It may be pointed out that not all kinds of deviations can be corrected while the chemical plant is in operation. To give an example, a pump can (i) have fluctuations in electrical input, (ii) develop extra friction at the rotor pin, (iii) face a loss of suction because there is no material to pump, and (iv) face a loss of suction due to clogging. Out of these, (ii) and (iv) require plant shutdown whereas controllers can be employed against faults (i) and (iii).

4.6 CONTROLLERS AND THEIR SEMANTIC EQUATIONS

As pointed out in Chapter 1, a process stream can be characterized by the specification of pressure, temperature concentration, phase and flow rates. Among these, pressure, temperature and flow rates can be controlled very conveniently. Controllers are devices which sense deviations at one point and produce corrections at some other. Figure 4.7 shows the schematic drawings of level, pressure, temperature and flow controllers. Semantic equations for these are given below:

A pressure controller is modelled by noting its function. If there is a deviation in pressure from its set value, the controller corrects it by regulating the exit stream flow rate by a suitable valve. Consequently, its semantic equation is given by

$$\overline{P} \rightarrow Q \qquad (4.6.1)$$

where

$$\left.\begin{aligned} P &= \text{there is no deviation in pressure from the set value} \\ Q &= \text{one regulates the exit stream flow rate by a valve} \end{aligned}\right\} \qquad (4.6.2)$$

A temperature controller eliminates fluctuations in temperature. If there is any deviation in temperature from the set value, the heat exchanging system, HES, regulates it. The HES could be a mere heat exchanger, a coal or oil fired furnace or an electric heater. Its semantic

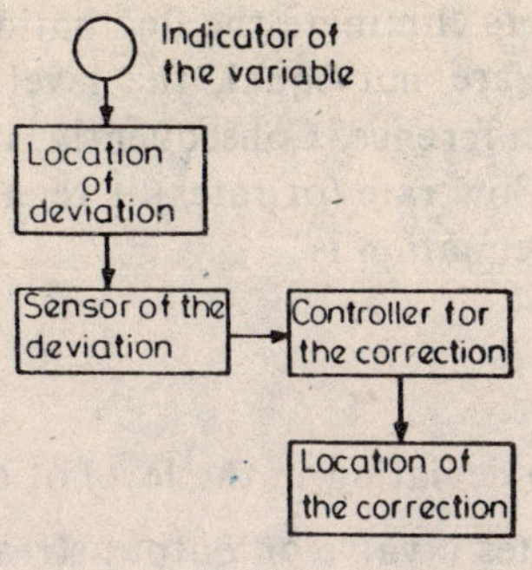

(a) Elements of Controllers

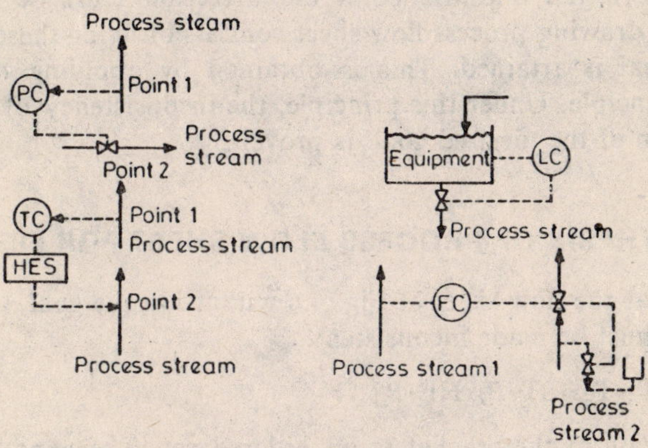

(b) Some Controllers

Fig. 4.7 Schematic representation of controllers

equation therefore is

$$\bar{T} \rightarrow HES \tag{4.6.3}$$

where

$$\left.\begin{aligned} T &= \text{there is no deviation in temperature from the set value} \\ HES &= \text{one regulates the heat exchanging system} \end{aligned}\right\} \tag{4.6.4}$$

The function of a flow controller is described as follows. If there is any deviation in flow rate, one puts a bypass system to work. The bypass system is a mere pipeline placed parallel to the process stream so that the adjustment in flow rate can be made through it. Its semantic equation is given by

$$\bar{Q} \rightarrow BP \tag{4.6.5}$$

where

$$\left.\begin{aligned} Q &= \text{there is no deviation in flow rate} \\ BP &= \text{one regulates the bypass system} \end{aligned}\right\} \tag{4.6.6}$$

Lastly, level controllers eliminate the fluctuations in liquid level. If input and output flow rates are not equal, the level would either rise or fall depending upon the difference. Consequently, if there is any deviation in the level, the output flow rate (or rates) is regulated by a valve. In view of this, the semantic equation is

$$\bar{\mathrm{L}} \rightarrow \mathrm{V} \tag{4.6.7}$$

where

$$\left.\begin{array}{l}\mathrm{L} = \text{there is no deviation in the level of the liquid} \\ \mathrm{V} = \text{one regulates a valve on output streams}\end{array}\right\} \tag{4.6.8}$$

The examination of Fig. 4.7 reveals that in placing controllers, the point of deviation and the point of correction must be known. The problem of drawing process flow sheet consists of fixing these points such that the goal is attained. This is obtained by applying the Theorem Proving Principle. Under this principle, the inconsistency of 'facts' with the negation of the 'desired goal' is proven.

4.7 SYNTHESIS OF PROCESS FLOW SHEET FOR FIG. 4.4

To show that the flow sheet of Fig. 4.4 will attain the goal, the following statement must be made inconsistent:

$$[\mathrm{HT} \cdot \mathrm{P_1} \cdot \mathrm{PL} \cdot \mathrm{ST} \cdot \mathrm{P_2} \cdot \mathrm{HE} \cdot \mathrm{R}] \cdot \bar{\mathrm{G}} \tag{4.7.1}$$

The expression in the bracket is an ordered set. The expression for G, R, HE, $\mathrm{P_2}$, ST, PL, $\mathrm{P_1}$ and HT from Sec. 4.4 are substituted to obtain

$$\begin{aligned} F = {} & [(Q_{\mathrm{HTi}} \cdot Q_{\mathrm{HTO}}) \rightarrow M_{\mathrm{HT}}] \cdot [M_{\mathrm{P1}} \rightarrow Q_{\mathrm{P1}}] \cdot [(LK_{\mathrm{PL}} \cdot B_{\mathrm{PL}}) \rightarrow Q_{\mathrm{PL}}] \\ & \cdot [(Q_{\mathrm{STi}} \cdot Q_{\mathrm{STO}}) \rightarrow L_{\mathrm{ST}}] \cdot [(Q_{\mathrm{STi}} \cdot Q_{\mathrm{STO}} \cdot L_{\mathrm{ST}}) \rightarrow C_{\mathrm{WST}}] \\ & \cdot [M_{\mathrm{P2}} \rightarrow Q_{\mathrm{P2}}] \cdot [(Q_{\mathrm{Ei}} \cdot T_{\mathrm{Ei}} \cdot Q_{\mathrm{H}} \cdot T_{\mathrm{H1}} \cdot T_{\mathrm{H2}}) \rightarrow T_{\mathrm{EO}}] \\ & \cdot [(P_{\mathrm{R}} \cdot T_{\mathrm{R}} \cdot T_{\mathrm{Ri}} \cdot Q_{\mathrm{Ri}} \cdot C_{\mathrm{WR}}) \rightarrow Q_{\mathrm{P}}] \cdot \bar{Q}_{\mathrm{P}} \end{aligned} \tag{4.7.2}$$

Since this is an ordered set, one must proceed exactly in the order written above. As a result, the negation of goal with the reactor is considered first, i.e.,

$$F_{\mathrm{R}} \equiv [(P_{\mathrm{R}} \cdot T_{\mathrm{R}} \cdot T_{\mathrm{Ri}} \cdot Q_{\mathrm{Ri}} \cdot C_{\mathrm{WR}}) \rightarrow Q_{\mathrm{P}}] \cdot \bar{Q}_{\mathrm{P}} \tag{4.7.3}$$

Simplifying the implication sign

$$F_{\mathrm{R}} = [\overline{(P_{\mathrm{R}} \cdot T_{\mathrm{R}} \cdot T_{\mathrm{Ri}} \cdot Q_{\mathrm{Ri}} \cdot C_{\mathrm{WR}})} + Q_{\mathrm{P}}] \cdot \bar{Q}_{\mathrm{P}} \tag{4.7.4}$$

and then using the DeMorgon's rule, one obtains

$$F_{\mathrm{R}} = [(P_{\mathrm{R}} \cdot T_{\mathrm{R}} \cdot T_{\mathrm{Ri}} \cdot Q_{\mathrm{Ri}} \cdot C_{\mathrm{WR}}) \cdot \bar{Q}_{\mathrm{P}})] + (Q_{\mathrm{P}} \cdot \bar{Q}_{\mathrm{P}}) \tag{4.7.5}$$

The statement $(Q_{\mathrm{P}} \cdot \bar{Q}_{\mathrm{P}})$ is always false. Consequently, the truth values of F_{R} is decided by

$$F_{\mathrm{R}} = (\overline{P_{\mathrm{R}} \cdot T_{\mathrm{R}} \cdot T_{\mathrm{Ri}} \cdot Q_{\mathrm{Ri}} \cdot C_{\mathrm{WR}}}) \cdot \bar{Q}_{\mathrm{P}}$$

$$= (\bar{P}_{R} + \bar{T}_{R} + \bar{T}_{Ri} + \bar{Q}_{Ri} + \bar{C}_{WR})\cdot\bar{Q}_{P} \tag{4.7.6}$$

Using DeMorgan's rule once again,

$$F_R = (\bar{P}_R\cdot\bar{Q}_P) + (\bar{T}_R\cdot\bar{Q}_P) + (\bar{T}_{Ri}\cdot\bar{Q}_P) + (\bar{Q}_{Ri}\cdot\bar{Q}_P) + (\bar{C}_{WR}\cdot\bar{Q}_P) \tag{4.7.7}$$

To make F_R false it is necessary that each of the statements in the parentheses of the right hand side be made false since these are connected by disjunction.

$(\bar{P}\cdot\bar{Q}_P)$ is examined first. This would be false for all possible truth values of $\bar{Q}_P$ only when $\bar{P}_R$ is made false. From Eq. 4.6.1 it follows that this will be so only when a pressure controller is put upon it.

Figure 4.8 shows the resultant arrangement on the reactor. Function $(\bar{T}_R\cdot\bar{Q}_P)$ is considered next and noted that $\bar{T}_R$ should be made false for the function to be false. As $\bar{T}_R$ cannot be simplified any more, a temperature controller is placed on the reactor (Eq. 4.6.4) as shown in Fig. 4.8. This way, $(\bar{T}_R\cdot\bar{Q}_P)$ and $(\bar{P}_R\cdot\bar{Q}_P)$ are made false always. Thus the effect of temperature and pressure controllers makes the truth value of F_R depend only on the remainder of variables in Eq. 4.7.7 and only these are considered in F_R now

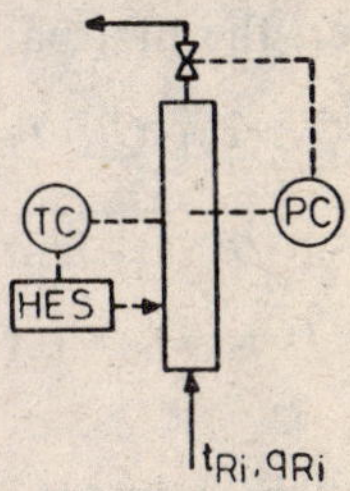

Fig. 4.8 Controller requirements for the reactor, R, of Fig. 4.4

$$F_R = (\bar{T}_{Ri}\cdot\bar{Q}_P) + (\bar{Q}_{Ri}\cdot\bar{Q}_P) + (\bar{C}_{WR}\cdot\bar{Q}_P)$$

[after controllers of Fig. 4.8] (4.7.8)

If $\bar{Q}_P$ is true, the truth value of F_R is entirely determined by truth values of $\bar{Q}_{Ri}$, $\bar{T}_{Ri}$ and $\bar{C}_{WR}$. As a result, $\bar{Q}_P$ is deleted from the above equation to obtain

$$F_R = \bar{Q}_{Ri} + \bar{T}_{Ri} + \bar{C}_{WR} = \overline{(Q_{Ri}\cdot T_{Ri}\cdot C_{WR})} \tag{4.7.9}$$

The next step in Eq. 4.7.1 is to consider the combination of the heat exchanger, and the reactor, i.e.,

$$F_{HE} = [(Q_{Ei}\cdot T_{Ei}\cdot Q_H\cdot T_{H1}\cdot T_{H2}) \rightarrow T_{EO}]\cdot\overline{(Q_{Ri}\cdot T_{Ri}\cdot C_{WR})} \tag{4.7.10}$$

Next, it is noted that the inlet temperature of the reactor is equal to the outlet temperature of the heat exchanger. Consequently, any deviation in them should also be equal, or

$$T_{EO} = T_{Ri} \tag{4.7.11}$$

Therefore Eq. 4.7.10 becomes

$$\begin{aligned} F_{HE} &= [(Q_{Ei}\cdot T_{Ei}\cdot Q_H\cdot T_{H1}\cdot T_{H2}) \rightarrow T_{EO}]\cdot\overline{(T_E\cdot Q_{RI}\cdot C_{WR})} \\ &= [\{(Q_{Ei}\cdot T_{Ei}\cdot Q_H\cdot T_{H1}\cdot T_{H2}) \rightarrow T_{EO}\}\cdot\bar{T}_{EO}] \\ &\quad + [\{(Q_{Ei}\cdot T_{Ei}\cdot Q_H\cdot T_{H1}\cdot T_{H2}) \rightarrow T_{EO}\}\cdot\bar{Q}_{Ri}] \\ &\quad + [\{(Q_{Ei}\cdot T_{Ei}\cdot Q_H\cdot T_{H1}\cdot T_{H2}) \rightarrow T_{EO}\}\cdot\bar{C}_{WR}] \end{aligned} \tag{4.7.12}$$

When the heat exchanger function, i.e., $[(Q_{Ei}\cdot T_{Ei}\cdot Q_{H}\cdot T_{H1}\cdot T_{H2})\rightarrow T_{EO}]$ operates over $\bar{Q}_{Ei}$ and $\bar{C}_{WR}$, no further simplification can result because there is no common variable between them. As the heat exchanger function describing its working is always true, the truth value of the combination is completely determined by $\bar{Q}_{Ri}$ and $\bar{C}_{WR}$. Thus,

$$[(Q_{Ei}\cdot T_{Ei}\cdot Q_{H}\cdot T_{H1}\cdot T_{H2})\rightarrow T_{EO}]\cdot\bar{Q}_{Ri}=\bar{Q}_{Ri} \tag{4.7.13a}$$

and

$$[(Q_{RI}\cdot T_{Ei}\cdot Q_{H}\cdot T_{H1}\cdot T_{H2})\rightarrow T_{EO}]\cdot\bar{C}_{WR}=\bar{C}_{WR} \tag{4.7.13b}$$

However, the first parenthesis of Eq. 4.7.12 can be simplified to

$$\begin{aligned}&[(Q_{Ei}\cdot T_{Ei}\cdot Q_{H}\cdot T_{H}\cdot T_{H1}\cdot T_{H2})\rightarrow T_{EO}]\cdot\bar{T}_{EO}\\&=[(\overline{Q_{Ei}\cdot T_{Ei}\cdot Q_{H}\cdot T_{H1}\cdot T_{H2}})+T_{EO}]\cdot\bar{T}_{EO}\\&=(\bar{Q}_{Ei}\cdot\bar{T}_{EO})+(\bar{T}_{Ei}\cdot\bar{T}_{EO})+(\bar{Q}_{H}\cdot\bar{T}_{EO})+(\bar{T}_{H1}\cdot\bar{T}_{EO})+(\bar{T}_{H2}\cdot\bar{T}_{EO})\end{aligned} \tag{4.7.14}$$

The last three terms in Eq. 4.7.4, $(\bar{Q}_{H}\cdot\bar{T}_{EO})$, $(\bar{T}_{H1}\cdot\bar{T}_{EO})(\bar{T}_{E2}\cdot\bar{T}_{EO})$, would become false provided Q_H, T_{H1} and T_{H2} are true. But it is noted that

$$Q_H = Q_P \tag{4.7.15}$$

and it has already been admitted that $\bar{Q}_P$ is true. Consequently from Eq. 4.7.15, statement $\bar{Q}_H$ is true and the statement $(\bar{Q}_{H}\cdot\bar{T}_{EO})$ is false only when $\bar{T}_{EO}$ is made false. As the statement $\bar{T}_{EO}$ cannot be further simplified, a temperature controller is used on this variable through a heat exchanging system, as shown in Fig. 4.9. The heat exchanging system, HES, of the temperature controller is placed between the reactor and the heat exchanger because the adjustment of the process stream temperature has to be done at this point.

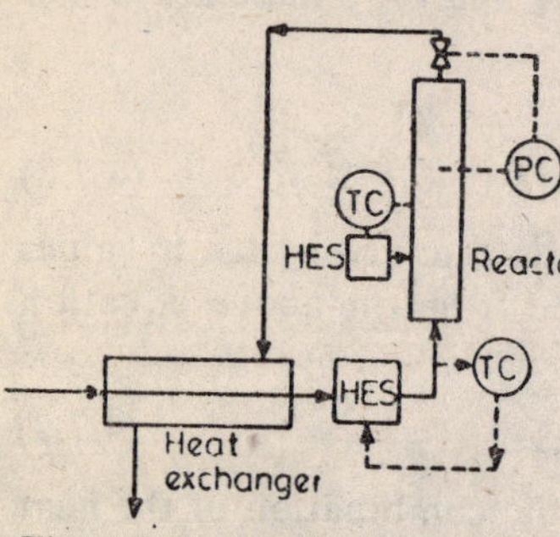

Fig. 4.9 Controller requirements for the combination of R and HE of Fig. 44.

When the statement $\bar{T}_{EO}$ is made false through the use of the heat exchanging system, HES, the statements $(\bar{Q}_{Ei}\cdot\bar{T}_{EO})$, $(\bar{T}_{Ei}\cdot\bar{T}_{EO})$, $(\bar{T}_{H1}\cdot\bar{T}_{EO})$ and $(\bar{T}_{H2}\cdot\bar{T}_{EO})$ all become false, and the statement in Eq. 4.7.14 is always false. This mathematically confirms the intuitive thinking that whatever deviation that might exist in the desired and the achieved final temperature out of the first exchanger can be taken care of by the HES irrespective of the cause. With the help of Eq. 4.7.13 and 4.7.14, Eq. 4.7.12 is written as

$$F_{HE}=\bar{Q}_{Ri}+\bar{C}_{WR}=(\overline{Q_{Ri}\cdot C_{WR}})\quad\text{[after controllers of Fig. 4.9]} \tag{4.7.16}$$

The next equipment on the flow sheet is pump P2, which has the dynamic equation given by Eq. 4.4.10. The combination of the pump, heat exchanger and reactor, F_{P2}, is given by

$$F_{P2} = \text{P2}\cdot(\text{HE}\cdot\text{R}\cdot\bar{\text{G}})$$
$$= (M_{P2} \to Q_{P2})\cdot(\overline{Q_{Ri}\cdot C_{WR}}) \tag{4.7.17}$$

Equation 4.7.17 can be simplified to

$$F_{P2} = [(M_{P2} \to Q_{P2})\cdot\bar{Q}_{Ri}] + [(M_{P2} \to Q_{P2})\cdot\bar{C}_{WR}] \tag{4.7.18}$$

In the second parenthesis, the function for the pump operates on an unrelated statement $\bar{C}_{WR}$. As $(M_{P2} \to Q_{P2})$ is always true, the truth value of its combination with $\bar{C}_{WR}$ is decided by the latter. Thus

$$(M_{P2} \to Q_{P2})\cdot\bar{C}_{WR} = C_{WR} \tag{4.7.19}$$

and Eq. 4.7.17 becomes

$$F_{P2} = [(M_{P2} \to Q_{P2})\cdot\bar{Q}_{Ri}] + \bar{C}_{WR}$$
$$= [(\bar{M}_{P2} + Q_{P2})\cdot\bar{Q}_{Ri}] + \bar{C}_{WR}$$
$$= [(\bar{M}_{P2}\cdot\bar{Q}_{Ri}) + (Q_{P2}\cdot\bar{Q}_{Ri})] + \bar{C}_{WR} \tag{4.7.20}$$

Once again to make F_{P2} false, $(\bar{M}_{P2}\cdot\bar{Q}_{Ri})$, $(Q_{P2}\cdot\bar{Q}_{Ri})$ and $\bar{C}_{WR}$ must individually be false. $(Q_{P2}\cdot\bar{Q}_{Ri})$ will always be false (or an invalid statement) provided

$$Q_{P2} = Q_{Ri} \tag{4.7.21}$$

and this can be achieved by placing a by-pass system of the flow controller on pump P2 which is actuated by the inlet flow rate of the reactor as shown in Fig. 4.10. Equation 4.7.21 is a semantic equation which

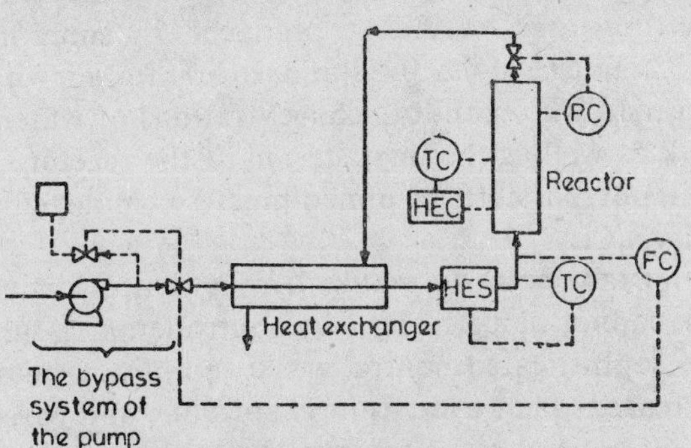

Fig. 4.10 Controller requirements for the combination of R, HE and P2 of Fig. 4.4

means that whenever a perturbation in flow rate at the reactor exists, the flow at the pump must be perturbed (intuitively to occur in the opposite direction).

After substituting Eq. 4.7.21, in Eq. 4.7.20, one has

$$F_{P2} = (\bar{M}_{P2}\cdot\bar{Q}_{P}) + \bar{C}_{WR} \text{ [after controllers of Fig. 4.10]} \tag{4.7.22}$$

As $\bar{Q}_P$ is true, the truth value of $(\bar{M}_{P2}\cdot\bar{Q}_P)$ is given by $\bar{M}_{P2}$. Thus

$$F_{P2} = (\bar{M}_{P2} + \bar{C}_{WRi}) \tag{4.7.23}$$

After this, the settling tank is combined with F_{P2} as

$$\begin{aligned} F_{ST} &\equiv (ST\cdot P2\cdot HE\cdot R)\cdot\bar{G} \\ &= [(Q_{STi}\cdot Q_{STO}) \rightarrow C_{WST}]\cdot(\overline{\bar{M}_{P2}\cdot C_{WR}}) \end{aligned} \tag{4.7.24}$$

It is noted that there can be enough material to pump if the level in the tank is maintained. Therefore these are equated

$$L_{ST} = M_{P2} \tag{4.7.25}$$

and substituting this in Eq. 4.7.24, one has

$$\begin{aligned} F_{ST} &= [(\overline{Q_{STi}\cdot Q_{STO}}) + C_{WST}]\cdot(\overline{L_{ST}\cdot C_{WR}}) \\ &= \{[(\overline{Q_{STi}\cdot Q_{STO}}) + C_{WST}]\cdot\bar{C}_{WR}\} + \{[(\overline{Q_{STi}\cdot W_{STO}}) + C_{WST}]\cdot\bar{L}_{ST}\} \end{aligned} \tag{4.7.26}$$

As the statement $[(\overline{Q_{STi}\cdot Q_{STO}}) + C_{WST}]$ is the semantic equation of the settling tank and is always true, the truth value of the combination is given by $\bar{L}_{ST}$. Therefore

$$F_{ST} = [(\overline{Q_{STi}\cdot Q_{STO}})\cdot\bar{C}_{WR}] + [C_{WST}\cdot\bar{C}_{WR}] + \bar{L}_{ST} \tag{4.7.27}$$

If by some mechanism, C_{WST} is related such that

$$C_{WST} = C_{WR} \tag{4.7.28}$$

Therefore, if $\bar{C}_{WR}$ is true, according to this relation, the concentration in the settling tank should also be perturbed simultaneously. From the control point of view if the deviation in the concentration of the inlet stream of the reactor is positive, a negative deviation in the concentration of water at the settling tank should be produced by some mechanism. As this regulation cannot be done by standard controllers of Fig. 4.7, this is assigned to a man who monitors concentrations of water at the exit of the settling tank as well as the inlet stream of the reactor. If the remote measurement is not possible, two men must be employed instead of one as shown in Fig. 4.11.

The most important point to realize here is that one employs men for activities which cannot be performed by controller. It is always possible to invent more sophisticated controllers to eliminate human functions. For economic reasons, however, this is not done and personnel requirements occur in a chemical plant because they have certain control operations to perform. In Fig. 4.11, for example, the activity of workers A and B would be the following. Worker B measures the concentration at the reactor and reports to the worker A. He finds the difference in the concentration at the exit of the settling tank and based upon this, the working of the settling tank is decided.

Equation 4.7.28 is substituted in Eq. 4.7.27 to obtain

$$F_{ST} = [(\overline{Q_{STi}\cdot Q_{STO}})\cdot\bar{C}_{WST}] + \bar{L}_{ST} \tag{4.7.29}$$

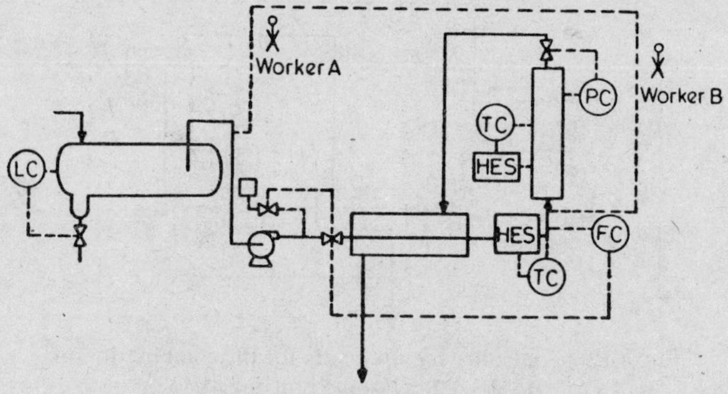

Fig. 4.11 Controller requirements for the combination of R, HE, P2 and ST of Fig. 4.4

To make the statement in the parenthesis above false for all truth values of C_{WST}, it is necessary to make the following statement false:

$$F_{ST} = (\overline{Q_{STi}\cdot Q_{STO}}) + \bar{L}_{ST} = \bar{Q}_{STi} + \bar{Q}_{STO} + \bar{L}_{ST} \tag{4.7.30}$$

As the statement L_{ST} cannot be further simplified, level controller of Fig. 4.7 is used on the settling tank to make it false. According to its semantic equation, both $\bar{L}_{ST}$ and $\bar{Q}_{STO}$, are simultaneously falsified. Thus, the truth value of F_{ST} is determined by Q_{STi}, or

$$F_{ST} = Q_{STi} \text{ [After controllers of Fig. 4.11]} \tag{4.7.31}$$

The pipeline is considered next. The statement, F_{PL}, formed by the conjunction of pipeline (PL), settling tank (ST), pump (P2), heat exchanger (HE) and the reactor (R) is

$$F_{PL} = [(Q_{PLi}\cdot LK_{PL}\cdot B_{PL}) \rightarrow Q_{PLO}]\cdot\bar{Q}_{STi} \tag{4.7.32}$$

As the location of statements Q_{PLO} and Q_{STi} is the same,

$$Q_{PLO} = Q_{STi} \tag{4.7.33}$$

and as a result of which,

$$\begin{aligned} F_{PL} &= [(\overline{Q_{PLi}\cdot LK_{PL}\cdot B_{PL}}) + Q_{PLO}]\cdot\bar{Q}_{PLO} \\ &= [(\bar{Q}_{PLi} + \overline{LK}_{PL} + \bar{B}_{PL})\cdot\bar{Q}_{PLO}] + [Q_{PLO}\cdot\bar{Q}_{PLO}] \\ &= (\bar{Q}_{PLi}\cdot\bar{Q}_{PLO}) + (\overline{LK}_{PL}\cdot\bar{Q}_{PLO}) + (\bar{B}_{PL}\cdot\bar{Q}_{PLO}) \end{aligned} \tag{4.7.34}$$

The above simplification is obtained by dropping $(Q_{PLO}\cdot\bar{Q}_{PLO})$. The second and third statements once again cannot be made false by any of controllers discussed in Fig. 4.7. Consequently, inspection and maintenance by a team of workers would ensure that it remains false for all time to come. This is depicted in Fig. 4.12 by a team of workers with the job to inspect the pipeline for any leakage and breakage. Consequently Eq. 4.7.34 becomes

$$F_{PL} = (\bar{Q}_{PLi}\cdot\bar{Q}_{PLO}) \text{ [after controllers of Fig. 4.12]} \tag{4.7.35}$$

If $\bar{Q}_{PLO}$ is false, it is immaterial what the truth value of $\bar{Q}_{PLi}$ is. If state-

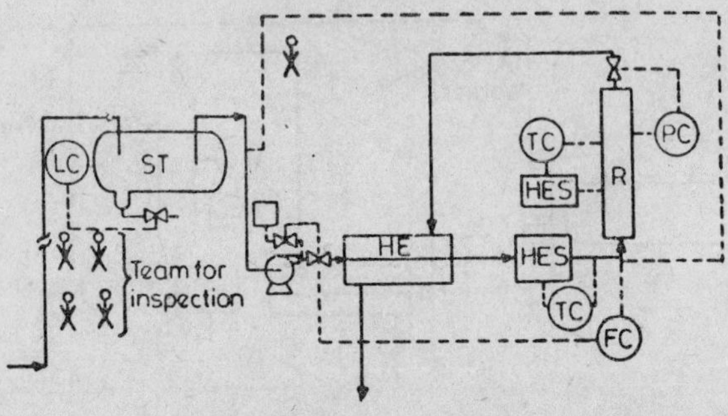

Fig. 4.12 Controller requirements for the combination of R, HE, P2, ST and PL of Fig. 4.4

ment $\bar{Q}_{PLO}$ is true, it is necessary to have $\bar{Q}_{PLi}$ false. Considering the situation when $\bar{Q}_{PLO}$ is true, it is required to make $\bar{Q}_{PLi}$ false, i.e.,

$$F_{PL} = \bar{Q}_{PLi} \tag{4.7.36}$$

After this, the combination of pump P1, pipeline PL, settling tank ST, pump P2, heat exchanger HE and the reactor R is considered. The statement for the combination of all these equipments, F_{P1}, is

$$F_{P1} = (M_{P1} \rightarrow Q_{P1}) \cdot F_{PL} = (M_{P1} \rightarrow Q_{P1}) \cdot \bar{Q}_{PLi} \tag{4.7.37}$$

As the location of statements Q_{P1} and Q_{PLi} is the same,

$$Q_{P1} = Q_{PLi} \tag{4.7.38}$$

Therefore Eq. 4.7.37 becomes

$$\begin{aligned} F_{P1} &= (M_{P1} \rightarrow Q_{P1}) \cdot \bar{Q}_{P1} \\ &= (\bar{M}_{P1} + Q_{P1}) \cdot \bar{Q}_{P1} \\ &= (\bar{M}_{P1} \cdot \bar{Q}_{P1}) + (Q_{P1} \cdot \bar{Q}_{P1}) \\ &= \bar{M}_{P1} \cdot \bar{Q}_{P1} \end{aligned} \tag{4.7.39}$$

If perturbation in the pump flow rate is allowed, $\bar{Q}_{P1}$ is true. Thus the truth value of F_{P1} is completely determined by $\bar{M}_{P1}$ or

$$F_{P1} = \bar{M}_{P1} \tag{4.7.40}$$

Lastly the storage tank is combined and the entire flow sheet of Fig. 4.4 is considered. The statement, F, of Eq. 4.7.1 with the help of Eq. 4.4.16 becomes

$$\begin{aligned} F &= [(Q_{HTi} \cdot Q_{HTO}) \rightarrow M_{HT}] \cdot F_{P1} \\ &= [(\overline{Q_{HTi} \cdot Q_{HTO}}) + M_{HT}] \cdot \bar{M}_{P1} \end{aligned} \tag{4.7.41}$$

Statements M_{HP} and M_{P1} are found to be identical, or

$$M_{HT} = M_{P1} \tag{4.7.42}$$

The statement F then becomes

$$F = [(\overline{Q_{HTi} \cdot Q_{HTO}}) + M_{HT}] \cdot \bar{M}_{HT}$$

$$= [(\overline{Q_{HTi} \cdot Q_{HTO}}) \cdot \overline{M}_{HT}] + [M_{HT} \cdot \overline{M}_{HT}]$$

$$= (\bar{Q}_{HTi} \cdot \overline{M}_{HT}) + (\bar{Q}_{HTO} \cdot \overline{M}_{HT}) \tag{4.7.43}$$

There are two ways by which F can be falsified:

(a) $\overline{M}_{HT}$ is falsified by putting a man at the storage tank who ensures the availability of material for all time, or

(b) one installs a flow controller between input and output stream, thus making $\bar{Q}_{HTi}$ and $\bar{Q}_{HTO}$ false permanently.

The first strategy is adopted when the raw material supply is intermittent whereas the second strategy is preferred when there is a continuous supply of the raw material. The final flow sheet for the control action (a) is given in Fig. 4.13.

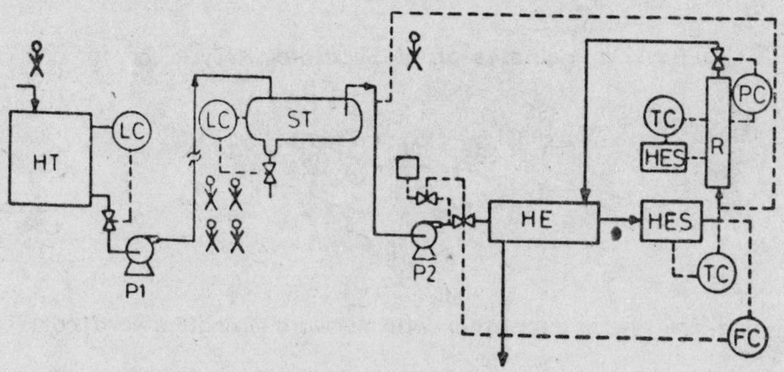

Fig. 4.13 Controller requirements for the entire flow sheet of Fig. 4.4

4.8 GENERALIZATION OF THE METHOD OF SYNTHESIS OF PROCESS FLOW SHEET

The method of obtaining the process flow sheet can now be generalized. First it is noted that the scanning is started from the end of the flow sheet and hierarchy assigned to equipment in the opposite direction to the process stream as shown in Fig. 4.14. In a single product flow sheet, there is a

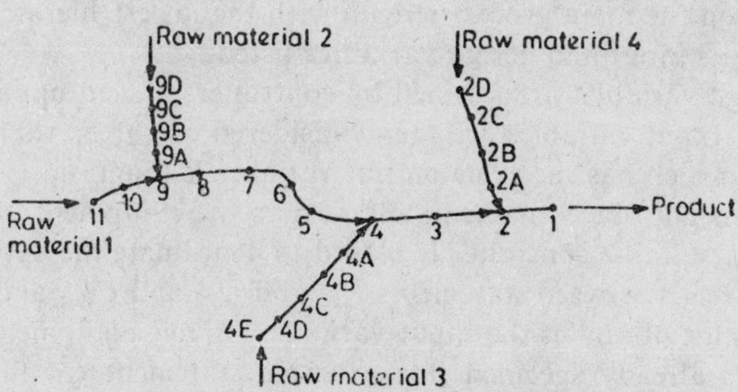

Fig. 4.14 Schematic representation of steady state flow sheet with one product and several raw materials

main process stream in which equipment is numbered in ascending order starting from the equipment where the final product is generated. The main process streams can branch at the separators and mixers and carry the number of the equipment along with it as shown. Two more special situations of recycle and forward mixing can arise as shown in Fig. 4.15.

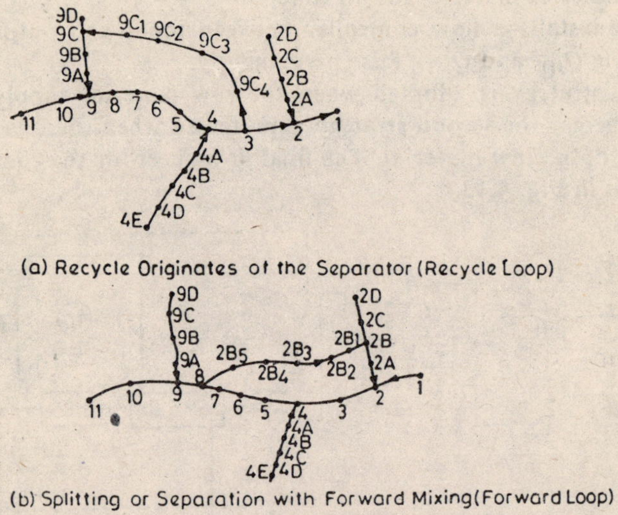

Fig. 4.15 Schematic representation of steady state flow sheet with recycle and forward mixing

Even in these cases the numbering is done in the opposite direction to the flow of the process stream. The hierarchy of the branch is once again decided by the equipment on the main stream.

The procedure of scanning in the opposite direction to the flow of the process stream is derived from the principle of theorem proving. The equipment on the main process stream with the lowest hierarchy is located and all its input and design variables listed.

The design variables are falsified by controllers placed upon the equipment itself. Input variables are then considered one after the other. The equipment which has the same output variable is found upstream. If this equipment is not placed immediately next to the equipment of the lowest hierarchy, a suitable controller is placed to coordinate the deviations between them (e.g., between statements $\bar{Q}_{Ri}$ and $\bar{Q}_{P2}$ and $\bar{C}_{WR}$ and $\bar{C}_{WST}$). If the truth value of any of the input variables of the equipment of higher hierarchy is already specified (for example statement $\bar{Q}_{II}$ for HE), the output variable must be controlled independently. These steps are given as a flow chart in Fig. 4.16.

The procedure of Fig. 4.16 has been applied to the steady flow sheet of Fig. 4.4 and the resultant process flow sheet given in Fig. 4.17. On comparison Fig. 4.13 and Fig. 4.17 are found to be identical.

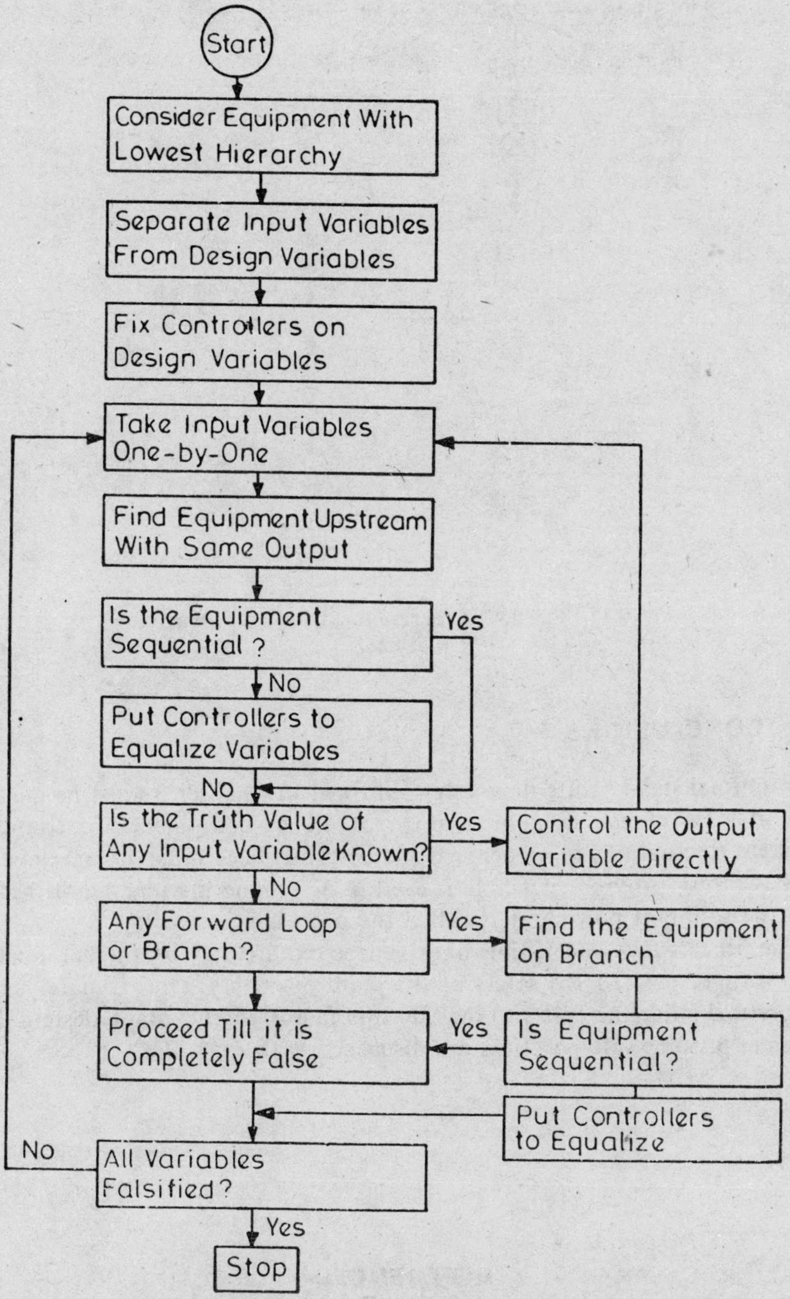

Fig. 4.16 Stepwise procedure for synthesis of process flow sheet

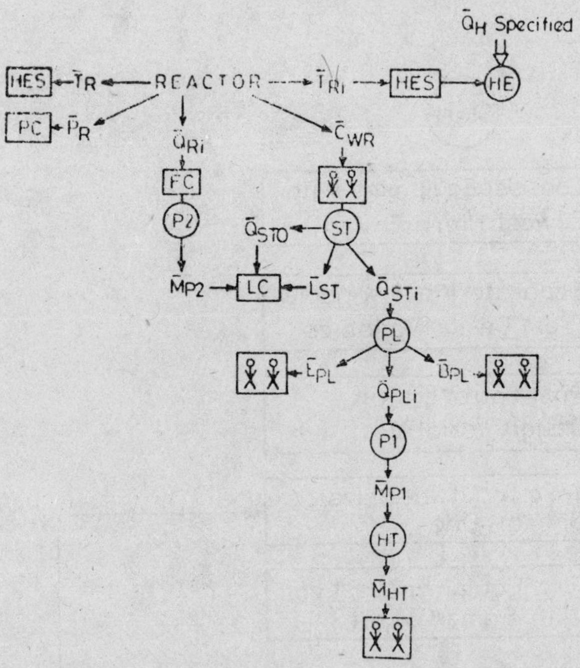

Fig. 4.17 The flow sheet resulting from the method of Fig. 4.16

4.9 CONCLUSIONS

The optimal steady state flow sheet obtained in Chapter 3 must be modified in view of the dynamic nature of all its equipment. To contain different fluctuations in a chemical plant, controllers must be employed. The theorem proving principle is used to determine the placing of these, and the resultant flow sheet is called the process flow sheet.

The process flow sheet, thus obtained, is required to be further modified with respect to the safety of the plant operation. Unfortunately, no theoretical guideline exists to include this factor in the plant design. In Chapter 5, some safe practices are discussed with examples.

REFERENCES

1. Kumar, A. and Gupta, S. K., *Fundamentals of Polymer Science and Engineering*, Tata McGraw-Hill, New Delhi, 1978.
2. Davis, J. J. and Kermode, R. I., *I. & E. C. Process Des. Dev.*, **14**, 459, 1975.

3. Lawley, H. G., *C.E.P.*, 70(4), 45, 1974.

4. Rudd, D. F. and Watson, C. C., *Strategy of Process Engineering*, 1st edn., Wiley, New York, 1968.

PROBLEMS

4.1 Prove the semantic relation of Eq. 4.4.15.

4.2 Consider the following equipment and write their semantic equations after determining their design variables:
(a) Distillation column
(b) CSTR
(c) Filter press
(d) Semibatch reactor
(e) Mixer
(f) Humidifier
(g) An oil fired furnace
(h) High pressure reactor
Also specify input and design variables for each case.

4.3 In the analysis of the dimerization plant of Fig. 4.4, a pump was assumed to give the desired flow rate if there was sufficient material. However, this is not quite correct and perhaps it would be apt to say that it would give the desired flow rate if there is sufficient material as well as proper input head. In view of this, modify the analysis and show the changes in the final flow sheet if any.

4.4 Using the algorithm of Fig. 4.16, synthesize the process flow sheet of the steady state flow sheet of Fig. 1.17.

4.5 Consider the steady state flow sheet of P. 1.16 and synthesize its process flow sheet.

4.6 Startup and shutdown of an equipment are very important operations and they are not simply switching the unit on or off. One should foresee problems arising during this, while designing the plant. Can you envisage the problem for the following equipment:
(a) reactor at 150°C
(b) pumps at the flow rate of 1000 l/min
(c) distillation columns at pressure 3 atm and 100°C
(d) heat exchangers at −30°C
(e) filter press for a 10% slurry at the total flow rate of 1000 l/h
(f) storage tank for petrol

4.7 Startup and shutdown of a plant requires coordination of activities of different equipment. Suggest the startup procedure of the dimerization plant discussed in this chapter.

4.8 Do the same as above for an ether plant.

4.9 It may be noted that during the startup, there exist deviations in variables (input as well as design) from steady state values and the basic goal of a startup procedure is to bring these to their s.s. values. Follow the same mathematical procedure given in this chapter and derive the strategies of plant startup. Please note that with every equipment, there is a finite delay associated.

4.10 The analysis of this chapter can also be extended to plant shutdown. Modify and derive similarly the strategies for this using the T.P. method.

5

Safety in Chemical Plant Design

5.1 INTRODUCTION

In Chapter 4, it was shown that a realistic plant design is the one that takes its dynamic nature into account. The process flow sheet arrived at in the previous chapter lays down a method of controlling the drift of variables from their steady state values. This analysis, however, is limited because the installed controllers are assumed to work with 100% probability of success.

In reality, controllers also have a finite probability of failure, which may lead to serious problems. To demonstrate this, one considers the flow sheet of Fig. 4.13 and analyzes a situation when the level indicator and controller on storage tank, HT, stop working and overfilling occurs. If unchecked, this would lead to spilling of hydrocarbons in adjacent areas. If the spilled material is not immediately disposed of, it will catch fire from accidental sparks propagating into a disastrous fire in the plant. Figure 4.13 does not account for this situation, even though the probability of such an event may be small. For similar reasons, spillage can also occur in the settling tank. The spillage can be avoided from these tanks if they are closed from the top. In this case, due to failure of level controllers, rise in level would cause excessive pressure build-up, ultimately leading to their rupture. Also because centrifugal pumps, P1 and P2, contain moving parts, they have finite probability of failures and it needs no imagination to see that the whole plant must be shut down when they fail.

The kinds of problems mentioned above are only a few of the several which are not accounted for by the analysis earlier. Since the goal of a chemical plant is not only to produce the chemical, but to do so safely, it is therefore part of the designing process to evolve a plant structure that can handle different problems that may arise from time to time. This situation amounts to a state of war, where lines of defence are drawn against all possible eventualities.

It is unfortunate that any systematic or theoretical guideline for safety does not exist. One has no choice but to analyze the case history of accidents in similar plants and base the design of a new one upon these. In the following a few safety considerations are given to familiarize readers with the dimension of the problem.

5.2 RELIABILITY OF EQUIPMENT

The problem of successful plant operation can be broken down as follows. It is important that the equipment used is reliable. Reliability is a difficult concept and its quantification will be attempted in Chapter 8, but it will suffice here to state that it is a measure of the inherent ability of the equipment in question to perform what it is expected to do. To illustrate this, a centrifugal pump is considered which has been guaranteed by the manufacturer to deliver 1000 l/min at a pressure of 5 atm. When this is placed in the plant where 1000 l/min of a liquid at 5 atm. is required, three possible situations can arise: firstly, it may not deliver the required flow rate because of some mechanical fault; secondly, it does so initially but later it develops some problem in its machinery; and lastly, it does not deliver the desired flow rate because of problems in equipment ahead of the pump in the plant. The first two situations measure the performance ability of the equipment and, consequently, it is a measure of reliability (or a measure of unreliability). Since this is an undesirable state of affairs, one likes to buy "reliable" equipment, i.e., equipment which is rated to perform for a long time "without developing troubles". It is only in this context that reputed manufacturers are preferred to non-reputed ones and one chooses to buy equipment from them even if it costs more.

There are several factors which affect the performance of equipment in an installation. The basis on which a new equipment is purchased are several and are given in Table 5.1. Any machinery which does not

Table 5.1[1] Some Factors Affecting the Decision of Purchasing an Equipment

A.	Single train application
B.	Unique or unprecedented application
C.	New and unproven design
D.	Proven design; little in-house experience
E.	A major scale-up of past design or experience
F.	Prior unsatisfactory performance
G.	Exceeds capacities of existing fabricating or handling facilities
H.	Requires scarce or unavailable skills
I.	Replacement parts and materials expensive or not readily available

have any standby is called a single train application, and if it has a higher chance of failure, it is necessary that it gets extra maintenance. Similarly, if an equipment has been used for the first time in a totally new situation (B, C and E of Table 5.1), it is once again necessary to watch its performance closely. If replacements are expensive or require repairs from specially trained people (H and I of Table 5.1), once again they are handled with extra caution to minimize operating costs. These considerations must be accounted for before the purchase is made. It may be appreciated that the cost of an equipment includes not only the money spent in buying it, but also the expenditure incurred on maintaining it.

5.3 PREVENTION OF ACCIDENTS

The second aspect of attaining successful plant operation is to prevent accidents. Chapter 6 elaborates on this subject at length, but for the discussion here, it would be sufficient to classify it broadly as minor and major. Minor accidents are sometimes called trouble shooting whereas major accidents in which there is loss of property as well as life would be called hazards. It is sometimes likely that trouble shooting conditions propagate into hazards; however, at present the distinction between the two is maintained.

By the very definition of the word, accidents are something which cannot be foreseen. However, in the limited context of chemical plant operation, they can be predicted with a favourable probability, and herein the experience of a designer plays a major role. Every plant accident that occurs must always be analyzed very thoroughly in terms of the reason of its occurrence. After having detected the cause, one alters the standards of the plant operation so that similar conditions may never occur to cause the same kind of disaster.[3] The occurrence of accidents can also be reduced with a better understanding of chemicals and equipment handled. With this understanding, the dangerous situations are handled more judiciously preventing this fault from developing into a major accident. In the following section, the nature of chemicals handled and equipment performance in dynamic conditions are given in brief, so that readers can appreciate their role in plant design.

5.4 FLAMMABILITY OF CHEMICALS

Chemicals used in plant operation can undergo a reaction under suitable conditions. The flammability is the property which measures their ability of reacting with oxygen in the air. In other words, flammability is a measure of the readiness with which chemicals can be oxidized.

Flammability of a chemical can be quantified by flash point and ignition points.[3] *Flash point* is the temperature at which a significant amount of vapour is generated in air. These vapours can burn but will not give rise to continuous burning when ignited by an outside agency. *Ignition point*, as opposed to this, is the temperature at which the chemical will continue to burn even without the aid of an outside ignition source. This ultimately leads to uncontrolled fire and explosion.

Flash point is only an indicative but not a complete measure of the fire hazard and explosiveness of a flammable liquid. A liquid will cause an explosion only when its vapour is mixed with air in a certain range of ratio. When the concentration of vapour is very low, the vapour will burn but the heat generated due to the combustion is not large enough to either maintain or propagate the flame. Similarly, when the vapour is present in large concentration, it tends to exclude oxygen, as a result of which the combustion does not produce enough heat for the maintenance of the flame. This range of flammability concentration is very specific to the particular vapour and is determined experimentally for it. Table 5.2 gives this concentration range for some systems.[4-6] It may be realized that if the flame remains localized, it would not cause a major disaster, and it is only in this context, that the speed of propagation of the flame front becomes extremely important. As a matter of fact, flames are classified as combustion and detonation waves depending upon their speeds: in the former, the flame front moves at a speed less than the velocity of sound and in the latter the front travels at the velocity of sound. Detonation flames are more dangerous as it is more difficult to control and causes more damage because of associated shock waves. A good example of a combustion flame would be an oxy-acetylene torch where oxygen and acetylene gases are passed at controlled rates and a steady flame is obtained at the torch head. When the gas mixture is throttled to the flashback point, the combustion wave starts travelling backward into the duct of the torch and develops into a detonation flame. It is only because of the flame arrester at the torch, that the flame gets extinguished. Further detail on this subject can be had in a more advanced text.[4] However, it might suffice here to say that an uncontrolled combustion flame has every likelihood of converting to a detonation flame.[7,8]

Table 5.2[4-6] Limits of Flammability of Gases and Vapour with Air Determined at 1 atm and Room Temperature

Compounds	Limits of flammability, per cent by volume	
	Lower	Upper
Paraffin hydrocarbons		
Methane	5.3	15.0
Ethane	3.0	12.5
Propane	2.2	9.5
Butane	1.9	8.5
Isobutane	1.8	8.4

Table 5.2 (*Contd.*)

Compounds	Limits of flammability, per cent by volume	
	Lower	Upper
Pentane	1.5	7.8
Isopentane	1.4	7.6
2,2-dimethyl propane	1.4	7.5
Hexane	1.2	7.5
Heptane	1.2	6.7
2,3-dimethyl pentane	1.1	6.7
Octane	1.0	6.0
Nonane	0.8	—
Decane	0.8	5.4
Olefins		
Ethylene	3.1	32.0
Propylene	2.4	10.3
Butene-1	1.6	9.3
Butene-2	1.8	9.7
Amylene	1.5	8.7
Acetylene	2.5	80.0
Aromatics		
Benzene	1.4	7.1
Toluene	1.4	6.7
o-Xylene	1.0	6.0
Cyclic hydrocarbons		
Cyclopropane	2.4	10.4
Cyclohexane	1.3	8.0
Methylcyclohexane	1.2	—
Alcohols		
Methyl alohol	7.3	36.0
Ethyl alcohol	0.3	19.0
Allyl alcohol	2.5	18.0
n-Propyl alcohol	2.1	13.5
Isopropyl alcohol	2.0	12.0
n-Butyl alcohol	1.4	11.2
Amyl alcohol	1.2	—
Isoamyl alcohol	1.2	—
Aldehydes		
Formaldehyde	7	73.0
Acetaldehyde	4.1	57.0
Crotonic aldehyde	2.1	15.5
Furfural	2.1	—
Ethers		
Diethyl ether	1.9	48.0
Divinyl ether	1.7	27.0
Ketones		
Acetone	3.0	13.0
Methyl ethyl ketone	1.8	10.0
Methyl propyl ketone	1.5	8.0
Methyl butyl ketone	1.3	8.0

Table 5.2 (*Contd.*)

Compounds	Limits of flammability, per cent by volume	
	Lower	Upper
Acids		
Acetic acid	5.4	—
Hydrocyanic acid	5.6	40.0
Esters		
Methyl formate	5.9	22.0
Ethyl formate	2.7	16.4
Methyl acetate	3.1	16.0
Ethyl acetate	2.5	9.0
Propyl acetate	2.0	8.0
Isopropyl acetate	1.8	8.0
Butyl acetate	1.7	7.6
Amyl acetate	1.1	—
Inorganic		
Hydrogen	4.0	75.0
Carbon monoxide + water vapour at 18°C	12.5	74.0
Ammonia	15.0	28.0
Cyanogen	6.0	32.0
Oxides		
Ethylene oxide	3.0	80.0
Propylene oxide	2.0	22.0
Dioxane	2.0	22.0
Sulphides		
Carbon disulphide	1.2	[illegible]
Hydrogen sulphide	4.3	[illegible]
Carbon oxysulphide	12.0	[illegible]
Fuels		
Jet fuel, JP-3	0.8	6.5
Kerosene	0.6	5.6
Petrol	1.3	6.0–7.6
Coal gas	5.3	32.0
Water gas	7.0	72.0
Naphtha, VM and P	0.9	6.0
Petroleum ether	1.4	5.9
Chlorides		
Methyl chloride	10.7	17.4
Ethyl chloride	3.8	14.8
Propyl chloride	2.6	11.1
Butyl chloride	1.8	10.1
Isobutyl chloride	2.0	8.8
Allyl chloride	3.3	11.1
Amyl chloride	1.6	8.6
Vinyl chloride	4.0	22.0
Ethylene dichloride	6.2	16.0
Propylene dichloride	3.4	14.5

Table 5.2 (*Contd.*)

Compounds	Limits of flammability, per cent by volume	
	Lower	Upper
Bromides		
Methyl bromide	13.5	14.5
Ethyl bromide	6.7	11.3
Allyl bromide	4.4	7.3
Amines		
Methyl amine	4.9	20.7
Ethyl amine	3.5	14.0
Dimethyl amine	2.8	14.4
Propyl amine	2.0	10.4
Diethyl amine	1.8	10.1
Trimethyl amine	2.0	11.6
Triethyl amine	1.2	8.0

The flammability property of vapours and gases is extremely important and in several instances governs the ultimate plant design to a large extent. A good example of this is the design of the reactor for oxidation of methanol to formaldehyde using iron-molybdenum oxide as the catalyst. From stoichiometry, one mole of pure oxygen is required for every two moles of methanol vapour to be oxidized. Since air is used industrially instead of pure oxygen, one would require about five moles of air for the reaction (air consists of about 20% oxygen). The mole fraction (same as volume fraction) of methanol in the feed to the reactor would be then 2/7 or about 29%. On checking Table 5.2, methanol is found to have lower and upper explosive limits as 7.3 and 36% and a single reactor with the feed at stoichiometric ratio would lead to explosion. In view of this, the flow sheet of the reaction section is modified by having a sequence of reactors in such a way that, at no moment, the concentration of methanol in air exceeds 7% by volume. Figure 5.1 shows a train of six tubular reac-

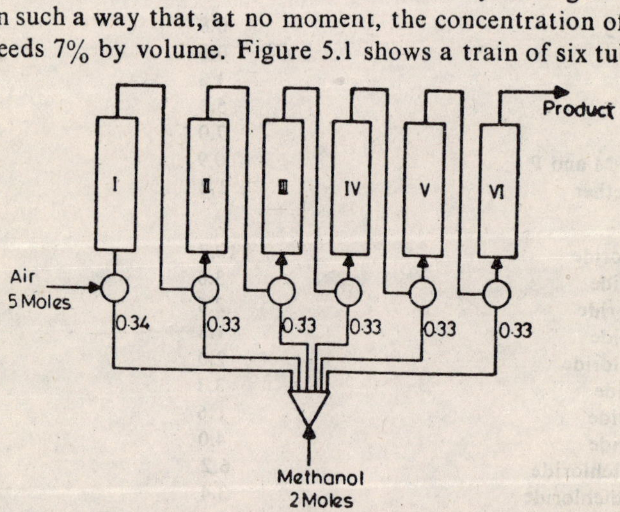

Fig. 5.1 Suggested flow sheet for oxidation of methanol to formaldehyde

tors in which the feed of one is the product of the previous reactor (except in the case of the first reactor) and methanol is added uniformly in all these reactors. In the calculation of 0.33 moles of methanol per reactor, it has been assumed that all methanol fed in a reactor gets completely consumed in it.

5.5 SAFETY CONSIDERATIONS IN PLANT LAYOUT

Plant layout, to a large extent, is decided by safety considerations. To give an example, the production of acetylene is considered. The gas is prepared by a chemical reaction between calcium carbide and water according to

$$CaC_2 + H_2O \rightarrow Ca(OH)_2 + CH \equiv CH \tag{5.5.1}$$

This reaction is instantaneous and goes to complete conversion. Since the gas does not dissolve in water, it is liberated which can be filled in cylinders for further distribution. The steady state flow sheet would consist of a storage space for calcium carbide, a means of transporting the raw material to the reactor where it is mixed with water. The gas liberated is sent to the distribution system. The plant installed based upon this flow sheet will not work because of the explosive nature of calcium carbide and acetylene. In this section it is shown that the plant can be made to work provided its layout is improved by including a suitable protection system.

The protection system mentioned above depends significantly upon the nature of chemicals handled. In the case of an acetylene plant, calcium carbide is discussed first. It is a stable solid in dry air. With moisture, it reacts instantaneously at the ambient conditions with liberation of acetylene. Calcium carbide is manufactured by a reaction between lime and coke in an electric furnace and is normally supplied in closed metal drums at the factory site.[8] As long as these drums are closed, there is no danger of fire provided it is not exposed to excessive heating. As a result, these drums must be stored in a separate warehouse which should be well ventilated. The warehouse should also be provided with fire-fighting arrangements so as to contain fire if at all it occurs. As water cannot be used, extinguishers containing either carbon dioxide or carbon tetrachloride are used. Analyses of previous accidents in acetylene plants have shown that the occurrence of fire can be minimized if a large quantity of calcium carbide in drums is not stored at *one* place. This is because small amount of the raw material at one place will cause only a small fire reducing chances of spreading in large areas. Thus, instead of one big warehouse, several small ones separated by a safe distance is a good idea. The second good practice would be to open one drum at a time and use its contents completely as soon as it is opened. The reason for doing this is that calcium carbide open to atmosphere can absorb moisture to form acetylene, which in turn can catch fire. Lastly, the point to remember is that all fires start with a source of ignition, which could be either an open flame or electric sparks. This implies that both should be avoided at all costs. The former is easy to control whereas the latter can be avoided by using special electrical wiring systems in which sparking is either minimal or covered.

The acetylene reactor is discussed next, where solid calcium carbide is contacted with water. This can be done in two ways. One is to maintain an excess of water to which calcium carbide solid is slowly added. Another method would be to add a limited amount of water to the carbide solid already charged into the reactor. Since the reaction between calcium carbide and water is highly exothermic, the first method is preferred because the excess water tends to dissipate the heat generated. The latter method of contacting has the disadvantage of causing localized heating. In view of this, the suggested reactor for acetylene generation is given in Fig. 5.2. The level of water in the tank is controlled by a float. Calcium carbide is charged into the hopper and the feed of solid is

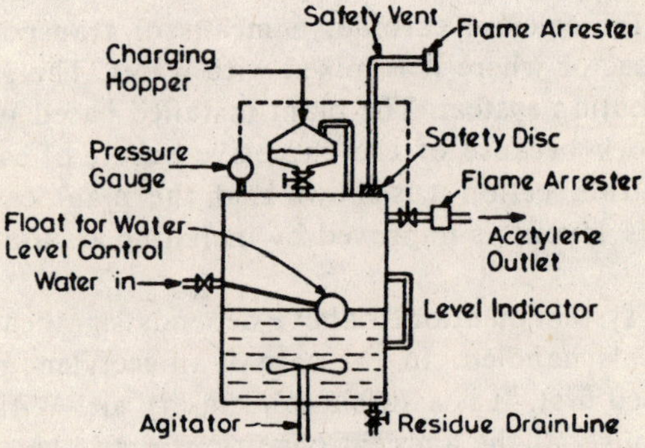

Fig. 5.2 Proposed reactor for generation of acetylene

gravitational because of the pressure equalizing line as shown. The withdrawal of acetylene gas is done at a rate such that the pressure in the tank remains constant. All the outlets of the reactor are equipped with flame arresters. For safeguarding against excessive pressure development, the reactor is provided with a rupture disc. When this occurs, the gas is released through the vent-line to the atmosphere. The vent-line is also provided with a flame arrester as shown in the figure. Lastly an agitator is provided in the reactor to maintain a well-mixed condition within.

The gas distribution system for an acetylene plant is shown in Fig. 5.3. The main supply line is branched to provide facility of filling at several ports as shown in the figure. Each branch can be isolated through a valve along with a special type of flame arrester, called flash arrester. This is required because of its special property as explained below.

Table 5.2 gives flammability limits of acetylene as 2.5% to 80% by

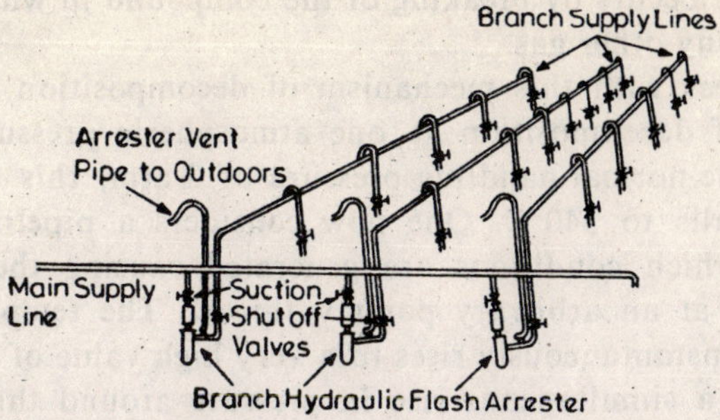

Fig. 5.3 Distribution system for acetylene. (Reprinted from Ref. 6 with the permission of Factory Mutual Engineering Corporation, Massachusetts.)

volume with air, which is a very broad range. This implies that air should be excluded at any cost. Not only this, it can also decompose spontaneously at high temperatures into its elements liberating large amounts of heat as follows:

$$CH \equiv CH \rightarrow 2C + H_2 \tag{5.5.2}$$

This decomposition is very fast and involves almost the whole body of the gas. According to the U.S. Bureau of Mines, it occurs at 635°C and with the increase in gas pressure, this flash temperature is reduced further. This reaction is attributed to the instability of the compound and when hydrocarbons are added to the gas, they are found to have a stabilizing effect causing the postponement of this explosive decomposition. This is shown in Table 5.3 in which the safe operating pressure is given against

Table 5.3[6] Amount of Various Hydrocarbons Required to Prevent Explosive Decomposition of Acetylene

Initial pressure atm (kPa)	Hydrocarbon in per cent by volume		
	Butane	Propane	Natural gas
1.02 (103)	8.4	9.2	13.8
1.70 (172)	11.0	13.0	21.0
3.40 (344)	17.0	20.0	31.0
5.10 (517)	19.0	24.0	36.0
6.80 (690)	21.0	26.0	39.0

(Reprinted from reference 6 with the permission of Factory Mutual Engineering Corporation, Massachusetts)

the amount of hydrocarbon added. By safe pressure, it is meant that at this pressure the spontaneous decomposition does not occur at any temperature. The decomposition according to Eq. 5.5.2 should be differentiated from combustion (which can on occasion be explosive) in the sense that the former occurs by breaking of the compound in which it requires no oxygen or any other gas.

Sargent agrees with this mechanism of decomposition but states the temperature of decomposition at one atmospheric pressure as 800°C.[10] However, at the normal handling pressure of 2 atm, this decomposition temperature falls to 540°C. One now considers a pipeline system for acetylene in which conditions are generated causing the spontaneous decomposition at an arbitrary point within it. The temperature at this point almost instantaneously rises to a very high value of about 3000°C. Also, there is a simultaneous rise in pressure around this region, thus further lowering the decomposition temperature. This would cause inward travel of waves. Depending upon the speed of travel it is either called deflagration flame or detonation flame.[10,11] In the former, the speed of propagation is less than the velocity of sound whereas in the latter, more than the velocity of sound. This causes production of shock waves and the static pressure shoots up to a very high value.

Due to devastations caused by detonation flames in acetylene pipelines, it is desirable to investigate factors responsible for its formation. It is found that deflagration flames are developed first which subsequently convert to detonation flames as it travels further down in the pipe. The distance between the point of formation and point of conversion is called the predetonation distance and it depends upon several factors including the pipe diameter. Experiments reveal that, for a given pipe diameter, there is a minimum acetylene pressure below which a deflagration flame cannot convert into detonation flame. Table 5.4 lists the minimum gas

Table 5.4[10] Minimum Pressure Levels Required for Acetylene Detonation in Various Pipes

Pipe diameter, cm	Minimum pressure, atm
30.48	0.82
20.32	1.02
10.16	1.50
5.08	2.18
2.54	3.27
1.27	4.90

(Reprinted from reference 10 with the permission of American Institute of Chemical Engineers, New York.)

pressure against the pipe diameter. Acetylene pipeline design is therefore governed by safety considerations. If the normal gas pressure in the pipeline is expected to be around 2 atm, the pipe diameter should lie anywhere between 1 and 2 inches and this value need not be the optimal diameter as given in Perry's Handbook[12]. If the distance to be traversed as well as the total flow rate is large, one is left with the following options: use several pipelines or use one pipeline with several intermediate booster pumps or use both.

Once the pressure is more than the minimum value for a given pipe diameter, it is found that detonation occurs and the predetonation distance is a function of initial acetylene pressure and independent of pipe diameter. Figure 5.4 gives the predetonation distance as a function of initial gas pressure and the value lies in a range as shown. It is also of great interest to know what final pressure would develop as a result of this detonation. For this the predetonation distance is read from Fig. 5.4; and from the length of the pipe the ratio of these distances is calculated. Then Fig. 5.5

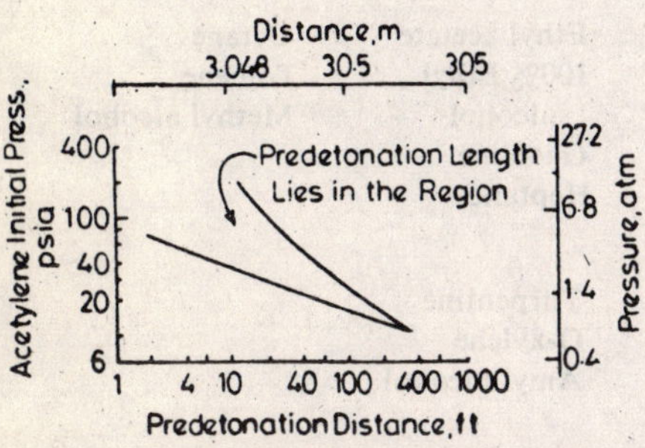

Fig. 5.4 Dependence of predetonation length on the initial pressure of acetylene.[10] (Reprinted from Chemical Engineering with the permission of McGraw-Hill, New York.)

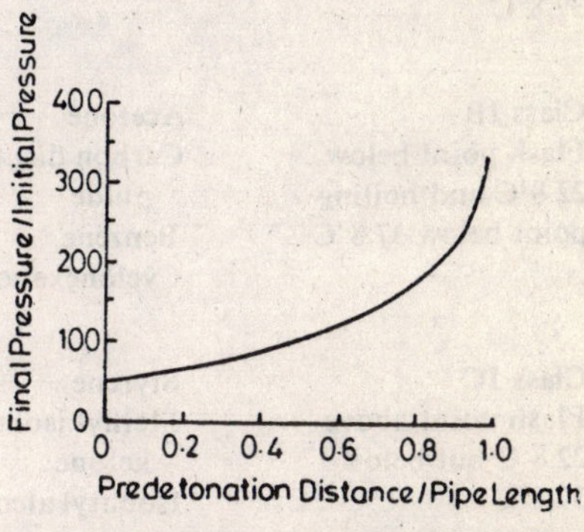

Fig. 5.5 Relation of final acetylene pressure with predetonation distance. (Reprinted from Ref. 11 with permission of the American Institute of Chemical Engineers, New York.)

is used to determine the ratio of the final pressure and the initial pressure. From the figure it is seen that the pressure increase due to the detonation is highest when the predetonation distance compares with the pipe length. It is important, therefore, for the best plant design, to be equipped against this eventuality which means that in addition to having a small pipe diameter (Table 5.4), one should have pipes thick enough to contain the detonation, if it occurs.[11]

5.6 CLASSIFICATION OF CHEMICALS AND HANDLING PROBLEMS

The example given in Sec. 5.5 also indicates that a good plant design requires a detailed understanding of different chemicals handled during the manufacturing process. The study of combustion and flame propagation for gases and liquids is a science by itself and is normally time consuming. For convenience, liquids have been roughly divided into four classes (I, II, IIIA and IIIB) according to their flash points and boiling points as given in Table 5.5. This division is not rigorously correct, but for almost all practical cases, this is good enough. For example, from the point of view of fire-fighting, fires resulting from liquids of the first group is devastating. Liquids from second groups can be easily ignited but they are less volatile, and liquids of the third group are safe to handle.

Table 5.5[6] Classification of Liquids

Classification	Representative flammable liquids		
Class IA Flash point below 22°C and boiling point below 37.8°C	Acetaldehyde Ethyl ether Ethyl chloride	Isoprene Methyl formate Pentane	Propylene oxide
Class IB Flash point below 22.8°C and boiling point below 37.8°C	Acetone Carbon disulphide Benzene Cyclohexane	Ethyl acetate 100% Ethyl alcohol Gasoline Heptane	Octane Toluene Methyl alcohol
Class IC Flash point above 22.8°C but below 37.8°C	Styrene Methyl isobutyl ketone Isobutyl alcohol	Turpentine O-xylene Amyl alcohol	
Class II Flash point above 37.8°C but below 60°C	No. 1 to 3 Fuel oils Kerosene n-Decane Hexyl alcohol	Glacial acetic acid Stoddard solvent	
Class IIIA Flash point above 60°C but below 93.4°C	Aniline Benzaldehyde Butyl cellosolve	Nitrobenzene Pine oil	
Class IIIB Flash point above 93.4°C	Animal oils Ethylene glycol Glycerine	Lubricating oil Quenching oil Transformer oil Triethanol amine	Benzyl alcohol Vegetable oil

(Reprinted from reference 6 with the permission of Factory Mutual Engineering Corporation, Massachusetts.)

As pointed out earlier, good practices tend to minimize hazards. General safeguards in handling flammable liquids are now discussed which are essential for safe plant operation. These are given in Table 5.6, and it can be seen that these recommendations are more on the conservative side. In Sec. 5.7 venting of tanks and vessels is discussed.

Table 5.6[6,12–14] General Consideration in Handling Flammable Liquids

1. Eliminate source of ignition; if not possible efforts should be directed towards isolating them.
2. Ventilate to prevent the formation of explosive mixture with air. If natural ventilation is not sufficient, the use of an exhaust is recommended.
3. Provide adequate fire protection.
4. Confine the liquid in closed containers, which could either be drums or storage tanks.
5. Isolate hazardous liquids with non-flammable walls. Outdoor location preferred.
6. Use rupture discs (or a relief valve) against excessive pressure build-up. The release through the rupture disc is discharged through a vent line to some distant place.
7. Break down the total task into smaller steps and follow them religiously.
8. Minimize chances of accidental spillage.
9. Instal a salvage tank to which contents of the plant can be drained in the event of emergency.
10. Do not repair any equipment on the plant site. It should be removed to the repair shop. Even the tightening of bolts is not recommended when the equipment is working in the plant.

5.7 VENTING OF TANKS AND VESSELS

Proper ventilation of the plant area as well as the storage area is extremely important. When one considers the storing of a liquid having a low flash point and if the ambient temperature is known, the rate at which the vapour is generated can be found out. The velocity of the air due to ventilation should be such that at the point of vapour generation, the composition of the air and vapour mixture is less than the lower explosive limits of the chemical. This situation can be easily modelled to calculate the velocity of the necessary draft of air. If this speed is more than that for natural ventilation, one instals a suitable exhaust system.

The situation discussed above arises when a liquid is stored in an open vessel. There is another very common situation which occurs in a chemical plant operation where proper venting is extremely necessary. In storage tanks, the process fluid (or raw material) is pumped in and out which need not occur at the same time. Two extreme situations are now considered: first, when the outflow is zero and the liquid is pumped in and second, when inflow is zero and the liquid is pumped out. In the former, the tank is pressurized and in the latter, there is a generation of vacuum. In addition to generating extra stresses on the tank, there will be great inconvenience when pumping in or pumping out the material because it has to work under a varying delivery head or suction pressure. As a consequence of this, breathing vents are provided on tanks, through which air can get in or leave to maintain a constant delivery head. It is of great interest as to how these must be designed if the maximum input or output flow rate of the tank are known. The recommended practice is to provide outbreathing vent capacity of 1 m^3/h of free air for each 0.47 m^3/h of the maximum tank filling rate for Class I liquids and 1 m^3/h of free air for each 0.94 m^3/h for Class II and III liquids. Similarly 1 m^3/h of inbreathing vent capacity of free air is provided for each 1 m^3/h of the maximum emptying rate for *all liquids*. In addition to this venting capacity, thermal venting capacity is provided. This depends upon a number of factors and recommendations of the American Petroleum Institute given in Table 5.7. From the table it is seen that the thermal venting capacity under pressure provided to liquids of high flash point is much lower compared to those of low flash points.

Table 5.7[6] Required Thermal Venting Capacity in m^3/h of Air

Tank capacity	Vacuum venting for all stock	Pressure venting, m^3/h	
m^3	m^3/h	Class I liquid	Class II and III liquid
160	28	30	17
320	57	60	34
480	85	90	51
640	113	120	68
800	142	150	85
1,600	280	300	170
2,400	420	450	255
3,200	570	600	340
4,000	680	720	420
4,800	790	840	480
5,600	880	930	540
6,400	960	1,020	590
7,200	1,050	1,110	650
8,000	1,130	1,200	680
9,600	1,250	1,320	760
11,200	1,360	1,440	820
12,800	1,470	1,560	880
14,400	1,590	1,680	960
16,000	1,700	1,800	1,020
19,200	1,930	2,040	1,160
22,400	2,120	2,250	1,270
25,600	2,320	2,460	1,420
28,800	2,550	2,700	1,530

(Reprinted from reference 6 with the permission of Factory Mutual Engineering Corporation, Massachusetts.)

In addition to this, all tanks are required to instal emergency relief valves. These are spring-loaded valves which open to release excessive vapours whenever there is a build-up of internal pressure. Recommendations for tanks are given in Table 5.8. It may be noted that the emer-

Table 5.8[6] Required Venting Capacity for Emergency Relief of Excessive Internal Pressure in Above Ground Storage Tanks

Wetted area of tank, m²	Total pressure-relief capacity at 100 kPa absolute and 157°C, m³/h
1.9	597
2.8	894
3.7	1,191
4.6	1,491
5.6	1,789
6.5	2,086
7.4	2,383
8.4	2,683
9.3	2,970
11.2	3,570
13.0	4,160
14.9	4,750
16.7	5,380
18.6	5,970
23.2	6,760
27.9	7,500
32.5	8,150
37.2	8,830
46.4	10,020
55.7	11,090
65.0	12,110
74.3	13,070
83.6	13,950
92.9	14,830
112	15,760
130	16,610
149	17,380
167	18,080
186	18,730
223	19,920
260	21,000

1. The wetted area of the tank is determined on the basis of 55% of the total exposed area of a sphere or spheroid, 75% of the total exposed area of a horizontal tank and the first 10 m above ground of the exposed shell area of a vertical tank.
2. For tanks operating at pressure less than 7 kPa and having exposed area exceeding 260 m², complete fire involvement is unlikely and overheating will probably cause loss of metal strength in the vapour space before the development of a maximum vapour-evolution rate. For such tanks, the maximum listed capacity is adequate.

(Reprinted from reference 6 with the permission of Factory Mutual Engineering Corporation, Massachusetts.)

gency vent is separately installed from the free vent. The free vent capacity is a sum of thermal venting capacity and that due to pumping of fluid in and out of the tank. As an example, consider an acetone storage tank of volume 380 m^3 (9.14 m dia and 5.8 m high) with 22.7 m^3/h of maximum pumping in rate (maximum delivery capacity of the pump) and 0.68 m^3/h of maximum outflow rate.[6] As acetone is a Class I liquid, the free venting requirement under pressure would be (27.7/0.47) or 48 m^3/h of air and under vacuum would be 68 m^3/h of air. Thermal venting for the tank for a tank of volume 480 m^3 is found to be 85 m^3/h of air and is used for this tank because interpolation is not recommended. The emergency venting capacity is found after calculating the shell area as 166.5 m^2 and from Table 5.8, it is found to be 18080 m^3/h. With this discussion, one now goes on to the design of relief valves in Sec. 5.8.

5.8 DESIGN OF SAFETY VALVES

Equipment like heat exchangers, distillation columns, reactors, steam boilers, etc. have shell structures and are treated as pressure vessels and to them, codes for mechanical design of unfired pressure vessels apply verbatim.[15-17] Since in this equipment, one has liquids as well as vapours, according to the codes, emergency safety valves are installed for liquids as well as vapours. According to the recommendations, the capacity of the safety valve for the liquid side is equal to the maximum input flow rate of liquid into the equipment and the capacity of the vapour relief valve is equal to the sum of maximum vapour input (including air and/or steam, if any) and the vapour generated in the equipment during its operation.

The use of relief valves is extremely important as it safeguards the equipment from hazards of explosion. According to codes of unfired pressure vessels, these valves should be set such that they start opening at 75% of the set pressure of the valve. Set pressure is the pressure at which the valve is meant to work and this should be lower than the maximum allowable pressure in the vessel. For the selection of proper valves, manufacturers are required to supply the following information on its body[16]:

(1) manufacturers' identification;
(2) nominal inlet and outlet size;
(3) set pressure and the normal temperature of operation, and
(4) certified capacity in kilograms of saturated steam per hour.

In the case of liquids, the certified capacity is expressed in litres of water per minute.

If the process stream is other than saturated steam or water, one suitably converts the rated capacity for the specific fluid to check if it is going to serve the purpose. For vapour relief valves, the following relation is used:

$$W = CKAp\sqrt{\frac{M}{T}} \tag{5.8.1}$$

where

W = rated capacity in kg/h, (5.8.2a)

C = a constant for gas or vapour which is a function of ratio of specific heats, γ (5.8.2b)

A = actual discharge area in mm²,

p = accumulation pressure which is equal to 1.10 times the set pressure of the valve in kgf/cm² (5.8.2c)

M = molecular weight of the vapour or gas (5.8.2d)

T = inlet temperature in °K (5.8.2e)

K = discharge coefficient, which depends upon the construction of the valve (5.8.2f)

The values of C for different gases and vapours are given in Table 5.9 and the procedure of calculation is as follows. From the specification of the rated capacity for saturated steam, the group of constant KA is calculated in Eq. 5.8.1. After substituting values of C, M and T for the process stream and using the calculated value of KA, the desired rated capacity is found.

Table 5.9[16] Constant C for Relief Valves as a Function of Specific Heat Ratio

γ	C	γ	C	γ	C
1.00	2.34	1.26	2.55	1.52	2.72
1.02	2.37	1.28	2.57	1.54	2.74
1.04	2.38	1.30	2.58	1.56	2.75
1.05	2.40	1.32	2.60	1.58	2.76
1.08	2.42	1.34	2.61	1.60	2.77
1.10	2.44	1.36	2.63	1.62	2.78
1.12	2.45	1.38	2.64	1.64	2.80
1.14	2.46	1.40	2.65	1.66	2.81
1.16	2.48	1.42	2.66	1.68	2.82
1.18	2.50	1.44	2.67	1.70	2.83
1.20	2.51	1.46	2.68	2.00	2.98
1.22	2.52	1.48	2.70	2.22	3.07
1.24	2.54	1.50	2.71		

(Reprinted from reference 16 with the permission of American Society of Mechanical Engineers, New York.)

Similar corrections are made for liquid relief valves using the following equation[16]

$$Q = 50.42\,KA\rho p \tag{5.8.3}$$

where

$$Q = \text{flow rate in m}^3/\text{min} \tag{5.8.4a}$$

$$p = \text{pressure drop in kgf/cm}^2 \tag{5.8.4b}$$

$$\rho = \text{specific gravity at inlet temperature} \tag{5.8.4c}$$

K and A in Eq. 5.8.3 have the same meaning as in Eq. 5.8.1. The group of constant KA is calculated as earlier from the relief capacity given by the manufacturer for water and from this information, the relief capacity of the valve is found for any other fluid.

5.9 SAFETY CONSIDERATIONS IN REACTOR DESIGN

In regular kinetic courses, a tubular reactor is shown as a mere tube and a stirred tank reactor as a vessel with a stirrer. In practice, however, they are much more complex in appearance as shown in Fig. 5.6.

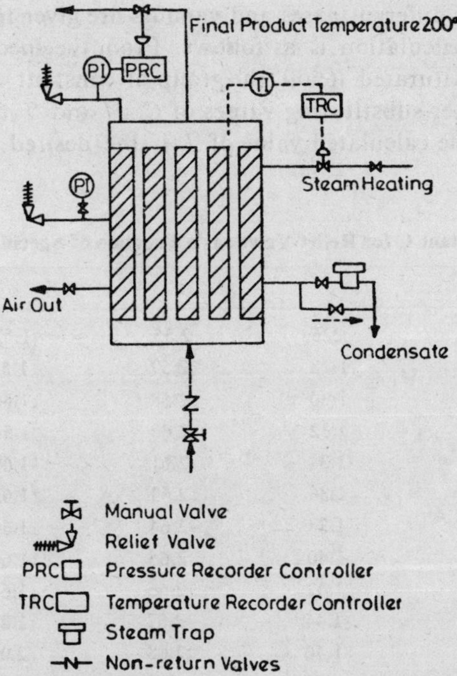

Fig. 5.6 Accessories of a safe reactor

If the desired temperature is 200°C, the reactor is nothing but a heat exchanger, with reaction occurring in the tube side with a suitable heating fluid in shell. If this temperature is to be obtained through steam heating, the arrangement of the reactor shown in Fig. 5.6 is justified as follows. Steam is passed through a pneumatically controlled valve, actuated by the temperature-recorder-controller, TRC. A manually controlled valve is placed beforehand so that the steam can be cut off during an emergency if it is desired to isolate the reactor completely. An emergency arises whenever a runaway reaction condition or failure of any of these controllers occurs. Steam is withdrawn as condensate through a steam trap at the bottom of the reactor. With every steam trap, it is almost necessary to instal a parallel bypass line through manual valves which is put to use when the temperature-recorder-controller (TRC) fails to open the valve on the steam input line completely. This situation would require opening of the bypass line to withdraw the extra steam in the exchanger. The bypass line is utilized also when the strainer of the steam trap gets choked and becomes inoperative. The 'air-out' line on the steam-side is necessary during the startup and shutdown process. Before steam is passed to heat up the reactor, it is necessary to displace the entire air of the shell. Since the steam trap line will not permit the escape of any vapour through it, the air-out line is utilized for this task. This task can also be achieved through a steam trap bypass line provided it is open to atmosphere; but this need not be so, as the condensate drain line is a long pipe dipping in the condensate collector. During the shutdown process, if the air-out line had not been there, it would produce a condition of vacuum in the shell side when steam flow is closed and the reactor cools down. In accordance with the code of pressure vessel design, a shell side relief valve is installed as shown so that the excessive pressure can be relieved through it when all other outlets get clogged, or do not open when desired. Now consider the tube side of the exchanger reactor. The pressure controller is placed very close to the exit stream and the relief valve installed in the same control loop. A manual valve and a non-return valve are necessary at the inlet for the following reason. Supposing an excessive pressure builds up in the reactor and there is no non-return valve at its inlet. If this pressure is more than the maximum delivery pressure of the feed pump, it will cause a back flow into the storage tank, placed upstream. The excessive build-up of pressure is normally indicative of explosion, fire or both and a non-return valve isolates the equipment, as it prevents the back-flow. The manual valve on the reactor input line, which is also called an emergency isolation valve, is utilized for the same function of isolation in the case of fire.

The safe design made in this section for a reactor can be extended to other chemical equipment. One now goes on to a more difficult problem of leakage and spillage which a chemical industry faces. These can be minimized as discussed in Sec. 5.10.

5.10 LEAKAGES

Leakages invariably occur in chemical plants and these are extremely dangerous. If flammable liquids are handled, their escape inevitably

Table 5.10[18] Olefin Plant Leakage Profile

Sl. No.	Item of equipment which might leak	Material	Temp °C	Press., psig	Inventory, tons	Leakage type A or B	EIV fitted?	Notes
1.	Furnaces	Naphtha	100	120	Small	—	Yes	Control valves used as EIV in event of tube failure and reasonably remote hand isolation.
2.	Feed pump	Naphtha (gassy)	15	140	70	B	No	No history of leaks material cold
3.	Bottoms pump	Fuel oil	210	30	100	A	Yes	Close to auto-ignition temperature. Similar pumps have ignited.
4.	Side stream pump	Distillate fuel oil	160	25	20	B	Yes, border line	—
5	Bottoms pump	Fuel oil	210	30	60	A	Yes	Inventory can be pumped out with other pump.
6.	Bottoms pump	Gasoline	85	20	25	—	No	Inventory can be pumped out with other pump.
7.	Bottoms pump	Fuel oil	220	30	1	A	No	Inventory small and feed to column can be isolated.
		Distillate fuel oil	180	25	1	—	No	
8.	Reflux pump	Ethylene	−30	270	30	A	Yes	Pump has leaked.
9.	Side stream pump	Ethylene	−30	270	10	A	Yes	Pump has leaked and ignited.
10.	Reflux pump	Propylene	40	250	50	B	Yes	—
11.	Cold Al. exchanger	Methane	−100	370	30	A	Yes	Aluminium exchanger more fragile than steel, therefore classified A. Isolation by a remote hand valve and one extra EIV.
12	Process gas compressor	Methane, Ethylene, Propylene etc.	40	30	30	—	No	Centrifugal machines—no history leaks except one bad leak on ethylene compressor. Most of inventory would be flared. Leaks would be gas.

13.	Compressor	Ethylene	−100	65	40	—	Yes, but mainly for process convenience	
14.	Compressor	Propylene	−40	40	100	—	No	
15.	Bottoms pump	Light gasoline	90	150	5	A	No	Congested area but remote manual valve accessible.
16.	Reflux pump	Methane	−100	370	5	A		Valve fitted remote by distance. Despite low temperature there is no local history of leaks.
17.	Reflux pump	Ethane/Ethylene	−14	350	5	A		Valves fitted remote by distance.
18.	Converter	Ethane/Ethylene	80/150	350	—	A	Yes	Valves fitted on inlet and blowdown lines to isolate and blowdown vessels to stop runway reactions.
19.	Reflux pump	Propylene	5	90	1	—	No	—
20.	Reflux pump	Propylene	30	175	5	—	No	—
21.	Bottoms pump	Butylene	125	50	15	—	No	—
22.	Reflux pump	Butylene	35	50	5	—	No	—

EIV: Emergency isolation valve
(Reprinted from reference 18 with the permission of American Institute of Chemical Engineers, New York.)

Table 5.11[18] Aromatics Plant Leakage Profile

No.	Items of equipment which might leak	Material	Temp. °C	Press., psig	Inventory, cu m	Leakage type A or B	EIV fitted?	Notes
1.	Reboiler circulation pumps	Gas oil	275	7/10	7	A	Yes	Material above autoignition point
2.	Wash oil circulation pumps	Gas oil	275	7/10	7	A	Yes	Take suction from source as previous item
3.	Bottoms pumps	C_9	215	11/17	115	A	Yes	Two pumps during normal operation. A fire damaged electric cable rendering all pumps unserviceable. Material above atmospheric pressure boiling point
4.	Bottoms pumps	C_6, C_7, C_8, C_9	180	37/39	65	B	Yes	Material above atmospheric pressure boiling point
5.	Feed pumps	Naphtha	120	50	85	B	Yes	—
6.	Feed pumps	Naphtha (C_6, C_7, C_8)	230	180	40	A	Yes	Material above atmospheric pressure boiling point
7.	Flash drum	Naphtha (C_6, C_7, C_8)	50	560	80	B	Yes	Air operated emergency valve fitted after pipe-work fractures
8.	Bottoms pumps	Raffinate	210	205	50	A	Yes	—
9.	Bottoms pumps	Mixed aromatics	120	8	50	A	Yes	Materials above atmospheric pressure boiling point
10.	Bottoms pumps	Mixed aromatics	175	25	70	A	Yes	-do-
11.	Bottoms pumps	C_9	195	22	760	A	Yes	-do-
12.	Intermediate reflux	Ethyl benzene/ xylenes	150	22	740	A	Yes	-do-

	Pumps							
13.	Bottoms pumps	Xylenes	190	32	730	A	Yes	Materials above atmospheric pressure boiling point
14.	Reactor circulating pumps	Cyclohexane	210	460	25	A	Yes	EIV is fitted to suction and delivery line
15.	Reaction circuits involving hydrocarbon and hydrogen		330	850	—	A	Yes	Air operated dump valves to flare system fitted to operate in the event of
			360	630	—	A	Yes	(a) temperature run away,
			475	225 & 400	—	A	Yes	(b) line or tube failure, or
			600	800	—	A	Yes	(c) failure of normal relief system.
			210	460	—	A	Yes	
16.	Pump	Benzene	30	2	20	—	No	—
17.	Pump	Benzene	30	2	30	—	No	—
18.	Pump	Solvent and aromatics	30	20	500	B	No	Low volatility (8/1 solvent/hydrocarbon)
19.	Pump	Benzene and toluene	195	110	10	A	No	Far from hot areas of the plant

EIV: Emergency isolation valve
(Reprinted from reference 18 with the permission of American Institute of Chemical Engineers, New York.)

would cause fire, explosion or both. Consequently, their occurrence must be reduced, if not completely eliminated. It is in this context, that it is worthwhile to analyze likely sources that can cause leakage in a plant.[18]

The most important factor that minimizes leakage is the correct choice of construction material. If the material is corroded with time, leaks will continue appearing and a trouble-free plant operation can never be achieved. Next it should be ensured that all the fittings used on any equipment are leakproof. Experience of working in a plant reveals that pumps are frequent sources. Other places which equally contribute to it are equipment joints, flanges and valves. Normally these joints and flanges are held togther with a gasket between them, and here too, the suitability of the latter must be ensured at the temperature and pressure of the operation. There are some valves in chemical plants which are used only for emergency situations. As a result of less usage they get rusted or clogged due to depositions and do not function when they are really required. Also rusting affects the seat of these valves thus not closing completely.

If locations of leakage in a given plant are known, their occurrence could be minimized, if not eliminated completely. To find them, one needs to analyze case histories of several plants. Tables 5.10 and 5.11 give some data on an olefin and aromatic plant respectively. The first column gives the likely location and subsequent columns specify the nature of material handled, its temperature, pressure and the quantity. In the sixth column, they have been broadly classified into the following two classes:

1. Type A: the equipment is likely to leak.
2. Type B: the equipment is less likely to leak, but if it does, a very large quantity of material will run out.

Tables 5.10 and 5.11 reveal that most leakages occur near and around pumps, specially those which handle either hot or cold fluids. As has been explained earlier, a leakage is never repaired while the plant is in operation. It is in view of this, that emergency isolation valves are installed before such equipment to isolate them. Ideally these valves should be placed before every equipment as they can be a source of leakage, but for economic reasons it is done on those which have history of leakage. It is because of this, that Nos. 2, 12, 13, 14 and 19 to 22 of Table 5.10 and 17 to 20 of Table 5.11 do not have an emergency isolation valve fitted on them. No. 16 of Table 5.11 involves reaction circuits and emergency isolation valves are used to isolate them against leakage as well as explosions.

Tables 5.10 and 5.11 suggest the following guidelines. Pumps are normally the source of trouble because they involve a moving rotor and are consequently subject to wear and tear. As a result of this, important pumps in a plant operation require a constant watch, and to avoid frequent shutdowns, a standby pump is used as shown in Fig. 5.7.

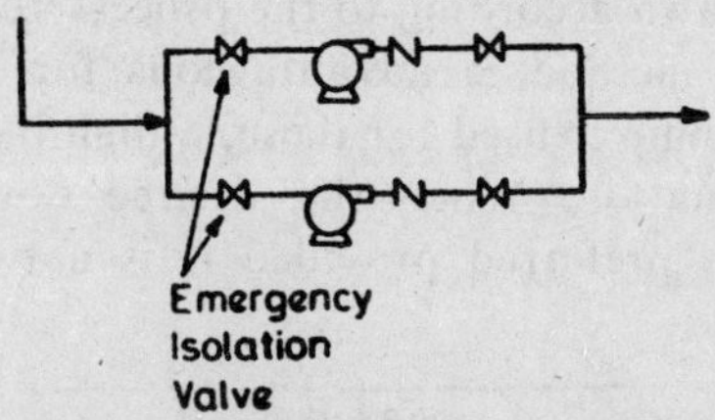

Fig. 5.7 Standby arrangement for a centrifugal pump. (Reprinted from Ref. 18 with permission of the American Institute of Chemical Engineers, New York.)

5.11 HANDLING OF FLUIDS

After having discussed safeguards against leakage, the handling of flammable liquid is discussed. The word handling is meant to include the storing, spillage and transfer of the material. Aspects of fluid storage have already been discussed in Sec. 5.7. Spillage normally occurs from open storage tanks specially when their level controller does not regulate the flow of the input pump. When the material has spilled into the adjoining area, it is necessary to get rid of it as soon as possible. In this context, dikes and drainage are provided. Figure 5.8 shows the method of providing a sioping floor in the vicinity of storage tanks for getting rid of the spillage.

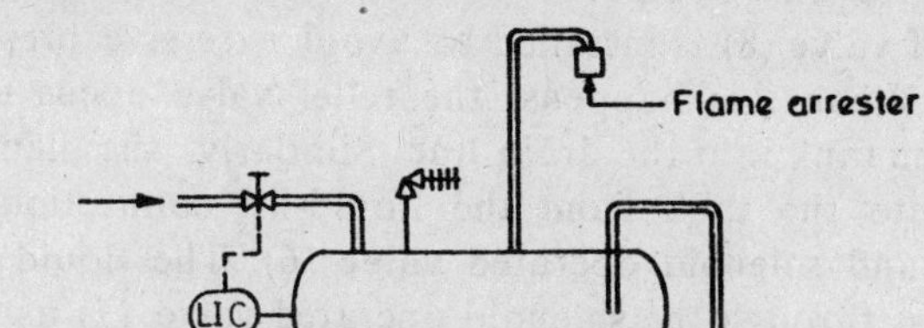

Fig. 5.8 Sloping ground to ensure quick drainage of spilled material

There are three recommended ways by which flammable liquids can be transferred from one point to another[8]:

1. gravity transfer,
2. inert gas or hydraulic transfer, and
3. transfer by pumping.

In the gravity transfer, an overhead tank is filled with liquid by a pump and material withdrawn according to the process requirement through a suitable valve. This method is advantageous for liquids having large volatility. When a pump is used for these, a high back pressure develops due to vapour formation. When the desired flow rate is very high, hydraulic transfer is preferred provided it is not miscible with water.

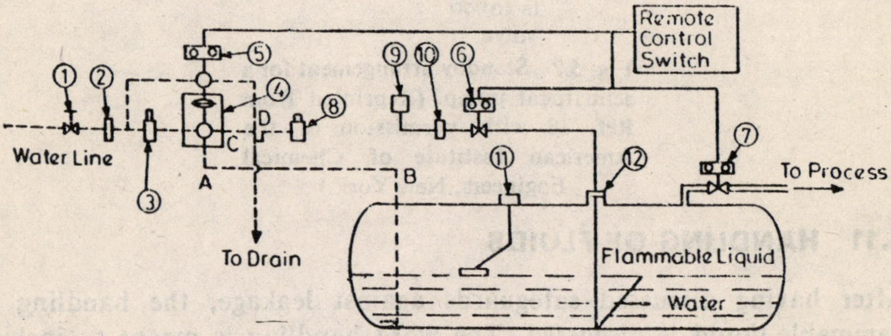

Fig. 5.9 Hydraulic transfer of a flammable material lighter than water. (Reprinted from reference 6 with the permission of Factory Mutual Engineering Corporation, Massachusetts.)

Figure 5.9 shows the approximate arrangement of controllers and indicators for the safe delivery of flammable liquids lighter than water. For heavier liquids a similar arrangement can be had. Water is pumped through a manual valve (1), a strainer (2), a pressure regulating valve (3), and a three-way two-port water control valve into the tank. The pressure relief valve (8) is installed to avoid excessive pressure build-up in the tank. When this happens, the relief valve opens to release the contents of the tank into the drain line. Similarly, the flammable liquid is pumped into the tank from the liquid-fill connection (9), through strainer (10) and solenoid operated valve (6). The liquid is withdrawn for the process through the solenoid operated valve (7) as shown in the diagram. The out-flow of liquid is controlled through the regulation of pressure from valve (3). If it is desired to isolate the storage tank for some reason, one uses the remote switch which operates solenoid valves (5), (6) and (7) simultaneously. On actuation, (5) opens the pilot valve (4) connecting the line BA to the drain line CD. The solenoid valve (6) opens up and valve (7) closes at the same time. During tank filling with the liquid, the same remote controlling switch is used making the same valve positions as above. As the liquid fills in, water drains through the pipe line BACD. The valve (12) closes the inflow of material when the tank is completely full with the liquid. The inert gas transfer system is not discussed here as it uses the same controllers and valves arrangement as in Fig. 5.9.

Transfer of fluid by use of pumps is a very common industrial practice because it does not require as elaborate control as that for the hydraulic transfer system. One employs either a centrifugal pump or a positive displacement pump. The latter has the advantage of complete flow stoppage when the pump is shut off. This does not occur in the case of centrifugal pumps as they normally take suction under head and on stopping the pump the material continues to flow due to the siphoning effect. While handling very flammable liquids, it must be ensured that the centrifugal pump does not operate dry, and if the plant operation is such that at times it does, one should have all moving parts of the pump that are of the non-sparking type. In view of the fact that centrifugal pumps can give a flow of fluids through siphoning when it is shut off, it is sometimes desirable that the pump takes the suction under lift (as against head). The priming of the pump in this case is done either through a vacuum pump or a priming tank.

It has already been noted that it is quite common for pumps to develop leakage. In view of this, it is preferred to locate pumps outdoors so that a fire in them does not damage other equipment in the vicinity. Sometimes to contain fire it is desirable to keep a separate pump room which houses all pumps of the chemical plant and provides a trapped floor drainage around it from which leaked fluid does not expose other facilities of the plant. As demonstrated through the example of a distillation column, this would change the complete plant layout.

Figure 5.10a shows a distillation column with feed pump FP1, cooler C, reflux drum RD, reflux pumps RD1, reboiler R and the reboiler pump BP1. As pumps FP1, RD1 and BP1 are handling hot fluids, they are likely to leak, as a result of which they have standby units FP2, RD2 and BP2 with isolation valves as shown. One way of arranging these pumps would be to place them at the height shown as this would minimize the total pipeline required. But this is not desirable because the leaked material at height can catch fire in the air. As distillation columns have tall structures, fire fighting at a height may be an impossible task. Consequently, they should be grounded and a leak trap built around each pump.

From the viewpoint of fire-fighting, it is desirable to have a prior knowledge of hazardous locations so that in the event of fire, these could be cut off immediately to prevent it from spreading. Figure 5.10(b) shows all pumps at one place which is a far better arrangement.

Lastly, in connection with the handling of flammable liquids, it is best to divide an operation into subtasks and instruct workers to adhere to the routine. This routine should be treated as standard operating procedure and under no circumstances should any 'shortcuts' be allowed. This is because invariably shortcuts lead to plant accidents.[13] As an example, a batch process of dilution of concentrated sulphuric acid is considered. It is not a flammable material, but demands similar careful handling because of large heat generated in mixing with water. Table 5.12 gives a breakdown of the steps required for dilution in a tank.

It is worthwhile to demonstrate that the steps given in Table 5.12 do not change whether the plant is semi-continuous or continuous.

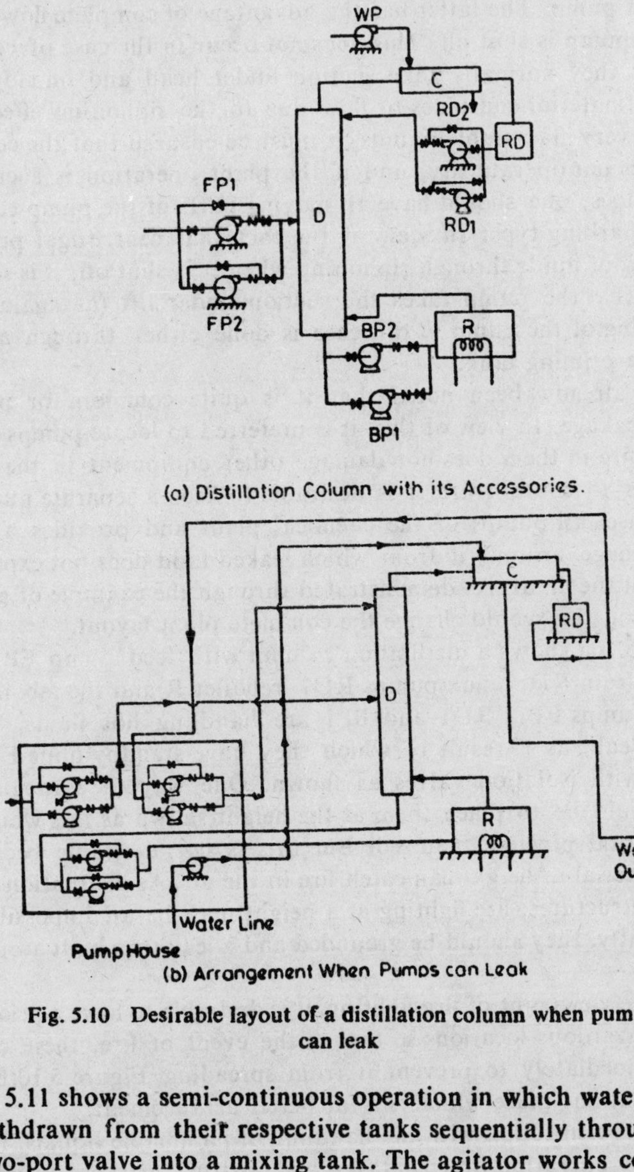

(a) Distillation Column with its Accessories.

(b) Arrangement When Pumps can Leak

Fig. 5.10 Desirable layout of a distillation column when pumps can leak

Figure 5.11 shows a semi-continuous operation in which water and acids are withdrawn from their respective tanks sequentially through a three-way two-port valve into a mixing tank. The agitator works continuously and the amount of water and the acid passed into it is measured through a meter. The density of the solution is checked and then transferred into another holding tank from which it is withdrawn continuously. The three-way two-port valve ensures that the acid and the water are not withdrawn simultaneously. In the flow sheet of Fig. 5.11, one has assigned the work of measurement of quantities of water and acid to the meter, sequencing of raw material to the three-way two-port valve and the transferring material to gravity tanks and a pump. In this flow sheet not only has the participation of a man been reduced but the number of steps required to

carry out the work remain unchanged. The same can be shown for a continuous plant which is left as an exercise to the reader.

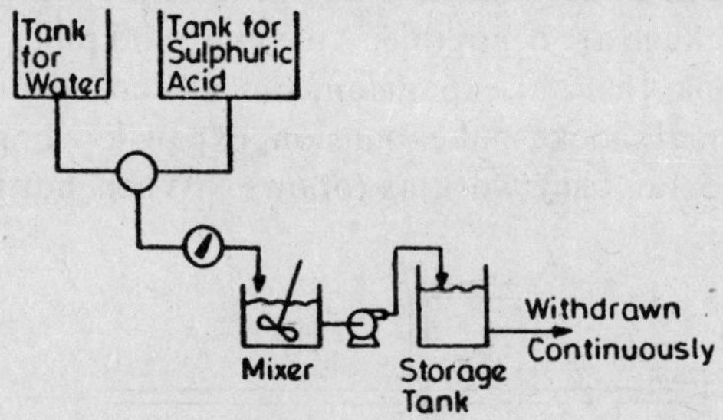

Fig. 5.11 Semicontinuous arrangement for dilution of sulphuric acid

Table 5.12[13] Breakdown of the Job of Dilution of Concentrated Sulphuric Acid

Part of Material	Operation
Sulphuric acid	Diluting acid with water
Important steps in the operation	Key points—Cracks—Hazards, Feel, Timing, Special information
Step 1 Add 100 lit of water to tank	Use dip stick to measure water depth before adding acid.
Step 2 Start agitator	Make certain, agitator is operating
Step 3 Open the valve on 98% sulphuric acid line and run into the measuring tank until 10 lit are in the tank	Wear safety protection such as acid goggles, rubber gloves and apron. Use face shield when checking contents of the tank.
Step 4 Transfer acid from measuring tank to the water in the receiving tank	Add acid to the water in the receiving tank slowly and keep the agitator operating. Always add the acid to water slowly. *Never* add water to acid because of reaction heat generated and danger of splashes.
Step 5 Make specific gravity test	Use Baume's scale.
Step 6 Run mix into holding tank	Make certain that there is room in holding tank and discharge valve is closed

(Reprinted from reference 13 with the permission of John Wiley & Sons Inc., New York.)

5.12 SAFETY IN PIPELINE LAYOUT

Safety considerations in deciding pipeline layout are considered in this section. Pipe layout in a building should be kept to the shortest possible and most of the pipeline must be kept preferably outdoors. Pipelines could be laid above or underground. Underground layout is preferred only when chances of leakage are negligible. Above-ground pipes must be guarded against vibrations, thermal expansion, shocks and sagging due to their own load. For thermal shocks and expansion, expansion loops are provided as shown in Fig. 5.12. They work as follows.[19] When hot process fluid is

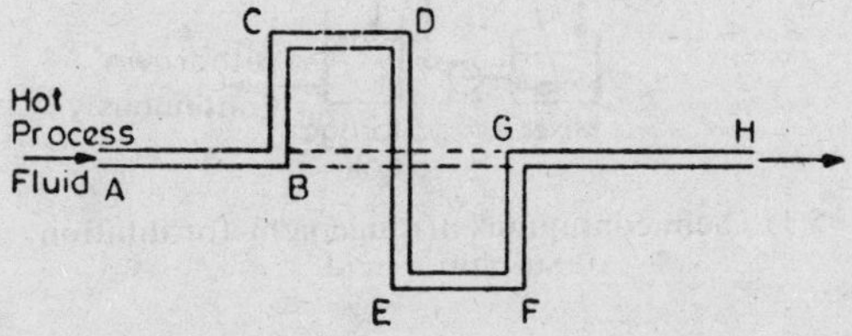

Fig. 5.12 Expansion loop to relieve thermal stresses

passed into the pipe line, there is thermal expansion, as a result of which internal stresses develop due to its constrained ends. In straight pipes, only compressive stresses exist, due to which it will bend to form an arc of a circle to accommodate the expansion. Since this is not desirable, by providing expansion loop in the pipeline, the change in length due to thermal expansion is contained within the loop. In addition to this, at high temperature, creep phenomena sets in which slowly relaxes the stresses developed. To guard pipelines against vibrations and sagging due to their own weight, they should be secured by fastening with suitable supports. If they have to pass under a building or railway tracks, they must be enclosed in a larger pipe.

5.13 DUST EXPLOSION[20–26]

In previous sections, the problem of designing with respect to the constraint of safety was considered. Since most devastations are preceded by explosions, this is analyzed now in detail. In this section, dust explosion is considered.

There are several chemical industries dealing with solid products. Handling of solids generates chemical dusts which can become a menace in three ways.[20] They are harmful to the respiratory system if the particle size is less than one micron. The second harmful aspect of dusts is that particles upto the size of 10 microns form a stable suspension in air leading to reduced visibility. Lastly, particles upto 700 microns can cause dust explosion, which can be severe enough to destroy the whole chemical plant. Industries where dust explosion can occur and where it has occurred in the past are grain elevators, flour mills, wood working plants, feed and cereal mills, manufacture of starch and corn products, cotton mills, paper mills, rubber factories, sugar refineries, pitch and resin plants, spice and coffee plants, drug and chemical plants, pulverized coal installations, malt houses, plastic industries and factories producing organic powders, and dusts.

Dust explosion occurs because of an uncontrollable chemical reaction at the surface of the particle. The second component is normally oxygen of the air, but this need not necessarily be so. There are some materials which can undergo dust explosion in an atmosphere of carbon dioxide. Experimental work shows that practically every material can be ignited in air to cause dust explosion. The necessary condition for its occurrence is that the material should be at a certain temperature, called ignition temperature and a suitable source of energy should be available to cause the ignition. Explosion would occur only when the energy generated due to the surface reaction at a given place is enough to heat up the neighbouring area to the ignition temperature. Dust explosion has been studied in great detail and following factors have been found to affect the phenomena:

1. chemical and physical properties of dust,
2. concentration of the dust cloud dispersion,
3. properties of the atmosphere in the explosion space,
4. ignition source, and
5. the shape of the explosion space.

Among the chemical properties of dust, the important ones are the rate of oxidation, specific heats of the dust and the final product, heat of combustion and oxygen required for complete burning. It can be seen that large heat of combustion and low specific heat of the dust would bring the neighbouring dust to its ignition temperature quicker. If the oxygen required for complete combustion is large, it is expected to have a much lower chance of igniting in an atmosphere deficient in oxygen. Among physical properties, the dust particle size is the most important one. In Fig. 5.13, minimum energy required to cause the explosion is given for cellulose acetate as a function of the average particle size and it is seen that the amount of energy required reduces almost linearly as the particle size reduces. The ignition temperature is also found to decrease with average particle size as shown in Fig. 5.14. This is not surprising if one recalls that dust explosion occurs because of a surface reaction, and the smaller the particle, the larger is its surface area per unit weight.

Concentration of dust particles is a very important parameter and it is found that the explosion does not occur below a certain minimum concentration, but unlike gas explosion, it does not have an upper limit of concentration. It is important to design the equipment to be mechanically strong such that it is not wrecked by it. To do so, the following two requirements are necessary:

1. maximum pressure reached in a closed container, and
2. the rate of pressure rise.

The experiment is normally carried out in a standard closed cubical box,

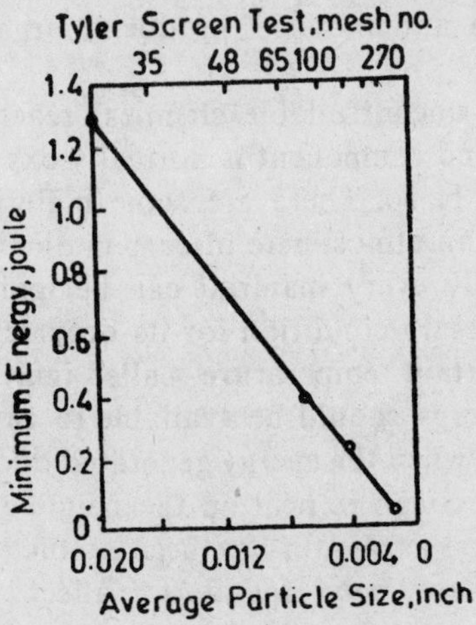

Fig. 5.13 Effect of fineness of dust upon the energy to cause ignition.[20] (Reprinted from I. Hartmann, *Ind. Eng. Chem.*, **40**, 752, 1948 with permission of the American Chemical Society, Washington.)

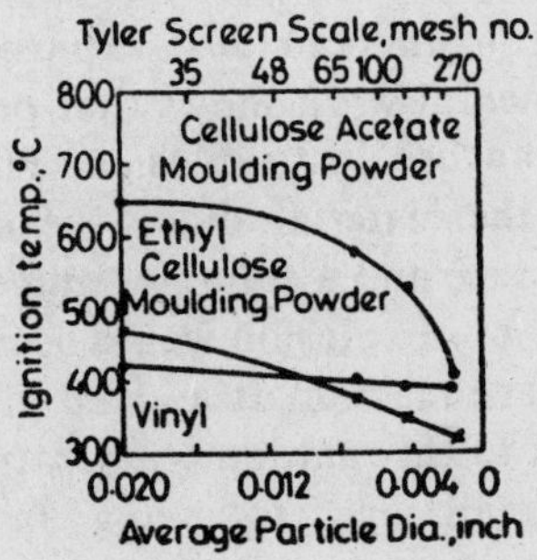

Fig. 5.14 Effect of fineness of dust upon the ignition temperature.[20] (Reprinted from I. Hartmann, *Ind. Eng. Chem.*, **40**, 752, 1948.)

called the Hartmann apparatus. Figure 5.15 gives experimental results for magnesium powder undergoing explosion. The maximum pressure developed in the apparatus has been plotted as a function of dust concentration. It is found that there is minimum concentration for explosion. The curve undergoes a maximum at about the stoichiometric concentration of magnesium with oxygen and then flattens out for large dust

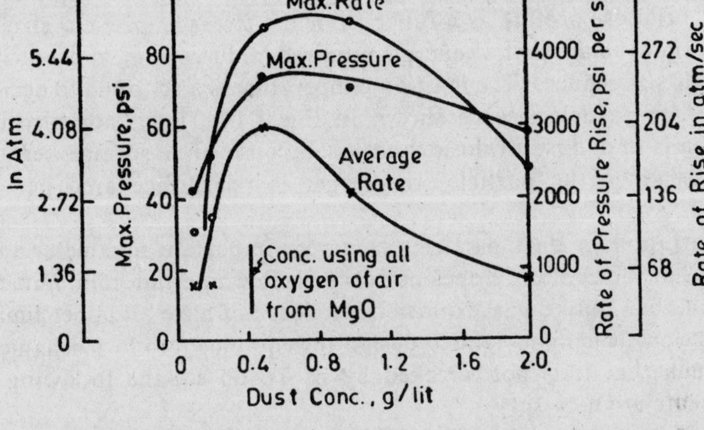

Fig. 5.15 Dependence of pressure and rate of pressure rise upon dust concentration in an explosion. (Reprinted from I. Hartmann, *Ind. Eng. Chem.*, **40**, 752, 1948 with permission of the American Chemical Society, Washington.)

concentrations as shown in the figure. On measurement of the rate of pressure rise, (dp/dt) is found to be time dependent. In Fig. 5.15, the maximum as well as the average (dp/dt) are given. Both undergo a maximum with respect to the dust concentration, but contrary to what is found for gas explosion, it does not go to zero for large concentrations even though reduced considerably.

The effect of oxygen concentration on dust explosion for 200 mesh coal particles is shown in Fig. 5.16. The oxygen concentration is changed by adding carbon dioxide as a diluent and the temperature at which its ignition occurs is measured. Experiments show that below a certain oxygen level, there is no ignition whereas the ignition temperature decreases very rapidly with increasing concentration.

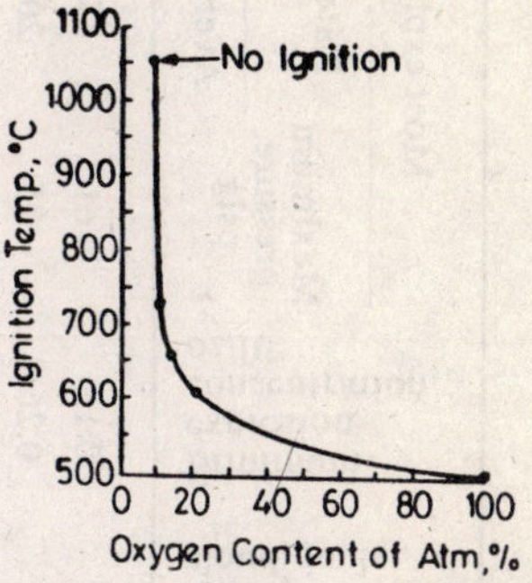

Fig. 5.16 Effect of environment upon ignition temperature (Reprinted from I. Hartmann, *Ind. Eng. Chem.*, **40**, 752, 1948 with permission of the American Chemical Society, Washington.)

Lastly, the explosion is found to depend upon the shape of the space which remains an unsolved problem to date. No satisfactory study on the subject exists because of the following reason. It has already been pointed out that dust particles can be fairly big and need not form a stable suspension. In industrial operations, there exists a variation in the dust concentration with respect to height due to gravity settling. This concentration variation pattern is time dependent. An ignition source is assumed to exist somewhere at the top. The combustion front, after the initiation of explosion, moves downwards and encounters larger and larger concentrations of dust. Consequently, it would require a larger heat to bring these particles to ignition temperature and as a result there is a chance that the combustion front might be extinguished. In addition, the dust explosion causes excessive turbulence which shakes up a lot of settled particles, giving rise to the formation of a cloud of dust.

Table 5.13 gives some of the data on dust explosion from references 6 and 22, which have been taken on the Hartmann apparatus. The last column gives the maximum pressure, the rate of pressure rise and the dust composition that will give that. The fourth column gives the ignition temperature and this is found to vary significantly if the dust is distributed "layer-like" or "cloud-like". Refer to the example of a container in which dust concentration variation exists. Supposing the dust is ignited. Due to turbulence, a dust cloud would be raised, which would increase the ignition temperature further. If the heat of combustion is not large enough to bring the neighbouring area to this temperature, the explosion will be extinguished.

Thus, dust explosion cannot be modelled satisfactorily and in the

Table 5.13[6,22] Data on Dust Explosion

Material	Explosibility index	Ignition sensitivity	Explosion severity	Ignition temp. °C		Minimum ignition energy cloud, joule	Minimum explosion concentration, oz/ft³	Most explosive mixture			
				Cloud	Layer			Maximum pressure, psig	Rate of pressure rise		oz/ft³
									Average	Maximum	
Alfalfa meal	0.1	0.1	0.8	470	220	0.37	0.16	61	350	800	0.5
Almond shell, hard	<0.1	0.1	<0.1	440	200	0.22	0.26	75	200	400	1.0
Almond shell, soft	0.3	0.9	0.3	450	210	0.08	0.065	101	700	1400	1.0
Apricot pit	1.9	1.6	1.2	440	230	0.08	0.035	109	1500	4000	1.0
Cereal grass, dehydrated	<0.1	<0.1	0.1	550	220	0.8	0.25	58	250	600	1.0
Cherry pit	0.4	2.0	2.2	430	220	0.08	0.03	113	1700	4400	1.0
Cinnamon	5.8	2.5	2.3	440	230	0.03	0.06	114	1400	3900	0.5
Citrus peel, dehydrated	1.0	1.1	0.9	730	270	0.045	0.065	51	600	1000	0.5
Cloverseed	0.4	0.9	0.4	470	270	0.08	0.06	76	450	1000	0.5
Cocoa bean shell	13.7	3.6	3.8	470	370	0.03	0.04	69	1100	3300	0.5
Cocoa with milk, sugar	0.5	0.4	1.2	420	390	0.18	0.065	70	500	1100	1.0
Coconut shell	4.2	2.0	2.1	470	220	0.06	0.035	115	1700	4200	1.0
Coffee, fully roasted	<0.1	0.2	0.1	720	270	0.16	0.085	38	100	150	1.0
Coffee, instant, spray dried	<0.1	0.1	0.1	410	350	*	0.28	68	200	500	1.0
Pectin from dried apple pulp	10.3	2.2	4.7	410	200	0.035	0.075	132	1900	7000	1.0
Sucrose, chemically pure	3.3	1.1	3.0	420	470	0.10	0.045	71	1000	2500	0.5
Tea	—	—	—	—	—	—	—	105	—	—	—
Wheat flour, commercial	4.1	1.5	2.7	440	440	0.06	0.05	97	900	2800	1.0
Starch, edible	17.7	5.2	3.4	430	—	0.025	0.045	100	1600	6500	0.5

Rubber, crude, hard	7.4	4.6	1.6	350	—	0.05	0.025	80	1200	3800	0.5
Adipic acid	1.9	1.7	1.1	550	—	0.06	0.035	76	1200	2700	0.5
Casein	0.4	1.4	0.3	520	—	0.06	0.045	66	400	1000	0.5
Fumaric acid, dried	1.6	1.3	1.2	520	—	0.035	0.085	89	1100	3000	1.0
Para formaldehyde	>10.0	6.1	6.9	410	—	0.02	0.04	102	2500	13000	0.5
Phthalic anhydride	>10.0	13.8	1.6	650	—	0.015	0.015	72	1300	4200	0.5
Salicylic acid	—	—	—	590	—	—	0.025	46	—	—	0.5
Naphtha	—	—	—	211	—	—	0.8(vol%)	—	—	—	—
Aluminium	—	—	—	700	—	0.02	0.04	58	1050	2250	—
Terephthalic acid	6.9	3.0	2.3	680	—	0.02	0.05	65	1900	8000	0.02
Urea crystals, ground	≪0.1	≪0.1	—	900	—	*	**	†	—	—	—
Cellulose	2.8	1.0	2.8	480	270	0.08	0.055	119	1800	8500	0.05
Cellulose flock	1.4	1.1	1.3	460	290	0.06	0.065	87	1300	2900	0.05
Cork	9.7	2.7	3.6	470	—	0.045	0.035	80	2300	7000	0.2
Wood flour	>10.0	2.8	3.8	450	—	0.04	0.4	97	2300	7500	0.5
Pittsburgh coal	1.0	1.0	1.0	610	—	0.06	0.055	83	800	2300	0.5
HEXANE, Quiescent	1880.0	855.0	2.2	235	—	0.0002	0.02	100	2000	4250	—

* no ignition upto 8.32 j, the highest tried. ** no ignition upto 2.0 oz/ft^3, the highest tried.

† no ignition obtained by standard igniting sources.

absence of any theoretical studies, one uses three empirical parameters: ignition sensitivity, explosion severity and index of explosivity as defined in Table 5.14.[23] These parameters have been calculated for different dusts and presented in the first three columns of Table 5.13. Explosions are characterized as weak, moderate, strong and severe depending upon the values of these parameters, as given in Table 5.14.

Table 5.14 Classification of Dust Explosion

Type of explosion	Ignition sensitivity	Explosion severity	Index of explosivity
Weak	0.2	0.5	0.1
Moderate	0.2 to 1.0	0.5 to 1.0	0.1 to 1.0
Strong	1.0 to 5.0	1.0 to 2.0	1.0 to 10.0
Severe	5.0	2.0	10.0

Ignition sensitivity

$$= \frac{\text{(ign. temp.} \times \text{min. ign. energy} \times \text{min. expl. conc.) of Pittsburgh coal dust}}{\text{(ign. temp.} \times \text{min. ign. energy} \times \text{min. expl. conc.) sample dust}}$$

Explosion severity

$$= \frac{\text{max. expl. press.} \times \text{max. rate of press. rise) of sample dust}}{\text{(max. expl. press.} \times \text{max. rate of press. rise) of Pittsburgh coal dust}}$$

Index of explosivity = ignition sensitivity × explosion severity

5.14 GAS EXPLOSION

Gas explosion is found to have similar characteristics as dust explosion, except for the fact that the former has an upper limit of concentration (Table 5.2). Several experimental as well as theoretical studies exist upon this subject and can be seen in reference 4. In Tables 5.13 and 5.14, the simplest model is described which establishes the basic relation between different parameters affecting it.

One assumes that there is a spherical container with an ignition source at its centre. If the combustible gas is ignited, a flame is generated at the centre travelling outwards. As long as the flame front is not travelling at a speed equal to or more than the velocity of sound, the pressure within the container at a given time can be assumed to be uniform everywhere. As the combustion flame travels outwards, the pressure in the vessel increases and is maximum when the flame reaches the outside surface. If this final pressure is P_f, reference 4 derives the following relation

$$p = \frac{KS_r^3 t^3}{V} P_f \tag{5.14.1}$$

where S_r is the radial flame velocity, V is the volume of the container, t is the time after the inception of explosion and K is a constant. To apply this equation to any container other than a spherical one is mathematically incorrect, but experimentally it is found that the functionality remains essentially the same. Consequently one has following three rules for *any shape of the vessel*:

(1) For a given gaseous fuel, the final pressure P_f is found to be almost independent of the volume of the vessel, and the pressure in it varies as

$$p = k_1 \frac{t^3}{V} \tag{5.14.2}$$

where k_1 is constant for a given fuel. For a spherical container, it is equal to $(KS_r^3P_f)$, but for any other shape of the vessel, k_1 has to be found experimentally. The above equation (Eq. 5.14.2) is rearranged to find the time required to reach a given pressure p as

$$t = (p/k_1)^{1/3}V^{1/3} \tag{5.14.3}$$

(2) For two vessels of volumes V_1 and V_2, one assumes the time taken to reach a given pressure level p as t_1 and t_2 respectively. These two times are related as

$$\left(\frac{t_1}{t_2}\right) = \left(\frac{V_1}{V_2}\right)^{1/3} \tag{5.14.4}$$

(3) For different gaseous fuels in a container of known volume at the same time

$$p = k_2S_r^3 \tag{5.14.5}$$

where k_2 is a constant. If it is strictly a spherical container, it is equal to (Kt^3P_f/V).

Equations 5.14.2 to 5.14.5 constitute the basic scale-up rules for gas explosion. From laboratory studies, one can evaluate the constants in these equations and use the result for the vessel of interest.

When a vessel is provided with relief valves, the nature of explosion is completely changed. However, these scale-up rules are found to still hold good. More information on it can be had from reference 26. Even though dust explosion behaves similarly in a closed vessel, scale-up rules are slightly changed as discussed below in Sec. 5.15.

5.15 DUST EXPLOSION IN VESSEL WITH A VENT

One considers dust explosion in a vessel equipped with a safety relief valve or a rupture disc. When excessive pressure builds up in it, the pressure first rises in the vessel according to Eq. 5.14.2. When the relief valve opens (or rupture disc bursts), the combustion front encounters a larger volume because of the opening. This volume is not infinity, but some effective volume which keeps increasing as the flame front travels upwards out of the vessel.

On opening of the rupture disc the material is forced out of the vent. After this has occurred, the pressure in the container first increases to some maximum value as the flame front accelerates upwards subsequently falling as shown in Fig. 5.17a. It undergoes a maximum at p_{red} whereas for large times, it falls to a considerably lower value, p_{st}.

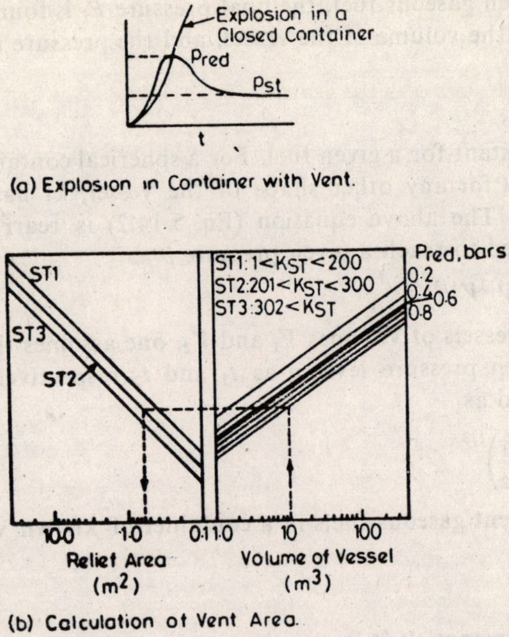

Fig. 5.17 Dust explosion in a container with a relief vent. (Reprinted from reference 24 with permission of the American Institute of Chemical Engineers, New York.)

This is found to be more than the pressure of the container before explosion has occurred as shown in the figure. Increase in the container pressure is explained by treating the vent opening as an orifice. As the flame front accelerates, the flow through the orifice continues increasing with a maximum flow rate when the front emerges from the opening. Because the pressure drop across an orifice increases with increasing flow rate, the pressure in the container starts decreasing when the front travels out of it, thus undergoing a maximum.

Vent area F required for the dust explosion of a specified dust has been experimentally found to depend upon:

(1) Shape of the container, S,
(2) Volume of the container, V,
(3) Maximum pressure, p_{red},
(4) Static pressure, p_{st}, and
(5) The rate of pressure rise, $\frac{dp}{dt}$.

The rate of pressure rise $\frac{dp}{dt}$ has been further shown to depend upon V

as follows:

$$\left(\frac{dp}{dt}\right)V^{1/3} = K_{st} \tag{5.15.1}$$

where K_{st} is a constant independent of container volume for a given dust; however, for different dust, it has different values. Thus the functionality of the vent area F can be written as

$$F = f(p_{red}, p_{st}, K_{st}) \tag{5.15.2}$$

Since experimental results are generated in the Hartmann apparatus (a cubical container with a volume of 1 m^3), it is therefore necessary to develop necessary scale-up rules for designing. The scale-up rule which is found to apply is called 'cubic law' and is given by[23-26]

$$\frac{V_2}{V_1} = \left(\frac{F_2}{F_1}\right)^{3/2} \tag{5.15.3}$$

where V_2 and V_1 are the volumes of the vessels and F_2 and F_1, relief vent areas provided on them.

Equations 5.15.2 and 5.15.3 have been combined in the form of a nomogram given in Fig. 5.17b. For this figure p_{st} has been taken as constant value of 0.1 bar. Different dust particles have been divided into three classes, ST1, ST2 and ST3, depending upon K_{st} values. Most violent explosions belong to ST3 class. The relief area provided should be such that p_{red} does not exceed a gauge pressure of 2 bar. To find the relief area F for a container of volume V, one reads the volume on the abscissa, goes upto the line corresponding to the specified p_{red}, travels horizontally to the adjoining diagram upto the line corresponding to the specified dust and reads the relief area on the abscissa. The procedure of reading the nomogram is also shown in the diagram. For volumes more than 400 m^3 one uses the scale-up rule of Eq. 5.15.3.

This nomogram is strictly true for cubical containers on which the relief area can be fitted anywhere on its body. In chemical engineering, however, the common shape is cylindrical vessels. They are further subdivided for the purpose of providing relief areas as

(1) short cylindrical vessels for which $H \leq 5D$, and

(2) long cylindrical vessels whose height H is more than five times the diameter, D.

While designing the vent area for long vessels, it may be noted that the nomograms given in Fig 5.17(b) are valid for vessel volumes upto 1000 m^3. The relief area calculated from the figure must not be undercut under any circumstances. This means that the diameter of the vessel defines the maximum volume that can effectively be protected and its length must accordingly be limited.

In this section the phenomena of dust explosion was analyzed. The preventive methods of dust explosion are to minimize dust and at times use inert atmosphere. Providing a suitable relief vent on equipment (e.g., dust collectors, driers, filters, etc.) is the best preventive measure.

5.16 DESIGN OF FLARE SYSTEM[27-30]

It was argued in Sec. 5.15 that the use of relief valves for the safeguard of equipment as well as plant personnel is necessary when excessive pressure builds up. Because one cannot release these in the plant area, suitable drains for liquids and a flare system for gases are provided.

A flare system consists of a stack with a pilot flame. The purpose of the pilot flame is to ignite the released vapour so that nothing escapes to the atmosphere. One considers the situation when the pilot flame fails during the discharge and the stack height is short. The released gases are normally heavier than air and would ultimately settle on the vicinity of the plant area, which can catch fire from any ignition source nearby. It may be appreciated that a fire, in an environment of this kind, would spread very fast causing more widespread devastation. There is also a likelihood that the gas might ignite as soon as it comes out of a short stack. A flame over the stack would radiate a large amount of thermal energy which is hazardous to personnel working nearby. Therefore a flare stack is better than an unflared one, and a tall flare stack is ideal because it not only minimizes hazard from radiation energy but dilutes the discharge at the ground level.

While designing the stack height of the flare system, it may be remembered that there is always the possibility that the pilot flame may extinguish. At the same time, a discharge from one of several relief valves of the plant may cause a dangerous situation leading to fire. The pilot flame can be extinguished because of nonavailability of the burning gas, choking of burners due to carbon deposition or due to the hydrodynamic conditions existing at the flame. One would like to minimize the chance of its occurrence, but once it occurs it will cause a major plant fire. The fire can be prevented by making the stack tall as shown in Fig. 5.18.

When a gas heavier than air escapes from a chimney (or stack), it first goes up a certain distance due to its escaping velocity as shown schematically in Fig. 5.18. Subsequently, wind carries it in the forward direction and due to buoyancy it settles downwards, and also there is mixing with air. This mixing is not instantaneous, but takes a finite amount of time—plumes of discharged gases (nonflammable) are known to have settled on the ground level and stayed for several days before they completely mix with air. Intermingling with air occurs through the process of eddy mixing, and it is desired that this be as fast as possible. This problem has been solved mathematically by Bosanquet and Pearson and they obtained the following equation for the ground level concentration of the discharged gas:[28]

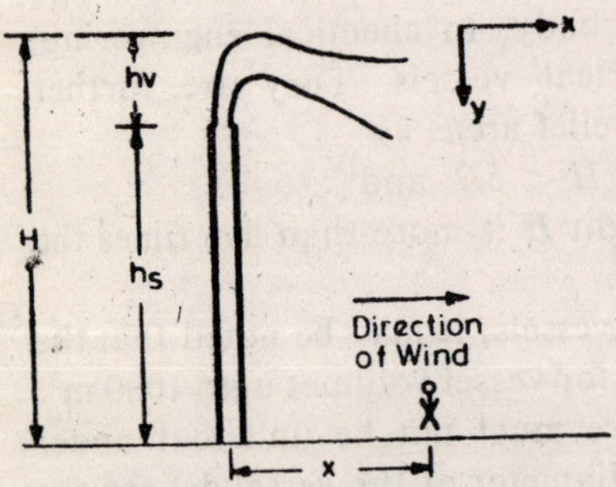

Fig. 5.18 Release of gases heavier than air through a stack

$$C_0 = \frac{(Q_m \times 10^6)e^{-H/px}}{(2\pi)^{1/2} pq\, u\, x^2} \tag{5.16.1}$$

where C_0 is the average ground level concentration in ppm by volume, Q_m is the emission rate of the contaminant at atmospheric temperature and pressure in m^3/s, p and q are vertical and horizontal atmospheric dispersion coefficients (dimensionless numbers), x is the downwind distance from the emission source in m and u is the mean horizontal wind velocity in m/sec. As shown in Fig. 5.18, the effective emission height H is the sum of the height of the stack h_s and the plume rise h_v, i.e.,

$$H = h_s + h_v \tag{5.16.2}$$

p and q are two constants which are difficult to measure experimentally. Table 5.15 gives their values under three different atmospheric conditions.

Table 5.15[28, 29] Values of Dispersion Coefficients Under Varying Atmospheric Conditions

	Low Turbulence	Average Turbulenee	Moderate Turbulence
p	0.02	0.05	0.10
q	0.04	0.08	0.16
x_m	$25H$	$10H$	$5H$

(Reprinted from F. T. Bodurtha, *Chemical Engineering*, Dec. 15, 1958 with the permission of McGraw-Hill Publishing Co., New York.)

It may be recalled that the process of mixing due to eddy motion is not deterministic, but stochastic in nature. This implies that if there is a man positioned at ground level at a distance X from the stack (as shown in Fig. 5.18) and measures concentration of the contaminant, he would find it to vary with time. It is in view of this, Bosanquet et al. take C_0 as the concentration at the ground level averaged for a time of 30 minutes or more. As a matter of fact, Bodurtha claims that peak 1-minute-concentration would be about ten times the value of C_0 and peak a-few-second-concentration would be about 50 times C_0[29]. The height of plume rise has been experimentally determined by Bodurtha by simulating a stack in a wind tunnel for different gases and he suggests the following empirical equation:

$$h_v = 5.44 \times 0.30D\left(\frac{V_s}{u}\right)^{0.75} \tag{5.16.3}$$

when h_v is in m, D is the stack diameter in m, V_s is the stack gas velocity in m/s and u is the horizontal wind velocity.

C_0 in Eq. 5.16.1 undergoes a maximum with respect to the horizontal distance x, which can be found by setting dC_0/dx equal to zero. The distance, x_m, at which this occurs is given by

$$x_m = H/2p \tag{5.16.4}$$

and the corresponding maximum concentration, $C_{0,\max}$, is given by

$$C_{0,\max} = \frac{(Q_m \times 10^6)e^{-1/2}(p/q)}{(2\pi)^{1/2}H^2} \tag{5.16.5}$$

As the stack height h_s increases, the effective emission height correspondingly increases and according to Eq. 5.16.5 $C_{0,\max}$ reduces. The correct stack height is the height which makes $C_{0,\max}$ less than the lower explosive limit at the ground level for the contaminant. For the reason of air pollution, it is best to burn any contaminant, but for the worst situation when the pilot flame fails, the stack should be high enough such that the contaminant does not catch fire at the ground level.

5.17 CONCLUSION

A chemical plant is dynamic in nature and this should be properly accounted for while designing it. In this chapter, important safety considerations in plant design have been given. In the absence of any theoretical guidelines, the performance of previous plants is used as the basis of all future designs.

The design of the plant should be such that it minimizes accidents. In Chapter 6, a systematic procedure is developed by which one can analyze conditions that lead to accidents. A prior knowledge of these helps one to arrive at a rational design.

APPENDIX 5.1

Standard Formats of Equipment Specifications

In steady state plant design, one normally uses principles of unit operations and stage processes and specifies equipment accordingly. It was noted in this chapter that such specifications alone are not sufficient, as one needs to know some of its important mechanical as well as electrical aspects.

In a chemical plant one very commonly comes across process vessels (storage tanks and pressure vessels), heat exchangers (evaporators, steam boilers, coolers, shell-and-tube exchangers, etc.), distillation columns, pumps, compressors, valves, piping, tees and elbows. In this appendix, a standard format for specifying these are given.

A5.1.1 Standard Codes Used by Mechanical Design Group

1. *Material specifications*
   American Society of Mechanical Engineers (ASME), Section II
   American Society of Testing and Materials (ASTM), Part A for

Ferrous materials, Part B for non-ferrous materials and Part C for welding rods, electrodes and filler materials

2. *Unfired pressure vessels*
   ASME, Section VIII, Division 1 and 2—Pressure vessels
   ASME, Section X, Fibre glass reinforced plastic pressure vessels
3. *Fired pressure vessels*
   ASME, Section IV & VI Heating boilers
   ASME, Section I & VII Power Boilers
4. *Atmospheric storage tanks* API 650
5. *Heat exchangers* Tubular Exchanger Manufacturers' Association (TEMA) Class B, C and R
6. *Welding procedures* ASME, Section IX

A5.1.2 Process Vessel Specifications

1. Item No.
2. Type of process vessel
   (a) Vertical or horizontal, dished ends or flat bottom, coded or non-coded pressure vessels, or API storage tanks, etc.
   (b) Specification of code type
      (i) Coded pressure vessels cover vessels having internal pressures from 1 atm to 200 atm gauge and all vacuums
      (ii) Non coded vessels cover upto pressure of 1 atm gauge
      (iii) API storage tanks
3. Location
   Outdoor or indoor
4. Service
   Material to be stored, lethal or non-lethal, corrosive, etc.
5. Material of construction
6. Capacity
7. Overall size
   Diameter × overall height or length
8. Operating pressure
9. Operating temperature
10. Specific gravity of liquid
11. *Design specifications*
    11.1 Design pressure
    11.2 Design temperature
    11.3 Maximum allowable working pressure
    11.4 Minimum design temperature
    11.5 Construction code
    None, API 650 or ASME Section VIII
    11.6 Material code
    Specify ASTM (ASME Section II) or equal for shell and dished ends

11.7 Vessel thickness
Specify shell and end thicknesses.
11.8 Corrosion allowance
11.9 Special requirements
Loyds inspection required or not required
Foam chamber, none or required
11.10 Air and Hydro tests
11.11 Radiography
11.12 Joint efficiency
11.13 Stress relief
11.14 Support details
Skirt, lugs, saddles, etc.
11.15 Attachments
Require insulation support rings, clips for grounding ladders, platforms, etc.
11.16 Painting and protective coating
None or specify for inside surface, red oxide primer for outside surface, etc.

12. Connections
Refer sketch No. (attach sketch to show overall dimensions, quantity required, nozzle schedule, and orientation; manways and flange ratings, etc.).
Provide nozzles for liquid inlet and outlet; vent, drain and sampling; level gauge glass, pressure and temperature indicator; safety and emergency relief valves; adequate spares.
13. Weight empty and full of water
Necessary to compute dynamic loadings

A5.1.3 Heat Exchanger Specifications

1. Name and item no.
2. Type of heat exchanger
   (a) Horizontal or vertical; shell and tube; U-bend, floating head or fixed tube sheet; single or double pass shell, split or divided flow type, etc.
   (b) Specification of code type (TEMA, Class B, C and R)
3. Location
   Outdoor or indoor at ground level or on a structure, etc.
4. Service
   Fluid flowing, lethal or non-lethal, corrosive slurry, etc.
5. Material of construction for shell as well as tube
6. *Performance data*

	Shell Side	*Tube Side*
6.1 Material entering		
Name of components		

		Shell side	*Tube side*
	Phase		
	Individual flow rates		
	Total flow rate		
6.2	Material leaving		
	Name of components		
	Phase		
	Individual flow rates		
	Total flow rates		
6.3	Fluid composition for		
	Material entering		
	Material leaving		
6.4	Heat transferred		
	Sensible heat-vapour or gas		
	Condensing or vaporizing heat		
	Sensible heat-liquid		
	Total		
6.5	Operating conditions		
	Pressure		
	Temperature—Inlet		
	—Outlet		
	Bubble point		
	Dew point		
	Maximum allowable pressure drop		
	Fouling		

7. *Fluid properties*

		Shell side	*Tube side*
7.1	Arithmetic mean temperature		
7.2	Molecular weight		
7.3	Density—liquid		
	—vapour or gas		
7.4	Specific heat—liquid		
	—vapour or gas		
7.5	Latent heat		
7.6	Thermal conductivity—liquid		
	—vapour or gas		
7.7	Viscosity—liquid		
	—vapour or gas		

8. Restrictions if any
   Specify restrictions like size, maximum or minimum shell and the side velocities, tube diameters, etc.
9. Overall size and area
   Specify shell ID, length or height, number of tubes, tube OD × BWG, pitch, number and type of baffles, and spacing etc.; and surface area/shell.

10. *Design specifications*

		Shell side	*Tube side*
10.1	Design pressure		
10.2	Design temperature		
10.3	Maximum allowable working pressure		
10.4	Minimum design temperature		
10.5	Construction code TEMA, Class R, C or B and ASME, Section VIII, etc.		
10.6	Material code Specify ASTM (ASME Section II) or equivalent for shell, shell cover, tube, tube-sheet, head, head cover, baffles, tie rods, spacers, gaskets, packings, and bolting, etc.		
10.7	Corrosion allowance Specify for shell only		
10.8	Special requirements None, tubes welded to tube-sheet, expansion joint on shell, Lloyds inspection required or not required, etc.		
10.9	Air and hydro tests		
10.10	Radiography		
10.11	Joint efficiency		
10.12	Stress relief		
10.13	Support details		
10.14	Attachments		
10.15	Painting		

11. Connections
Refer Sketch No. (attach sketch to show overall dimensions, quantity required, nozzle schedule, orientation and flange ratings).
Provide nozzles for inlet, outlet, vent, drain, safety valve and spares
12. Drawings
Specify drawing numbers or vendor to supply

A5.1.4 Distillation Equipment

Distillation equipment consist of columns installed with trays or packings. The column is usually designed and specified as a coded pressure vessel. The trays are, however, covered by proprietary designs. Some of the well-known tray names are Hydronyl, Nutter, Glitach, Koch, Linde, etc.

(a) *Distillation Column*

1. Name and Item No.
2. Type of column
   Vertical column with dished heads, etc.
3. Location
   Outdoor or indoor
4. Service
   Type of process material (organic, hydrocarbon, etc.), lethal or non-lethal, corrosive, etc.
5. Material of construction
   Specify for each tray section and heads
6. Operation conditions
   Pressure
   Temperature
   Flow rates—top section
   —bottom section
   Densities—top section
   —bottom section
7. Tray details
   Actual trays—top section
   —bottom section
   Tray spacing—top section
   —bottom section
   Side takeoff trays
8. Overall size
   Column ID
   Maximum space above top tray
   Maximum space below bottom tray
   Height/tray section
   Overall height
9. *Design Specifications*
   9.1 Design pressure
   9.2 Design temperature
   9.3 Maximum allowable working pressure
   9.4 Minimum design temperature
   9.5 Construction code
   9.6 Material code
   9.7 Vessel thickness
   9.8 Corrosion allowance
   9.9 Special requirements
   Lloyds inspection, none, etc.
   9.10 Air and hydro tests
   9.11 Radiography
   9.12 Joint efficiency
   9.13 Stream relief

9.14 Support details
Specify column support like skirt, lugs, tray support details, etc.

9.15 Attachments
Require insulation support, ladder, platforms, etc.

9.16 Painting
Internal none, external red oxide primer, etc.

10. Connections —
Refer sketch no. (attach sketch to show overall dimensions, quantity required, nozzle schedule and orientation, manways and flange ratings).
Provide nozzles for feed, reflux, vapour outlet, bottom liquid outlet, kettle vapour return, differential pressure, pressure, temperature, and level gauges, gauge glass, side takeoffs, sampling, safety valves, and spares.
Provide manways for each tray section.

11. Weight empty and with liquid
Specify empty weight of the column, maximum height of liquid on trays (usually assumed 3-4 inches/tray) and its specific gravity, maximum height of liquid in the bottom head and its specific gravity. If used as an extractor, specify weight full of liquid.
This data is required to calculate dynamic loadings of the column.

12. Drawings
Specify drawing number or vendor to supply.

(b) *Distillation trays*

1. Name and Item No.
2. Tray type
Sieve, valve, bubble cap, etc. with construction—Package, sectional, single piece, etc.
3. Service
Hydrocarbon, organic, lethal, non-lethal, corrosive, etc.
4. Column diameter
Specify column ID or tray vendor to specify.
5. Number of trays
Specify tray numbers above and below feed nozzles.
Also specify side takeoff tray numbers.
6. Tray spacing
Specify spacing or tray vendor to specify.
7. Tray support
Ring type, specify size or vendor to supply.
8. *Performance data*

	Top section	*Bottom section*
Vapour rate		
Liquid rate		
Vapour density		
Liquid density		

Top section *Bottom section*

Liquid molecular weight
Vapour molecular weight
Liquid viscosity
Vapour viscosity
Surface tension
Operating pressure
Operating temperature
Specify tray loadings for other locations in the column.
Specially if the flows are significantly different like side stream take-off trays. Specified flow rates are based on maximum required flow rates. Flow specified are vapour to the tray and liquid from the tray.

9. Foaming characteristics
Specify non-foaming, slight, moderate, heavy foaming, etc.
10. Fouling tendency
Specify fouling problems like polymer, scale, precipitation etc.
11. Maximum allowable pressure drop
Specify reasonable pressure drop/tray. It is desirable to allow a pressure drop/tray of at least 3 mm Hg. Tray pressure drops less than 2 mm Hg are usually impractical.
12. Turn down ratio:
Maximum loading
Minimum loading
Specify flexibility of operation as % of design.
Vapour and liquid loads at constant L/V ratio
13. Corrosion allowance
14. Material of construction
Tray deck
Downcomer
Valves or caps
Fasteners
Support material and hardware
Specify material and thicknesses or request vendor to specify.
15. Special requirements
Specify special requirements like adjustable weir, preferred weir height, top removeable preferred for top section trays and bottom removeable preferred for lower section trays.
16. Drawings
Attach and specify drawing numbers of the column showing full details of nozzle locations, tray support details if available, manway locations and sizes, etc.
17. Vendor's supplies to include
17.1 Complete set of trays with downcomers, weirs, receiving and seal pans, distributors, supporting members tray components, clamps, washers, gaskets, packing, fasteners and all other

necessary appurtenances. Vendor to ensure that tray components will pass through the specified manway size.

17.2 Performance guarantee at maximum and minimum loadings of the specified design rates.

17.3 Complete tray installation drawings

17.4 Essential tray data, such as
Flow paths—no. of paths or passes, cross flow, etc.
Bubbling area
Blanking area
Design pressure drop
% flooding at design rates
Downcomer type, size, area and clearances
Type, size and number of valves, caps, sieves, etc.
Weir heights
Seal pan heights
Drain holes, etc.

A5.1.5 Pumps

The most common types of pumps installed in chemical plants are centrifugal and positive displacement pumps. Centrifugal pumps have rotating impellers in a fixed casing to develop the head either by radial or axial flow of liquid through the impeller. Depending upon the service, the impeller of a centrifugal pump may be closed or open type. Generally, an open impeller is specified if the liquid to be pumped has polymerizing tendencies or has suspended solids. Positive displacement pumps are of two types—rotary or reciprocating. Rotary pumps may be of gear, screw or vane type. Reciprocating pumps are of piston, plunger or diaphragm type and are used for high viscosity liquids or where high heads are necessary. The centrifugal pumps are driven by either an electric motor or steam turbine and the reciprocating pumps are usually driven by steam or air.

The following are the typical methods of specifying a centrifugal and reciprocating piston pump:

(a) *Centrifugal Pump*

1. Name and Item No.
2. Type of pump
   Specify horizontal or vertical, open or close impeller, etc.
3. Service
   Clear or viscous, non-corrosive, non-abrasive or corrosive, having suspended solids, continuous or non-continuous, etc.
4. Location
   Outdoor or indoor.
5. *Performance data*
   5.1 Capacity
   5.2 Total head of liquid

5.3 NPSH available of liquid
5.4 Pressure at suction (normal/maximum)
5.5 Discharge pressure
5.6 Material handled
Specify fluid, composition if fluid is a mixture, character of fluid, suspended solids, particle size and percentage by weight of solids, etc.
5.7 Specific gravity (normal/maximum)
5.8 Viscosity (normal/maximum)
5.9 Fluid temperature at suction
5.10 Atmospheric boiling point
5.11 Brake horsepower
(Vendor to specify)
5.12 Pump speed
(Vendor to specify)

6. *Driver specifications*
Type of driver
Electric motor—specify enclosure like Class I, Group D, Division 2, and supply voltage like 440 V, 3 Phase, 50 Cycles AC
Steam turbine—specify type like back pressure type and steam quality, supply and exhaust pressures as dry/saturated steam
7. Material of construction
Specify for casing, impeller, shaft, gland and bolting
8. *Construction*

8.1	Casing type	radially split preferred
8.2	Casing vent and drain	required/not required
8.3	Bearing type	ball or roller, water cooled required/not required
8.4	Jacketed stuffing box	required/not required
8.5	Shaft seal	packing or mechanical
8.6	Mechanical seal	single or double, water cooled, flush across seal face etc.
8.7	Shaft	solid or sleeve type preferred
8.8	Coupling	direct coupled, flexible type preferred
8.9	End connections	flange type per ANSI standard
8.10	Self priming	required/not required

9. Vendor's supplies to include
9.1 Complete pump, driver, base plate, couplings, etc. in semi-assembled condition
9.2 Characteristic curves showing total head, efficiency, brake horsepower, and NPSH required, as a function of volume pumped
9.3 Complete information of driver, pump rating, bearings, lubrication, coupling, shaft seal, materials of construction, peak head and brake horsepower; foundation and installation drawings and details

9.4 Operating manual, parts catalogue and recommended spare parts for 2 years' operation

9.5 Performance guarantee for operation at design conditions

(b) *Reciprocating Piston Pump*

1. Name and Item No.
2. Type of pump
   Specify simplex, duplex, double acting, horizontal type etc.
3. Servicc
   Specify type and nature of liquid.
4. Location
   Outdoor or indoor
5. Capacity
6. Total head
7. Suction pressure
8. Discharge pressure
9. NPSH available
10. Material handled
11. Specific gravity (normal/maximum)
12. Viscosity (normal/maximum)
13. Fluid temperature
14. Steam or air pressure
    Inlet
    Exhaust
15. Construction

Size	vendor to specify liquid and steam/air cylinder diameter × stroke × normal and maximum piston speeds
Steam cylinder	cast iron, lubricated, standard type, ratings
Liquid cylinder	twin, submerged piston, cap and valve plate type with renewable type linear material vendor to specify
Liquid piston	body and follower type
Piston rings and rods	standard type
Cylinder valves	standard force feed lubricator
Flanges	required per ANSI standard

16. Vendor's supplies to include
    16.1 Complete pump and base plate
    16.2 Speed versus delivery curve per cylinder
    16.3 Steam/air consumption at design pumping rate
    16.4 Operating manual, parts catalogue, recommended spare parts, foundation and installation drawings
    16.5 Performance guarantee for operation at design conditions

A5.1.6 Compressors

Compressors are used for pumping gases and can be broadly classified as positive displacement compressors and dynamic compressors. Positive displacement compressors are of two types, reciprocating or rotary. Dynamic compressors are classed as centrifugal, axial or mixed flow compressors. The most common installations in chemical plants are reciprocating and centrifugal compressors. Depending upon the size, service and location, a reciprocating or a centrifugal compressor is selected.

A typical method of specifying a reciprocating and a centrifugal compressor is stated as follows:

(a) *Reciprocating Compressor*

1. Name and Item No.
2. Type of machine
   Specify lubricated or non-lubricated, horizontal or vertical, single acting or double acting, single or multistage, balanced opposed or tandem arrangement, etc.
3. Service
   Specify service as continuous/intermittent, corrosive, dirty, or clean.

5. *Performance requirements*
   - 5.1 Gas condition
     Specify the gas to be compressed, its exact condition like clean, fouling, superheated, saturated, etc.
     Specify composition if gas is a mixture.
   - 5.2 Number of stages
     Specify number of stages as required by process condition, otherwise vendor will specify
   - 5.3 Gas molecular weight
   - 5.4 Suction condition at each stage:
     - 5.4.1 Humidity or dew point
     - 5.4.2 (C_p/C_v) value
     - 5.4.3 Compressibility
     - 5.4.4 Intake pressure
     - 5.4.5 Intake temperature
     - 5.4.6 Flow density
     - 5.4.7 Capacity, actual
   - 5.5 Discharge pressure at each stage
   - 5.6 Discharge temperature at each stage
     Specify temperature limit, otherwise vendor will specify
   - 5.7 Capacity control
     Specify range and mode of control like external by-pass, 0–50–75% capacity control by valve unloaders, clearance volume pockets, etc.
   - 5.8 Safety protection
     Specify protection to the compressor, like high discharge pressure and temperature to trip the motor, low lube oil pressure to trip the motor, high bearing temperature alarm, etc.

5.9 Accessories

Intercoolers	required/not required
Liquid separators	-do-
Volume bottles	-do-
Pulsation dampers	specify % peak to peak residual pulsation, if required

5.10 Vibration and torsional analysis
Specify required or not required

6. Brake horsepower and speed
Vendor to specify bhp required, driver hp and compressor speed
7. Driver
Electric motor—specify induction or synchronous, area classification like suitable for Class I, Group D, Division 2 location, supply voltage, phase, frequency, power factor, etc.
Steam turbine—specify steam quality, inlet and exhaust pressures
8. Drive preferred
Specify type like V-belt, speed reducer gears, flanged coupling, etc.
9. Utilities available
Cooling water—specify close cycle, inhibited, or raw river water, fouling factor, supply and return pressures, supply temperatures and allowed temperature rise
Plant air or nitrogen—specify supply pressure
10. Mechanical specification
Vendor to specify the following:
Frame size and rating
Number of stages and cylinders/stage
Cylinder size and stroke
Cylinder design pressures
Cylinder clearances
Cylinder volumetric efficiencies
Piston displacement/cylinder
Piston speed
Piston rod diameter and maximum pin load
Type of valves
Number of valves/cylinder
Valve gas velocity
Packing type
Flange sizes and ratings
11. Lubrication system
11.1 Frame lubrication—specify splash or pressure type
11.2 Cylinder and packing—specify force feed motor driven or compressor shaft driven, etc.
11.3 Lubricating oil type
12. Material of construction
Vendor to specify materials for cylinders, liners, pistons, piston rod, piston rings, valve seat, rings/plates, and springs, bottles,

intercooler shell, tube, tubesheet, separators, etc.

13. Test and inspection
Specify hydrostatic test pressures of cylinders, bottles, etc.; and requirements of no-load or on-load tests and inspection during fabrication, etc.
14. Piping and instruments
Specify requirements such as complete interstage piping and valves, pressure and temperature gauges at suction and discharge of each stage, safety valve at the discharge of each stage, etc.
15. Foundation details
Vendor to specify
16. Overall size and weight
Vendor to specify space required and gross weight of the machine
17. Scope of supply
17.1 Complete compressor unit with driver and all accessories, piping, instruments, etc.
17.2 Performance guarantee
17.3 Foundation design
17.4 Foundation bolts
17.5 Barring device
17.6 Special tools
17.7 Spare parts (2 years' operation)
17.8 Instruction book
17.9 Installation drawings

(b) *Centrifugal compressor*

1. Name and Item No.
2. Type of compressor
Specify single or multistage, horizontally or vertically split, etc.
3. Service
Specify nature of gas to be compressed like hydracarbon gas, carbon dioxide, etc.; clean/dirty, saturated or superheated, corrosive or non-corrosive, continuous dependable service, etc.
4. Location
Outdoor or indoor
5. *Performance requirements*
5.1 Gas composition
5.2 Gas molecular weight
5.3 (C_p/C_v) value—attach (C_p/C_v) versus temperature curve
5.4 Relative humidity or dew point
5.5 Compressibility factor
5.6 Suction pressure
5.7 Suction temperature
5.8 Flow density
5.9 Flow volume, actual
5.10 Final discharge pressure
5.11 Final discharge temperature (specify limit if any)

5.12 Interstage flow stream (if any)
Specify flow rate, pressure and temperature, C_p/C_v value, flow density, compressibility etc.

5.13 Intercoolers and separators
Specify requirements and allowable pressure drops

6. *Compressor details*

6.1	Exact type, size and description	vendor to specify
6.2	Number of stages	-do-
6.3	Compression ratio/stage	-do-
6.4	Operating speeds (maximum/normal/minimum)	-do-
6.5	Critical speeds	-do-
6.6	Rotation from driver end	-do-
6.7	Maximum allowable working pressure	-do-
6.8	Maximum allowable working temperature	-do-
6.9	Shaft horsepower	-do-

7. Utilities available
Cooling water—closed cycle, inhibited etc., specify supply and return pressures, supply temperature and allowable temperature rise.

8. Driver
Electric motor—vendor to specify horsepower, speed and type. Specify supply voltage, phases, frequency and power factor. Specify electrical area classification, etc.

9. *Gear speed increaser*

9.1	Horsepower rating	vender to specify
9.2	Gear ratio	-do-
9.3	Service factor	-do-
9.4	Low/high speed shaft	-do-
9.5	Low/high speed couplings	-do-

10. Construction
Generally in accordance with API 617, centrifugal compressors for refinery service

11. *Material of construction*
11.1 Casing material
11.2 Diffuser
11.3 Guide vanes
11.4 Shaft material
11.5 Impeller material
11.6 Bolting material

12. End connections
Vendor to specify nozzle location, size and rating

13. Seal type
Vendor to specify—purge type or oil film type (labyrinth), etc.

14. Lube and seal oil system
Complete lube and seal oil system for compressor, driver, gear speed increaser in accordance with API 617 to be specified by vendor

15. *Controls required*
    - 15.1 Constant discharge pressure control
    - 15.2 Minimum flow control (Antisurge)
    - 15.3 Excessive vibration control
    - 15.4 Seal and lube oil failure control
    - 15.5 Bearing temperature control
16. Bearing
    Vendor to specify
17. *Testing and inspection*
    Specify the requirements as applicable:
    - 17.1 Static and dynamic balancing of impeller and completely assembled rotor
    - 17.2 Ultrasonic inspection of the compressor shaft
    - 17.3 Hydrostatic test
    - 17.4 Speed test
    - 17.5 Vibration level test
    - 17.6 No-load or on-load tests
18. Overall size and weight
    Vendor to specify overall dimension and weight of the compressor
19. Scope of supply
    - 19.1 Complete unit with compressor, gears, driver and all accessories including seal and lube oil system, safety devices, interstage piping and instruments, etc.
    - 19.2 Foundation design and foundation bolts
    - 19.3 Installation drawings
    - 19.4 Head capacity and other performance curves
    - 19.5 Performance guarantee
    - 19.6 Instruction books and recommended spare parts for 2 years' operation
    - 19.7 Special tools

A5.1.7 Specification of Pipes and Valves

(a) *Pipes*

Specify:
1. Size
2. Material and code
3. Schedule
4. Construction (seamless, ERW, butt welded, cold drawn, etc.)

(b) *Flange*

Specify:
1. Size
2. Rating and code
3. Material and code

4. Construction (screwed, slip-on, welding neck, etc.)
5. Facing (flat face, raised face, ring joint, etc.)
6. Flange type (straight, reducing, orifice, blind, etc.)
7. Bore and class

(c) *Elbow, Tee and Reducer*

Specify:
1. Size
2. Material and code
3. Schedule
4. Construction (seamless, welded, etc.)
5. Type (90°/45° long radius equal or reducing elbow, equal or reducing tee, concentric and eccentric reducers, etc.)

(d) *Gaskets*

Specify:
1. Flange size and rating
2. Types and class:
   Compressed asbestos ANSI B 16.21
   Soft iron oval ring ANSI B 16.20
   Spiral wound gaskets API 601

(e) *Valves*

Specify:
1. Size
2. Type (gate, globe, check, plug valves, etc.)
3. Rating and code
4. End connection (flat face, raised face, ring joint, screwed, butt welding, etc.)
5. Material for body, disc, seat and trim
6. Packing (asbestos, teflon, etc.)
7. Gasket (compressed asbestos, spiral wound, soft iron ring, metallic flat or corrugated, etc.)
8. Bonnet (bolted, screwed, inside screw, outside screw and yoke, etc.)
9. Prefix details with valve numbers from piping and valve specification sheets

(f) *Bolting*

Specify:
1. Size × length
2. Material and code

Common standards used in piping and valve specifications are:

ASTM	American Society of Testing Materials
API	American Petroleum Institute

ASI	American Standards Association
ANSI	American National Standards Institute
AAR	Association of American Railroad
MSS	Manufacturer's Standardization Society of Valves and Fittings Industry

REFERENCES

1. Neale, D. F., *C.E.P.*, **70**(10), 53, 1974.
2. Kletz, T. A., Second International Symposium on Loss Prevention and Safety Promotion in Process Industries, Heidelberg, Sept. 6–9, 1977.
3. Stainiar, W., *Plant Engineering Handbook*, 1st edn, McGraw-Hill, N.Y., 1950.
4. Lewis, B. and von Elbe, G., *Combustion, Flames and Explosion of Gases*, 1st edn, Academic Press, New York, 1961.
5. Coward, H. F. and Jones, G. W., *U.S. Bureau of Mines Bull.*, No. 503, 1952.
6. Factory Mutual Engineering Corporation, *Loss Prevention Data*, Massachusetts (these data sheets supersede their earlier publication: Staff of Factory Mutual Insurance, *Handbook of Industrial Loss Prevention*, 1st edn., McGraw-Hill, N.Y., 1959).
7. van Eijnatten, A. L. M., *C.E.P.*, **73**(9), 69, 1977.
8. Freeman, R., *C.E.P.*, **71**(11), 71, 1975.
9. Wallace, R. D. and McKinney, R. W., *Ind. Eng. Chem.*, **44**, 1508, 1952.
10. Sargent, H. B., *Chem. Eng.*, **64**, 250, Feb. 1957.
11. Sutherland, M. E. and Wegert, H. W., *C.E.P.*, **69**(4), 48, 1973.
12. Perry, R. H. (ed.), *Chemical Engineers' Handbook*, 5th edn., McGraw-Hill, N.Y., 1975.
13. Fawcett, H. H. and Wood, W. S. (eds.), *Safety and Accident Prevention in Chemical Operations*, 1st edn., Wiley, N.Y., 1965.
14. Handley, W., *Industrial Safety Handbook*, 1st edn., McGraw-Hill, London, 1969.
15. Code for Unfired Pressure Vessels (IS : 2825–1969), Indian Standards Institution, New Delhi, 1969.
16. Rules for Construction of Unfired Pressure Vessels, Section VIII, The American Society of Mechanical Engineers, N.Y., 1962.
17. A.P.I. Codes for Unfired Pressure Vessels.
18. Kletz, T. A., *C.E.P.*, Vol. **71**(9), 63, 1975.
19. Robinson, E. L., *Trans. ASME*, **77**, 1147, 1955.
20. Hartmann, I., *Ind. Eng. Chem.*, **40**, 752, 1948.
21. Hartmann, I. and Nagy, *Ind. Eng. Chem.*, **49**, 1734, 1957.
22. Maisy, H. R., *Chem. & Process Eng.*, **46**, 662, 1965.
23. Gibson, N. and Harris, G. F. P., *C.E.P.*, **72**(11), 62, 1976.
24. Bartknecht, W., *C.E.P.*, **73**(9), 45, 1977.

25. Cubbage, P. A. and Simonds, W. A., Gas Council Research Communications, Nos. GC 23 (1955) and GC 43 (1943).
26. Stull, D. R., *Fundamentals of Fire and Explosion*, No. 10, Vol. **73**, AIChE Monograph Series, 1977.
27. Bodurtha, F. T., Palmer P. A. and Walsh, W. H., *C.E.P.*, **69**(4), 37, 1973.
28. Bosanquet, C. H. and Pearson, J. L., *Trans. Farad. Soc.*, **32**, 1249, 1936.
29. Bodurtha Jr., F. T., *Chem. Eng.*, **65**, No. 25, 177, Dec. 17, 1958.
30. Bodurtha Jr., F. T., *J. Air Poll. Control Assoc.*, **11**(9), 431, 1961.
31. Basu, A., Hindustan Lever, Bombay, Private Communication, 1974.
32. Ludwig, E. E., *Applied Process Design for Chemical and Petrochemicals Plants*, Vols. 1 to 3, 1st edn., Gulf Publishing Co., Houston, Texas, 1964.

PROBLEMS

5.1 There are several problems in the handling of different intermediates and raw materials in the manufacture of ether. Envisage them for
(a) ethanol, and
(b) diethyl ether.
Show how you improve the flow sheet from the safety point of view. What safety precautions would you take on reactor and bottle filling sections. Design a suitable stack for this plant.

5.2 Consider the strategy of storing diethyl ether in an open storage tank and it is desired to evaluate its intrinsic safety. The concentration of ether at the surface is that of a pure component, however, as one goes up, it keeps reducing. The distance between the upper and lower concentration limits constitutes the dangerous zone. Find this treating it as a one dimensional problem.

5.3 Solve the problem above if a gentle wind is blowing across the tank.

5.4 All liquids have a critical handling temperature above which they form an explosive mixture with air. Find this temperature for
(a) acetaldehyde
(b) diethyl ether
(c) acetone
(d) ethanol.
Hint: Use the Clausius-Clapeyron equation to get the temperature dependence of the vapour pressure.

5.5 Design a safe pipeline system for acetylene when one requires 1000 kg/h of gas at 2 atm from an acetylene plant situated 300 km away. Write down different ways one can do this and find the best alternative.

5.6 Design a suitable acetylene reactor system giving 1000 kg/h gas *continuously*. The proposed reactor given in Fig. 5.2 has following steps of operation:
(1) Bringing CaC_2 from storage area to the reactor (10 min)
(2) Filling of the reactor with water (10 min)
(3) Charging hopper (10 min)
(4) Feeding CaC_2 (fed continuously during the reaction time, 40 min)
(5) Draining residue from it (10 min)
(6) Agitation (agitated continuously).

It may be noted that it is a batch reactor and to get acetylene continuously, one must have more than one reactor. How many reactors would give continuous supply of acetylene ? Give complete details of sequencing.

5.7 Assume that the slurry of calcium hydroxide consists of 10% (by weight) solid in P. 5.6. Determine
(a) the volume of the reactor,
(b) H.P. of the pump required to feed water,
(c) the pump required to drain the slurry,
(d) the drain line size (optimum dia),
(e) the diameter of the safety vent line, and
(f) the horse power required for the agitator.

5.8 Design the safety valve for the reactor of P. 5.6. Take the value of K as one.

5.9 Design the flare system for the
(a) acetylene plant,
(b) diethyl ether plant (throughput of 1000 kg/h ether),
(c) acetaldehyde plant (throughput 1000 kg/h acetaldehyde).

5.10 In a urea plant there is a prilling tower to make granules from the solution. One wishes to dry the solution as fast as possible in order to suppress biurette formation. Its diameter is 4 m and height 10 m. However, one deals with dry solid powder in this tower and therefore it is likely to undergo a dust explosion. Find the vent area using Fig. 5.18 assuming that it is a ST1 type explosion.

5.11 As has been pointed out in this chapter that to build in safety into plant design, it is necessary that one foresees all the problems that one would face in any chemical plant operation. It is a well-known fact that the plant layout affects the intrinsic safety of the plant.

Ideal plant layout would be one which minimizes total pipeline length. However, for safety purposes it is desired to put all pumps together. Consider the other plant once again and on a linear graph show all equipment 10 ft away from each other. Assume all pipelines arranged in a rectangular grid. Find
(a) the minimum pipe length
(b) the length of the pipe when all pumps are together.

6
Trouble-Shooting and Hazard Analysis

6.1 INTRODUCTION

A chemical plant consists of equipment, controllers and working personnel, and as explained earlier, the combination should be viewed in totality. The synthesis of process flow sheet in Chapter 4 and safety considerations discussed in Chapter 5 are to achieve a coordination of different activities in the production of the chemical with an effort to minimize *danger*. It may be remembered that the total life of a plant depends to a large extent upon both the number and nature of accidents. A realistic plant design is the one that minimizes accidents. The subject of this chapter is to establish a systematic method of locating and analyzing accidents.

It was shown in Chapter 4 that there are always deviations from the steady state in a plant operation. Accidents occur whenever these deviations become unbounded due to incorrect interaction between man and machine. They can be roughly classified as[1]

(1) nuisance,
(2) marginal,
(3) critical, and
(4) catastrophic.

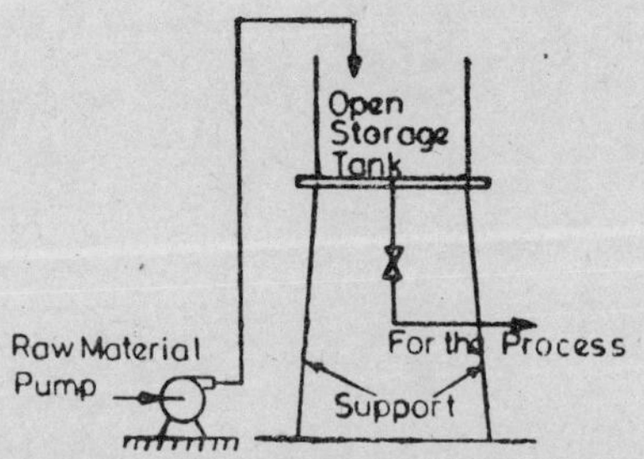

Fig. 6.1 A raw material storage tank

This classification is qualitative and is based upon the relative loss of properties and the life of personnel working in the plant. To illustrate the difference between them, an open storage tank of raw material shown in Fig. 6.1 is considered as an example. The raw material is assumed to be supplied intermittently through an outside agency (for example through a tank-car) is pumped into it using a large capacity pump. Supposing one is handling water. Since there is no mechanism to stop the input of water in the tank, there is a likelihood of spillage due to overflow. Since water is neither flammable nor corrosive, its spillage is totally unharmful. However, it is undesirable and is regarded as a nuisance because it does not lead to any loss of property or life.

Instead of water, it is now assumed that the raw material is a non-corrosive oily liquid with a very high flash point. Once again, its spillage does not cause any loss of life or property, but unlike water, it does not dry up by itself. In addition, slippery areas adjacent to the tank can lead to serious injuries and also the cleaning of oil is not an easy task. This situation is more serious than the above one and, therefore, is regarded as a marginal accident.

If the material to be pumped is concentrated sulphuric acid, the spillage will possibly lead to loss of life due to acid burns. The acid will also corrode the structure supporting the storage tank and with time it may collapse leading to loss of property also. Such an accident is regarded as critical because one can hardly afford to lose acid from spillage.

Lastly the pumping of acetone is considered which is class IB flammable material (*see* Table 5.5). On its spillage, it quickly evaporates forming an explosive mixture with air and can explode if a nearby ignition source is available. In addition, the fire can propagate upwards to the storage tank causing a very severe fire ultimately leading to loss of both life and property. This is called catastrophic because of its speed of propagation and the widespread devastation. The difference between critical and catastrophic is mainly in terms of the degree and the speed of spreading.

In this chapter, a method of identifying and analyzing accidents is given. Trouble shooting analysis, presented here, is a systematic but approximate procedure to identify possible sources of accidents and actions required to overcome them. This may be sufficient for nuisance and marginal accidents only. Critical and catastrophic accidents require a much more careful analysis of all factors that can lead to them. This is necessary for effective control. Fault tree method of analysis for critical and catastrophic situations is presented later in this chapter and it is shown how the chance of attaining the final goal can be improved by re-arrangement of plant structure with its help.

6.2 TROUBLE-SHOOTING ANALYSIS OF EQUIPMENT AND CHEMICAL PLANTS

Since any unbounded deviation from steady state value leads to accidents, hence its analysis and pinpointing the source of error, commonly known as trouble-shoooting analysis, must always be carried out. This can be performed for an equipment as well as for the whole plant. In the following, it is done for an equipment first.

In a trouble-shooting analysis one considers the equipment as a whole, which includes its mechanical as well as electrical aspect. This makes the analysis very difficult which is shown through an example of a centrifugal pump. The detailed drawing of its mechanical aspects is shown in Fig. 6.2. A pump can be roughly divided into three major sections: pump housing the impeller, driver (which is normally an electrical motor) and the pump driving shaft.

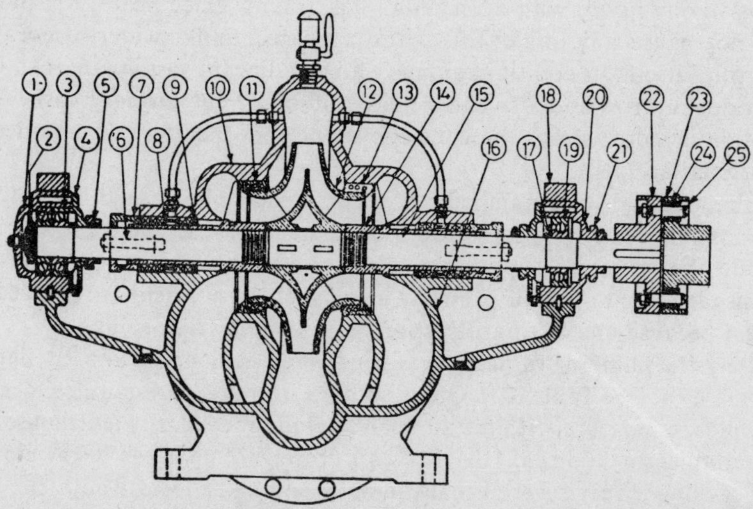

Fig. 6.2 Mechanical details of a centrifugal pump: 1. Outboard bearing house, 2. Bearing locknut and washer, 3. Ball radial and thrust bearing, 4. Bearing end cap, 5. Water slinger, 6. Glandswing bolt, 7. Stuffing box gland halves, 8. Water seal ring, 9. Stuffing box bottom, 10. Upper case, 11. Case wearing rings, 12. Impeller, 13. Impeller wearing rings, 14. Shaft sleeve and impeller, 15. Shaft, 16. Lower case, 17. Shaft collar, 18. Bearing cap, 19. Ball radial bearing, 20. Inboard bearing housing, 21. Grease guard, 22. Pump half coupling, 23. Motor half coupling, 24. Coupling bushing. (Reprinted from Reference 2 with the permission of the Factory Mutual Corporation, Massachusetts.)

In any pump, the design of the stuffing box is crucial because it provides a seal on the process fluid and prevents it from leaking out. At the same time, it should be possible for the driving shaft to rotate freely. On occasions, the process fluid gets hot, and under these circumstances it becomes necessary to cool the stuffing box.

Table 6.1 analyzes the pump for ten effects and different factors that will give rise to these. Readers are asked to go through the table and understand it completely. It may be realized that it is worthwhile to monitor some mechanical and electrical variables (like rotating speed, bearing temperature, rotor axial position, rotor axial piston, rotor radial vibration, balance piston pressures, electric motor power consumption and winding temperatures and gear acoustics), process variables (like suction and discharge pressures, process temperatures, process flows, suction liquid levels) and auxiliary conditions (like lube and seal oil

Table 6.1 Pump Troubles and Causes[a]

SUCTION

S.No.	Causes	(1) Discharge pressure too low for gpm discharge	(2) Insufficient water discharge	(3) Pump loses suction after start up	(4) Discharge pressure not constant for same flow rate	(5) Too much power required	(6) Pump is noisy or vibrates	(7) No water discharge	(8) Pump unit will not start	(9) Pump or driver overheats	(10) Excessive leakage at stuffing box
1.	Suction lift too high	×	×	×	×		×	×			
2.	Foot valve too small, partially obstructed, or of inferior design causing excessive suction head loss	×	×	×	×		×				
3.	Air drawn into suction connection through foot valve not submerged enough	×	×	×	×		×				
4.	Air drawn into suction connection through leak	×	×	×	×			×			
5.	Suction connection obstructed	×	×	×				×			
6.	Air pocket in suction pipe	×	×	×	×			×			
7.	Hydraulic cavitation from too high suction life	×	×				×				
8.	Well collapsed or serious misalignment					×	×		×	×	
9.	Stuffing box too tight or packing improperly installed, worn, defective, too tight or incorrect type	×			×	×	×		×	×	×
10.	Water seal or pipe to seal obstructed	×	×	×	×					×	

	S.No.	Causes	(1) Discharge pressure too low for gpm discharge	(2) Insufficient water discharge	(3) Pump loses suction after start up	(4) Discharge pressure not constant for same flow rate	(5) Too much power required	(6) Pump is noisy or vibrates	(7) No water discharge	(8) Pump unit will not start	(9) Pump or driver overheats	(10) Excessive leakage at stuffing box
SUCTION	11.	Air leak into pump through stuffing boxes	×	×	×	×						
	12.	Impeller obstructed	×	×			×	×	×		×	
	13.	Wearing rings worn	×	×			×					
	14.	Impeller damaged	×	×				×				
	15.	Wrong diameter impeller	×	×			×				×	
	16.	Actual net head lower than rated	×	×								
	17.	Casing gasket defective, permitting internal leakage (multistage pumps only)	×	×			×					
	18.	Pressure gauge on top of pump casing	×									
	19.	Incorrect impeller adjustment (vertical pump only)		×			×			×	×	
PUMP AND/OR DRIVER	20.	Impeller locked								×		
	21.	Pump frozen								×		
	22.	Pump shaft or shaft sleeve scored, bent or worn					×	×			×	×
	23.	Pump not primed							×		×	
	24.	Seal ring improperly located in stuffing box, preventing water from entering space to form seal			×						×	

PUMP AND/OR DRIVER	25.	Excess bearing friction due to lack of lubrication wear, dirt, rusting, failure or improper installation			×	×	×		
	26.	Rotating element binds against stationary element			×	×		×	
	27.	Pump and driver misaligned. Shaft running off centre or misalignment			×	×		×	×
	28.	Foundation not rigid			×	×		×	
	29.	Engine cooling system obstructed. Heat exchanger or cooling system too small. Cooling pump faulty						×	
DRIVER	30.	Faulty driver					×		
	31.	Lack of lubrication			×	×	×	×	
	32.	Speed too low	×	×					
	33.	Wrong direction of rotation	×	×	×			×	
	34.	Speed too high			×			×	
	35.	Rated motor voltage different from line voltage	×	×	×			×	
	36.	Faulty electric circuit, obstructed fuel system or obstructed steam pipe or dead battery					×		

(Reprinted by permission from NFPA 20, Standards for the Installation of Centrifugal Fire Pumps, Copyright © 1978, National Fire Protection Association, Boston, USA)

pressures and temperatures, tube and coupling filter pressure drops, lube and seal oil tank levels, oil drain trap levels, strainer differential pressures).[3] Monitoring of these may not be important for small pumps, but for large ones, all these are monitored and an alarm or an automatic shutdown system is installed for the safeguard of the equipment. Lastly the performance equation of the pump given in Chapter 4 should be compared with Table 6.1. Except for effects indicated in columns (6), (9) and (10), all others are covered by it. Effects (6) and (9) arise due to mechanical faults and are symptoms which would lead to the pump failure in terms of reduced efficiency or complete breakdown. Effect (10) also arises because of faulty packing at the stuffing box and excessive loss of material due to leakage which may ultimately lead to reduced output flow rates.

Now the trouble-shooting analysis of a complete plant is considered. Theoretically one needs to do the detailed analysis of every equipment as was done for a centrifugal pump in Table 6.1 and the information so gathered combined together for the whole plant. Evidently this is not a simple task because this will require a complete understanding of mechanical as well as electrical details of all equipment. In the short cut method suggested below, such detailed understanding is not necessary and results arrived at are detailed enough for removing small snags for a trouble free production of the final product.

As emphasized earlier, a chemical plant is dynamic in nature and by this, it is meant that deviations will occur in variables from their steady state values. Table 6.2 gives five classes of deviations, each called by its guide word NONE, MORE OF, LESS OF, PART OF and MORE THAN. The last guide word OTHERS encompasses all other factors which are

Table 6.2[4] Meaning and Types of Process Deviations Generated

Guide word	Explanation
NONE	No forward flow when there should be, i.e., no flow or reverse flow.
MORE OF	More of any relevant physical property than should be, e.g., higher flow (rate or total quantity), higher temperature, higher pressure, higher viscosity, etc.
LESS OF	Less of any relevant physical property than there should be, e.g., lower flow (rate or total quantity), lower temperature, lower pressure, etc.
PART OF	Composition of system different from what it should be, e.g., change in ratio of components, components missing etc.
MORE THAN	More components present in the system than there should be, e.g., extra phase present (vapour, solid), impurities present (air, water, acids, corrosion products) etc.
OTHERS	What else can happen apart from normal operation, e.g., startup, shutdown, uprating, low rate running, alternative operation mode, failure of plant services, maintenance, catalyst change etc.

(Reprinted from reference 4 with the permission of Gulf Publishing Co., Houston, Texas)

important but not covered in the first five. The procedure of application of this method is to isolate a section of any given plant and consider these deviations one by one. For a given deviation, one considers all causes that would give rise to the deviation and against each cause, writes down the action taken to overcome it. The whole chemical plant is in this way analyzed section-by-section starting from any one end. The set of actions, thus generated, would be the operating procedure of the plant and constitutes its trouble shooting analysis. This procedure is demonstrated through an example below.

One considers the dimerization plant of Fig. 6.3. According to the procedure suggested above, the whole plant is divided into smaller subsections. The size of the section chosen can be small enough to consist of one equipment and big enough to contain the whole plant. However, the latter is a trivial situation, whereas the former is the same as the trouble-shooting analysis of an equipment discussed earlier. For example, among the ten effects of pump malfunctions studied in Table 6.1, effects (7) and (8) fall under the guide word NONE, effects (5), (9) and (10) under the guide word MORE OF, effects (1), (2) and (4) under the guide word LESS OF and effects (3) and (6) under the guide word OTHERS. Guide words PART OF and MORE THAN have no relevance to the operation of a pump.

The size of the subsection should be chosen such that it reveals the pitfalls in the running of the plant. Equipment whose operations are crucial for the running of a plant as a whole are considered individually whereas 'non-critical' equipment are combined together. For example, the flow sheet of Fig. 6.3 can be divided into two sections: *subsection 1* consisting of storage tank HT, pump p1, $\frac{1}{2}$-mile pipeline and settling tank ST and *subsection 2* consisting of heat exchanger HE, pump P2 and reactor R. Trouble-shooting analyses of each of these subsections have been made and summarized in Tables 6.3 and 6.4.

Readers should familiarize themselves with these tables. It may be realized in general that a chemical plant will fail to produce the product if any of its equipment develops a fault. Also an equipment will fail to work if any of its components stops functioning and facilities are provided to overcome these (for example, bypass line, shutdown, use of standby or repair after isolation, etc.). Under no circumstances, the faulty equipment is repaired while the plant is running and in view of this, it is not necessary for a chemical engineer to know complete mechanical details of chemical equipment. However, an awareness of approximate constructional detail helps.

In these Tables possibility of human error has been considered, like for example the closure of valves incorrectly. If a given valve is not mechanically operated, but is closed or opened automatically through the use of a controller air pressure, wrong position of the valve could develop due to faults in electrical circuit of the controller, failure of air pressure or excessive friction in the valve. In such situations it is recommended to use some kind of an alarm to help focus attention on them.

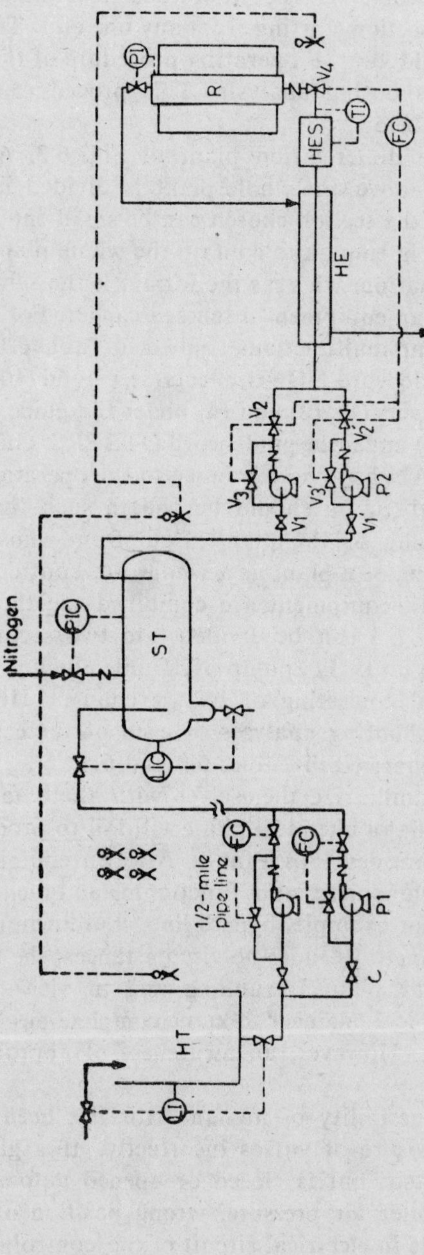

Fig. 6.3 Process flow sheet of a dimerization unit (see Chapter 4 also)

Table 6.3[5] Trouble Shooting Analyses of the Combination of Storage Tanks HT, Pump P1, Pipe Line PL and the Settling Tank ST

Guide word	Deviation	Possible causes	Consequence	Action required
NONE	No flow	(1) No hydrocarbon available in the storage tank, HT	Loss of feed to the reaction section and reduced output. Polymer formed in heat exchanger under no flow condition	(a) Ensure good communication with the storage tank operator (b) Install low level alarm on the settling tank
		(2) Pump P1 fails (motor fault, loss of drive, impeller corroded away etc.)	As for (1)	Covered by (b)
		(3) Line blockage, isolation valve closed in error, or LCV fails shut	As for (1)	Covered by (b)
			Pump P1 over-heats	(c) Install kickback on P1 (d) Check design of P1 pump strainer
		(4) Line fracture	As for (1)	Covered by (b)
			Hydrocarbon discharged into adjacent areas	(e) Institute regular patrolling and inspection of pipeline PL
MORE OF	More flow	(5) LCV fails open or LCV bypass open in error	Settling tank overfills	(f) Install high level alarm on LIC and check sizing of relief opposite liquid overfilling (g) Institute locking off procedure for LCV bypass when not in use
			Incomplete separation of water phase in tank leading to problems in the reaction section	(h) Extend pump P2 suction line to 25 cm above tank base

Guide word	Deviation	Possible causes	Consequence	Action required
	More pressure	(6) Isolation valve closed in error or LCV closes with pump P1 running	Transfer line, PL, subjected to full pump delivery pressure	(i) Covered by (c) except when blocked or isolated. Check line and flange ratings and reduce stroking speed of pump if necessary. Install a pressure gauge upstream of LCV and an independent pressure gauge on the settling tank
		(7) Thermal expansion in an isolated valved section due to fire or strong sunlight	Line fracture or flange leak	(j) Install thermal expansion relief valve on valved section (relief discharge route to be decided later)
	More temperature	(8) High storage tank temperature	Higher pressure in pipeline, PL, and settling tank, ST	(k) Check whether there is adequate warning of high temperature at storage tank. If not, install
LESS OF	Less flow	(9) Leaking flange or valved stub not blanked and leaking	Material loss to adjacent areas	Covered by (e) and checks in (i)
	Less temperature	(10) Winter conditions	Water sump and drain line freeze	(l) Lag water sump down to drain valve, and steam trace drain valve and drain line downstream
PART OF	High water concentration in stream	(11) High water concentration in storage tank	Water sump fills more quickly. Increased chance of water phase passing to reaction section	(m) Arrange for frequent draining of water from storage tank. Install high interface level alarm on sump

	High concentration of lower alkanes or alkenes in stream	(12) Disturbance in input hydrocarbon concentration	Higher system pressure	(n) Check that design of settling tank and associated pipe work, including relief valve sizing, will cope with sudden ingress of more volatile hydrocarbons
MORE THAN	Organic acid present	(13) As for 12	Increased rate of corrosion of tank base, sump and drain line	(o) Check suitability of materials of construction
OTHERS	Maintenance	(14) Equipment failure, flange leak, etc.	Line cannot be completely drained or purged	(p) Install low-point drain and N_2 purge point downstream of LCV. Also N_2 vent on settling tank

(Reprinted from reference 5 with the permission of American Institute of Chemical Engineers, New York)

Table 6.4[5] Trouble Shooting Analysis of Section 2 Consisting of Pump P2, Heat Exchanger HE and Reactor R

Guide word	Deviation	Possible causes	Consequences	Action required
NONE	No flow	(1) No hydrocarbon in settling tank	Loss of feed to reaction section and reduced output. Polymer formed in heat exchanger under no flow condition	(a) Covered by low alarm on settling tank (action b in Table 6.3)
		(2) Pump P2 fails (motor fault, loss of drive, impeller corroded away or detached, etc.)	As for (1), but also see 'reverse flow' below for other consequences	(b) Install low flow alarm on reactor feed controller, FC
		(3) Main line blockage, delivery line isolation valve V'1 closed in error, or V3 and V2 both closed simultaneously	As for (1)	Covered by (b)
			Pump P2 over-heats	(c) Check design/suitability of pump P2 strainer (d) Ensure valves V1, V'1' and V4 are locked open (e) Arrange the FC valve V2 and V'2' to remain open when air supply to controller fails
			Line section subjected to full delivery pressure	(f) Check line and flange ratings for withstanding maximum delivery pressure
		(4) Heat exchanger blockage (reactor feed side)	As for (1) and (3)	(g) Consequences covered (b), (c) and (f), but re-route reactor feed through tubes of exchanger and product through shell to facilitate cleaning (feed more prone to polymerization than product on hot

			surfaces; tubes easier to clean than shell side). Consider the need of relief valve on rerouted feed side of exchanger (h) Install pressure gauges upstream and downstream of exchanger to give warning of fouling up (j) Consider the need for a by-pass on feed side as well as product side of the exchanger (to be locked shut when not in use)
	(5) Line fracture	As for (1)	Covered by (b) and (f)
		Major spillage of hydrocarbon in plant area	(k) Partly covered by (f) but check plant layout and drainage with respect to fire hazard
Reverse flow	(6) On-line pump P2 fails	Sudden backflow of hot material from high pressure section into settling tank via pump P2 by-pass line or via P2 if its non-return valve has also failed	(l) Install two non-return valves in series immediately downstream of reactor feed control valve (a single non-return valve was deemed to be insufficiently unreliable) (m) Ensure that bypass lines of pump P2 should be of sufficient diameter to cope up with the maximum possible back flow (n) Provided for regular testing of all valves and non-return valves which forms part of safety protection system (general point for the whole project)

(*Contd.*)

Guide word	Deviation	Possible causes	Consequences	Action required
MORE OF	More flow	(7) Reactor feed control valve V2 does not sufficiently close or valve V3 remains completely open	Incomplete conversion of olefin in reaction section (due to shorter residence time) leading to lower efficiencies and to problems later in the process	(o) Install high flow alarm on reactor feed controller (p) The automatic valve V3 should be installed such that it is completely closed without air pressure
	More pressure or temperature	(8) Line or heat exchanger blockage, closure of valve V2 or V4; high reactor product temperature; or local plant fire	Line fracture, major spillage, lost output, possibly, major plant fire	(q) Covered by (f) and (k) but reconsider the need for relief protection on reactor feed side in (g), and also need for remote isolation. Also check design of the exchanger shell with respect to the maximum pressure and temperature
LESS OF	Less flow	(9) Flange leak, leak at valved stub or at pump P2 seals	Less of material into plant area and possible fire	(r) Covered by (f) and (k), but important to keep the number of flanges in the high-pressure section to a minimum, and to fit permanent blanks on all valved stubs (s) Check design of P2 pump seals, and consider the need for a permanent steam or water quench
		(10) Holed heat exchanger tube	Reactor product contaminated with reactor feed, leading to reduced chemical efficiencies and to problems later in the process	(t) Install appropriate sample points in reactor product line upstream and downstream of exchanger

Guide word	Deviation	Possible causes	Consequences	Action required
	Less pressure	(11) Deterioration in pump P2 efficiency	Reduced feed rate to reaction and lower output	(u) Check pump P2 design for delivery at 20 atm, and for compatibility with reactor control system
	Less temperature	(12) Fouling of heat exchanger leading to reduced heat transfer	Increased heat load on the heat exchanging system (HES1) and increased cooling requirement on reactor product after-cooler	(v) Consider installation of thermowells inlet and exit exchanger to permit periodic checks on performance
PART OF	High water concentration in reactor feed	(13) Incomplete separation of suspended water in settling tank	Reduced conversion and efficiency on reaction side	Covered by action (n) in Table 6.3
	High concentration of lower alkanes or alkenes in the reactor feed	(14) Variation in the available raw material	Higher system pressure	(w) Check that the design of the heat exchanger and associated pipework, including the sizing of any relief valves, will cope with sudden ingress of more volatile hydrocarbons
	Polymer formation in heat exchanger	(15) Low feed rate to the reaction section	Increased fouling of the heat exchanger and reduced reaction efficiencies and output	(x) Covered by (b), but install sample point in reactor feed line down-stream of exchanger to permit periodic checks on polymer formation. This sample point could also be used for general analysis, e.g., for checks on suspended water
OTHERS	Maintenance or inspection	(16) Fault or leak in the settling tank or on associated pipework. Statutory tank inspection	Tanks cannot be inspected or maintained unless all connecting lines are positively isolated	(y) Make provision for slip plating of the nitrogen inlet and exit lines on the settling tank. Also consider the need for emptying the tank prior to maintenance

(Reprinted from reference 5 with the permission of American Institute of Chemical Engineers, New York).

Leakage and spillage of material must be avoided not only because of financial loss due to loss of material, but also because of fire hazards due to it. In view of this, number of flanges, specially in high pressure regions, must be minimized. Actions like (a) and (e) of Table 6.3 (and similar actions of both these Tables) are, by any stretch of imagination, not trivial because they define the interaction of men and machines. Only an analysis of this kind can reveal the nature of this kind of interaction.

Lastly, any good plant design should accommodate the nature of the fluid handled. For example in the dimerization plant, the reaction of dimerization is known to be followed with polymerization, which is not desired. This can be minimized by carefully controlling reactor temperature and residence time. In spite of this, polymerization occurs, and it does so not only in the reactor but also in the heat exchanger. This has been pointed out in Table 6.4, and the only method of surmounting this process difficulty is to monitor it. One discusses next the fault tree analysis for critical and catastrophic accidents.

6.3 FAULT TREE ANALYSIS OF ACCIDENTS

Fault tree analysis is a method of analyzing an event, for example spillage of liquid from the storage tank (Fig. 6.1). An event which is desired to be analyzed is called a *head event*. As opposed to this, *basic events* are the ones which occur at the elemental level. By the word element, one implies the smallest subdivision of the analysis. This method of analysis consists of breaking a head event into its basic events. Basic events are further classified as primary and secondary events. The former is the one caused by the characteristics inherent in the component itself whereas the latter is the one caused by an external source. In the following the method of drawing a fault tree is illustrated through examples in order of increasing difficulty.

Figure 6.4 shows the schematic drawing of a furnace. Air and gaseous fuel are sent to burners placed suitably within it. Their flow rates are controlled through valves which may be manual or automatic. The purpose of these is to maintain a constant fuel to air ratio for smooth burning of the fuel in the furnace. A pilot light is provided within the furnace so that the ignition of burners can be obtained automatically by opening air and fuel valves. One considers the head event of explosion in the furnace and the fault tree analysis of this event can be carried out as follows.[6]

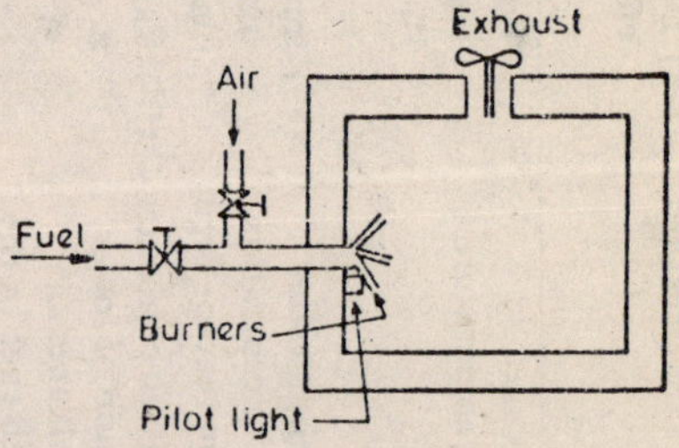

Fig. 6.4 Schematic diagram of a furnace

Explosion in the furnace will occur when the gas and air are in a ratio lying between the upper and lower explosive limits of the gaseous fuel. It is also necessary that there is an ignition source nearby. In calculating the gas and air ratio, the whole furnace volume and the unburnt gas in it must be considered. In a normal operation, as all incoming gas is totally burnt, explosion can never occur. However, if the operational procedure is incorrect or there is some fault in the furnace, it can. Unburnt gas can build up in its volume under three possible situations. The first situation giving unburnt gas is when the gas line within the furnace is leaking or the flow control valve outside does not close completely when shut. The second situation giving unburnt gas in the furnace can arise when:

(1) There is no combustion air. This occurs when the fan (or a compressor) supplying air fails. The flame in this case extinguishes and there will be lot of unburnt gas in the furnace.

(2) The ratio of gas to air flow rate is more than what it should be. This can occur if the air fan (or compressor) starts working inefficiently. This can also occur when the gas valve does not control the flow rate or the air inlet line is partially blocked.

(3) There is a rise in the gas line pressure and it remains undetected due to faulty pressure indicator.

Lastly the event giving rise to unburnt gases arises when the flame in the furnace gets extinguished and this can occur under the following situations:

(4) The ratio of gas to air flow rates becomes very low. When this occurs the flame cools down ultimately extinguishing, and in spite of the pilot light, unburnt fuel accumulates in the furnace. This condition can occur when the fuel control valve does not function properly.

(5) The flame will also be extinguished if there is no combustion air for the same reason as in (1).

(6) The ratio of gas to air flow rate becomes high for the same reason as in (2).

(7) The gas line pressure falls and remains undetected due to faulty pressure indicator leading to low ratio of gas to air flow.

(8) There is also a likelihood of the pilot flame failure and unburnt fuel would accumulate in the furnace during the startup.

It is desired to express these reasonings of fault conditions pictorially. Table 6.5 gives convention adopted for showing it. Primary events are linked to secondary or head events through logical connectors, AND and OR, and they are symbolized as shown. Sometimes, it is not possible to develop an event upto its primary constituents due to lack of sufficient information. They are shown by diamond boxes. It has already been indicated that personnel working should be treated as an integral part of the plant and in view of this, a man fails to perform the assigned task (due to negligence or deliberately in an act of sabotage) exactly like the failure of a controller due to mechanical or electrical fault within it. This kind of an event is primary in nature and symbolized as

K-failure. Lastly there are occasions when it is necessary to force an equipment (or a part) to shut down when the controlled variable drifts out of bounds. This is called a command failure, as shown in Table 6.5.

Table 6.5 Symbols Used for Drawing Fault Tree[1, 7]

Symbols	*Events*
(a) Connectors (or gates)	
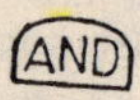	When two or more events occur simultaneously for the secondary (or head) event to occur
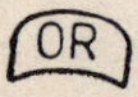	When any one of the two or more events occurs for the secondary (or head) event to occur
(b) Events	
	Primary events
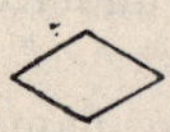	An event which is not developed further, which is either not worthwhile to develop more or cannot be because of lack of data
	Event which has to be developed further (secondary or head)
(c) Use of connectors	

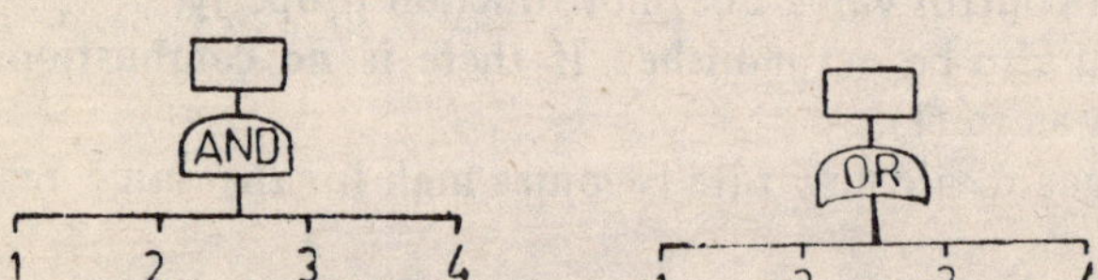

(d) Classification of primary events

P : The primary failure caused by inherent failure of the machine

K : The primary failure induced by human alteration of design set points, valve positions etc., deliberately or otherwise

COMMAND : The component is 'commanded' to perform an event by the built-in control system

With the heip of this table, the analysis of explosion is shown pictorially in Fig. 6.5. The fault tree diagram for explosion consists of sixteen primary events. The event of the availability of an ignition source is not developed completely, and is shown in a diamond box. Since the furnace considered in Fig. 6.4 has only flow controllers and no operators,

K and COMMAND events of Table 6.5 do not occur in it.

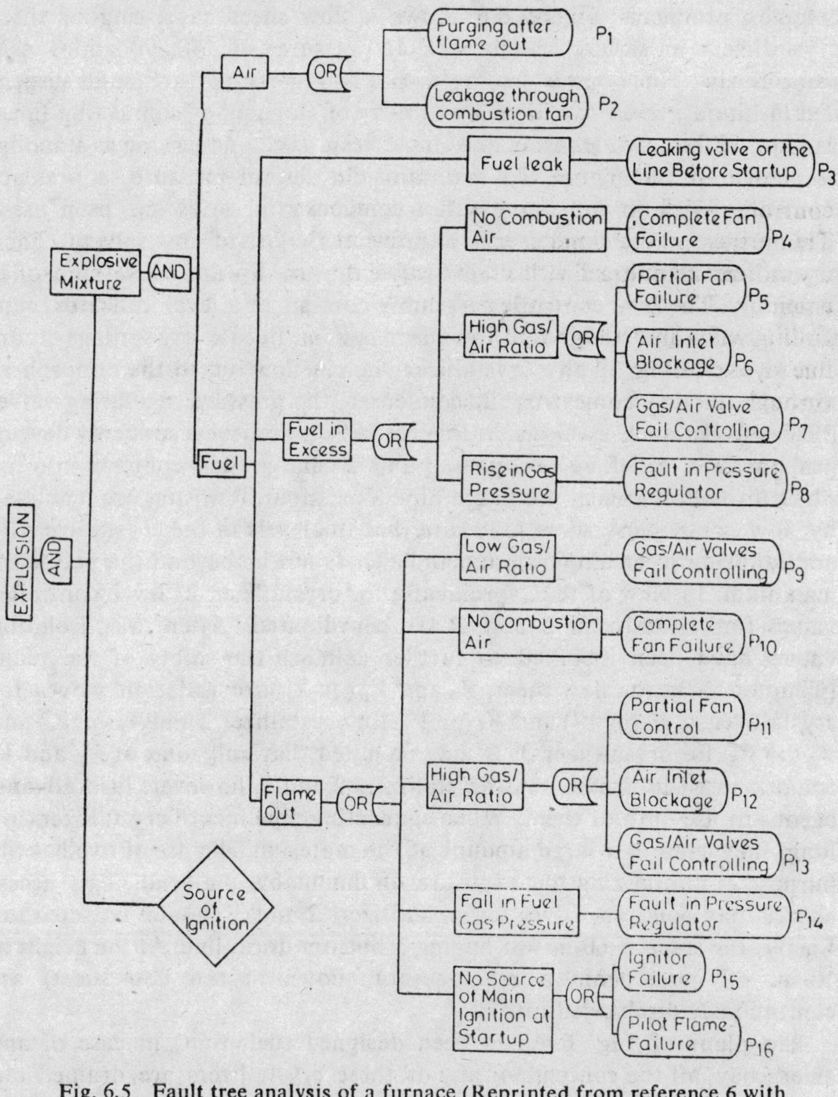

Fig. 6.5 Fault tree analysis of a furnace (Reprinted from reference 6 with permission of the American Institute of Chemical Engineers, New York.)

On going through all primary events of Fig. 6.5, it is realized that they are all independent, i.e. the occurrence of one does not affect the occurrence of any other. However, they are not mutually exclusive, i.e. they have overlapping domains. As a matter of fact, P_4 and P_{10}, P_6 and P_{12}, P_8 and P_{14}, P_7 and P_{13}, and P_9 and P_{13} are exactly the same. It is possible to determine mathematically whether different events are overlapping or mutually exclusive. If they are of the former type, it is important to reduce them into mutually exclusive ones. This mathematical process is called simplification and will be dealt with later in this chapter.

The method of drawing the fault tree can easily be extended to more complex problems. Figure 6.6 shows a flow sheet consisting of three crystallizers in series, which work at pressures of 140, 70 and 1 psig respectively[5]. Since one is dealing with a two phase mixture (solid suspended in liquid), there is a high probability of clogging of connecting lines. In view of this, two transfer lines have been used, one serving as standby, as shown in the figure. To maintain the desired pressure, a pressure controller (with an indicator) with a condenser in series has been used. The purpose of the condenser is to prevent the loss of any solvent. Each crystallizer is equipped with a relief valve, rupture disc and a level controller assembly. The level controller assembly consists of a level indicator, controlling valve and a high-and-low-alarm system. In case any spillage occurs due to overfilling of any crystallizer, it can flow out to the atmosphere through the line connecting the condenser, the pressure regulating valve, PRC and the stack as shown in Fig. 6.6. Also to relieve a suddenly developed pressure, relief valves open up and discharge their contents into the stack through a special discharge pipe line. Input flow rates are regulated by flow controllers, so as to ensure that the levels in the crystallizers do not fall below a certain minimum and also do not go beyond the permitted maximum. In view of this, specifically for crystallizer 2, level controller valves for crystallizers 2 and 3 are coordinated. After this, isolation valves have been installed to further enhance the safety of the plant (Chapter 5). In the flow sheet, V_0 and V_1 (or V_1') are isolation valves for crystallizer 1, V_2 (or V_2') and V_3 (or V_3') for crystallizer 2 and V_4 (or V_4') and V_5 (or V_5') for crystallizer 3. It may be noted that only one of V_4 and V_5 could be used to isolate both crystallizers 2 and 3; however, it is advantageous to use both of them. When connecting pipelines of crystallizers are long, they contain a large amount of the material. The use of two has the purpose of minimizing the exposure of flammable material. This necessitates that isolation valves for crystallizers 2 and 3 should be separate. Lastly, the stack is 60 m tall having a bottom drain line. At the height of 10 m, off gases from a reactor (not shown in the flow sheet) are continuously discharged into it.

The plant of Fig. 6.6 has been designed such that, in case of any emergency, all the contents of any of these crystallizers are drained out into the stack. The nature of the discharge is solid suspended in a solvent. There is no possibility of explosion in crystallizers, but it can occur above (or around) the stack at the moment of discharge. It is desired to investigate conditions which will lead to the slurry discharge into atmosphere through the stack and is considered as the head event.

From a given crystallizer (No. 1, 2 or 3), discharge can occur when the level controller on it does not maintain the desired level. The rise in level will continue beyond the maximum permitted height when.

(1) the high level alarm does not work, or

(2) the high level alarm works, but the worker does not intervene by operating the manual control valve

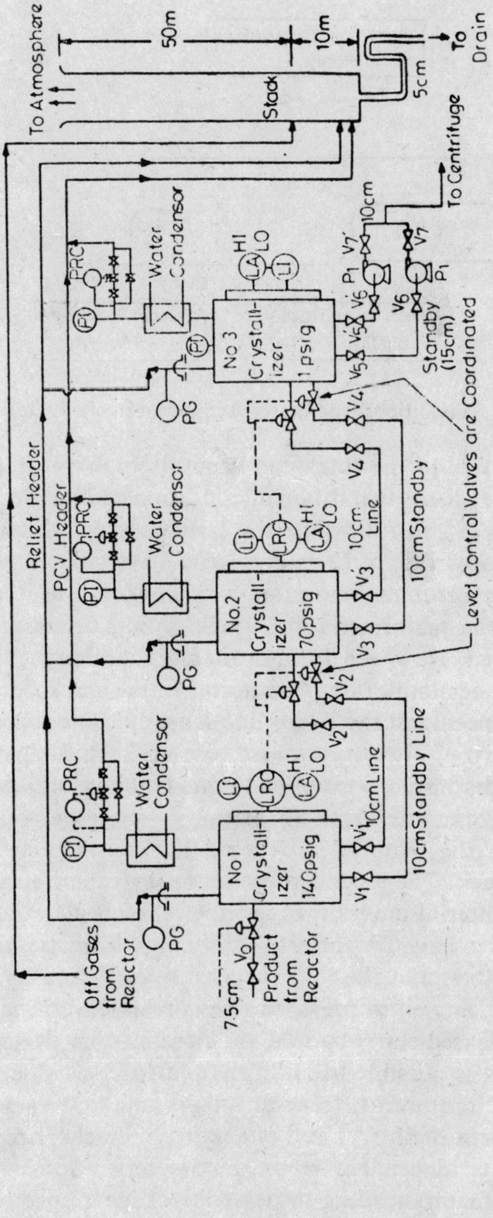

Fig. 6.6 A flow sheet of three crystallizers (Reprinted from reference 5 with permission of the American Institute of Chemical Engineers, New York.)

The event of rising level in a crystallizer is shown in Fig. 6.7 as a fault tree. The discharge of slurry into the stack is now considered. It is noted that the discharge in the stack will occur *only when* the level of liquid in any of the crystallizers keeps rising unabated. The slurry discharge to *atmosphere* would occur only when the discharged slurry in the stack rises upto the height of 10 m where it is entrained by off-gases from the reactor.

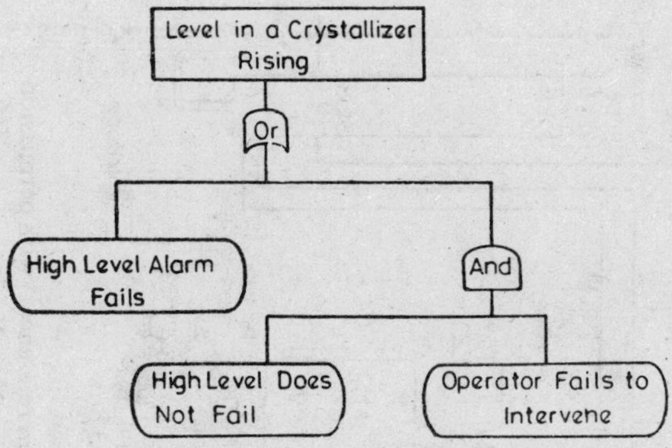

Fig. 6.7 Conditions leading to rise in level in crystallizers

Discharge of slurry to the stack can occur at high pressure as well as at low pressure and these two differ in conditions which lead to spillage in the atmosphere in the following way. When the discharge occurs at low pressure, the flow rate of the slurry is relatively small and in normal circumstances, the incoming material will flow out of the drain line. Accumulation of the material in the stack would occur only when the drain line is clogged. However, in high pressure discharge, clogging is not conditional to the accumulation of material in the stack because it comes out at very high speed and the drain line is insufficient to remove it. As a result, unpressurized discharges must be investigated separately.

Unpressurized discharge is examined first. If the crystallizer 3 overfills, the slurry can move to the stack through two possible routes. The first one will be when the material moves up through condenser, the PCV header into the stack. The second route is through the bursting disc when it ruptures and material moves through the relief header into the stack. It is also possible to have low pressure discharge from crystallizers 1 and 2 through PRC valves and the PCV header line. When PRC valves are working normally, any extra pressure rise is released through it and the pressure of the released slurry would be close to one atmosphere. With this information it is possible to build the fault tree as done in Fig. 6.8. It may be realized that events (reactor off-gas rate to the stack sufficiently high for entrainment of slurry) and (stack drain blocks) are not primary events as they are dependent upon parameters affecting the reactor performance and factors leading to drain blockage respectively. In view of this, they are shown in diamond boxes in Fig. 6.8. Lastly, conditions that lead to rise in level of slurry in all three crystallizers are exactly the same and this fact is represented by a common dotted box in the figure.

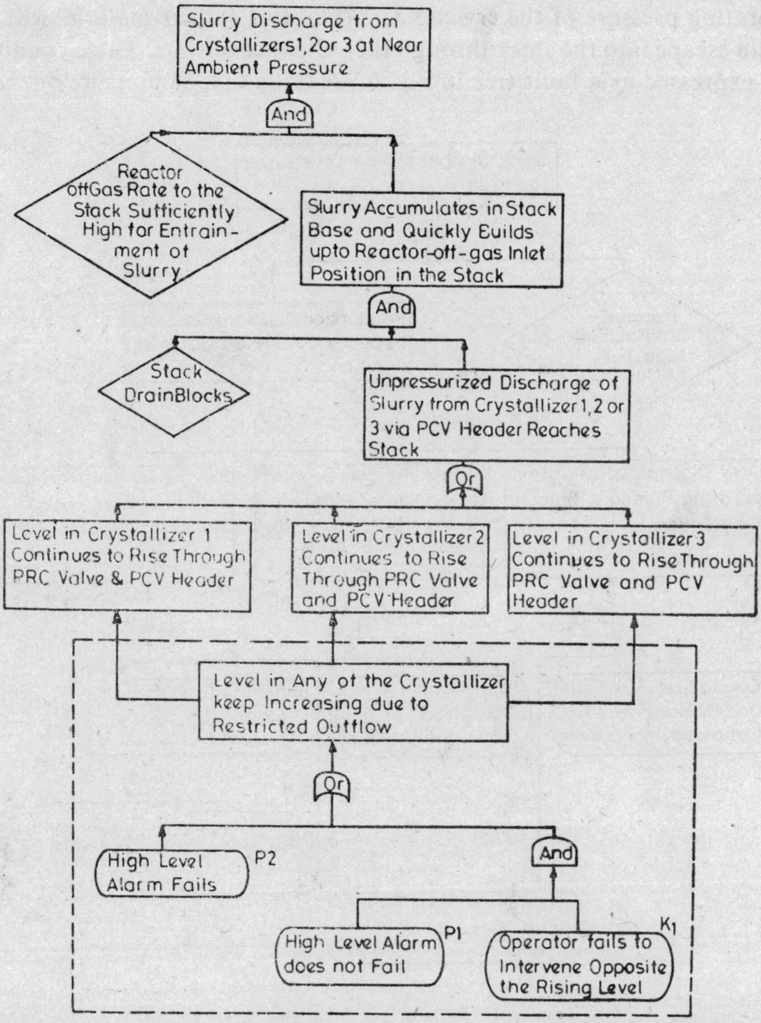

Fig. 6.8 Fault tree analysis of low pressure discharge (Reprinted from reference 5 with permission of the American Institute of Chemical Engineers, New York.)

Now the high pressure discharge to the stack is considered. Naturally this can occur from crystallizers 1 and 2 because of their high working pressure. Once again the prerequisite to any discharge is overfilling of crystallizers either due to the failure of level alarm or the absence of operator's intervention in spite of the alarm (Fig. 6.7). Due to high liquid pressure, the relief valve opens up discharging contents of the crystallizers (1 and 2) into the stack through relief header. This discharge can also occur if PRC controller valves are not functioning properly: If they open excessively, in the case of overflow, material escapes at high pressure. In addition, there is also a likelihood of failure of the rupture discs at the operating pressure of the crystallizers (No. 1 or 2). The material will once again escape into the stack through the relief header line. These conditions are expressed as a fault tree in Fig. 6.9 leading to high pressure discharge.

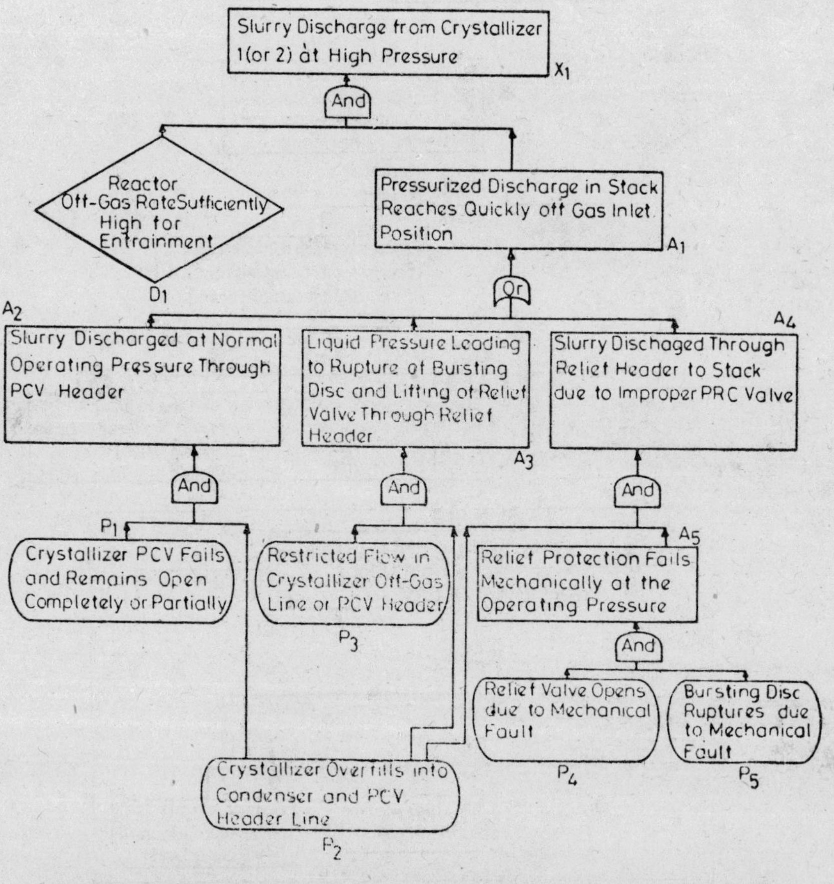

Fig. 6.9 Fault tree analysis of high pressure discharge (Reprinted from reference 5 with the permission of American Institute of Chemical Engineers, New York.)

Fault tree analysis is a method to determine the cause-and-effect sequence of any dangerous situation systematically. A given hazardous condition is reduced to a set of primary events. Information of this kind is extremely crucial for safe plant design. Also for an existing plant, fault tree analysis can improve the reliability of the process.

It is possible to carry out fault tree analysis for a single equipment (like a furnace) or a small chemical plant. However, for a large plant where several hazardous incidents can occur simultaneously this becomes a very cumbersome process, as a result of which a computer-aided fault tree analysis has been suggested.[8,9]

Mathematically fault tree analysis represents any head event as a function of primary events joined by connectives, AND and OR. The Boolean function so obtained needs to be simplified with respect to overlapping domains and the method of simplification is given in the Sec. 6.4.

6.4 SIMPLIFICATION OF BOOLEAN FUNCTIONS

Once a fault tree analysis has been carried out, it is quite straightforward to write the corresponding Boolean function for the head event. As an example, the fault tree of high pressure discharge given in Fig. 6.9 is considered. In the figure itself, the variables are defined against each box. The head event X_1 is equal to the conjunction of D_1 and A_1, (dot for conjunction not shown for convenience), i.e.

$$X_1 = D_1 A_1 \tag{6.4.1}$$

The event A_1 is a disjunction of A_2, A_3 and A_4, or

$$A_1 = A_2 + A_3 + A_4 \tag{6.4.2}$$

A_2 is a conjunction of P_1 and P_2, A_3 is a conjunction of P_2 and P_3 and P_4 is a conjunction of A_5 and P_2, i.e.,

$$A_2 = P_1 P_2 \tag{6.4.3a}$$

$$A_3 = P_2 P_3 \tag{6.4.3b}$$

and

$$A_4 = P_2 A_5 \tag{6.4.3c}$$

But event A_5 is a conjunction of P_4 and P_5,

$$A_5 = P_4 P_5 \tag{6.4.4}$$

A_5 is eliminated in Eq. 6.4.3c by substituting Eq. 6.4.4 and then substituting these in Eq. 6.4.2, A_1 is found to be

$$A_1 = P_1 P_2 + P_2 P_3 + P_2 P_4 P_5 \tag{6.4.5}$$

Thus the head event function is obtained by eliminating A_1 in Eq. 6.4.1 as

$$X_1 = D_1(P_1 P_2 + P_2 P_3 + P_2 P_4 P_5) \tag{6.4.6}$$

While drawing the fault tree in Section 6.3, primary event was assumed to be independent, which means the occurrence of one is not correlated with the other. Since variables P_1 to P_5 represent primary events, these are independent but their combinations have overlapping domains and in Eq. 6.4.6, they are counted more than once. It is desired to write the function, X_1, in such a way that these domains are counted only once. This is called simplification of Boolean functions.

It may be reemphasized that in Eq. 6.4.6 one is dealing with Boolean functions which are different from ordinary known mathematical functions. In Sec. 6.4.1, mapping method of simplification is given.

6.4.1 Mapping Method of Simplification[1]

Two Boolean variables M and N, which are independent and mutually exclusive are considered. In the following, their simplest combinations (i.e. conjunctive and disjunctive) are mapped.

$$X = MN \tag{6.4.7a}$$

and

$$Y = M + N \tag{6.4.7b}$$

It is noted that Boolean functions X and Y can either be true or false. If zero denotes the false condition and one the true condition, it is possible to express the truth value of a function through these numbers. In terms of these numbers the truth table (as done in Chapter 1) for X can be written as

$$X = MN = (11, 10, 01, 00) \tag{6.4.8}$$

where 11 stands for the fact that both M and N are true, 10 for M is true and N is false, 01 for M is false and N is true whereas 00 for both M and N are false. According to Eq. 6.4.8, X has four states and it is true only in state 11. This can be pictorially represented as in Fig. 6.10. If variables M and N are independent and mutually exclusive, conjunction of these variables would denote the area (or domain) common to both *M and N* and in Fig. 6.10 it is shown to be the third column.

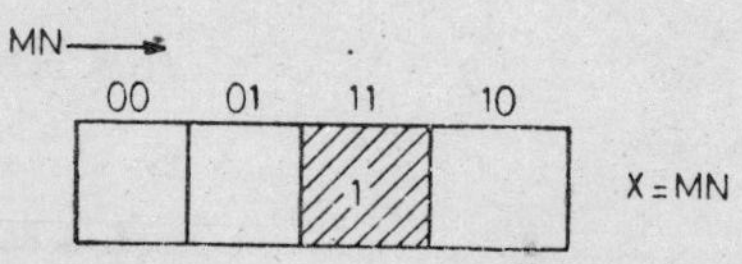

Fig. 6.10 Mapping of $X = MN$

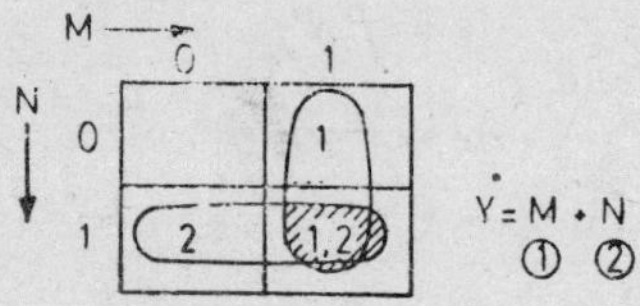

Fig. 6.11 Mapping of $Y = M + N$

The function Y defined by Eq. 6.4.7b is now considered. Y is plotted in a matrix of Fig. 6.11 by inserting entries for the truth values of M and N on it. It is seen that the block of the second row and the second column is common to both M and N and in the function Y, the common area is counted twice. In view of it, the function written in Eq. 6.4.7b is not at its simplest form because of the redundancy.

The methodology given above can be extended to more complex functions to find the overlapping areas. As an example function, Z, as defined below, is considered

$$Z = (A + B)(C + D) \tag{6.4.9}$$

Its right hand side is expanded by multiplying term-by-term

$$Z = AC + AD + BC + BD \tag{6.4.10}$$

From this, Z is found to involve four Boolean variables A, B, C and D and the product of any two of them (for example, AB, CD etc.) can have four possible states (11, 10, 01, 00) as given in Eq. 6.4.8. Consequently, the entire domain covered by the four variables can be represented by a 4 × 4 square grid as shown in Fig. 6.12. In the first column for any row in the

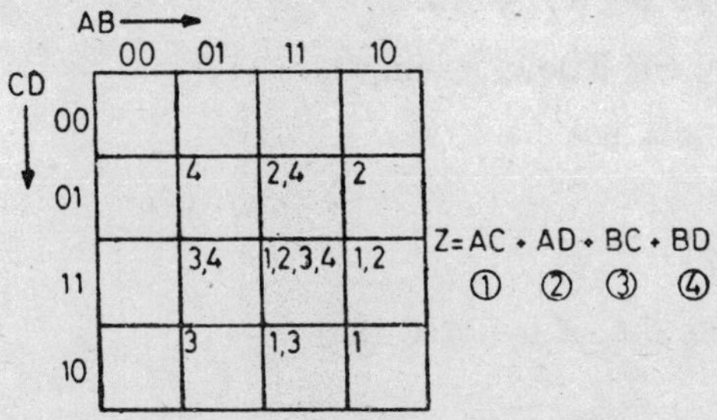

Fig. 6.12 Mapping of $Z = (A + B)(C + D)$

figure, both A and B are false, in the second column for any row, A is false and B is true and so on. Similarly, in the first row for any column, both the variables C and D are false, in the second row for any column, C is false and D is true and so on. Specifically, any square in this grid represents one of the states of the function $ABCD$. For example the square formed by the second row and third column represents the state 1101 in which A is true, B is true, C is false and D is true. Supposing it is desired to express AC. As Boolean variables B and D in it can take any truth value; one looks for columns in which A is true only and rows in which C is true only. A is found to be true in 3rd and 4th columns and C is true in 3rd and 4th rows and the Boolean function AC is true in the region formed by these rows and columns as shown in Fig. 6.12. Functions AD, BC and BD can be similarly expressed in the matrix and the resultant map for the function Z is shown in the figure. Once again, there are several overlapping regions: the square (3rd row, 3rd column), has been counted four times, the squares (2nd row, 3rd column), (3rd row, 2nd column), (3rd row, 4th column) and (4th row, 3rd column) have been counted twice. From this mapping, it is thus seen that the function Z as written in Eq. 6.4.10, is not in the simplest form.

After understanding the method of Boolean function mapping, one goes on to the method of simplifying it. The method of simplification is dependent upon rules which can be derived from basic principles of semantic mathematics given in Chapter 1. Some of these rules (some repeated for convenience) are given in Table 6.6. Rule 1 is evident because the domain encompassed by a function formed by either conjunction or disjunction with itself is the same as that for the variables. (XX should not be confused with X^2 because the latter has no meaning in Boolean algebra). It may be reemphasized that the easiest way of proving an equality in Boolean algebra is to construct a truth table. This is normally very cumbersome for complex functions, and as a result, algebraic manipulations are done. As an example, to prove rule 3 of Table 6.6, function A is defined as

$$A = (X + Y)(X + Z) \tag{6.4.11}$$

which can be expanded to

$$A = XX + XZ + XY + YZ \tag{6.4.12}$$

On using rule 1 of the Table, it simplifies to

$$A = X + XB + YZ \tag{6.4.13}$$

where

$$B = (Y + Z) \tag{6.4.14}$$

From rule 2, Table 6.6, it is noted that

$$X = X\blacksquare \tag{6.4.15}$$

and

$$B + \bar{B} = \blacksquare \tag{6.4.16}$$

Therefore

$$X = X(B + \bar{B}) \tag{6.4.17}$$

Substituting Eq. 6.4.17 in Eq. 6.4.13, one has

$$A = X(B + \bar{B}) + XB + YZ \tag{6.4.18}$$

On expanding the parenthesis, A becomes

$$A = XB + X\bar{B} + XB + YZ \tag{6.4.19}$$

which, on use of rule 1 of Table 6.6, reduces to

$$A = XB + X\bar{B} + YZ \tag{6.4.20}$$

On application of Eq. 6.4.17, the above relation becomes the same as written in Table 6.6 for rule 3. Rules 4, 5 and 6 can be easily obtained from the first three and their derivation is left as an exercise to readers.

On carefully going through Table 6.6, the process of simplification is found to consist of eliminating terms which appear more than once. Since by the method of mapping, the areas of overlapping are known, the process of simplification is an easy task after the map has been prepared.

Table 6.6 Simplification Rules for Boolean Functions

1.	(a) $XX = X$	(b) $X + X = X$
2.	(a) $X\bar{X} = \square$	(b) $X + \bar{X} = \blacksquare$ (c) $X\blacksquare = X$
3.	(a) $XY + XZ = X(Y + Z)$	(b) $(X + Y)(Z + Z) = X + YZ$
4.	(a) $X \rightarrow Y = \bar{X} + Y$	(b) $X \rightarrow Y = \bar{Y} \rightarrow \bar{X}$
5.	(a) $YX + X\bar{Y} = X$	(b) $(X + Y)(X + \bar{Y}) = X$
6.	(a) $X + XY = X$	(b) $X(X + Y) = X$
7.	(a) $X = \bar{X}Y = X + Y$	(b) $X(\bar{X} + Y) = XY$

$\square$: invalid statement (or null space).
$\blacksquare$: valid statement (or encompasses the entire domain).

Since in mapping method, each square is associated with a well defined truth value of all Boolean variables, any given function can therefore be converted into the one without any overlap by simply counting all those squares where the function spreads only once. As an example, Y defined in Eq. 6.4.7b can be written as

$$Y = M\bar{N} + MN + \bar{M}N \tag{6.4.21}$$

Similarly Z of Eq. 6.4.10 can be written as

$$\begin{aligned} Z = \bar{A}B\bar{C}D + AB\bar{C}D + A\bar{B}\bar{C}D + \bar{A}BCD + A\bar{B}CD + ABCD \\ + \bar{A}BC\bar{D} + ABC\bar{D} + A\bar{B}C\bar{D} \end{aligned} \tag{6.4.22}$$

Boolean functions Y and Z, written in Eqs. 6.4.21 and 6.4.22 are at their simplest form because they do not have any redundancy. However, these Boolean expressions are not unique because they can be written in more than one way without having any overlapping. For example, it will also be equal to

$$Z = AC = \bar{A}BC + A\bar{C}D + \bar{A}B\bar{C}D \tag{6.4.23}$$

Similarly Z can also be written as

$$Z = \bar{A}BC + A\bar{C}D + ABC + A\bar{B}C + \bar{A}B\bar{C}D \tag{6.4.24}$$

In view of multiplicity of functional forms, the following convention for the simplification of Boolean functions is normally adopted. Instead of counting single squares, as done in Eq. 6.4.22, several of them are lumped together for convenience. In case of the function Z, the domain for Z can be covered by a group formed by the intersection of 2nd, 3rd and 4th rows with the 2nd, 3rd and 4th columns, and this can best be represented by Eq. 6.4.24.

As a more complex example of simplification, one considers function P defined by

$$\underset{1}{P = AB} + \underset{2}{BC} + \underset{3}{B\bar{D}} + \underset{4}{A\bar{B}D} + \underset{5}{AB\bar{C}D} \tag{6.4.25}$$

and it is mapped in Fig. 6.13. Areas covered by different sub-functions are shown by their numbers and the overlapping is evident. In the figure,

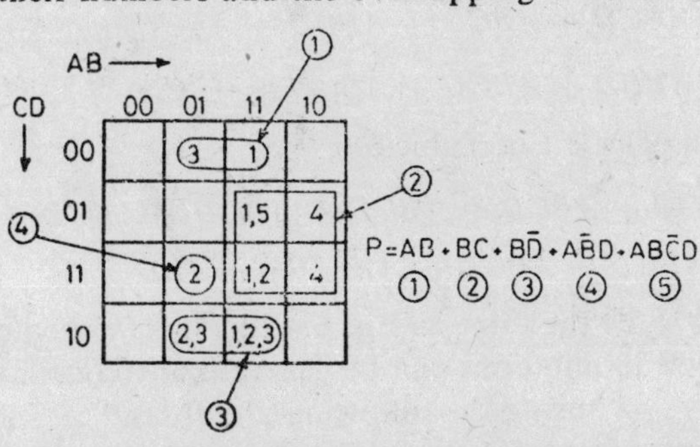

Fig. 6.13 Mapping of $P = AC + BC + B\bar{D} + A\bar{B}D + AB\bar{C}D$

these squares have been regrouped such that their domains do not intersect, and then functions corresponding to these areas are written. P would be the disjunctive combination of these functions, i.e.,

$$P = \underset{1}{B\bar{C}\bar{D}} + \underset{2}{AD} + \underset{3}{BC\bar{D}} + \underset{4}{\bar{A}BCD} \tag{6.4.26}$$

6.4.2 Analytical Method of Simplification

The discussion so far was restricted to a maximum of four Boolean variables. Mapping method has been extended upto maximum of six variables, but for more than six variables, this cannot be used because of inability to represent the grid in a two-dimensional plane. However, it is possible to generalize the mapping method into an analytical scheme. It is recalled that a single square in 4×4 grid (Fig. 6.12), stands for a conjunctive combination of all these four Boolean variables (for example, $\bar{A}\bar{B}\bar{C}\bar{D}$ $A\bar{B}\bar{C}\bar{D}$, $AB\bar{C}\bar{D}$, $ABC\bar{D}$ etc.). Such combinations are called *minterms*. Thus $\bar{A}\bar{B}\bar{C}\bar{D}$, $A\bar{B}\bar{C}\bar{D}$, etc. each are minterms of a function involving four Boolean variables A, B, C and D. The property of the minterm is that it represents a single 'square' in the domain of the function. Consequently, if a function is expanded in minterms it is possible to eliminate overlapping areas (or redundancy). The method of expansion in minterms is suggested through an example of expanding function Z defined by Eq. 6.4.10. The function involves terms AC, AD, BC and BD. To AC, one multiplies with $(B + \bar{B})\cdot(D + \bar{D})$, to $A \cdot D$ with $(B + \bar{B})(C + \bar{C})$ and so on. As $(B + \bar{B})$, $(C + \bar{C})$ etc. are valid statements, multiplication by these does not change the truth value of function Z. Therefore, Z can be rewritten as

$$\begin{aligned} Z = {} & AC(B + \bar{B})(D + \bar{D}) + AD(B + \bar{B})(C + \bar{C}) \\ & + BC(A + \bar{A})(D + \bar{D}) + BD(A + \bar{A})(C + \bar{C}) \end{aligned} \tag{6.4.27}$$

These can be multiplied term by term to obtain

$$\begin{aligned} Z = {} & (ABC\bar{D} + A\bar{B}CD + A\bar{B}C\bar{D} + ABCD) \\ & + (ABCD + AB\bar{C}D + A\bar{B}CD + A\bar{B}\bar{C}D) \\ & + (ABCD + ABC\bar{D} + \bar{A}BC\bar{D} + \bar{A}BCD) \\ & + (ABCD + AB\bar{C}D + \bar{A}BC\bar{D} + \bar{A}B\bar{C}D) \end{aligned} \tag{6.4.28}$$

On application of rule 1 of Table 6.6,

$$\begin{aligned} Z = {} & ABCD + ABC\bar{D} + A\bar{B}CD + A\bar{B}C\bar{D} + AB\bar{C}D \\ & + A\bar{B}\bar{C}D + \bar{A}BCD + \bar{A}B\bar{C}D + \bar{A}\bar{B}CD \end{aligned} \tag{6.4.29}$$

which is found to be the same as Eq. 6.4.22. This method of expanding a Boolean function in minterms can be generalized to one involving n variables $X_1, X_2, \ldots, X_n$ using the following algorithm[10].

(a) Each term of the function is examined; if it is a minterm it is retained and one continues to the next term.

(b) If it is not a minterm; one scans for the Boolean variables which do not occur in it; for each X_i which does not occur, one multiplies the specific with $(X_i + \bar{X}_i)$.

(c) One multiplies out all products and eliminates redundant terms using rule 1 of Table 6.6.

The fault tree of Fig. 6.9 is now simplified as an example. The function X_1 representing the fault tree has been given in Eq. 6.4.6 and is rewritten below after expanding the parenthesis

$$X_1 = D_1P_1P_2 + D_1P_2P_3 + D_1P_2P_4P_5 \tag{6.4.30}$$

X_1 above involves six Boolean variables D_1, P_1, P_2, P_3, P_4 and P_5 and none of the terms are minterms. The second step is carried out to convert them into minterms as follows

$$\begin{aligned} X_1 = {} & D_1P_1P_2(P_3+\bar{P}_3)(P_4+\bar{P}_4)(P_5+\bar{P}_5)+D_1(P_1+\bar{P}_1)P_2P_3(P_4+\bar{P}_4)(P_5+\bar{P}_5) \\ & +D_1(P_1 + \bar{P}_1)P_2(P_3 + \bar{P}_3)P_4P_5 \end{aligned} \tag{6.4.31}$$

$$\begin{aligned} X_1 = {} & D_1P_1P_2P_3P_4P_5+D_1P_1P_2P_3\bar{P}_4P_5+D_1P_1P_2\bar{P}_3P_4P_5+D_1P_1P_2\bar{P}_3\bar{P}_4P_5 \\ & +D_1P_1P_2P_3P_4\bar{P}_5+D_1P_1P_2P_3\bar{P}_4\bar{P}_5+D_1P_1P_2\bar{P}_3P_4\bar{P}_5+D_1P_1P_2\bar{P}_3\bar{P}_4\bar{P}_5 \\ & +D_1P_1P_2P_3P_4P_5+D_1P_1P_2P_3\bar{P}_4P_5+D_1\bar{P}_1P_2P_3P_4P_5+D_1\bar{P}_1P_2P_3\bar{P}_4P_5 \\ & +D_1P_1P_2P_3P_4\bar{P}_5+D_1P_1P_2P_3\bar{P}_4\bar{P}_5+D_1\bar{P}_1P_2P_3P_4\bar{P}_5+D_1\bar{P}_1P_2P_3\bar{P}_4\bar{P}_5 \\ & +D_1P_1P_2P_3P_4P_5+D_1\bar{P}_1P_2P_3P_4P_5+D_1P_1P_2\bar{P}_3P_4P_5+D_1\bar{P}_1P_2\bar{P}_3P_4P_5 \end{aligned} \tag{6.4.32}$$

On application of rule 1 of Table 6.6, the simplified form is found to be

$$\begin{aligned} X_1 = {} & D_1P_1P_2P_3P_4P_5+D_1P_1P_2\bar{P}_3P_4P_5+D_1P_1P_2P_3\bar{P}_4P_5+D_1P_1P_2\bar{P}_3\bar{P}_4P_5 \\ & +D_1\bar{P}_1P_2P_3P_4P_5+D_1\bar{P}_1P_2P_3\bar{P}_4P_5+D_1P_1P_2P_3P_4\bar{P}_5+D_1P_1P_2P_3\bar{P}_4\bar{P}_5 \\ & +D_1P_1P_2\bar{P}_3P_4\bar{P}_5+D_1P_1P_2\bar{P}_3\bar{P}_4\bar{P}_5+D_1\bar{P}_1P_2P_3P_4\bar{P}_5+D_1\bar{P}_1P_2\bar{P}_3P_4P_5 \end{aligned} \tag{6.4.33}$$

The form of the function resulting from the use of the above algorithm is sometimes called *canonical sum of products* or *a disjunctive normal expression*. It may be clarified here that the simplification in a given function is obtained in terms of removal of redundancies and *not* in the reduction of the total number of terms. As a matter of fact between Eq. 6.4.30 and 6.4.33, the number of terms has been increased. The importance of canonical form of a function would become evident when the probability of Boolean function is discussed.

In this section, both mapping and analytical methods of simplification were discussed. The former has the advantage that the function can be represented pictorially and the simplification is carried not only in terms of removal of redundancies, but also in reduction in the total number of terms. However, this has limited application because it cannot be applied for functions consisting of six and beyond Boolean variables. For complex problems, the analytical method of simplification becomes very handy and can also be carried out on a computer. However, the canonical form resulting from this invariably leads to a larger number of terms even though overlapping areas are removed from the function.

6.5 PROBABILITY OF BOOLEAN FUNCTION

The example of Fig. 6.6 consisting of three crystallizers in series is considered once again. A good plant design would be the one which minimizes the occurrence of unwanted events, if not eliminates them completely. However, from experience it is known that complete elimination is an impossible task and in spite of all precautions, things do go wrong. It is only in this context that the occurrence of any event is described stochastically and a probability value is assigned to it.

In Sec. 6.3, a fault tree method was discussed in which a given event was broken down in terms of its constituents called primary events. It is now desired to find the probability of the head event in terms of probabilities of primary events because the latter can be determined experimentally.

A function X which is a conjunctive combination of A and B is considered

$$X = AB \tag{6.5.1}$$

and it is desired to find the probability of X. To do so, the definition of conditional probability is used[1]

$$P(A/B) = \frac{P(AB)}{P(B)} \tag{6.5.2}$$

where $P(A/B)$ is the probability of event A when event B is specified. $P(AB)$ and $P(A)$ stand for probability of events AB and A respectively. On rearrangement, the above relation becomes

$$P(X) = P(B)P(A/B) \tag{6.5.3}$$

which can be equivalently written as

$$P(X) = P(A)P(B/A) \tag{6.5.4}$$

When two Boolean functions A and B are independent of each other, mathematically it means that

$$P(A/B) = P(A) \tag{6.5.5a}$$

and

$$P(B/A) = P(B) \tag{6.5.5b}$$

On substitution of Eq. 6.5.5(b) in Eq. 6.5.4,

$$P(X) = P(A)P(B) \tag{6.5.6}$$

This relation can be generalized to conjunctive combination of n independent functions $A_1, A_2, \ldots, A_n$ as

$$P(A_1A_2A_3 \ldots A_n) = P(A_1)P(A_2)P(A_3) \ldots P(A_n)$$

$$= \prod_{i=1}^{n} P(A_i) \tag{6.5.7}$$

If they are not independent, the above relation is modified by extending Eq. 6.5.3 to

$$P(A_1A_2 \ldots A_n) = P(A_1)P(A_2/A_1)P(A_3/A_1A_2) \ldots P(A_n/A_1A_2 \ldots A_{n-1}) \tag{6.5.8}$$

For a function Y which is a disjunctive combination of A and B

$$Y = A + B \tag{6.5.9}$$

if events A and B are independent, then

$$P(Y) = P(A) + P(B) - P(AB) \tag{6.5.10}$$

It was shown in Fig. 6.11 that the function $(A + B)$ has an overlapping area and in view of this, $P(AB)$ is subtracted from the sum of probabilities of A and B. If functions A and B are mutually exclusive, the common area between them is zero. Thus

$$P(Y) = P(A) + P(B) \tag{6.5.11}$$

when A and B are mutually exclusive. This result can be generalized to a function formed by disjunctive combination of n mutually exclusive Boolean functions $A_1, A_2, \ldots, A_n$ as follows:

$$P(A_1 + A_2 + \ldots + A_n) = \sum_{i=1}^{n} P(A_i) \tag{6.5.12}$$

These results can now be applied to the fault tree of Fig. 6.9 to find the probability of the head event in terms of probabilities of primary events. The Boolean function representing the fault tree (Eq. 6.4.6) is first converted into mutually exclusive terms as in Eq. 6.4.33. On its application, $P(X_1)$ is obtained as

$$\begin{aligned} P(X_1) = {} & P(D_1P_1P_2P_3P_4P_5) + P(D_1P_1P_2P_3\bar{P}_4P_5) + P(D_1P_1P_2\bar{P}_3P_4P_5) \\ & + P(D_1P_1P_2P_3P_4P_5) + P(D_1P_1P_2P_3P_4P_5) + P(D_1\bar{P}_1P_2P_3\bar{P}_4P_5) \\ & + P(D_1P_1P_2P_3P_4P_5) + P(D_1P_1P_2P_3P_4P_5) + P(D_1P_1P_2P_3P_4\bar{P}_5) \\ & + P(D_1P_1P_2P_3P_4P_5) + P(D_1P_1P_2P_3P_4P_5) + P(D_1\bar{P}_1P_2\bar{P}_3P_4P_5) \end{aligned} \tag{6.5.13}$$

As D_1, P_1, P_2, P_3 and P_4 are all independent, Eq. 6.5.7 is used to obtain

$$\begin{aligned} P(X_1) \\ = {} & P(D_1)P(P_1)P(P_2)P(P_3)P(P_4)P(P_5) + P(D_1)P(P_1)P(P_2)P(P_3)P(P_4)P(P_5) \\ & + P(D_1)P(P_1)P(P_2)P(P_3)P(P_4)P(P_5) + P(D_1)P(P_1)P(P_2)P(P_3)P(P_4)P(P_5) \\ & + P(D_1)P(P_1)P(P_2)P(P_3)P(P_4)P(P_5) + P(D_1)P(\bar{P}_1)P(P_2)P(P_3)P(\bar{P}_4)P(P_5) \\ & + P(D_1)P(P_1)P(P_2)P(P_3)P(P_4)P(P_5) + P(D_1)P(P_1)P(P_2)P(P_3)P(\bar{P}_4)P(\bar{P}_5) \\ & + P(D_1)P(P_1)P(P_2)P(\bar{P}_3)P(P_4)P(\bar{P}_5) + P(D_1)P(P_1)P(P_2)P(\bar{P}_3)P(\bar{P}_4)P(\bar{P}_5) \\ & + P(D_1)P(\bar{P}_1)P(P_2)P(P_3)P(P_4)P(\bar{P}_5) + P(D_1)P(P_1)P(P_2)P(P_3)P(P_4)P(P_5) \end{aligned} \tag{6.5.14}$$

Thus the probability of the occurrence of the head event is obtained in terms of probabilities of the occurrence of primary events. As will be shown in Chapter 8 they can be determined experimentally with a fair amount of precision. Further discussion on probability of primary events will not be done here and in Sec. 6.6, it is assumed that these are known.

6.6 EFFECT OF CONFIGURATION OF EQUIPMENT ON PRODUCTION

In any plant operation, if the head event is defined as "successful production of the product", the information on its probability would be of great significance. In this section, the arrangement is shown to have considerable effect upon it.

Equipment in a chemical plant cannot be arranged in any fashion one wishes because it has to satisfy the constraints given in Chapter 1. This normally leads to their sequential arrangement, as shown in Fig. 6.3.

Even though it is conceptually the simplest, the sequential arrangement faces the following major problem. If any of the equipment in the train breaks down (or develops fault), it would require the entire plant shutdown. The flow sheet of Fig. 6.6 is next considered where the three crystallizers are arranged in a sequence, but the connecting pipelines and pumps are not. This is redrawn in a block diagram of Fig. 6.14 in which

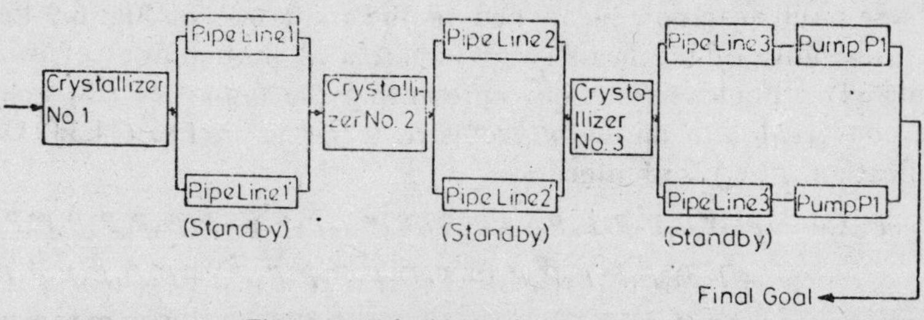

Fig. 6.14 Block diagram of Fig. 6.6

the stack, relief header and PCV header lines have not been shown as these have been provided for safety considerations and are not directly required in the processing of raw materials. As explained below the flowsheet of Fig. 6.6 will produce the desired chemical with higher probability.

On examination of the block diagram of Fig. 6.14, it is observed that the final goal will not be attained if any one of the crystallizers develops a fault or breaks down. However, when a pipeline or a pump stops functioning due to choking or fracture, its standby can immediately take over. This way the system as a whole can tolerate some failures. Plant configuration is thus seen to play a major role and is based upon the discussion above and can be classified as[11]

(a) a pure series configuration, which is defined as the one in which the final goal cannot be attained when any *one* of the train of equipment fails;

(b) a pure parallel configuration, which is defined as the one in which the final goal is attained if any *one* of the equipment of the configuration works properly; and

(c) a mixed configuration of these two.

The pure parallel configuration needs to be distinguished from the one with standby redundancy (Fig. 6.6) because in the former all units are working simultaneously whereas in the latter the standby unit is idle and is actuated to work only when there is failure. The pure series and parallel arrangements of n equipment are shown in Fig. 6.15. It is desired

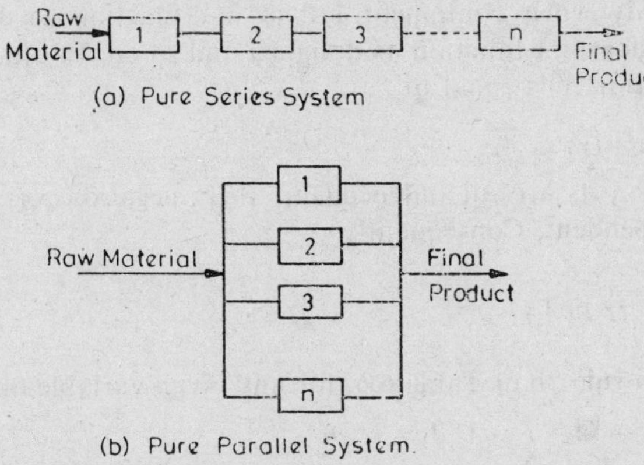

Fig. 6.15 Pure series and parallel configurations of a system

to find out the probability of attaining the goal successfully if the probability of successful operation of each equipment is known. The pure series configuration is considered first. In the series system, it is necessary that equipment 1 operates as designed *and* equipment 2 operates as designed *and* equipment 3 operates as designed *and* so on for the goal to be attained. If the head event, X, consists of carrying out the desired goal successfully,

$$X = A_1 A_2 A_3 \ldots A_n \tag{6.6.1}$$

where

$$A_1 = \text{Equipment 1 is working as required} \tag{6.6.2a}$$

$$A_2 = \text{Equipment 2 is working as required} \tag{6.6.2b}$$

$$\vdots$$

$$A_n = \text{Equipment } n \text{ is working as required} \tag{6.6.2c}$$

As working of each equipment is independent of any other, variables A_1 to A_n are all independent and the probability of X is given by Eq. 6.5.7:

$$P(X) = \prod_{i=1}^{n} P(A_i) \tag{6.6.3}$$

In analyzing the pure parallel configuration of Fig. 6.15b it is once again noted that the head event is attained when either equipment 1 functions as designed *or* equipment 2 functions as designed *or* equipment

3 functions as designed and so on, i.e.,

$$X = A_1 + A_2 + \ldots + A_n \tag{6.6.4}$$

where definitions of these variables are the same as given in Eq. 6.6.2. The probability of function X can be found by converting the right hand side into mutually exclusive terms according to the method given in Sec. 6.5. However, the same result can be obtained by using the following simplified approach. It is noted that the final goal will *not* be attained when and only when equipment 1 does not function as designed and equipment 2 does not function as designed and so on. In other words, the Boolean function $\bar{X}$ is equal to

$$\bar{X} = \bar{A}_1\bar{A}_2\bar{A}_3 \ldots \bar{A}_n \tag{6.6.5}$$

As $A_1, A_2, \ldots, A_n$ are all independent, their negations $\bar{A}_1, \bar{A}_2, \ldots, \bar{A}_n$ are also independent. Consequently

$$P(\bar{X}) = \prod_{i=1}^{n} P(\bar{A}_i) \tag{6.6.6}$$

Further, from rule 2b of Table 6.6, for any given variable one has

$$A_i + \bar{A}_i = \blacksquare, \quad i = 1, 2, \ldots, n \tag{6.6.7}$$

As $\blacksquare$ maps the entire Boolean space, the probability of its occurrence is one. Also A_i and $\bar{A}_i$ are mutually exclusive because the domain of A_i is excluded from $\bar{A}_i$ by definition. Therefore

$$P(\bar{A}_i) + P(A_i) = P(\blacksquare) = 1, \quad i = 1, 2, \ldots, n \tag{6.6.8}$$

Similarly, results can be found for X also:

$$P(X) + P(\bar{X}) = 1 \tag{6.6.9}$$

Eliminating $P(\bar{X})$, $P(A_i)$ etc. from Eq. 6.6.5,

$$P(X) = 1 - \prod_{i=1}^{n} (1 - P(A_i)) \tag{6.6.10}$$

which is the final expression for the probability of the head event in pure parallel configuration.

There are several examples of a parallel configuration which one comes across in chemical plants. Supposing it is desired to control the outlet temperature of a process stream through a heat exchanger. This can be conveniently done by sensing the outlet temperature using one thermocouple, and any deviation in the signal from its steady state value actuates the flow regulating valve of the heating (or cooling) fluid. Due to several reasons, these thermocouples can stop sensing and the outlet temperature becomes out of control. When the controlling has to be done with high probability of success (as required in nuclear plants), a single sensing unit is not a good design. To minimize the danger of failure, several thermocouples are used in parallel and any *one* good thermocouple would actuate the flow regulating valve. Another example of parallel configuration would arise when it is desired to make sufficient profit from a partially understood reaction carried out in a batch reactor. The chemical reaction may not succeed if it is very sensitive to impurities, whose concentration cannot be monitored. The success of such reactions is stochastic in nature and a probability of succes, p, can be associated with a given batch.

One considers a set of N reactors and assumes the cost of running a single batch as C_R (unit money per batch), which includes the cost of the raw material and the utility. When there are N reactors, the total cost is NC_R. Out of these some would fail to produce the desired chemical. The probability of all the reactors failing, $P(\bar{X})$, can be found with the help of Eq. 6.6.10 as

$$P(\bar{X}) = (1-p)^N \tag{6.6.11}$$

The probability of any n reactors ($n \leqslant N$) succeeding out of the total N can be found as

$$P(n) = \frac{N!}{n!\,(N-n)!}\, p^n \cdot (1-p)^{N-n} \tag{6.6.12}$$

The first term represents the number of combinations of n out of the total N. If C_f is the revenue per batch obtained through sale, the expected profit, S, is found to be

$$S = \sum_{n=0}^{N} (nC_f)\left(\frac{N!}{n!\,(N-n)!}\right)p^n(1-p)^{N-n} \tag{6.6.13}$$

On rearrangement of Eq. 6.6.13, one has

$$S = C_f Np \sum_{n=1}^{N} \frac{(N-1)!}{(n-1)!\,(N-n)!}\, p^{n-1}(1-p)^{N-n} \tag{6.6.14}$$

The term under the summation is a binomial expansion of $[p+(1-p)]^{N-1}$ which is equal to 1. Thus,

$$S = NpC_f \tag{6.6.15}$$

and the net profit, R_e, would be

$$R_e = NpC_f - NC_R = N(pC_f - C_R) \tag{6.6.16}$$

Thus, the reaction of this kind should not be run if the probability of success p is less than (C_R/C_f). Depending upon the cost of the final product relative to the cost of running a batch reactor, the failure tolerance is decided. If one is dealing with a very expensive final product, a couple of batches failing would not mean much, because a single successful batch would more than compensate the loss. The total number of reactors, N, is decided by the expected profit level, R_e.

Another interesting case of parallel configuration in a chemical plant is found in the control of a tubular reactor where an explosive reaction is carried out. A typical example would be that of ethylene oxide reactor where ethylene and oxygen are reacting. Figure 6.16 shows a simplified version of a tubular reactor in which the pressure is monitored very closely by placing several sensors (a total of N in number) and each of them independently operates the valve V_1 placed on the outlet stream. The valve V_1 is assumed to be an on-off type and if there is anything wrong with these sensing devices, the output signal from it is too weak to open it. In case none of them works, the pressure in the reactor builds upto a level where the safety valve opens draining out all its content into the stack. This leads to the entire plant shutdown because the reactor becomes inoperative.

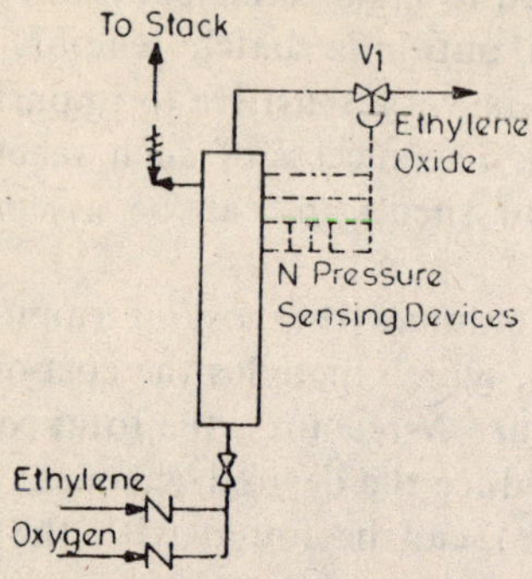

Fig. 6.16 Pressure sensing devices arranged in parallel on an ethylene oxide reactor

Thus it is seen that the event of all sensing devices failing amounts to a plant shutdown. If C_f is the cost associated with bringing the plant to working condition once again, then the cost, F_C, associated with this event can be written as

$$F_C = C_f(1 - p_1)^N \tag{6.6.17}$$

where p_1 is the probability of one sensing device measuring the pressure surge successfully in a given time and N is the total number of sensing devices installed on the reactor. As seen from Eq. 6.6.11, $(1 - p_1)^N$ is the probability of having all sensors in failed condition. If the cost of operating a single sensor is C_o, then the total operating cost of N sensors is NC_o and the total cost, C_t, is given by

$$C_t = C_f(1 - p_1)^N + NC_o \tag{6.6.18}$$

For a given value of p_1 ($p_1 \leqslant 1$), as N is increased, the first term reduces whereas the second term increases. Consequently an optimum value of sensing devices is expected which would minimize the total cost. This can be obtained by setting (dC_t/dN) equal to zero

$$\frac{dC_t}{dN} = C_f(1 - p_1)^{N^*} \ln (1 - p_1) + C_o = 0 \tag{6.6.19}$$

where N^* is the optimal number of sensing devices which can be solved from the above equation as

$$N^* = \frac{\ln \left\{ \dfrac{(C_o/C_f)}{\ln \left[\dfrac{1}{(1 - p_1)} \right]} \right\}}{\ln (1 - p_1)} \tag{6.6.20}$$

As an example, if the value of p_1 is 0.5 and the ratio of the operating cost of a single sensing device to the cost of shutdown is

$$(C_o/C_f) \simeq 0.01 \tag{6.6.21}$$

N^* is found from Eq. 6.6.20 as

$$N^* = \ln\left\{\frac{1}{100 \ln 2.0}\right\} \Big/ \ln 0.5 \simeq 6 \tag{6.6.22}$$

Thus, by having six sensing devices the total operating cost of the tubular reactor is minimized.

A standby equipment is the one which does not work during the regular plant operation, but is started immediately on detection of failure. It will be shown now that by having standby units the probability of attaining the final goal is improved from what it would be in a series configuration.

One is referred to the block diagram of Fig. 6.14. Equipment like crystallizer 1, pipeline 1, crystallizer 2, pipeline 2, crystallizer 3, pipeline 3 and pump P1 constitute a pure series system. By the system's failure here, it is meant that the whole plant is shut down. The problem dealt with now is different from that solved in Sec. 6.3. Earlier, conditions leading to spillage of slurry into the atmosphere were analyzed, whereas in this section the successful crystallization is of interest.

Table 6.7 Probabilities of Successful Operation of Different Equipment

Equipment	Probability of success, p_i	Numerical value
Crystallizer 1	p_1	0.90
Pipeline 1	p_2	0.75
Crystallizer 2	p_3	0.90
Pipeline 2	p_4	0.75
Crystallizer 3	p_5	0.90
Pipeline 3	p_6	0.75
Pump P1	p_7	0.75
Pipeline 3 and pump P1	$p_8 = p_6 p_7$	0.563

Table 6.7 gives probabilities of success for each equipment of Fig. 6.14. When there are no standbys, Eq. 6.5.7 is used to find the probability of success of the series system, R_s, as

$$R_s = p_1 p_2 p_3 p_4 p_5 p_6 p_7 = 0.308 \tag{6.6.23}$$

when the plant with standbys is operated, several chains of events can occur. For example, pipeline 1 gets choked and the standby pipeline 1 is made operative whereas the remainder of equipment of the plant are working properly. Similarly, pipeline 2 can be choked and the standby pipeline 2′ made operative with the remainder of the equipment working properly. In case of a third standby, it is a combination of a pipeline and a pump in series, and it succeeds only when both work properly. Table 6.8 gives all sequences of events that lead to successful plant operation.

Different events of Table 6.8 are generated by considering different combinations of failures. It is assumed that the probability of success of

Table 6.8 Different Chain of Events Leading to Successful Plant Operation of Fig. 6.14

No.	Chain of Events	Probability
1.	Equipment 1 to 7 all succeeding	$p_1p_2p_3p_4p_5p_6p_7$
2.	Pipeline 1 failing, pipeline 1′ succeeding	$p_1(1-p_2)p_2p_3p_4p_5p_6p_7$
3.	Pipeline 2 failing, pipeline 2′ succeeding	$p_1p_2p_3(1-p_4)p_4p_5p_6p_7$
4.	Combination of pipeline 3 and pump P1 failing and the standby succeeding	$p_1p_2p_3p_4p_5(1-p_6p_7)p_6p_7$
5.	Pipeline 1 and 2 failing, with their standbys succeeding	$p_1(1-p_2)p_2p_3(1-p_4)p_4p_5p_6p_7$
6.	Pipelines 1 and 3 failing, with their standbys succeeding	$p_1(1-p_2)p_2p_3p_4p_5(1-p_6p_7)p_6p_7$
7.	Pipelines 2 and 3 failing, with their standbys succeeding	$p_1p_2p_3(1-p_4)p_4p_5(1-p_6p_7)p_6p_7$
8.	Pipelines 1, 2 and 3 failing with their standbys succeeding	$p_1(1-p_2)p_2p_3(1-p_4)p_4p_5(1-p_6p_7)p_6p_7$

a given standby unit is the same as that of the unit which it is supposed to replace in the case of failure. Therefore the probability of event in which pipeline 1 fails and its standby succeeds is $(1-p_2)p_2$. If the remainder of the equipment of the plant are working as designed, the probability of the system's success is $p_1(1-p_2)p_2p_3p_4p_5p_6p_7$. In this way, the probability of each event of Table 6.8 is written and the total probability of success, P_s, of the whole plant with standbys is sum of all these terms, i.e.,

$$\begin{aligned} P_s = p_1p_2p_3p_4p_5p_6p_7\,[1 &+ (1-p_2) + (1-p_4) + (1-p_6p_7) \\ &+ (1-p_2)(1-p_4) + (1-p_2)(1-p_6p_7) + (1-p_4)(1-p_6p_7) \\ &+ (1-p_2)(1-p_4)(1-p_6p_7)] \end{aligned} \tag{6.6.24}$$

On substitution of numerical values, P_s is found to be 0.753, and is found to increase by 225% compared to the pûre series configuration. Thus by adding standby units to a chemical plant, the chance of success is improved considerably. If this result is extended to all the equipment of a chemical plant, the probability of successful operation can be maximized by having a standby unit to every equipment. However, this is not practical because of economic reasons and a dividing line is drawn depending upon the rate of failure and the fixed cost of providing an extra unit. Intuitively it is felt that equipment having high failure rates is provided with a standby.

There are several occasions where configurations cannot be classified as a pure series or a pure parallel one. These are called mixed configurations and Fig. 6.17 gives some involving one, two, three and four units. For two units, it can be either pure parallel or pure series. For units more than two, the number of combinations increases considerably as shown in the figure.

It may be understood that these configurations are not merely to satisfy one's mathematical curiosity but have considerable practical importance.

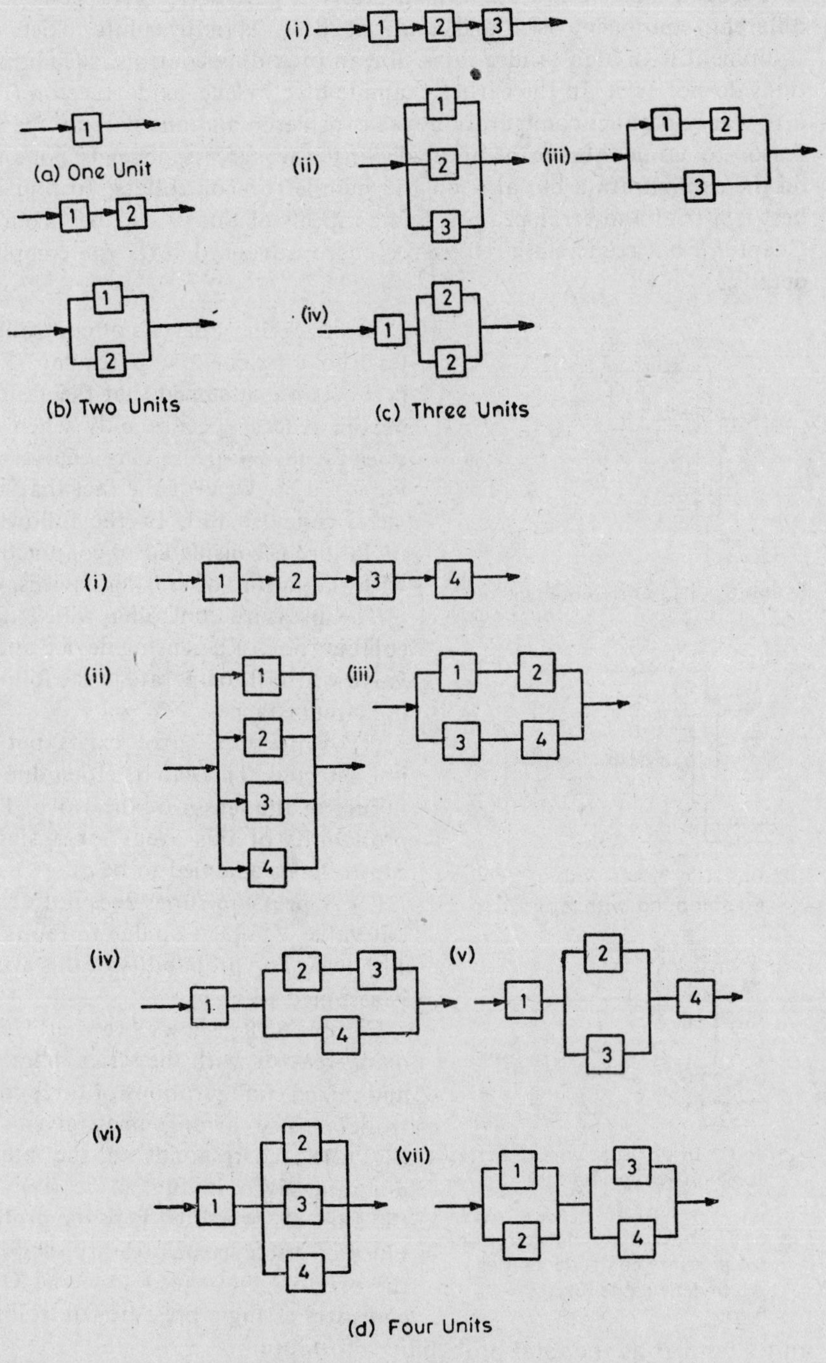

Fig. 6.17 Some mixed configurations involving 1, 2, 3 and 4 equipment

It is true in a given flow sheet, invariably, a pure series arrangement of different equipment is found and, at best, standby units exist for equipment with high failure rates. But in providing controls, such limitations do not exist. In the earlier example of ethylene oxide reactor (Fig. 6.16), only parallel configuration was considered and one has no specific reason to assume this to be optimal. Since the success depends not only on the configuration but also on the number of controllers, to find the best is a trial-and-error process. Some aspects of this would be taken in Chapter 8 but readers are referred to more advanced texts for complete details.[8]

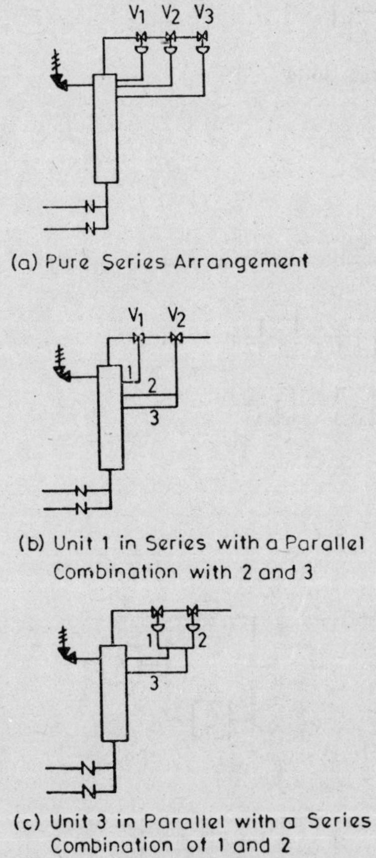

Fig. 6.18 Different configurations of three pressure controllers on an ethylene oxide reactor

In the earlier analysis of controllers on the ethylene oxide reactor (Fig. 6.16), it was assumed that the failure of the reactor occurs only when all sensors have failed. This analysis is idealized in view of the fact that the valve can also fail. In the following its failure is considered in conjunction with the failure of sensing devices.

The pressure controller which is a combination of a sensing device and a valve will fail to operate in the following circumstances.

(1) A pressure surge exists but is not detected. This can be either due to failure of the sensor or the valve. The probability of this event for a single controller is assumed to be q_1.

(2) A pressure surge does not exist, but valve V1 opens up due to faults in circuitry. The probability of this event is assumed to be q_2.

Figure 6.18 shows the ethylene oxide reactor with the series, parallel and mixed configurations of three controllers. The arrangement given in Fig. 6.18c corresponds to the block diagram given in Fig. 6.17c (iv). In the analysis presented here the probability of existence of pressure surge in the reactor is assumed to be α. One considers a single pressure controlling unit, and if P_1 is the total probability of failure,

$$P_1 = \alpha q_1 + (1 - \alpha) q_2 \tag{6.6.25}$$

where the first term corresponds to the failure of detection in the case of pressure surge and the second term accounts for the false detection. The probability of a system's failure can easily be found by determining the mutually exclusive ways by which it can fail. In the configuration of Fig. 6.17c (iv), failure will occur in the following mutually exclusive ways:

(a) When the pressure surge exists, failure will occur if

- (i) units 1, 2 as well as 3 fail to open. The probability is αq_1^3
- (ii) units 1 and 2 fail to open but 3 opens. The probability is $\alpha q_1^2(1-q_1)$
- (iii) units 1 and 3 fail to open but 2 opens. The probability is $\alpha q_1^2(1-q_1)$
- (iv) unit 1 fails to open but 2 and 3 open. The probability is $\alpha q_1(1-q_1)^2$
- (v) units 2 and 3 fail to open but 1 opens. The probability is $\alpha q_1^2(1-q_1)$

(b) When the pressure does not exist and there is false detection, the circuit will remain by error if:

- (i) 1, 2 and 3 are open. The probability is $(1-\alpha)q_2^3$.
- (ii) 1, 2 open but 3 is closed. The probability is $(1-\alpha)q_2^2(1-q_2)$.
- (iii) 1, 3 open but 2 is closed. The probability is $(1-\alpha)q_2^2(1-q_2)$.

As all these are mutually exclusive events, the total probability of failure, P_3, would be

$$P_3 = \alpha[q_1^3 + 3q_1^2(1-q_1) + q_1(1-q_1)^2] + (1-\alpha)[q_2^3 + 2q_2^2(1-q_2)] \quad (6.6.26)$$

6.7 CONCLUSIONS

In this chapter, accidents have been analyzed using trouble-shooting and fault tree approaches. In the fault tree analysis, a head event is broken down into primary events. Since the probability of occurrence of an accident is of great interest, this is written in terms of probabilities of primary events. If the head event is defined as the successful plant operation, the overall realizability of the plant can be calculated in terms of the probability of successful *operation* of individual equipment. It was then noted that a chemical plant is normally a pure series configuration and the probability of manufacturing the final product successfully is increased considerably by using standby units for equipment which have higher frequency of failure. In addition to pure series, there are pure parallel and mixed configurations which one comes across normally in the arrangement of controls on an equipment.

When a chemical plant has been designed and installed with respect to considerations outlined upto this chapter, the following problems are faced during its operation:

(a) monitoring, and
(b) maintenance.

All factors which are involved in these are discussed in Chapters 7 and 8.

REFERENCES

1. Brown, D. S., *Systems Analysis and Design for Safety*, 1st edn, Prentice-Hall, New Jersey, 1976.
2. Factory Mutual Engineering Corporation, *Loss Prevention Data*, Massachusetts. [These data supersede their earlier publication: Factory Mutual Insurance, *Handbook of Industrial Loss Prevention*, 1st edn, McGraw-Hill, N.Y., 1959.]
3. Neale, D. F., *C.E.P.*, **70**(10), 53, 1974.
4. Lawley, H. G., *Hydrocarbon Processing*, **55**(4), 247, 1976.
5. Lawley, H. G., *C.E.P.*, **70**(4), 45, 1974.
6. Kletz, T. A., *C.E.P.*, Loss Prevention Series, **6**, 15, 1972.
7. Browning, R. L., *C.E.P.*, **72**(6), 72, 1976.
8. Powers, G. J. and Lapp, S. A., *C.E.P.*, **72**(4), 89, 1976.
9. Shaelwitz, J. A., Lapp, S. A. and Powers, G. J., *I. and E. C. Proc. Des. Dev.*, **16**, 529, 1977.
10. Kohavi, Zvi, *Switching and Finite Automata Theory*, 1st edn, McGraw-Hill, New York, 1970.
11. Rudd, D. F. and Watson, C. C., *Strategy of Process Design*, 1st edn, Wiley, N.Y. 1968.
12. Lloyd, D. K. and Lipow, M., *Reliability : Management, Methods and Mathematics* 1st edn, Prentice-Hall, N.Y., 1962.

PROBLEMS

6.1 Carry out the fault tree analysis of a storage tank with spillage as the head event.

6.2 Consider the evaporation of ethanol in a vessel electrically heated. Enumerate different conditions which would lead to fire. Arrange them into a fault tree.

6.3 Supposing there is a batch reactor run at 100°C at a pressure of 2 atm and it is desired that its temperature does not go beyond 150°C under any case. First draw a fault tree treating this as the head event and then design a suitable reactor for this. Include the operator, if any, as a part of the fault tree.

6.4 It is desired to use plate and frame filter to separate solids from a slurry coming out of a reactor continuously. As you know, the filter press operates batchwise. Suggest a scheme to obtain the continuous separation. Treat the head event as continuous operation and draw the tree structure of events for two presses.

6.5 Write the Boolean function for the head event of Fig. 6.8. Expand it in minterms and find the probability of the head event.

6.6 Find the expression for probability of explosion in the furnace with the help of Fig. 6.3.

6.7 In the dimerization plant discussed in Chapter 4, water coming with the hydrocarbon remains as an immiscible phase and is not only detrimental to the conversion but makes the operation of the heat exchanger very difficult. Because water boils at 100°C, it expands suddenly at 260°C. This makes the flow of hydrocarbon bumpy thus making the control of the outlet temperature very difficult. Prepare a detailed fault tree for the head event 'excessive water with hydrocarbon'. Write then the Boolean function in minterms.

6.8 Note that the dimerization plant consists of two standby pumps. Show by how much the probability of successful operation of the plant is improved because of these pumps.

6.9 Examine the plant for manufacturing diethyl ether from ethanol. Point out areas which have higher chances of fire or explosion. Prepare a detailed fault tree for the reactor section with explosion as the head event. Subsequently write the Boolean function in minterms.

6.10 Diethyl ether also serves as an anaesthetic and in view of this, in the plant manufacturing it, its odour has to be strictly controlled. The most vulnerable area in the plant would be the bottle filling section. Prepare a fault tree with the head event "excessive ether in vapour" for the bottle filling section.

Hint: Find out the vapour pressure data of diethyl ether. Then suggest how you are going to fill bottles before you draw the fault tree.

6.11 In an acetaldehyde plant, the catalyst gets deactivated in 24 hours and it takes one hour to activate it. To obtain a continuous operation, one employs two reactors. Is the plant flow sheet a pure series, a pure parallel or a mixed configuration?

7

Fault Diagnosis in Chemical Plants Through Data Analysis and Coaptation

7.1 INTRODUCTION

It is important that all equipment in a chemical plant functions as designed to produce the desired chemical at the optimal level. Since a chemical plant is dynamic in nature suitable controllers are used to force it to operate around the steady state. Unfortunately Parkinson's famous law, "if a thing can go wrong, it will", fouls every human design: even controllers have finite probability of failure. It is in view of this that the plant as a whole must be monitored.

A flow sheet consisting of a storage tank, pump, heat exchanger and reactor is shown in Fig. 7.1 to illustrate what is involved in monitoring.

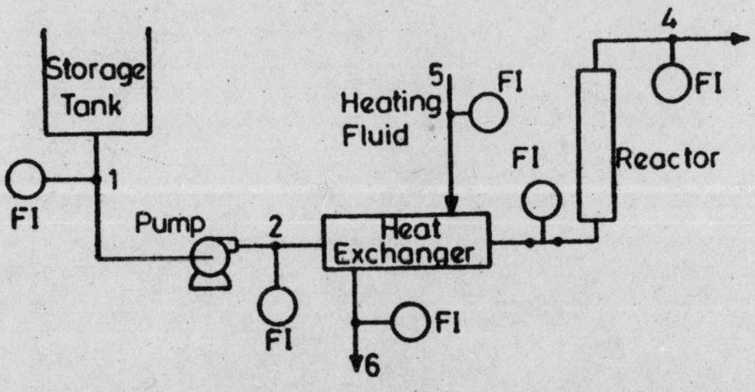

Fig. 7.1 An arbitrary flow sheet to demonstrate the importance of measurement

The storage tank with a level controller maintains sufficient raw material for the process. There is a flow controller on the pump and temperature controllers on the heat exchanger and the reactor. In the flow sheet, there are six streams involved as shown. One considers the total mass flow rate x_i of stream *i*. As long as the system is working at the steady state without leaks, out of these (x_1 to x_6) only one is independent and the measurement can be made anywhere within the system. However, as the system can develop a leak anywhere with equal likelihood, the flow rate at one point must be treated as independent of the one at any other. It is desirable, therefore, to measure them all and record them. Maintaining a record of these is called *log book-keeping* and this information is extremely important for the plant operation. In this chapter, it is shown that a proper analysis of this data can lead to the fault diagnosis in a running plant.

A steady state mass balance for different equipment gives the following for the flow sheet of Fig. 7.1.

$$x_2 - x_1 = 0 \tag{7.1.1a}$$

$$x_3 - x_2 = 0 \tag{7.1.1b}$$

$$x_4 - x_3 = 0 \tag{7.1.1c}$$

$$x_6 - x_5 = 0 \tag{7.1.1d}$$

If flow rates, x_1 to x_6, are measured and substituted in Eq. 7.1.1, they do not balance out *even when* the plant is functioning at the steady state level. This disbalance occurs because of

(1) Unsteady nature: It has already been stated that the plant can never be in true steady state, and as a result, steady state equations are not satisfied. When variables, measured under dynamic conditions, are substituted in these equations, there will always be a disbalance.

(2) Sudden influx or efflux in an equipment: This can arise because of several reasons; a more common one would be a sudden development or removal of barriers in the flow path.

(3) Leak development: In case a leak develops in an equipment, balances on equipment prior to it are likely to be satisfied by *measured* flow rates, but the equation for that equipment is not satisfied.

(4) Measurement errors: There are several kinds of measuring equipment and they are discussed in great detail in ref. 1. Given a measuring equipment, however sophisticated, there is always an associated measurement error, which can arise because of reasons classified in Table 7.1. In the analysis presented here, systematic error is assumed to be absent.

As true values of mass flow rates, x_1 to x_6, are of great value, it is desired to estimate them from the measured values. The best estimate of these is defined as the one which minimizes the total error, and is found through *data analysis*.

Data analysis is statistical in nature, and redundancy in measurement is known to improve the accuracy of estimation.[2–4] It is in this context that it is desirable to measure all streams even when the plant has no leak which gives the maximum redundancy. However, because of costs

Table 7.1 Classification of Measurement Errors

S. No.	Nature of Error
I.	Systematic or fixed errors
	A. Calibration error
	B. Error due to individual bias
	C. Errors due to faulty instruments
	D. Error of technique
	E. Errors due to leaks, depositions etc.
II.	Random errors
	A. Judgement error
	B. Errors due to fluctuations of conditions
	C. Errors due to vibration, pick up of extraneous electrical noise etc.
III.	Illegitimate errors
	A. Blunder or mistake
	B. Computational error
	C. Deliberate error

involved, only some of them are measured, the remaining estimated with the constraint of minimum error. The process of estimating unmeasured streams is called *coaptation.*

In this chapter, the principles of statistical analysis are given first which are later used in data analysis and coaptation. The discussion is basically concerned with process variables of flow rate and enthalpy, which are governed by mass and energy balance equations. Estimations of true process conditions are done by the method of least square with respect to constraints of these balance equations. In the end, a general fault detection scheme has been presented which utilizes the result of the data analysis.

7.2 PRINCIPLES OF STATISTICS[2-4]

Sample space is defined as the domain which includes all possible outcomes of an experiment. For example, an experiment could be a single throw of a dice. A possible outcome of the experiment is the occurrence of either 1, 2, 3, 4, 5 or 6, and therefore this constitutes the sample space.

A *random variable* is a numerical-valued variable defined in the sample space. For example the occurrence of any of the numbers (1 to 6) in the throw of a dice is a random variable. A random variable can be discrete in nature (as in a single throw of a dice) or continuous (as in the measurement of temperature).

To explain these concepts, an example of continuous variable is considered in which study habit of a single engineering student is recorded for a period of two years of engineering studies.[5] One normally starts the formal education at the age of five and finishes when he is about twenty-five. This twenty year period is the sample space and the two years which have been studied is one sample.

Table 7.2 gives the finding of this study in terms of number of days, n_1, the student works between 0 to 2 hours (this does not include number

Table 7.2[5] Study Habit of a Given Student

Time put in, hrs, x	Range Δx	Number of days out of 730 days, n_i	n_i/n	$n_i/n\ \Delta x$
0–2	2	120	0.126	0.083
2–4	2	65	0.090	0.045
4–6	2	186	0.256	0.128
6–8	2	146	0.200	0.100
8–10	2	71	0.100	0.050
10–12	2	90	0.122	0.061
12–14	2	42	0.056	0.028
14–16	2	10	0.010	0.005
16 and above		0	0.000	0.000
Total		730	1.000	0.500

(Reprinted from A. Kumar and S. K. Gupta, *Fundamentals of Polymer Science and Engineering*, with the permission of Tata McGraw-Hill Publishing Co., New Delhi.)

of hours he spends in attending classes), number of days, n_2, he works between 2 to 4 hours and so on. The number of hours of self-study here is the random variable, x, and it is assumed to vary by two units ($=\Delta x$). The graphical representation of the same data is called a histogram. The $n_i/(n\Delta x)$ is a very important quantity and is called *distribution function*, $f(x)$, and for a continuous variable, it is mathematically defined as

$$f(x) = \lim_{\Delta x \to 0} \left(\frac{n_i}{n\ \Delta x} \right) \tag{7.2.1}$$

The distribution function, $f(x)$, has following properties for a given random variable, X.

(i) $f(X) \geqslant 0$ for all X, $-\infty < x < \infty$ (7.2.2a)

(ii) $\int_{-\infty}^{\infty} f(X)\, dX = 1$ (7.2.2b)

(iii) $\Pr(X \leqslant A) = \int_{-\infty}^{A} f(X)\, dX$ (7.2.2c)

The first relation (7.2.2a) states that the distribution function is positive in the entire sample space. The second relation (7.2.2b) stands for the fact that it is distributed over the sample space and when it is summed, it is equal to one. The third relation (7.2.2c) follows from Eq. 7.2.1 and therefore may be regarded as the definition of the distribution function: the probability of the random variable X less than or equal to some number A in the sample space written as $\Pr(X \leqslant A)$ is equal to the integral of $f(x)$. These concepts can be easily extended to an ensemble of several random variables, but before doing it, a physical example is given where a distribution function of more than one random variable is used.

There is a steel ball, completely spherical in nature, whose diameter is to be determined. Evidently, measurements are subject to random errors irrespective of the instrument used for the measurement and the true diameter can be estimated through statistical analysis only. The collection of measurement data can be done in following two *distinct* ways:

1. A succession of readings taken under identical conditions (same observer and same instrument of measurement) except for time. This is called *single sample test*.

2. Repeated measurement of the diameter of the same steel ball using different instruments of measurement (e.g., a ruler, vernier callipers, travelling microscope, etc.) and different observers. Change of instruments and observers both are found to change the distribution of error and therefore one has a set of a random variables, x_i, y_i, z_i, etc. in this measurement. This is called *multisample test* and in situations of this kind, a distribution function of more than one random variable is utilized.

A set of n random variables are defined as follows:

$$X_1 = x_{1,i}, \quad i = 1, 2, \ldots \tag{7.2.3a}$$

$$X_2 = x_{2,i}, \quad i = 1, 2, \ldots \tag{7.2.3b}$$

$$\vdots \qquad \vdots$$

$$X_n = x_{n,i}, \quad i = 1, 2, \ldots \tag{7.2.3c}$$

The distribution function $f(X_1, X_2, \ldots, X_n)$, defined in similar fashion, is a function with the following properties:

$$\text{(i) } f(X_1, X_2, \ldots, X_n) \geqslant 0 \tag{7.2.4a}$$

$$\text{(ii) } \int_{-\infty}^{\infty}\int_{-\infty}^{\infty} \cdots \int_{-\infty}^{\infty} f(X_1, X_2, \ldots, X_n)\, dX_1\, dX_2 \ldots dX_n = 1 \tag{7.2.4b}$$

$$\text{(iii) } \Pr(X_1 \leqslant A, X_2 \leqslant B, \ldots, X_n \leqslant Q)$$

$$= \int_{-\infty}^{A}\int_{-\infty}^{B} \cdots \int_{-\infty}^{Q} f(X_1, X_2, \ldots, X_n)\, dX_1\, dX_2 \ldots dX_n \tag{7.2.4c}$$

where $\Pr(X_1 \leqslant A, X_2 \leqslant B, \ldots, X_n \leqslant Q)$ is the probability of an event where $X_1 \leqslant A$, $X_2 \leqslant B, \ldots, X_n \leqslant Q$. A distribution function of any subset of these n random variables is called a marginal distribution function. For example $f(X_1, X_2, \ldots, X_p)$ will be one such provided p is less than n and this can be found from

$$f(X_1, X_2, \ldots, X_p) = \int_{-\infty}^{\infty}\int_{-\infty}^{\infty} \cdots \int_{-\infty}^{\infty} f(X_1, X_2, \ldots, X_n)\, dX_{p+1}\, dx_{p+2} \ldots dX_n \quad \text{for all } p < n \tag{7.2.5}$$

Random variables $X_1, X_2, \ldots, X_n$ are all *independent* if and only if

$$f(X_1, X_2, \ldots, X_n) = f_1(X_1) f_2(X_2) \ldots f_n(X_n) \tag{7.2.6}$$

where $f_1(X_1), f_2(X_2), \ldots, f_n(X_n)$ are distribution functions for individual random variables. The concept of independence used in Chapter 6 is identical to that being used here. In most of the physical situations that one comes across, the condition of independence is satisfied.

The distribution function reported in the form of a graph is normally found very cumbersome and in practice one calculates its moments. Both these representations are mathematically equivalent. The kth moment, m_k, of a single discrete random variable, x_i, is defined as

$$m_k = \sum_{i=1}^{n} x_i^k f(x_i) \quad k = 1, 2, \ldots \tag{7.2.7}$$

and for a continuous variable, it is

$$m_k = \int_{-\infty}^{\infty} x^k f(x)\, dx, \quad k = 1, 2, 3, \ldots \tag{7.2.8}$$

As it is possible to expand any continuous function in terms of its moments, a given set of experimental data can always be reduced to a set of moments. Of all these moments, the ones for $k = 1$ and 2 have a great physical significance and are discussed below.

The first moment, m_1, often called the mean, $\bar{x}$, of the random variable, is nothing but the weighted average about the origin with $f(x_i)$ as the weight:

$$m_1 \equiv \bar{x} = \sum_{i=1}^{n} x_i f(x_i) \tag{7.2.9}$$

The second moment, m_2, is a measure of variation and by definition it is

$$m_2 \equiv \sum_{i=1}^{n} x_i^2 f(x_i) \tag{7.2.10}$$

The variation is normally measured around the origin. When it is measured around the mean, $\bar{x}$, it is called variance, σ_i, and it is equal to

$$\sigma_i^2 \equiv m_{2,\bar{x}} = \sum_{i=1}^{n} (x_i - \bar{x})^2 f(x_i) \tag{7.2.11}$$

It is possible to relate variance to m_2 (defined by Eq. 7.2.10) and the mean $\bar{x}$ as follows

$$\begin{aligned}\sigma_i^2 &= \sum_{i=1}^{n} [x_i^2 - 2x_i\bar{x} + \bar{x}^2] f(x_i) \\ &= \sum_{i=1}^{n} x_i^2 f(x_i) - 2\bar{x} \sum_{i=1}^{n} x_i f(x_i) + (\bar{x})^2 \sum_{i=1}^{n} f(x_i)\end{aligned} \tag{7.2.12}$$

From the definition of distribution function, mean $\bar{x}$ and the second moment m_2 (Eqs. 7.2.2, 7.2.9 and 7.2.10), Eq. 7.2.12 simplifies to

$$\begin{aligned}\sigma_i^2 &= m_2 - 2\bar{x}\cdot\bar{x} + \bar{x}^2 \\ &= m_2 - \bar{x}^2\end{aligned} \tag{7.2.13}$$

which is a convenient relation to determine variance σ_i from experimental data. Higher moments are calculated in similar fashion; however, they are normally not necessary. This is because two parameters ($\bar{x}$ and σ) are invariably found to be sufficient in describing distribution functions.

There are several theoretical distribution functions that exist in literature and are summarized in Table 7.3. Out of these, normal distribution

Table 7.3[2-4] Some Theoretical Distribution Functions for Random Variables

1. Normal distribution	$f(x_i) = \frac{1}{(2\pi)^{1/2}\sigma} \exp\{-\frac{1}{2}((x - \bar{x})/\sigma)^2\}, -\infty < x < \infty$
2. Binomial distribution	$f(x) = \frac{n!}{x!\,(n-x)!}\, p^x q^{n-x}$
3. Poisson distribution	$f(x) = \frac{e^{-\mu}}{x!}\, \mu^x$
4. Lognormal distribution	$f(x) = \frac{1}{x\sigma\sqrt{2\pi}} \exp\{-(\log x - \bar{x})/2\sigma^2\}, \quad 0 < x < \infty$
5. Weibull distribution	$f(x) = \frac{\beta x^{\beta-1}}{\lambda^\beta} \exp\{-(x/\lambda)^\beta\}, \quad x \geqslant 0, \lambda, \beta > 0$
6. Gamma distribution	$f(x) = \frac{x^{\beta-1}}{\lambda^\beta \Gamma(\beta)} \exp\{-(x/\lambda)\}, \quad x \geqslant 0, \beta > 0$
7. Exponential distribution	$f(x) = \frac{1}{\lambda} \exp\{-(x/\lambda)\}, \quad x \geqslant 0$

finds greater application. Reference 3 gives normal approximations to Poisson and binomial distribution and it is found that as the sample size increases, this approximation becomes better and better. It is in view of this, that the discussion from now on will be restricted to normal distribution.

One next introduces the concept of moment generating function $M_x(\theta)$ defined for discrete random variable

$$M_x(\theta) = \sum_{x=0}^{\infty} e^{\theta x} f(x) \tag{7.2.14}$$

Expanding $e^{\theta x}$ in an infinite series and rewriting the above equation as

$$M_x(\theta) = \sum_{x=0}^{\infty} \left[1 + \theta x + \frac{\theta^2 x^2}{2!} + \frac{\theta^3 x^3}{3!} + \ldots\right] f(x)$$

$$= \sum f(x) + \theta \sum x f(x) + \frac{\theta^2}{2!} \sum x^2 f(x) + \frac{\theta^3}{3!} \sum x^3 f(x) + \ldots \tag{7.2.15}$$

Using Eqs. 7.2.2 and 7.2.7,

$$M_x(\theta) = 1 + m_1\theta + m_2 \frac{\theta^2}{2!} + m_3 \frac{\theta^3}{3!} + \ldots \tag{7.2.16}$$

Conversely the kth moment, m_k, can be written from the above relation as

$$m_k = \frac{d^k M_x(\theta)}{d\theta^k} \tag{7.2.17}$$

For a continuous variable, the generating function is defined as

$$M_x(\theta) = \int_{-\infty}^{\infty} e^{\theta x} f(x)\, dx \tag{7.2.18}$$

and the same result as in Eq. 7.2.17 can be derived. If there is some arbitrary function $g(x)$ of the random variable, x, its moment generating function $M_{g(x)}(\theta)$ can be written in complete analogy with Eq. 7.2.18 as

$$M_{g(x)}(\theta) = \int_{-\infty}^{\infty} e^{\theta g(x)} f(x)\, dx \tag{7.2.19}$$

The following two particular forms of $g(x)$ are of great interest:

$$g_1(x) = cx \tag{7.2.20a}$$

and

$$g_2(x) = x + d \tag{7.2.20b}$$

The moment generating function $M_{g_1(x)}(\theta)$ is

$$M_{g_1(x)}(\theta) = M_{cx}(\theta) = \int_{-\infty}^{\infty} e^{\theta(cx)} f(x)\, dx$$
$$= \int_{-\infty}^{\infty} e^{(c\theta)x} f(x)\, dx \tag{7.2.21}$$

and on comparison with Eq. 7.2.18, one obtains

$$M_{cx}(\theta) = M_x(c\theta) \tag{7.2.22}$$

Similarly

$$M_{g_2(x)}(\theta) = M_{x+d}(\theta) = \int_{-\infty}^{\infty} e^{\theta(x+d)} f(x)\, dx = e^{\theta d} M_x(\theta) \tag{7.2.23}$$

The moment generating function of a normal distribution function is now found for $g(x)$ defined by

$$g(x) = (x - \bar{x}) \tag{7.2.24}$$

Its moment generating function $M_{(x-\bar{x})}(\theta)$ is given by

$$M_{(x-\bar{x})}(\theta) = \frac{1}{(2\pi\sigma^2)^{1/2}} \int_{-\infty}^{\infty} \exp\{\theta(x - \bar{x})\} \exp\{-\tfrac{1}{2}(x - \bar{x})/\sigma^2\}\, dx \tag{7.2.25}$$

Defining

$$x = \frac{x - \bar{x}}{\sigma} \tag{7.2.26}$$

and writing Eq. 7.2.25 in terms of it as

$$M_{(x-\bar{x})}(\theta) = \frac{1}{(2\pi)^{1/2}} \int_{-\infty}^{\infty} \exp\{(\theta\sigma z - z^2/2)\}\, dz \tag{7.2.27}$$

or

$$M_{(x-\bar{x})}(\theta) = \frac{1}{(2\pi)^{1/2}} \exp(\tfrac{1}{2}\theta^2\sigma^2) \int_{-\infty}^{\infty} \exp\{-\tfrac{1}{2}(z - \theta\sigma^2)\}\, dz \tag{7.2.28}$$

On substituting

$$t = (z - \theta) \tag{7.2.29}$$

in Eq. 7.2.28, it becomes

$$M_{(x-\bar{x})}(\theta) = \frac{1}{(2\pi)^{1/2}} \exp\left(\tfrac{1}{2}\theta^2\sigma^2\right) \int_{-\infty}^{\infty} \exp(-t^2/2)\, dt$$
$$= \exp\left(\tfrac{1}{2}\theta^2\sigma^2\right) \tag{7.2.30}$$

The value of the integral can be found from any standard mathematical handbook[8] and is $(2\pi)^{1/2}$. One now considers the function

$$\bar{x} = \frac{x_1 + x_2 + x_3 + \ldots + x_n}{n} \tag{7.2.31}$$

$\bar{x}$ is nothing but an average of n independent random variables, x_1 to x_n and it is desired to find out its moment generating function. The use of Eq. 7.2.22 yields

$$M_{\bar{x}}(\theta) = M_{(x_1+x_2+\ldots+x_n)/n}(\theta) = M_{(x_1+x_2+\ldots+x_n)}(\theta/n) \tag{7.2.32}$$

As $x_1, x_2, \ldots, x_n$ all are independent, $M_{\bar{x}}$ is obtained by using Eq. 7.2.6,

$$M_{\bar{x}}(\theta) = \int_{-\infty}^{\infty}\int_{-\infty}^{\infty}\cdots\int_{-\infty}^{\infty} \exp\{\theta(x_1 + x_2 + \ldots x_n)/n\}$$
$$\times f(x_1, x_2, \ldots, x_n)\, dx_1\, dx_2 \ldots dx_n$$
$$= \int_{-\infty}^{\infty} \exp(\theta x_1/n)\, f(x_1)\, dx_1 \int_{-\infty}^{\infty} \exp(\theta x_2/n)\, f(x_2)\, dx_2 \ldots$$
$$\int_{-\infty}^{\infty} \exp(\theta x_n/n)\, f(x_n)\, dx_n$$
$$= M_{x_1}(\theta/n) M_{x_2}(\theta/n) \ldots M_{x_n}(\theta/n) \tag{7.2.33}$$

As $x_1, x_2, \ldots, x_n$ all are normally distributed with same mean and variance, one has (Eq. 7.2.23 and 7.2.30)

$$M_{x_1}(\theta/n) = M_{x_2}(\theta/n) = \ldots = M_{x_n}(\theta/n)$$
$$= \exp\{\bar{x}(\theta/n)\} \cdot \exp\{\tfrac{1}{2}\sigma^2(\theta/n)^2\} \tag{7.2.34}$$

Substituting this in Eq. 7.2.33, one obtains

$$M_{\bar{x}}(\theta) = [\exp \bar{x}(\theta/n) \exp(\tfrac{1}{2}\sigma^2\theta^2/n^2)]^n \tag{7.2.35}$$

or

$$M_{\bar{x}}(\theta) = \exp(\bar{x}\theta) \exp\{\tfrac{1}{2}\theta^2(\sigma^2/n)\} \tag{7.2.36}$$

On comparison with Eq. 7.2.30, it is seen that $\bar{x}$ is normally distributed with a variance of (σ^2/n)!. This is an extremely important result and is called the central limit theorem which is formally stated as follows:

Central Limit Theorem: If in a *single sample test* with a normally distributed sample space of mean μ and variance σ, a random sample of size n is drawn, the mean $\bar{x}$ of the *test sample* will be normally distributed with mean μ and variance $(\sigma/\sqrt{n})$.

Now it is explained how the testing of a statistical hypothesis is done and a confidence limit of the test is calculated.

A statistical hypothesis is an assumption concerning the distribution function of a random variable. A *test for statistical hypothesis* is a procedure for deciding whether to accept or reject the hypothesis. To give an example, a random variable is considered. The statistical hypothesis could be that it is normally distributed which is to be tested against, say, it is Poisson distributed. To test the hypothesis would consist of designing experiments on the random variable to confirm the hypothesis of the normal distribution function. Another example which is of great interest is when the distribution for the random variable is known on the basis of earlier theoretical or experimental work. One wishes to evaluate parameters involved in the distribution function and desires to find out how good the estimation is based upon a given test sample.

As an example of hypothesis testing, a batch reactor for polymerization reaction of phenol and formaldehyde is considered. To get the same degree of polymerization (which is found out in practice by a viscosity test), the reaction is carried out for a batch time t. This time t is a random variable because:

1. The temperature of the reactor can never be perfectly controlled. As the reaction mass is an organic fluid and has low thermal conductivity, there will necessarily be temperature distributions within the reactor.
2. The polymerization reaction is exothermic in nature. The heat liberation occurs uniformly over the entire reaction mass whereas the heat removal occurs either at the wall of the reactor or the cooling coil. This causes spatial inhomogeneity of temperature in the reaction mass.
3. This temperature inhomogeneity leads to concentration inhomogeneity because at places of higher temperatures there is a higher reaction rate leading to lower concentration.
4. As the reaction progresses the viscosity of the reaction mass is increased leading to problems of mixing. This difficulty further promotes the concentration inhomogeneity.

It may be recalled that the time of reaction can be deterministically evaluated *only* when the temperature and concentrations of species in the reaction mass are precisely homogeneous. Since this is rarely the case, the time of batch reaction for the same conversion (or the molecular weight of the polymer, which is related to conversion) is expected to be a random variable.

One assumes that extensive laboratory experiments have been carried out to confirm that the batch time t is normally distributed with an average, μ, of 10 hours and variance, σ, of 2 hours. As would be evident these μ and σ belong to the sample space. When the same reaction is carried out industrially in 10 batch reactors, the average batch time is reported to have gone up to 12 hours even though the operator *claims* that the reaction has been carried out under exactly the same conditions as in the laboratory. It is desired to check the claim of the operator. Mathematically this amounts to hypothesis testing on the average batch time $\bar{t}$ and hypothesis, H_0, that $\bar{t}$ remains the same as that of the sample space is compared with respect to the alternative hypothesis, H_1, that $\bar{t}$ has been increased. This is written as

$$H_0 : \bar{t} = 10 \tag{7.2.37a}$$

$$H_0 : \bar{t} \geqslant 12 \tag{7.2.37b}$$

According to the central limit theorem, the *test sample* of size 10 for $\bar{t}$ will have mean $\mu_{\bar{t}} = 10$, variance $\frac{2}{\sqrt{10}}$ or 0.632 and would be normally distributed. The probability of occurrence of $\bar{t}$ equal to 12 hours or greater, $\Pr(\bar{t} \geqslant 12)$ would be given by

$$\Pr(\bar{t} \geqslant 12) = 1 - \int_{-\infty}^{z} \frac{1}{\sigma\sqrt{2\pi/n}} \exp\left[-\tfrac{1}{2}\{(\bar{t} - 10)/\sigma/\sqrt{n}\}^2\right] d\bar{t}$$

$$= 1 - \Phi(z) \tag{7.2.38}$$

where

$$z = \frac{12 - 10}{(\sigma/\sqrt{n})} = 3.17 \tag{7.2.39}$$

and

$$\Phi(z) = \int_{-\infty}^{z} \frac{1}{\sigma\sqrt{2\pi/n}} \exp\left[-\tfrac{1}{2}\{(\bar{t} - 10/\sigma/\sqrt{n}\}^2\right] d\bar{t} \tag{7.2.40}$$

Values of $\Phi(z)$ are given in Appendix 7.1 from which one reads its value for $z = 3.17$. It is found to be 0.99924 and probability of $\bar{t}$ is obtained by substituting in Eq. 7.2.38 as

$$\Pr(\bar{t} \geqslant 12) = 1 - 0.99924 = 0.00076 \tag{7.2.41}$$

which is found to be very low. This means that the claim of the operator that the operating conditions are exactly the same as in the laboratory experiments is *not correct*. This indicates that the industrial reactor should be re-examined very carefully.

Supposing there is a random variable whose mean has been determined through a test sample. Sometimes it is desirable to find the upper and lower bounds $\underline{\theta}$ and $\underline{\theta}$ around the mean $\bar{x}$ so that most of the data occur in this range with a probability of 95%. In other words

$$\Pr[\underline{\theta} < z < \bar{\theta}] = (1 - \alpha) \equiv 0.95 \tag{7.2.42}$$

where $(1 - \alpha)$ is called a confidence coefficient. Values of $\underline{\theta}$ and $\underline{\theta}$ can be found from Appendix A7.1 such that the probability is 95%. This gives

$$\underline{\theta} = -1.96 \tag{7.2.43a}$$

and

$$\bar{\theta} = +1.96 \tag{7.2.43b}$$

From these, the upper and lower bounds on $\bar{t}$, $\bar{t}_u$ and $\bar{t}_l$, are calculated as

$$\bar{t}_u = 10 + 1.96 \times 0.632 = 11.24 \text{ h} \tag{7.2.44a}$$

$$t_l = 10 - 1.96 \times 0.632 = 9.76 \text{ h} \tag{7.2.44b}$$

Lastly, multisample test involving n random variables $X_1, X_2, \ldots, X_n$ is considered which are normally distributed with their means $\mu_1, \mu_2, \ldots \ldots \mu_n$ and variances $\sigma_1, \sigma_2, \ldots, \sigma_n$ respectively. These may or may not be independent. Random variable, Z, is a linear combination of these

$$Z = a_1X_1 + a_2X_2 + \ldots + a_nX_n \tag{7.2.45}$$

and one wishes to perform hypothesis testing on Z exactly as done for the case of single sample test. Z is assumed to be normally distributed with mean μ_z (which is some function of $\mu_1, \mu_2, \ldots, \mu_n$) and variance σ_z (which is some function of $\sigma_1, \sigma_2, \ldots, \sigma_n$). If $f(X_1, X_2, \ldots, X_n)$ is the distribution function, μ_z is given by

$$\mu_z = \int_{-\infty}^{\infty}\int_{-\infty}^{\infty}\cdots\int_{-\infty}^{\infty}(a_1X_1 + a_2X_2 + \ldots + a_nX_n)\, f(X_1, X_2, \ldots, X_n)\, dX_1\, dX_2 \ldots dX_n \tag{7.2.46}$$

On expanding, Eq. 7.2.46 simplifies to

$$\begin{aligned}\mu_z &= \int_{-\infty}^{+\infty}\int_{-\infty}^{\infty}\cdots\int_{-\infty}^{\infty} a_1X_1 f(X_1, X_2, \ldots, X_n)\, dX_1\, dX_2 \ldots dX_n \\ &+ \int_{-\infty}^{\infty}\int_{-\infty}^{\infty}\cdots\int_{-\infty}^{\infty} a_2X_2 f(X_1, X_2, \ldots, X_n)\, dX_1\, dX_2 \ldots dX_n \\ &+ \ldots \\ &= (a_1\mu_1 + a_2\mu_2 + \ldots + a_n\mu_n)\end{aligned} \tag{7.2.47}$$

i.e., the mean of the random variable Z is the weighted mean of individual random variables. To find the variance σ_z^2 it is noted that (Eqs. 7.2.45 and 7.2.47)

$$\begin{aligned}(Z - \mu_z) &= a_1(X_1 - \mu_1) + a_2(X_2 - \mu_2) + \ldots + a_n(X_n - \mu_n) \\ &= \sum_{i=1}^{n} a_i(X_i - \mu_i)\end{aligned} \tag{7.2.48}$$

and

$$\begin{aligned}(Z - \mu_z)^2 &= \left[\sum_{i=1}^{n} a_i(X_i - \mu_i)\right]^2 \\ &= \sum_{i=1}^{n} a_i^2(X_i - \mu_i)^2 + 2\sum_{i<j}\sum a_ia_j(X_i - \mu_i)(X_j - \mu_j)\end{aligned} \tag{7.2.49}$$

The variance σ_z^2 is given by

$$\begin{aligned}\sigma_z^2 &= \sum_{i=1}^{n}\left[\int_{-\infty}^{\infty}\int_{-\infty}^{\infty}\cdots\int_{-\infty}^{\infty} a_i^2(X_i - \mu_i)^2 f(X_1, X_2, \ldots, X_n) dX_1\, dX_2 \ldots dX_n\right] \\ &+ 2\sum_{i<j}\sum\left[a_ia_j\int_{-\infty}^{\infty}\int_{-\infty}^{\infty}\cdots\int_{-\infty}^{\infty}(X_i - \mu_i)(X_j - \mu_j)\, f(X_1, X_2, \ldots, X_n)\cdot dX_1\, dX_2 \ldots dX_n\right] \\ &= \sum_{i=1}^{n} a_i^2\sigma_i^2 + \sum_{i<j}\sum a_ia_j \operatorname{cov}(X_i, X_j)\end{aligned} \tag{7.2.50}$$

where cov (X_i, X_j) represents the covariance of random variables X_i and X_j defined by

$$\text{cov}(X_i, X_j) = \int_{-\infty}^{\infty}\int_{-\infty}^{\infty}\cdots\int_{-\infty}^{\infty}(X_i - \mu_i)(X_j - \mu_j) \, f(X_1, X_2, \ldots, X_n)\, dX_1\, dX_2 \ldots dX_n \tag{7.2.51}$$

Covariance is equal to zero if random variables X_i and X_j are independent of each other. Sometimes it is written as

$$\text{cov}(X_i, X_j) = \rho_{ij}\sigma_i\sigma_j \tag{7.2.52}$$

where ρ_{ij} is called correlation coefficient and in terms of this, Eq. 7.2.50 can be rewritten as

$$\sigma_z^2 = \sum_{i=1}^{n} a_i^2\sigma_i^2 + \sum\sum_{i<j} a_i a_j \rho_{ij}\sigma_i\sigma_j \tag{7.2.53}$$

Now one is in a position to discuss data analysis and coaptation as given in Sec. 7.3 and 7.4.

7.3 DATA ANALYSIS[7-9]

It has already been explained earlier that because of several reasons the measured values are significantly different from the true process conditions existing at the point. The basic problem of data analysis is to estimate process variables from measured values and in doing so, one desires to minimize the total error.

In a chemical plant, there are equipment and process streams and with each stream there is a process variable associated (flow rate or enthalpy, as the case may be). In the following, total flow rate as the process variable has been analyzed and results can be conveniently extended for enthalpy. Measured flow rates are denoted as v_j and the true flow rates as μ_j in stream j. From these measured values v_j, $\hat{v}_j$ is estimated as the value of the variable. Adjustment, x_j, and the error, e_j, are given by

$$x_j = \hat{v}_j - v_j \tag{7.3.1a}$$

and

$$e_j = (v_j - \mu_j) \tag{7.3.1b}$$

One assumes that there are n equipment and m streams in the chemical plant. To have a situation when the number of streams is less than the number of equipment is meaningless because this would amount to having some equipment completely unconnected. In view of this, the following inequality always holds good

$$m > n \tag{7.3.2}$$

Each measuring instrument has its own accuracy of measurement and it is assumed that *all* measurements are statistically independent and normally distributed. Consequently with each error term, e_j, a standard deviation (or variance) σ_j is associated (the mean is zero).

Because each instrument works independent of the other, the total set of measurements generated corresponds to the situation of the multi-sample test. Also, the measurement error is given by its absolute value, $|e_j|$ (or e_j^2) and the estimated flow variables $\hat{v}_j$ are found in such a fashion that the total weighted error, e_t^2, defined by

$$e_t^2 = m_1 e_1^2 + m_2 e_2^2 + m_3 e_3^2 + \ldots + m_m e_m^2 \tag{7.3.3}$$

is minimum. In the above equation, $m_1, m_2, \ldots, m_m$ are weighting factors. If all these are equal to one, e_t^2 is nothing but the sum of squares of all errors. Taking all weighting factors equal to one is not correct because measurements with larger standard deviations (consequently with higher probability of error) would contribute to e_t^2 larger than the measurements with lower standard deviations. In view of this, errors must be discounted accordingly. It is customary to take m_i as[9]

$$m_i = \frac{1}{\sigma_i^2}, \quad i = 1, 2, \ldots, m \tag{7.3.4}$$

and substituting this in Eq. 7.3.3, one gets

$$\begin{aligned} e_t^2 &= \frac{e_1^2}{\sigma_1^2} + \frac{e_2^2}{\sigma_2^2} + \ldots + \frac{e_m^2}{\sigma_m^2} \\ &= \sum_{i=1}^{m} \frac{e_i^2}{\sigma_i^2} \end{aligned} \tag{7.3.5}$$

It is desired to minimize e_t^2 with respect to the constraints that estimated flow rates, $\hat{v}_j$, satisfy mass balance equations. Since there are n equipment, therefore there are n constraints written as

$$a_{11}\hat{v}_1 + a_{12}\hat{v}_2 + \ldots + a_{1m}\hat{v}_m = 0 \tag{7.3.6a}$$

$$a_{21}\hat{v}_1 + a_{22}\hat{v}_2 + \ldots + a_{2m}\hat{v}_m = 0 \tag{7.3.6b}$$

$$\vdots \qquad \vdots \qquad \qquad \vdots$$

$$a_{n1}\hat{v}_1 + a_{n2}\hat{v}_2 + \ldots + a_{nm}\hat{v}_m = 0 \tag{7.3.6c}$$

Constants a_{ij} can take a value of $+1$, 0 or -1 depending upon whether the stream j is directed to, not connected or leaving the equipment i respectively. Or

$$a_{ij} = +1 \text{ when stream } j \text{ is directed to equipment } i \tag{7.3.7a}$$

$$a_{ij} = -1 \text{ when stream } j \text{ leaves equipment } i \tag{7.3.7b}$$

$$a_{ij} = 0 \text{ when stream } j \text{ is not connected to equipment } i \tag{7.3.7c}$$

Equations 7.3.5 and 7.3.7 are best written in matrix form as follows

$$e_t^2 = (\hat{\mathbf{v}} - \mathbf{v})^T \mathbf{Q} (\hat{\mathbf{v}} - \mathbf{v}) \tag{7.3.8a}$$

and

$$\mathbf{A}\hat{\mathbf{v}} = 0 \tag{7.3.8b}$$

where

$$(\hat{\mathbf{v}} - \mathbf{v}) = \begin{bmatrix} \hat{v}_1 - v_1 \\ \hat{v}_2 - v_2 \\ \vdots \\ \hat{v}_m - v_m \end{bmatrix}_{m\times 1} \tag{7.3.9a}$$

$$\hat{\mathbf{v}} = \begin{bmatrix} \hat{v}_1 \\ \hat{v}_2 \\ \vdots \\ \hat{v}_m \end{bmatrix}_{m\times 1} \tag{7.3.9b}$$

$$\mathbf{Q} = \begin{bmatrix} 1/\sigma_1^2 & 0 & 0 & 0 & \cdots \\ 0 & 1/\sigma_2^2 & 0 & 0 & \cdots \\ 0 & 0 & 1/\sigma_3^2 & 0 & \cdots \\ & & \cdot & \cdot & \cdot \\ 0 & 0 & 0 & 0 & 1/\sigma_m^2 \end{bmatrix}_{m\times m} \tag{7.3.9c}$$

$$\mathbf{A} = \begin{bmatrix} a_{11} & a_{12} & \cdots & a_{1m} \\ a_{21} & a_{22} & \cdots & a_{2m} \\ \vdots & \vdots & \cdots & \vdots \\ a_{n1} & a_{n2} & \cdots & a_{nm} \end{bmatrix}_{n\times m} \tag{7.3.9d}$$

The matrix **Q** is diagonal whereas matrix A is a very special one because it has its elements as either +1, 0 or −1 and is called *incidence matrix*. Minimization of e_t^2 with respect to the constraint of Eq. 7.3.8b can be done by defining Lagrange multipliers $\lambda_1, \lambda_2, \ldots, \lambda_n$ and writing a combined function, F, as

$$F = e_t^2 + \lambda_1(a_{11}\hat{v}_1 + a_{12}\hat{v}_2 + \ldots + a_{1n}\hat{v}_m) + \lambda_2(a_{21}\hat{v}_1 + a_{22}\hat{v}_2 + \ldots + a_{2n}\hat{v}_m) + \ldots \tag{7.3.10}$$

To optimize F with respect to variables $\hat{v}_1, \hat{v}_2, \ldots, \hat{v}_m$ and Lagrange multipliers λ_1 to λ_n, one has

$$\frac{\partial F}{\partial \hat{v}_i} = 0, \quad i = 1, 2, \ldots, m \tag{7.3.11a}$$

and

$$\frac{\partial F}{\partial \lambda_i} = 0, \quad i = 1, 2, \ldots, n \tag{7.3.11b}$$

On differentiation of F with respect to λ_i for different values of i, the set of equations written in Eq. 7.3.6 is obtained. Differentiation with respect to $\hat{\mathbf{v}}_i$, however, gives

$$\frac{\partial F}{\partial \hat{v}_i} = \frac{2(\hat{v}_i - v_i)}{\sigma_i^2} + (\lambda_1 a_{1i} + \lambda_2 a_{2i} + \ldots + \lambda_n a_{ni}) = 0 \tag{7.3.12}$$

or

$$\frac{2}{\sigma_i^2}\hat{v}_i + \sum_{j=1}^{n} \lambda_j a_{ji} = \left(\frac{2v_i}{\sigma_i^2}\right), \; i = 1, 2, \ldots, m \tag{7.3.13}$$

In $(m + n)$ equations in Eqs. 7.3.6 and 7.3.13, there are $(m + n)$ unknowns: $\hat{v}_1$ to $\hat{v}_m$ and λ_1 to λ_n and these can be solved uniquely provided measured variables v_1 to v_m are available. Linear equations are best solved by the matrix method as given below.

Equation 7.3.13 can be written in matrix form as

$$\mathbf{Q}_1\hat{\mathbf{v}} + \mathbf{A}^{\mathrm{T}}\boldsymbol{\lambda} = \mathbf{a} \tag{7.3.14}$$

where

$$\mathbf{Q}_1 = 2\mathbf{Q} \tag{7.3.15a}$$

and

$$\mathbf{a} = \begin{bmatrix} 2v_1/\sigma_1 \\ 2v_2/\sigma_2 \\ \vdots \\ 2v_m/\sigma_m \end{bmatrix}_{m\times 1} \tag{7.3.15b}$$

and $\mathbf{A}^{\mathrm{T}}$ is the transpose of the matrix $\mathbf{A}$ given in Eq. 7.3.9. $\hat{\mathbf{v}}$ and $\boldsymbol{\lambda}$ are $m \times 1$ and $n \times 1$ vectors. Equations 7.3.8b and 7.3.14 are combined together to obtain

$$\begin{bmatrix} \mathbf{Q}_1 & \mathbf{A}^{\mathrm{T}} \\ \mathbf{A} & \mathbf{O} \end{bmatrix}_{(m+n)\times(m+n)} \begin{bmatrix} \hat{\mathbf{v}} \\ \boldsymbol{\lambda} \end{bmatrix}_{(m+n)\times 1} = \begin{bmatrix} \mathbf{a} \\ \mathbf{O} \end{bmatrix}_{(m+n)\times 1} \tag{7.3.16}$$

One multiplies both the sides by $\begin{bmatrix} \mathbf{Q}_1 & \mathbf{A}^{\mathrm{T}} \\ \mathbf{A} & \mathbf{O} \end{bmatrix}^{-1}$ to solve $\hat{v}$. However, computations are considerably simplified if both sides are multiplied by matrix $\begin{bmatrix} \mathbf{Q}_1^{-1} & \mathbf{O} \\ -\mathbf{A}\mathbf{Q}_1^{-1} & \mathbf{I} \end{bmatrix}$. Here $\mathbf{I}$ is an identity matrix whose diagonal elements are all unity and non-diagonal elements are zero. On multiplication with this matrix, Eq. 7.3.16 becomes

$$\begin{bmatrix} \mathbf{Q}_1^{-1} & \mathbf{O} \\ -\mathbf{A}\mathbf{Q}_1^{-1} & \mathbf{I} \end{bmatrix} \begin{bmatrix} \mathbf{Q}_1 & \mathbf{A}^{\mathrm{T}} \\ \mathbf{A} & \mathbf{O} \end{bmatrix} \begin{bmatrix} \hat{\mathbf{v}} \\ \boldsymbol{\lambda} \end{bmatrix} = \begin{bmatrix} \mathbf{Q}_1^{-1} & \mathbf{O} \\ -\mathbf{A}\mathbf{Q}_1^{-1} & \mathbf{I} \end{bmatrix} \begin{bmatrix} \mathbf{a} \\ \mathbf{O} \end{bmatrix} \tag{7.3.17}$$

or

$$\begin{bmatrix} \mathbf{Q}_1\mathbf{Q}_1^{-1} & \mathbf{Q}_1^{-1}\mathbf{A}^{\mathrm{T}} \\ \mathbf{A}\mathbf{Q}_1\mathbf{Q}_1^{-1} + \mathbf{A}, & -\mathbf{A}\mathbf{Q}_1^{-1}\mathbf{A}^{\mathrm{T}} \end{bmatrix} \begin{bmatrix} \hat{\mathbf{v}} \\ \boldsymbol{\lambda} \end{bmatrix} = \begin{bmatrix} \mathbf{Q}_1^{-1}\mathbf{a} \\ \mathbf{A}\mathbf{Q}_1^{-1}\mathbf{a} \end{bmatrix} \tag{7.3.18}$$

or

$$\begin{bmatrix} \mathbf{I} & \mathbf{Q}_1^{-1}\mathbf{A}^{\mathrm{T}} \\ \mathbf{0} & -\mathbf{A}\mathbf{Q}_1^{-1}\mathbf{A}^{\mathrm{T}} \end{bmatrix} \begin{bmatrix} \hat{\mathbf{v}} \\ \boldsymbol{\lambda} \end{bmatrix} = \begin{bmatrix} \mathbf{Q}_1^{-1}\mathbf{a} \\ -\mathbf{A}\mathbf{Q}_1^{-1}\mathbf{a} \end{bmatrix} \tag{7.3.19}$$

From this it follows

$$\boldsymbol{\lambda} = (\mathbf{A}\mathbf{Q}_I^{-1}\mathbf{A}^T)^{-1}\mathbf{A}\mathbf{Q}_I^{-1}\mathbf{a} \tag{7.3.20a}$$

$$\begin{aligned}\hat{\mathbf{v}} &= \mathbf{Q}_I^{-1}\mathbf{a} - (\mathbf{Q}_I^{-1}\mathbf{A}^T)\boldsymbol{\lambda} \\ &= \mathbf{v} - \mathbf{Q}_I^{-1}\mathbf{A}^T(\mathbf{A}\mathbf{Q}_I^{-1}\mathbf{A}^T)^{-1}\mathbf{A}\mathbf{v}\end{aligned} \tag{7.3.20b}$$

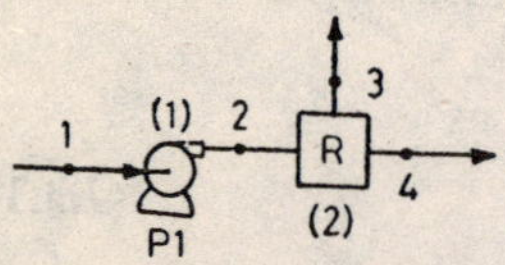

Fig. 7.2 A flow sheet involving two equipment and four streams

To demonstrate the use of the above equation, a combination of a pump and a reactor shown in Fig. 7.2 is considered. The system as a whole involves four streams. It is assumed that all streams have been measured and have variances σ_1^2 to σ_4^2 as 1, 2, 3 and 4. The incidence matrix $\mathbf{A}$ is given by

$$\mathbf{A} = \begin{array}{c} \text{Stream} \rightarrow \\ \text{Equipment} \downarrow \end{array} \begin{array}{c} \\ 1 \\ 2 \end{array}\begin{array}{cccc} 1 & 2 & 3 & 4 \end{array} \atop \begin{bmatrix} +1 & -1 & 0 & 0 \\ 0 & +1 & -1 & -1 \end{bmatrix}_{2\times 4} \tag{7.3.21}$$

and the variance matrix $\mathbf{Q}_I$ is

$$\mathbf{Q} = \begin{bmatrix} 2 & 0 & 0 & 0 \\ 0 & 1 & 0 & 0 \\ 0 & 0 & 2/3 & 0 \\ 0 & 0 & 0 & 1/2 \end{bmatrix}_{4\times 4} \tag{7.3.22}$$

Substituting these in Eq. 7.3.20,

$$\begin{bmatrix} \lambda_1 \\ \lambda_2 \end{bmatrix} = \frac{4}{23}\begin{bmatrix} 9/2(v_1 - v_2) + (v_2 - v_3 - v_4) \\ (v_1 - v_2) + \frac{3}{2}(v_2 - v_3 - v_4) \end{bmatrix} \tag{7.3.23}$$

and

$$\begin{bmatrix} \hat{v}_1 \\ \hat{v}_2 \\ \hat{v}_3 \\ \hat{v}_4 \end{bmatrix} = \begin{bmatrix} v_1 - \lambda_1 \\ v_2 + \lambda_1 - \lambda_2 \\ v_3 - \lambda_2 \\ v_4 - \lambda_2 \end{bmatrix} \tag{7.3.24}$$

Once values of v_1 to v_4 are available, λ_1 and λ_2 are calculated from Eq. 7.3.23 and substitute in Eq. 7.3.24 to calculate the best estimates $\hat{v}_1$ to $\hat{v}_4$. Intermediate steps in arriving at these equations are given in Table 7.4. As seen from the table, one needs to carry out only matrix multiplication and inversion operation. As these can be easily done on a computer, the algorithms of Eq. 7.3.20 are best suited for computer calculations.

The result presented in Eq. 7.3.20 is general and is applicable even when $\mathbf{A}$ is not an incidence matrix. The method can be used as long as the matrix $(\mathbf{A}\mathbf{Q}_I^{-1}\mathbf{A}^T)$ is non-singular. In Appendix 7.2 one considers the case where $\mathbf{A}$ is not an incidence matrix which will arise when

Table 7.4 Intermediate Steps of Estimation Problem of Fig. 7.2

1. $\mathbf{Q}_1^{-1} = \begin{bmatrix} 1/2 & 0 & 0 & 0 \\ 0 & 1 & 0 & 0 \\ 0 & 0 & 3/2 & 0 \\ 0 & 0 & 0 & 2 \end{bmatrix}_{4\times 4}$

2. $\mathbf{Q}_1^{-1}\mathbf{a} = \begin{bmatrix} v_1 \\ v_2 \\ v_3 \\ v_4 \end{bmatrix}$

3. $\mathbf{A}\mathbf{Q}_1^{-1} = \begin{bmatrix} 1 & -1 & 0 & 0 \\ 0 & 1 & -1 & -1 \end{bmatrix} \begin{bmatrix} 1/2 & 0 & 0 & 0 \\ 0 & 1 & 0 & 0 \\ 0 & 0 & 3/2 & 0 \\ 0 & 0 & 0 & 2 \end{bmatrix}_{4\times 4} = \begin{bmatrix} 1/2 & -1 & 0 & 0 \\ 0 & 1 & -3/2 & -2 \end{bmatrix}$

4. $\mathbf{Q}_1^{-1}\mathbf{A}^T = \begin{bmatrix} 1/2 & 0 & 0 & 0 \\ 0 & 1 & 0 & 0 \\ 0 & 0 & 3/2 & 0 \\ 0 & 0 & 0 & 2 \end{bmatrix} \begin{bmatrix} 1 & 0 \\ -1 & 1 \\ 0 & -1 \\ 0 & -1 \end{bmatrix} = \begin{bmatrix} 1/2 & 0 \\ -1 & 1 \\ 0 & -3/2 \\ 0 & -2 \end{bmatrix}$

5. $\mathbf{A}\mathbf{Q}_1^{-1}\mathbf{A}^T = (\mathbf{A}\mathbf{Q}_1^{-1})\mathbf{A}^T = \begin{bmatrix} 1/2 & -1 & 0 & 0 \\ 0 & 1 & -3/2 & -2 \end{bmatrix} \begin{bmatrix} 1 & 0 \\ -1 & 1 \\ 0 & -1 \\ 0 & -1 \end{bmatrix} = \begin{bmatrix} 3/2 & -1 \\ -1 & 9/2 \end{bmatrix}$

6. $(\mathbf{A}\mathbf{Q}_1^{-1}\mathbf{A}^T)^{-1} = \begin{bmatrix} 18/23 & 4/23 \\ 4/23 & 6/23 \end{bmatrix}$

7. $\mathbf{A}\mathbf{Q}^{-1}\mathbf{a} = \begin{bmatrix} 1/2 & -1 & 0 & 0 \\ 0 & 1 & -3/2 & -2 \end{bmatrix} \begin{bmatrix} 2v_1 \\ v_1 \\ 2v_3/3 \\ v_4/2 \end{bmatrix} = \begin{bmatrix} v_1 - v_2 \\ v_2 - v_3 - v_4 \end{bmatrix}$

component mass balance is made and the problem is solved once again using the algorithm of Eq. 7.3.20.

7.4 COAPTATION

From Sec. 7.3, it may be clear that it is desirable to measure *all* stream flow rates. However, due to the costs involved, only some are measured and the remainder are estimated based upon these measurements. The process of estimating unknown variables based upon measured ones is called coaptation. In this connection, it is of interest to find the least number of measurements. The second question to be answered in relation to this is what should the procedure of data analysis be in the case when all measurements are not available.

The first question can be answered by examining the set of mass balance equations in Eq. 7.3.6. There are n equations corresponding to each equipment in the flow sheet and there are m unknown variables. To solve these linear equations, it is necessary that $(m - n)$ of them are measured and it is only after that, that one can solve the remaining n unknown variables. Therefore if s is the number of unmeasured streams, s should be less than n. Or

$$s \leqslant n \tag{7.4.1}$$

In the analysis presented below, the condition set by the above equation is assumed to be satisfied.

The s unspecified variables are denoted by $u_1, u_2, \ldots, u_s$ to distinguish from measured variables $v_1, v_2, \ldots, v_{(m-s)}$. The variables $\hat{v}_1, \hat{v}_2, \ldots, \hat{v}_{m-s}$ are their best estimates and in terms of these, mass balance equations become

$$a_{11}\hat{v}_1 + a_{12}\hat{v}_2 + \ldots + a_{1,(m-s)}\hat{v}_{m-s} + a_{1,(m-s+1)}u_1 + \ldots + a_{1m}u_s = 0 \tag{7.4.2a}$$

$$a_{21}\hat{v}_1 + a_{22}\hat{v}_2 + \ldots + a_{2,(m-s)}\hat{v}_{m-s} + a_{2,(m-s+1)}u_1 + \ldots + a_{2m}u_s = 0 \tag{7.4.2b}$$

$$\vdots \qquad\qquad \vdots \qquad\qquad \vdots \qquad\qquad \vdots$$

$$a_{n1}\hat{v}_1 + a_{n2}\hat{v}_2 + \ldots + a_{n,(m-s)}\hat{v}_{m-s} + a_{n,(m-s+1)}u_1 + \ldots + a_{nm}u_s = 0 \tag{7.4.2c}$$

In matrix notation, these equations are written as

$$\mathbf{A}_1\hat{v} + \mathbf{A}_2\mathbf{u} = 0 \tag{7.4.3}$$

Matrices $\mathbf{A}_1$ and $\mathbf{A}_2$ are obtained by partitioning $\mathbf{A}$ of Eq. 7.3.8. The total error e_t^2 would be defined as

$$e_t^2 = \frac{e_1^2}{\sigma_1^2} + \frac{e_2^2}{\sigma_2^2} + \ldots + \frac{e_{m-s}^2}{\sigma_{m-s}^2} \tag{7.4.4}$$

because the remaining s variables have not been measured. To minimize e_t^2, n Lagrange multipliers $\lambda_1, \lambda_2, \ldots, \lambda_n$ are defined to obtain function, F

$$F = \sum_{i=1}^{(m-s)} \frac{(\hat{v}_i - v_i)^2}{\sigma_i^2} + \sum_{j=1}^{n} \lambda_j \left(\sum_{i=1}^{m-s} a_{ji}\hat{v}_i + \sum_{i=m-s+1}^{m} a_{ji}u_{i-m+s} \right) \tag{7.4.5}$$

which is to be minimized with respect to the choice of $\hat{v}_i$, u_{i-m+s} and λ_j. This would similarly give rise to m equations which can be solved for $\mathbf{u}$ and $\hat{v}$ exactly as done in Eq. 7.3.20. Results are identical except for the fact that vector $\mathbf{a}$ has to be rewritten for this specific case.

Even though the problems of estimation of process variables and coaptation can in principle be solved using Eq. 7.3.20, it may be clear that these matrix manipulations for large problems would be messy and sometimes unmanageable for large chemical plants. In view of this, it would be desirable to further simplify the calculation procedure.

The simplification in the calculation becomes possible because of the special form of the incidence matrix **A**. One considers the flow sheet of Fig. 7.3, in which R1 is a pyrolysis reactor where several products are generated. Gaseous products are allowed to escape whereas liquid products are sent to distillation columns S1 and S2. The two lighter fractions, thus separated, react in R2 to get the desired product. The process flow

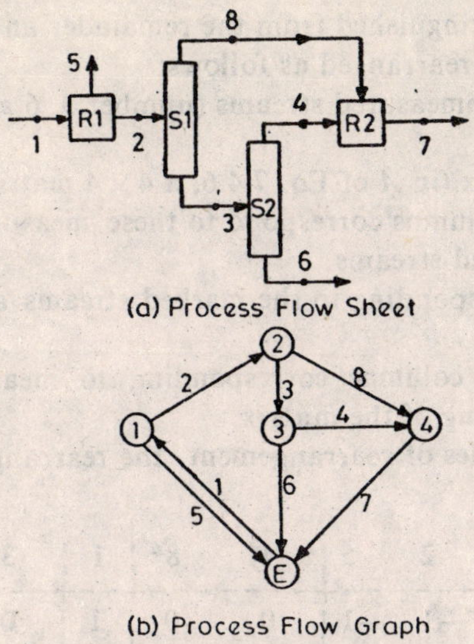

Fig. 7.3 A flow sheet involving two reactors and two distillation columns with its flow graph

graph of Fig. 7.3a is drawn as follows. An environment node, E, is defined from which all raw material streams emanate and products directed towards it. All equipment, whether a reactor or a distillation column, are represented by a small circle as shown in Fig. 7.3b and they are connected by streams exactly as in the flow sheet. These small circles are called *nodes* and in the process flow graph they are not distinguished from each other. The purpose of introducing environment node E is to ensure that every stream is connected to two nodes, which can be verified from the figure. The introduction of the fictitious node E does not change incidence matrix A because while writing the matrix, the presence of E in the process graph is ignored. The matrix A remains as

$$
\begin{array}{ll}
 & \text{Stream}\rightarrow \quad 1 \quad\; 2 \quad\; 3 \quad\; 4 \quad\; 5 \quad\; 6 \quad\; 7 \quad\; 8 \\
\mathbf{A} = & \begin{array}{c} \text{nodes no.} \\ \downarrow \end{array}
\begin{array}{c} 1 \\ 2 \\ 3 \\ 4 \end{array}
\begin{bmatrix}
1 & -1 & 0 & 0 & -1 & 0 & 0 & 0 \\
0 & 1 & -1 & 0 & 0 & 0 & 0 & -1 \\
0 & 0 & 1 & -1 & 0 & -1 & 0 & 0 \\
0 & 0 & 0 & 1 & 0 & 0 & -1 & 1
\end{bmatrix}_{4\times 8}
\end{array}
\qquad (7.4.6)
$$

It can be checked by evaluating any 4×4 determinant of A that the matrix is non-singular, and it is indeed because of this reason, that the node E has been ignored.

The problem of coaptation discussed here is a special case of the data analysis: in the latter there are zero number of unmeasured streams. For the purpose of discussion, *streams* 3, 6 *and* 7 are assumed as the unmeasured ones. It is observed that these connect nodes 2, 3, 4 and E as shown in Fig. 7.3b. Between these nodes, it is found that streams 4 and 8 are measured ones and do not connect any other node besides these. Such streams are distinguished from the remainder and are called *marked streams*. Matrix A is rearranged as follows:

(1) Columns for unmeasured streams (number 3, 6 and 7) are placed in the end.

(2) An $n \times n$ matrix (in A of Eq. 7.4.6, a 4×4 matrix) is marked. The $(n - s)$ number of columns correspond to these measured streams which are other than marked streams.

(3) Columns corresponding to the marked streams are placed next to the $n \times n$ matrix.

(4) Remainder of columns corresponding to measured streams are placed in the beginning of the matrix.

Following these rules of rearrangement, the rearranged **A** is obtained as

$$\begin{array}{ll} & \text{Stream} \rightarrow \quad \begin{array}{cc:cc:c:ccc} 2 & 5 & 4^* & 8^* & 1 & 3 & 6 & 7 \end{array} \\ & \text{Nodes} \downarrow \\ \mathbf{A} = & \begin{array}{c} 1 \\ 2 \\ 3 \\ 4 \end{array} \left[\begin{array}{cc:cc:c:ccc} -1 & -1 & 0 & 0 & 1 & 0 & 0 & 0 \\ 1 & 0 & 0 & -1 & 0 & -1 & 0 & 0 \\ 0 & 0 & -1 & 0 & 0 & 1 & -1 & 0 \\ 0 & 0 & 1 & 1 & 0 & 0 & 0 & -1 \end{array} \right] \end{array} \tag{7.4.7}$$

Columns 3, 6 and 7 correspond to the unmeasured streams whereas columns 4 and 8 are for marked streams and to indicate this, they have been shown with asterisks.

What has been done for the process flow graph of Fig. 7.3 can be done for any graph involving m streams and n equipment (or nodes) as follows[8]:

$$\begin{array}{ll} & \text{Stream} \rightarrow \begin{array}{c:c:c:c} m-n-k & k & n-s & s \end{array} \\ \mathbf{A} = & \left[\begin{array}{c:c:c:c} \mathbf{A}_{11} & \mathbf{A}_{12} & \mathbf{A}_{13} & \mathbf{A}_{14} \end{array} \right]_{n \times m} \end{array} \tag{7.4.8}$$

where k is the number of marked streams. Vector $\hat{\mathbf{v}}$ is split into three subvectors $\hat{\mathbf{v}}_1$, $\hat{\mathbf{v}}_2$ and $\hat{\mathbf{v}}_3$ corresponding to the splitting of **A** above. Thus if **u** is the vector for unmeasured variables the mass balance equation becomes

$$\mathbf{A}\hat{\mathbf{v}} = [\mathbf{A}_{11} \,\vdots\, \mathbf{A}_{12} \,\vdots\, \mathbf{A}_{13} \,\vdots\, \mathbf{A}_{14}] \begin{bmatrix} \hat{\mathbf{v}}_1 \\ \hat{\mathbf{v}}_2 \\ \hat{\mathbf{v}}_3 \\ \mathbf{u} \end{bmatrix} = 0 \tag{7.4.9}$$

A matrix **T** is defined as

$$\mathbf{T} = [\mathbf{A}_{13} \,\vdots\, \mathbf{A}_{14}] \tag{7.4.10}$$

and Eq. 7.4.9 is multiplied by $\mathbf{T}^{-1}$ to obtain

$$\mathbf{T}^{-1}[\mathbf{A}_{11} \mid \mathbf{A}_{12} \mid \mathbf{A}_{13} \mid \mathbf{A}_{14}]\begin{bmatrix} \hat{\mathbf{v}}_1 \\ \hat{\mathbf{v}}_2 \\ \hat{\mathbf{v}}_3 \\ \mathbf{u} \end{bmatrix} = 0 \tag{7.4.11}$$

The multiplication with $\mathbf{T}^{-1}$ is always possible because $\mathbf{T}$ is a non-singular matrix and on doing so, the following forms of the incidence matrix $\mathbf{A}$ is obtained:

$$\mathbf{T}^{-1}[\mathbf{A}_{11}\mathbf{A}_{12}\mathbf{A}_{13}\mathbf{A}_{14}] = \begin{array}{c} \\ n-s \\ s \end{array}\begin{array}{c} \begin{array}{cccc} (m-n-k) & k & (n-s) & s \end{array} \\ \left[\begin{array}{cccc} \mathbf{K}_1 & 0 & \mathbf{I}_{n-s} & 0 \\ \hline \mathbf{K}_2 & \mathbf{K}_3 & 0 & \mathbf{I}_s \end{array}\right] \end{array} \tag{7.4.12}$$

The proof of the above relation is beyond the scope of this text, but to demonstrate that one indeed gets this form, the incidence matrix of Fig. 7.3b is considered. Matrix $\mathbf{T}$ would be

$$\mathbf{T} = \begin{array}{c} \\ 1 \\ 2 \\ 3 \\ 4 \end{array}\begin{array}{c} \begin{array}{cccc} 1 & 3 & 6 & 7 \end{array} \\ \begin{bmatrix} 1 & 0 & 0 & 0 \\ 0 & -1 & 0 & 0 \\ 0 & 1 & -1 & 0 \\ 0 & 0 & 0 & -1 \end{bmatrix} \end{array} \tag{7.4.13}$$

and its inverse is given by

$$\mathbf{T}^{-1} = \begin{bmatrix} 1 & 0 & 0 & 0 \\ 0 & -1 & 0 & 0 \\ 0 & -1 & -1 & 0 \\ 0 & 0 & 0 & -1 \end{bmatrix} \tag{7.4.14}$$

All intermediate steps in calculating the inverse of $\mathbf{T}$ are given in Table 7.5 with the purpose to show that the determinant of an incidence matrix or any of its cofactors is either 1, 0 or -1. The value zero arises when it is singular. Also an incidence matrix under inversion remains an incidence matrix. This result is true for any matrix size and its proof can be had in reference 10. One next multiplies $\mathbf{T}^{-1}$ with $\mathbf{A}$ (Eq. 7.4.7) to obtain

$$\mathbf{T}^{-1}\mathbf{A} = \begin{bmatrix} 1 & 0 & 0 & 0 \\ 0 & -1 & 0 & 0 \\ 0 & -1 & -1 & 0 \\ 0 & 0 & 0 & -1 \end{bmatrix} \times \begin{bmatrix} -1 & -1 & 0 & 0 & 1 & 0 & 0 & 0 \\ 1 & 0 & 0 & -1 & 0 & -1 & 0 & 0 \\ 0 & 0 & -1 & 0 & 0 & 1 & -1 & 0 \\ 0 & 0 & 1 & 1 & 0 & 0 & 0 & -1 \end{bmatrix} \tag{7.4.15}$$

Table 7.5 Calculation of the Inverse of an Incidence Matrix T

S.No.	Matrix
1.	$\text{Det } T = 1 \times \begin{vmatrix} -1 & 0 & 0 \\ 1 & -1 & 0 \\ 0 & 0 & -1 \end{vmatrix} = -1$
2	A_{ij} are different cofactors of matrix T
	$A_{11} = \begin{vmatrix} -1 & 0 & 0 \\ 1 & -1 & 0 \\ 0 & 0 & -1 \end{vmatrix} = -1 \qquad A_{12} = \begin{vmatrix} 0 & 0 & 0 \\ 0 & -1 & 0 \\ 0 & 0 & -1 \end{vmatrix} = 0$
	$A_{13} = \begin{vmatrix} 1 & -1 & 0 \\ 0 & 1 & 0 \\ 0 & 0 & -1 \end{vmatrix} = 0 \qquad A_{14} = \begin{vmatrix} 0 & -1 & 0 \\ 0 & 1 & -1 \\ 0 & 0 & 0 \end{vmatrix} = 0$
	$A_{21} = \begin{vmatrix} 0 & 0 & 0 \\ 1 & -1 & 0 \\ 0 & 0 & -1 \end{vmatrix} = 0 \qquad A_{22} = \begin{vmatrix} 1 & 0 & 0 \\ 0 & -1 & 0 \\ 0 & 0 & -1 \end{vmatrix} = 1$
	$A_{23} = \begin{vmatrix} 1 & 0 & 0 \\ 0 & 1 & 0 \\ 0 & 0 & -1 \end{vmatrix} = +1 \qquad A_{24} = \begin{vmatrix} 1 & 0 & 0 \\ 0 & 1 & -1 \\ 0 & 0. & 0 \end{vmatrix} = 0$
	$A_{31} = \begin{vmatrix} 0 & 0 & 0 \\ -1 & 0 & 0 \\ 0 & 0 & -1 \end{vmatrix} = 0 \qquad A_{32} = \begin{vmatrix} 1 & 0 & 0 \\ 0 & 0 & 0 \\ 0 & 0 & -1 \end{vmatrix} = 0$
	$A_{33} = \begin{vmatrix} 1 & 0 & 0 \\ 0 & -1 & 0 \\ 0 & 0 & 1 \end{vmatrix} = +1 \qquad A_{34} = \begin{vmatrix} 1 & 0 & 0 \\ 0 & -1 & 0 \\ 0 & 0 & 0 \end{vmatrix} = 0$
	$A_{41} = \begin{vmatrix} 0 & 0 & 0 \\ -1 & 0 & 0 \\ 1 & -1 & 0 \end{vmatrix} = 0 \qquad A_{42} = \begin{vmatrix} 1 & 0 & 0 \\ 0 & 0 & 0 \\ 0 & -1 & 0 \end{vmatrix} = 0$
	$A_{43} = \begin{vmatrix} 1 & 0 & 0 \\ 0 & 0 & 0 \\ 0 & -1 & 0 \end{vmatrix} = 0 \qquad A_{44} = \begin{vmatrix} 1 & 0 & 0 \\ 0 & -1 & 0 \\ 0 & 1 & -1 \end{vmatrix} = 1$
	$T^{-1} = \dfrac{1}{\text{Det T}} \begin{bmatrix} A_{11} & A_{21} & A_{31} & A_{41} \\ A_{12} & A_{22} & A_{32} & A_{42} \\ A_{13} & A_{23} & A_{33} & A_{43} \\ A_{14} & A_{24} & A_{31} & A_{44} \end{bmatrix} = \begin{bmatrix} +1 & 0 & 0 & 0 \\ 0 & -1 & 0 & 0 \\ 0 & -1 & -1 & 0 \\ 0 & 0 & 0 & -1 \end{bmatrix}$

or

$$
\begin{array}{c}
\quad\quad (m-n-k) \quad\quad k \quad\quad n-s \quad\quad s \\
\mathbf{T}^{-1}\mathbf{A} =
\begin{array}{c|cc|cc|c|ccc}
 & 2 & 5 & 4^* & 8^* & 1 & 3 & 6 & 7 \\
\hline
1 & -1 & -1 & 0 & 0 & 1 & 0 & 0 & 0 \\
\hline
2 & -1 & 0 & 0 & 1 & 0 & 1 & 0 & 0 \\
3 & -1 & 0 & 1 & 1 & 0 & 0 & 1 & 0 \\
4 & 0 & 0 & -1 & -1 & 0 & 0 & 0 & 1
\end{array}
\end{array}
\tag{7.4.16}
$$

which is found to be exactly of the same form as given in Eq. 7.4.12. The form of the matrix in Eq. 7.4.12 leads to further simplification in the set of linear equations as follows: One considers the set of equations

$$
\begin{bmatrix} \mathbf{K}_1 & 0 & \mathbf{I}_{n-s} & 0 \\ \mathbf{K}_2 & \mathbf{K}_3 & 0 & \mathbf{I}_s \end{bmatrix} \begin{bmatrix} \hat{\mathbf{v}}_1 \\ \hat{\mathbf{v}}_2 \\ \hat{\mathbf{v}}_3 \\ \mathbf{u} \end{bmatrix} = 0 \tag{7.4.17}
$$

which on expansion gives

$$\mathbf{K}_1\hat{\mathbf{v}}_1 + \hat{\mathbf{v}}_3 = 0 \tag{7.4.18a}$$

and

$$\mathbf{K}_2\hat{\mathbf{v}}_1 + \mathbf{K}_3\hat{\mathbf{v}}_2 + \mathbf{u} = 0 \tag{7.4.18b}$$

This shows that the number of equations that have to be solved simultaneously has been reduced, thus reducing the total computation time.

At the first instance it might appear that matrices $\mathbf{K}_1$, $\mathbf{K}_2$ and $\mathbf{K}_3$ themselves are to be found through intricate matrix manipulation and whatever time saving that may occur in solving Eq. 7.4.18 is more than compensated in calculating $\mathbf{K}_1$, $\mathbf{K}_2$ and $\mathbf{K}_3$. However, this is not so because these can be directly determined from the process flow graph and the procedure of doing it is given below.[7,8]

(1) From the process flow sheet, the process flow graph is drawn after introducing an extra environment node, E.

(2) All streams which have not been measured are found (in Fig. 7.3, these are streams 3, 6 and 7).

(3) All *nodes* connecting unmeasured streams are found (nodes 2, 3 and 4 in Fig. 7.3).

(4) These nodes are redrawn on a separate graph showing *only* unmeasured streams. If in the resultant graph, there is any *cyclic path*, it is broken by removing one of the streams such that it becomes a tree involving all the *nodes*. These terms are explained a little later in detail. The graph thus generated is a subgraph (subgraph 1) of the process flow graph.

(5) Subgraph 2 is drawn by combining nodes in such a fashion that all unmeasured streams disappear. For example, by combining nodes 2 and

3, stream 3 disappears and so on.

(6) The incidence matrix of subgraph 2 is written and is called B_1 with the streams as columns and nodes as rows.

(7) A table consisting of connections of nodes and measured and unmeasured streams is prepared.

(8) With the help of the above table incidence matrix, B_2, is written giving connection between streams of matrix B_1 (i.e., all measured streams) and all those nodes from which unmeasured streams emanate.

(9) Incidence matrix B_3 is written which gives connection between unmeasured streams and the nodes connecting them.

Then it is stated without proof that

$$\begin{array}{c} \\ n-s \\ \\ s \end{array}\left[\begin{array}{c|c|c|c} m-n-k & k & n-s & s \\ \mathbf{K}_1 & \mathbf{0} & \mathbf{I}_{n-s} & \mathbf{0} \\ \hline \mathbf{K}_2 & \mathbf{K}_3 & 0 & \mathbf{I}_s \end{array}\right]\begin{bmatrix} \hat{\mathbf{v}} \\ \mathbf{u} \end{bmatrix}$$

$$= \begin{array}{c} \\ n-s \\ \\ s \end{array}\left[\begin{array}{cc} m-s & s \\ \mathbf{B}_1 & \mathbf{0} \\ \hline \mathbf{B}_2 & \mathbf{B}_3 \end{array}\right]\begin{bmatrix} \hat{\mathbf{v}} \\ \mathbf{u} \end{bmatrix} = 0 \tag{7.4.19}$$

On comparison it is found that

$$\mathbf{B}_1 = (n-s)\,[\mathbf{K}_1 \quad \mathbf{0} \quad \mathbf{I}_{n-s}]_{(n-s)\times(m-s)} \tag{7.2.20a}$$

and

$$\mathbf{B}_3^{-1}\mathbf{B}_2 = [\mathbf{K}_2 \quad \mathbf{K}_3 \quad \mathbf{0}] \tag{7.4.20b}$$

Before proceeding to demonstrate the validity of the above relation, cyclic path and trees are defined formally. A *cyclic path* is the one where it is possible, starting from a given equipment, to reach the same equipment following connecting streams. Figure 7.4a shows a few cyclic graphs and by removing some of the streams from these, their cyclic nature can be broken. A tree is any graph in which there is no cyclic path. Any cyclic graph, when broken in such a way that *no isolated node occurs*, would lead to a tree as shown in Fig. 7.4b.

Now one goes back to the example of Fig. 7.3. Step (1) consists of drawing a process flow graph which is given in Fig. 7.3b. Streams 3, 6 and 7 are unmeasured ones and nodes 2, 3, 4 and E are the ones connected by them. Since this does not have any cyclic path, it is called subgraph 1 and is shown in Fig. 7.5. To generate subgraph 2, nodes E, 2, 3 and 4 are merged as shown in the figure. Only node 1 is left after the merger and is connected to nodes E in Fig. 7.3b. These streams are shown in subgraph 2 in Fig. 7.5. The incidence matrix $\mathbf{B}_1$ of subgraph 2 would be

$$\begin{array}{lcl} & \text{Streams} \rightarrow & 1 \quad\; 2 \quad\; 4 \quad\; 5 \quad\; 8 \\ & \text{Node} & \\ \mathbf{B}_1 = & \downarrow 1 & [1 \quad -1 \quad 0 \quad -1 \quad 0]_{1\times 5} \end{array} \tag{7.4.21}$$

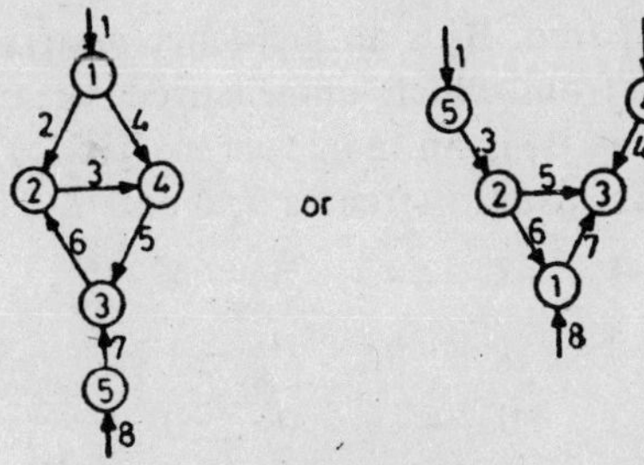

(a) Cyclic Graphs

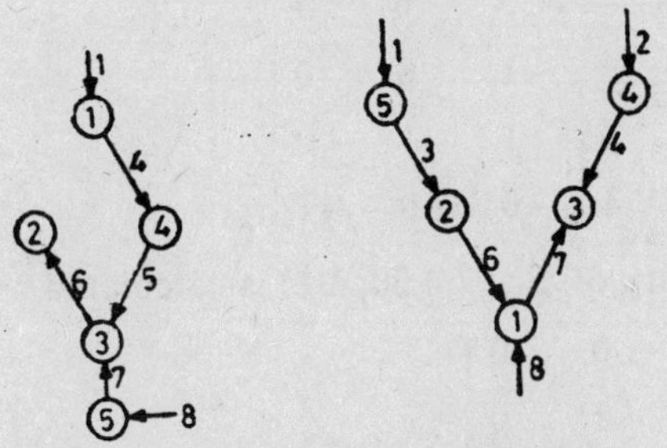

(b) Trees Obtained from Cyclic Graphs

Fig 7.4 Some examples of cyclic graphs with the trees obtained from them

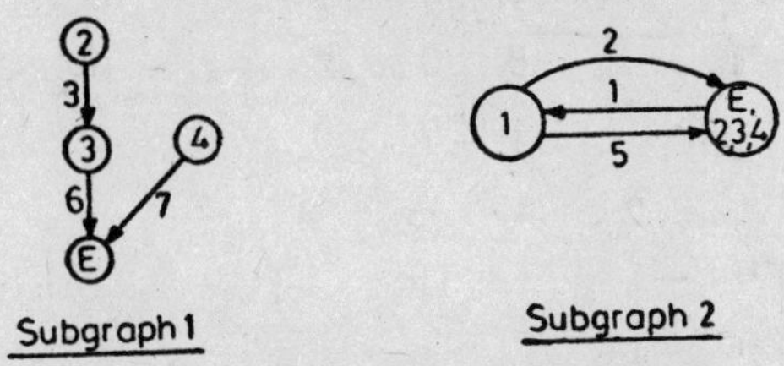

Fig. 7.5 Subgraphs 1 and 2 obtained from Fig. 7.3b

To find $\mathbf{B}_2$ and $\mathbf{B}_3$, a connection table (Table 7.6) of measured and un-

Table 7.6 Stream-Node Connection Table of Process Graph of Fig. 7.3

S.No.	Nature	Connection		
		Stream	From node	To node
1.	For unmeasured stream	3	2	3
		6	3	E
		7	4	E
2.	For measured stream	1	E	1
		2	1	2
		4	3	4
		5	1	E
		8	2	4

measured streams is prepared. $\mathbf{B}_2$ is an incidence matrix for three (equal to s) nodes 2, 3 and 4 (from which unmeasured streams emanate) and measured streams whereas $\mathbf{B}_3$ is an incidence matrix for three (equal to s) nodes 2, 3 and 4 and unmeasured streams 3, 6 and 7. Thus,

$$\begin{array}{ccc} & \text{Streams} \longrightarrow & \begin{array}{ccccc} 1 & 2 & 4 & 5 & 8 \end{array} \\ & \text{Node} & \\ \mathbf{B}_2 = & \downarrow & \begin{array}{c} 2 \\ 3 \\ 4 \end{array} \begin{bmatrix} 0 & 1 & 0 & 0 & -1 \\ 0 & 0 & -1 & 0 & 0 \\ 0 & 0 & 1 & 0 & 1 \end{bmatrix}_{3\times 5} \end{array} \tag{7.4.22a}$$

and

$$\begin{array}{ccc} & \text{Streams} \longrightarrow & \begin{array}{ccc} 3 & 6 & 7 \end{array} \\ & \text{Nodes} & \\ \mathbf{B}_3 = & \downarrow & \begin{array}{c} 2 \\ 3 \\ 4 \end{array} \begin{bmatrix} -1 & 0 & 0 \\ 1 & -1 & 0 \\ 0 & 0 & -1 \end{bmatrix}_{3\times 3} \end{array} \tag{7.4.22b}$$

To show the equality of Eq. 7.4.20, $\mathbf{B}_3^{-1}$ is calculated as

$$\mathbf{B}_3^{-1} = \begin{bmatrix} -1 & 0 & 0 \\ -1 & -1 & 0 \\ 0 & 0 & -1 \end{bmatrix} \tag{7.4.23}$$

and

$$\mathbf{B}_3^{-1}\mathbf{B}_2 = \begin{bmatrix} -1 & 0 & 0 \\ -1 & -1 & 0 \\ 0 & 0 & -1 \end{bmatrix} \begin{bmatrix} 0 & 1 & 0 & 0 & -1 \\ 0 & 0 & -1 & 0 & 0 \\ 0 & 0 & 1 & 0 & 1 \end{bmatrix}$$

$$= \begin{array}{c} \\ 2 \\ 3 \\ 4 \end{array} \begin{array}{c} \begin{array}{ccccc} 1 & 2 & 4 & 5 & 8 \end{array} \\ \begin{bmatrix} 0 & -1 & 0 & 0 & 1 \\ 0 & -1 & 1 & 0 & 1 \\ 0 & 0 & -1 & 0 & -1 \end{bmatrix} \end{array} \tag{7.4.24}$$

On comparison with Eq. 7.4.16, the validity of Eq. 7.4.20 is established.

In view of Eq. 7.4.19, the matrix equations are given by

$$\mathbf{B}_1\hat{\mathbf{v}} = 0 \tag{7.4.25a}$$

and

$$\mathbf{B}_2\hat{\mathbf{v}} + \mathbf{B}_3\mathbf{u} = 0 \tag{7.4.25b}$$

and $\hat{\mathbf{v}}$ can be calculated in terms of measured parameters $\mathbf{v}$, completely analogous to Eq. 7.3.20b, as

$$\hat{\mathbf{v}} = \mathbf{v} - \mathbf{Q}_1^{-1}\mathbf{B}_1^T(\mathbf{B}_1\mathbf{Q}_1^{-1}\mathbf{B}_1^T)^{-1}\mathbf{B}_1\mathbf{v} \tag{7.4.26}$$

From Eq. 7.4.25, $\mathbf{u}$ is obtained as

$$\begin{aligned} \mathbf{u} &= -\mathbf{B}_3^{-1}\mathbf{B}_2\hat{\mathbf{v}} \\ &= -\mathbf{B}_3^{-1}[\mathbf{B}_2\mathbf{Q}_1^{-1}\mathbf{B}_1^T(\mathbf{B}_1\mathbf{Q}_1^{-1}\mathbf{B}_1^T)^{-1}\mathbf{B}_1 - \mathbf{B}_2]\mathbf{v} \end{aligned} \tag{7.4.27}$$

Since matrix $\mathbf{B}_1$ is much smaller in size compared to A, the amount of simplification resulting through the procedure suggested here is evident.

Before concluding this section, a more difficult example involving a cyclic graph in subgraph 1[8] is solved. One considers the flow sheet of Fig. 7.6a and draws the process flow graph in Fig. 7.6b after introducing the environment node E. As has been explained earlier, the nature of equipment involved in the flow sheet is unimportant for data analysis and coaptation calculations.

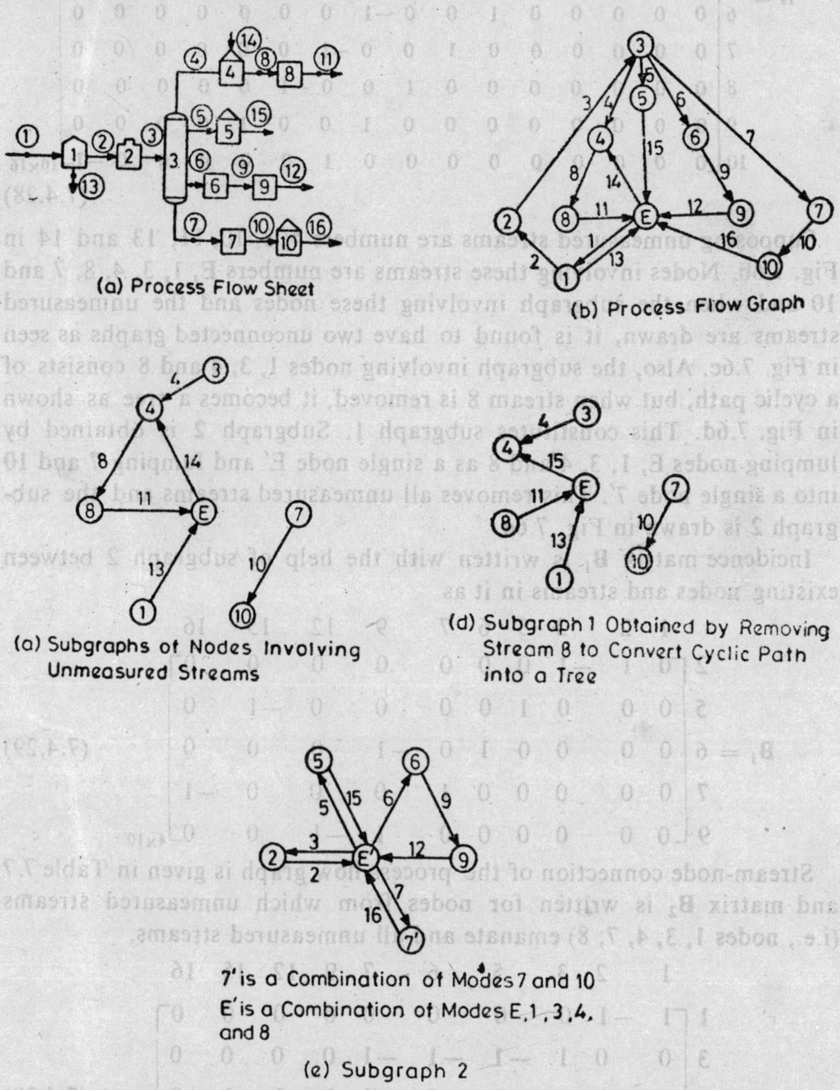

Fig. 7.6 Subgraphs 1 and 2 obtained from a flow sheet consisting of a cyclic graph.[8] (Reprinted from Mah et al., *I. & E. C. Process Design and Development*, **15**, 175, 1976 with permission of the American Chemical Society, Washington.)

The incidence matrix $\mathbf{A}$ of the flow graph is given by

$$
\begin{array}{c}
\text{Streams} \longrightarrow \\
\text{Nodes} \downarrow
\end{array}
\mathbf{A} =
\begin{array}{c|cccccccccccccccc|}
 & 1 & 2 & 3 & 4 & 5 & 6 & 7 & 8 & 9 & 10 & 11 & 12 & 13 & 14 & 15 & 16 \\
\hline
1 & 1 & -1 & 0 & 0 & 0 & 0 & 0 & 0 & 0 & 0 & 0 & 0 & -1 & 0 & 0 & 0 \\
2 & 0 & 1 & -1 & 0 & 0 & 0 & 0 & 0 & 0 & 0 & 0 & 0 & 0 & 0 & 0 & 0 \\
3 & 0 & 0 & 1 & -1 & -1 & -1 & -1 & 0 & 0 & 0 & 0 & 0 & 0 & 0 & 0 & 0 \\
4 & 0 & 0 & 0 & 1 & 0 & 0 & 0 & -1 & 0 & 0 & 0 & 0 & 0 & 1 & 0 & 0 \\
5 & 0 & 0 & 0 & 0 & 1 & 0 & 0 & 0 & 0 & 0 & 0 & 0 & 0 & 0 & -1 & 0 \\
6 & 0 & 0 & 0 & 0 & 0 & 1 & 0 & 0 & -1 & 0 & 0 & 0 & 0 & 0 & 0 & 0 \\
7 & 0 & 0 & 0 & 0 & 0 & 0 & 1 & 0 & 0 & -1 & 0 & 0 & 0 & 0 & 0 & 0 \\
8 & 0 & 0 & 0 & 0 & 0 & 0 & 0 & 1 & 0 & 0 & -1 & 0 & 0 & 0 & 0 & 0 \\
9 & 0 & 0 & 0 & 0 & 0 & 0 & 0 & 0 & 1 & 0 & 0 & -1 & 0 & 0 & 0 & 0 \\
10 & 0 & 0 & 0 & 0 & 0 & 0 & 0 & 0 & 0 & 1 & 0 & 0 & 0 & 0 & 0 & -1
\end{array}_{10\times 16}
\tag{7.4.28}
$$

Supposing unmeasured streams are numbers 4, 8, 10, 11, 13 and 14 in Fig. 7.6b. Nodes involving these streams are numbers E, 1, 3, 4, 8, 7 and 10 and when the subgraph involving these nodes and the unmeasured streams are drawn, it is found to have two unconnected graphs as seen in Fig. 7.6c. Also, the subgraph involving nodes 1, 3, 4 and 8 consists of a cyclic path, but when stream 8 is removed, it becomes a tree as shown in Fig. 7.6d. This constitutes subgraph 1. Subgraph 2 is obtained by lumping nodes E, 1, 3, 4 and 8 as a single node E′ and lumping 7 and 10 into a single node 7′. This removes all unmeasured streams and the subgraph 2 is drawn in Fig. 7.6e.

Incidence matrix $\mathbf{B}_1$ is written with the help of subgraph 2 between existing nodes and streams in it as

$$
\mathbf{B}_1 =
\begin{array}{c|cccccccccc|}
 & 1 & 2 & 3 & 5 & 6 & 7 & 9 & 12 & 15 & 16 \\
\hline
2 & 0 & 1 & -1 & 0 & 0 & 0 & 0 & 0 & 0 & 0 \\
5 & 0 & 0 & 0 & 1 & 0 & 0 & 0 & 0 & -1 & 0 \\
6 & 0 & 0 & 0 & 0 & 1 & 0 & -1 & 0 & 0 & 0 \\
7 & 0 & 0 & 0 & 0 & 0 & 1 & 0 & 0 & 0 & -1 \\
9 & 0 & 0 & 0 & 0 & 0 & 0 & 1 & -1 & 0 & 0
\end{array}_{4\times 10}
\tag{7.4.29}
$$

Stream-node connection of the process flow graph is given in Table 7.7 and matrix $\mathbf{B}_2$ is written for nodes from which unmeasured streams (i.e., nodes 1, 3, 4, 7, 8) emanate and all unmeasured streams.

$$
\mathbf{B}_2 =
\begin{array}{c|cccccccccc|}
 & 1 & 2 & 3 & 5 & 6 & 7 & 9 & 12 & 15 & 16 \\
\hline
1 & 1 & -1 & 0 & 0 & 0 & 0 & 0 & 0 & 0 & 0 \\
3 & 0 & 0 & 1 & -1 & -1 & -1 & 0 & 0 & 0 & 0 \\
4 & 0 & 0 & 0 & 0 & 0 & 0 & 0 & 0 & 0 & 0 \\
7 & 0 & 0 & 0 & 0 & 0 & 1 & 0 & 0 & 0 & 0 \\
8 & 0 & 0 & 0 & 0 & 0 & 0 & 0 & 0 & 0 & 0
\end{array}_{5\times 10}
\tag{7.4.30}
$$

Table 7.7 Stream-node Connection of Process Graph of Fig. 7.6b

S. No.	Nature	Connection		
		Stream	From node	To node
1.	For unmeasured streams	4	3	4
		8	4	8
		10	7	10
		11	8	E
		13	1	E
		14	E	4
2.	For measured nodes	1	E	1
		2	1	2
		3	2	3
		5	3	5
		6	3	6
		7	3	7
		9	6	9
		12	9	E
		15	5	E
		16	10	E

(Reprinted from *I. & E. C. Process, Design and Development* with permission of the American Chemical Society, Washington.)

The incidence matrix $\mathbf{B}_3$ is drawn directly from subgraph 1 and is

$$\mathbf{B}_3 = \begin{array}{c} \\ 1 \\ 3 \\ 4 \\ 7 \\ 8 \end{array} \begin{array}{c} \begin{array}{ccccc} 4 & 10 & 11 & 13 & 14 \end{array} \\ \begin{bmatrix} 0 & 0 & 0 & -1 & 0 \\ -1 & 0 & 0 & 0 & 0 \\ +1 & 0 & 0 & 0 & 1 \\ 0 & -1 & 0 & 0 & 0 \\ 0 & 0 & -1 & 0 & 0 \end{bmatrix}_{5\times 5} \end{array} \tag{7.4.31}$$

The absence of stream 8 should be noticed. Thus it is seen that the 10×16 **A** matrix has been partitioned to 4×10 $\mathbf{B}_1$, 5×10 $\mathbf{B}_2$ and 5×5 $\mathbf{B}_3$ matrices which considerably reduces the total matrix operation in Eqs. 7.4.26 and 7.4.27 in calculation of $\hat{\mathbf{v}}$ and $\mathbf{u}$.

In this section, the problem of data analysis and coaptation has been discussed. The former arises because of the presence of random errors in measured data whereas the latter occurs when all variables are not measured due to costs involved. These problems are solved by minimizing the total error in measurement with respect to the constraint that the estimated variables satisfy balance equations. A procedure based upon graph theory has been suggested by which the partitioned matrices can be determined directly from the process flow graph. In Sec. 7.5, the fault detection scheme based on logbook measurements is given.

7.5 FAULT DETECTION SCHEME[8]

A chemical plant whose process variables have been measured is considered. Due to several reasons, it can develop faults causing the fall in the production level. A systematic procedure needs to be evolved for locating the points where the faults exist with minimum trial and error. The scheme proposed here is an extension of the Sec. 7.5. In the following, the total flow rate as the process variable is considered and the method is equally applicable for variable enthalpy.

The plant is assumed to have *m* streams and *n* equipment (or nodes) and balance around each equipment leads to one equation and therefore for the plant as a whole, there are *n* linear equations. One defines E_i^* as the total deviation in the balance equation at node *i* as

$$a_{11}v_1 + a_{12}v_2 + \dots + a_{1m}v_m = E_1^* \tag{7.5.1a}$$

$$a_{21}v_1 + a_{22}v_2 + \dots + a_{2m}v_m = E_2^* \tag{7.5.1b}$$

$$\vdots \qquad \vdots \qquad \vdots \qquad \vdots$$

$$a_{n1}v_1 + a_{n2}v_2 + \dots + a_{nm}v_m = E_n^* \tag{7.5.1c}$$

It may be recalled that if $v_1, v_2, \dots, v_m$ are replaced by $\hat{v}_1, \hat{v}_2, \dots, \hat{v}_m$ in above equations, $E_1^*, E_2^*, \dots, E_n^*$ would all be equal to zero exactly according to Eq. 7.3.6. Since v_1 to v_m are random variables, E_1^* to E_n^* are also random variables. If v_1 to v_m are assumed as normally distributed, then E_1^* to E_n^* are also normally distributed.

The expected value of E_i^* is zero because on substitution of the best estimate in Eq. 7.5.1, one obtains zero. The variance $\sigma_{z,i}^*$ of E_i^* is obtained using Eq. 7.2.53 and is given by

$$\sigma_{zi}^{*2} = \sum_{j=1}^{m} a_{ij}^2\sigma_j^2 \tag{7.5.2}$$

Since measured data are all independent, the correlation coefficients ρ_{ij} are all zero in the above equation. Each node is statistically examined for the normal deviation N_i^* defined by

$$N_i^* = \frac{E_i^*}{\sigma_{zi}^{*2}}, \; i = 1, 2, \dots, n \tag{7.5.3}$$

and the evaluation of N_i^* would indicate the presence or absence of fault. Before proceeding further on the fault detection method, it may be pointed out that the dimension of the problem can be reduced when some of the streams are unmeasured through coaptation procedure. The simplification results from the partitioning of matrix *A*, and the following set of equations can be equivalently considered.

$$\mathbf{B}_1\hat{\mathbf{v}} = 0 \tag{7.5.4}$$

If $\hat{\mathbf{v}}$ are replaced by $\mathbf{v}$ in the above equation, vector $\mathbf{E}_l$ is generated which is given by

$$\mathbf{E}_l = \mathbf{B}_l \mathbf{v} \tag{7.5.5}$$

Its variance σ_{zi}^2 is determined by

$$\sigma_{zi}^2 = \sum_j b_{ij}^2 \sigma_j^2 \tag{7.5.6}$$

and the normal deviation function N_i is correspondingly evaluated as

$$N_i = \left(\frac{E_l}{\sigma_{zl}^2}\right) \tag{7.5.7}$$

The working of any equipment (or a node) can now be checked through making a statistical test on the variable E_i. If values of N_l fall within the tolerance limit (as for example in Eq. 7.2.43), it is deemed to be functioning properly, otherwise it is declared bad. Thus a node error term n_i can be generated corresponding to equipment i (or more appropriately to the specific balance equation corresponding to it) which can be either 0 or 1 depending on the result of the statistical test on E_i (Eq. 7.5.3 or 7.5.7). i.e.,

$$n_i = 1 \text{ if the equipment is malfunctioning} \tag{7.5.8a}$$

$$= 0 \text{ if it is working normally} \tag{7.5.8b}$$

From measured data, it is possible to find out n_i for every equipment and from this information, one wishes to deduce the following:

1. Which of the measurements are likely to be bad ?
2. Which of the equipment is possibly leaking and the measurement is good ?

Bad measurements and leaks differ in the following sense. The process flow graph of Fig. 7.7 is considered. From v_1, v_2, v_3, v_4 and v_5 measurements, n_1, n_2, n_3 and n_4 are generated by making a statistical test on E_1, E_2, E_3 and E_4. Supposing v_2 measurement is bad and this is indicated in the figure by a cross and called 'bad stream'. Since v_2 appears both in E_1 and E_2, n_1 and n_2 both would therefore be 1. When the combination of E_1 and E_2 is examined, it amounts to examining the mass balance equation of the combination of equipment 1 and 2 as shown by a dotted area in Fig. 7.7. Since in $(E_1 + E_2)$ given by

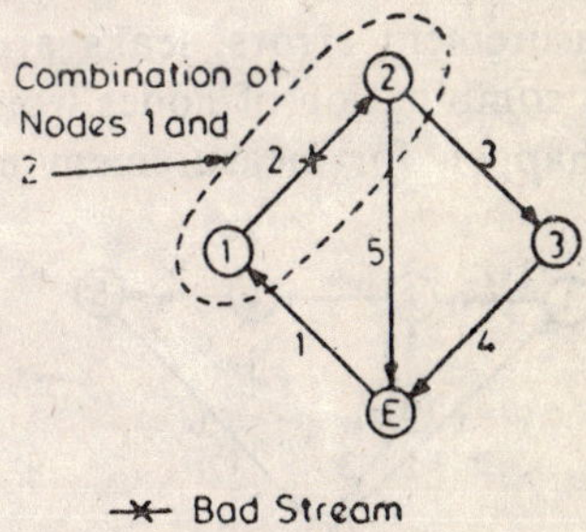

Fig. 7.7 An arbitrary process flow graph

$$(E_1 + E_2) = v_1 - v_3 - v_5 \tag{7.5.9}$$

v_2 does not appear, any error in the combination of nodes 1 and 2 will

not be detected. For the combination of equipment 1 and 2, error term $n_{(1,2)}$ is defined and for the measurement error (or bad v_2 stream) in v_2, it is equal to zero. One next defines a stream error term m_j as

$$m_j = 1 \quad \text{if } j\text{th stream is bad} \tag{7.5.10a}$$

$$= 0 \quad \text{if } j\text{th stream is good} \tag{7.5.10b}$$

and the "goodness" and "badness" of stream can be related to the "goodness" and "badness" of nodes (or equipment) in terms of the set of rules given below:

Rule 1 Any single *measurement* error would give rise to node error at both adjoining nodes.

Rule 2 If there are two nodes i_1 and i_2, such that

$$n_{(i_1, i_2)} = 0, \tag{7.5.11a}$$

$$n_{i_1} = 1, \tag{7.5.11b}$$

and

$$n_{i_2} = 1, \tag{7.5.11c}$$

then there is a measurement error in the connecting jth stream, or

$$m_j = 1 \tag{7.5.12}$$

If there is more than one connecting stream, at least one would have the measurement error.

Rule 3 If there is a single node or a combination of nodes for which the node error term n_I is the 1 and j is the only adjacent stream to it, then $m_j = 1$. The subscript I in n_I stands for the fact that it could be a combination of nodes as shown in Fig. 7.7.

One now considers the situation in Fig. 7.7 when there is a leak in the system. All measuring instruments are assumed to be working properly and equipment 1 develops a leak. If E_1, E_2, E_3 and E_4 are evaluated according to Eq. 7.5.1, only E_1 will be found to be affected. Unlike measurement errors, leaks are associated with a single node. Secondly, any combination of nodes would not remove the error, as it was shown to happen for measurement errors. Supposing both equipment 1 and 2 are leaking. This cannot be confused with measurement error because not only n_1 and n_2 would be 1, but $n_{(1,2)}$ will also be 1. Thus it is seen that leakage can be differentiated from measurement errors. To further confirm this claim, the flow sheet of Fig. 7.8 is considered in which a combination of bad streams and leaking nodes occurs. On determination of n_i for every node, it will be found that n_1, n_2, n_3, n_4 and n_5 are all equal to 1. One then examines combination of

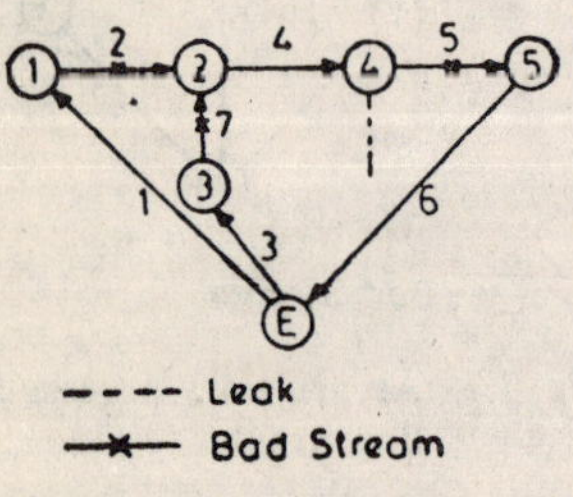

Fig. 7.8 Identification of leaks in a process flow graph

2 nodes, $n_{(1,2)}$, $n_{(2,3)}$, $n_{(2,4)}$ and $n_{(4,5)}$. They will all be 1 as can be seen from the figure. One then examines $n_{(1,2,3)}$ and $n_{(4,5)}$ and it will be found that the former equals zero whereas the latter is 1. Since $n_{(4,5)}$ does not disappear from any combination of nodes, it implies that there is a leak at node 4. It is also seen that adjacent to the combined node of 4 and 5, there is a combination of nodes 1, 2 and 3 for which node error parameter is zero. Based upon these observations it is possible to suggest a few more rules:

Rule 4 A leak causes an imbalance at *one* node.

Rule 5 If there is an aggregation I of nodes i_1, i_2, . . . , such that $n_I = 0$, it would imply that there is no leakage within the congregation considered. If one defines leak error parameter l_i such that

$$l_i = 1 \quad \text{if there is a leak} \tag{7.5.13a}$$

$$= 0 \quad \text{if there is no leak} \tag{7.5.13b}$$

then this observation can be stated by the following equation:

$$l_I = 0 \tag{7.5.14}$$

Also n_I can be zero only when all the adjacent streams to the aggregated nodes are good, i.e.,

$$m_{(j_1,\, j_2, \ldots)} = 0 \tag{7.5.15}$$

where j_1, j_2 etc. are adjacent streams.

Rule 6 If E is the environment node and $n_E = 0$, from Rule 5 it follows that no equipment in the entire chemical plant is leaking.

Rule 7 If there is a node or a combination of nodes I is bad (i.e., $n_I = 1$) and all its adjoining nodes (or any combination thereof) is good, then $l_I = 1$.

In view of these rules, the following systematic algorithms can be written for the detection of bad streams and leakages:

1. The flow sheet is converted into a process flow graph with an environment node included; with n equipment in the flow sheet, there are $(n + 1)$ nodes in the graph.
2. One node at a time is examined and this way all the nodes of the process graph are analyzed. All good nodes are removed. After this, a group of two connected nodes is examined in the remaining graph and good streams are removed once again. The process is repeated for groups of three nodes, etc. till one cannot find any connected graph.
3. An adjacent stream of the aggregate of nodes is then examined. If it has one stream, it means, according to Rule 3, that it must be bad. If it has no bad stream it means that the aggregate has a leak.

To demonstrate this algorithm, the process flow graph of Fig. 7.9 is considered. One starts considering a single node at a time. Since node 4 is good, n_4 would be found to be zero. This node is eliminated along

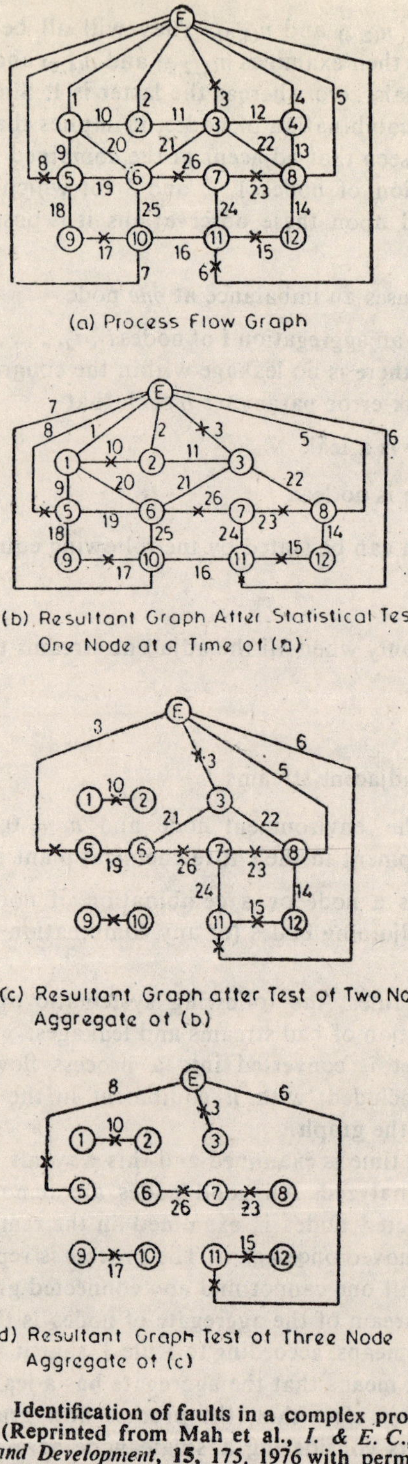

Fig. 7.9 Identification of faults in a complex process flow graph[8]. (Reprinted from Mah et al., *I. & E. C., Process Design and Development*, 15, 175, 1976 with permission of the American Chemical Society, Washington.)

with its streams 4, 12 and 13 in Fig. 7.9b. In the resultant graph, the combination of two nodes is considered and through statistical tests, it will be found that $n_{(1,2)}$ and $n_{(9,10)}$ are equal to zero. In Fig. 7.9c, good streams of combinations of nodes 1 and 2 (i.e., 1, 2, 9, 11 and 20) and 9 and 10 (i.e., 7, 16, 18 and 25) are removed. In consideration of three node aggregates, $n_{(6,7,8)}$ will be found equal to zero and in Fig. 7.9d, good streams of this combination (i.e., 5, 19, 21, 22, 24 and 14) are eliminated. When Rule 2 is applied to the combination of nodes (1, 2), (9, 10), (6, 7, 8) and (3, 5, 11, 12, E), it is discovered that streams, 3, 6, 8, 10, 15, 17, 23 and 26 are with bad measurements.

When equipment 7 is leaking, one node consideration yields Fig. 7.9b and two node analyses gives Fig. 7.9c. However, there is no more simplification after this and the leakage cannot be identified with equipment 7 leaking. This is because it is surrounded by measurement errors. In addition, it is also impossible to differentiate measurement error from the leakage now. Such a situation is hypothetical in nature in any plant operation. Normally at any given time, only a few measurement errors exist and it is always possible to distinguish leakage from them.

7.6 CONCLUSION

Since the basic purpose of any chemical plant activity is to synthesize the desired chemical, a two-pronged attack is maintained to attain the goal. The first one is at the design level and the second one is at the operational level.

During the operation of a plant, a log-book of measured data is maintained and its analysis is carried out. The data analysis is statistical in nature and it is possible to conclude about performance of an equipment through it. There can be two types of errors at a given point: measurement and leakage. An algorithm has been given in this chapter to distinguish between the two.

APPENDIX 7.1

Values of Normal Distribution Functions

$$z = \left(\frac{x - \bar{x}}{\sigma}\right)$$

$$f(z) = \frac{1}{(2\pi)^{1/2}} e^{-1/2z^2}$$

$$\phi(z) = \phi(-z) = \int_0^z f(z)\, dz$$

z	φ(z)	z	φ(z)
0.00	0.00000	1.80	0.46407
0.05	0.01994	1.85	0.46784
0.10	0.03983	1.90	0.47128
0.15	0.05962	1.95	0.47441
0.20	0.07926	2.00	0.47725
0.25	0.09871	2.05	0.47982
0.30	0.11791	2.10	0.48214
0.35	0.13683	2.15	0.48422
0.40	0.15542	2.20	0.48610
0.45	0.17364	2.25	0.48778
0.50	0.19146	2.30	0.48928
0.55	0.20884	2.35	0.49061
0.60	0.22575	2.40	0.49180
0.65	0.24215	2.45	0.49286
0.70	0.25804	2.50	0.49379
0.75	0.27337	2.55	0.49461
0.80	0.28814	2.60	0.49534
0.85	0.30234	2.65	0.49598
0.90	0.31594	2.70	0.49653
0.95	0.32894	2.75	0.49702
1.00	0.34134	2.80	0.49744
1.05	0.35314	2.85	0.49781
1.10	0.36433	2.90	0.49813
1.15	0.37493	2.95	0.49841
1.20	0.38493	3.00	0.49865
1.25	0.39435	3.05	0.49886
1.30	0.40320	3.10	0.49903
1.35	0.41149	3.15	0.49918
1.40	0.41924	3.20	0.49931
1.45	0.42647	3.25	0.49942
1.50	0.43319	3.30	0.49952
1.55	0.43943	3.35	0.49960
1.60	0.44520	3.40	0.49966
1.65	0.45053	3.45	0.49972
1.70	0.45543	3.50	0.49977
1.75	0.45994		

APPENDIX 7.2[11]

Data Analysis When A is not an Incidence Matrix

In Sec. 7.3, one deals with incidence matrix **A** because the total flow rate (or enthalpy) has been analyzed. An incidence matrix will not be obtained if component mass balance is made, even though the set of equations remains linear. Since in arriving at best estimate parameters $\hat{v}$, linear property of balance equations is used, Eq. 7.3.20 remains valid.

As an example, a mixer (a stirred tank) is considered in which two streams, v_1 and v_2, enter and two distinct streams, v_3 and v_4, leave. If each has three components, the three independent mass balance equations are

$$0.1\hat{v}_1 + 0.6\hat{v}_2 - 0.2\hat{v}_3 - 0.7\hat{v}_4 = 0 \tag{A7.2.1a}$$

$$0.8\hat{v}_1 + 0.1\hat{v}_2 - 0.2\hat{v}_3 - 0.1\hat{v}_4 = 0 \tag{A7.2.1b}$$

$$0.1\hat{v}_1 + 0.3\hat{v}_2 - 0.6\hat{v}_3 - 0.2\hat{v}_4 = 0 \tag{A7.2.1c}$$

The matrix **A** is not an incidence matrix and is given by

$$\mathbf{A} = \begin{bmatrix} 0.1 & 0.6 & -0.2 & -0.7 \\ 0.8 & 0.1 & -0.2 & -0.1 \\ 0.1 & 0.3 & -0.6 & -0.2 \end{bmatrix} \tag{A7.2.2}$$

Using measured values v_1, v_2, v_3 and v_4 in Table A7.2.1 the best estimate parameters can be obtained as

$$\begin{bmatrix} \hat{v}_1 \\ \hat{v}_2 \\ \hat{v}_3 \\ \hat{v}_4 \end{bmatrix} = \begin{bmatrix} 0.1676 \\ 4.8595 \\ 1.1730 \\ 3.8541 \end{bmatrix} \tag{A7.2.3}$$

Table A7.2.1[11] Measured Data for Solving Best Estimates $\hat{v}_1$ to $\hat{v}_4$

i	Measurements $v_i \times 1000$ mol/h	σ_i	$\hat{v}_i \times 1000$
1	0.1858	0.017	0.1676
2	4.7935	0.050	4.8595
3	1.2295	0.024	1.1730
4	3.8800	0.200	3.8541

REFERENCES

1. Beckwith, T. G. and Buck, N. I., *Mechanical Measurements*, 1st edn., Addison Wesley, Reading, Massachusetts, 1961.

2. Hald, A., *Statistical Theory with Engineering Applications*, Wiley, N.Y., 1952.
3. Hoel, P. G., *Introduction to Mathematical Statistics*, 3rd edn, Wiley, New York, 1962.
4. Graybill, F. A., *An Introduction to Linear Statistical Models*, McGraw-Hill, N.Y., 1961.
5. Kumar, A. and Gupta, S. K., *Fundamentals of Polymer Science and Engineering*, 1st edn, Tata McGraw-Hill, New Delhi, 1978.
6. Abromowitz, M. and Stegun, I. A., *Handbook of Mathematical Functions*, 1st edn., Dover, N.Y., 1968.
7. Vaclavek, C., *Chem. Eng. Sci.*, **24**, 947, 1969.
8. Mah, R. S., Stanley, G. M. and Downing, D. M., *I. & E. C. Proc. Des. Dev.*, **15**, 175, 1976.
9. Kuehn, D. R. and Davidson, H., *C.E.P.*, **57**(6), 44, 1961.
10. Deo, N., *Graph Theory with Applications to Engineering and Computer Science*, 1st edn, Prentice-Hall, 1974.
11. Nogita, S., *I. & E. C. Proc. Des. Dev.*, **11**, 197, 1972.

PROBLEMS

7.1 From the data of Table 7.2, find out the mean and standard deviation. What is the probability of finding a student studying 2 hours per day on an average.

7.2 Find out expressions of $\overline{X}$ and σ^2 for different distributions of Table 7.3.

7.3 Find the expressions for the moment generating function for the exponential distribution function.

7.4 The result of Eq. 7.2.53 was derived for the normal distribution function. Find a similar expression for other distribution functions of Table 7.3.

7.5 Supposing the distribution function is of the exponential type. How are you going to carry out hypothesis testing ? Suggest the procedure for any other distribution function.

7.6 Consider a chemical plant manufacturing acetaldehyde. Find out the minimum number of flow measurements that have to be made. For this extreme case, find B_1, B_2, and B_3.

7.7 Consider the process flow graph of Fig. 7.9. Assume that streams 7, 16, 21, 23 and 26 are not measured. Find B_1, B_2 and B_3.

7.8 Consider the flow sheet of Fig. 1.17. Convert it into its process flow graph. Supposing streams s_6, s_{11}, s_{13}, s_{29}, s_{38}, s_{31}. s_{23} and s_{34} are not measured. Write B_1, B_2 and B_3 for this process graph.

7.9 Consider the problem of Appendix 7.2. Solve this problem using the algorithm presented in the text.

7.10 In the process graph of Fig. 7.3, average values of flow rates and their standard deviations are as follows:

Stream No.	1	2	3	4	5	6	7	8
Average values	10	8	4	2	2	2	6	4
$3\sigma^2$	0.2	0.3	0.2	0.1	0.3	0.1	0.2	0.3

Actual measurements, v_i, are found to be:

Stream Nos.	1	2	3	4	5	6	7	8
v_i	10.5	3.1	3.7	1.9	2.1	1.9	4.0	3.8

Use the detection scheme suggested in the text to find where the measurement error exists and where the leak exists.

7.11 In the process graph of Fig. 7.6, the following are known:

Stream No.	1	2	3	4	5	6	7	8
Average flow rates	10	8	8	2	2	2	2	7
$3\sigma^2$	0.10	0.20	0.20	0.10	0.08	0.09	0.15	0.13
Measured values	10.20	8.10	7.4	1.81	1.83	1.81	1.81	7.2
Stream No.	**9**	**10**	**11**	**12**	**13**	**14**	**15**	**16**
Average flow rates	2	2	7	2	2	5	2	2
$3\sigma^2$	0.12	0.12	0.10	0.12	0.11	0.13	0.16	0.14
Measured values	1.83	1.79	7.1	1.83	1.81	5.3	1.80	1.82

From the measured values given in the third row, find the point of leakage. Also find the places of measurement errors.

8
Reliability Considerations in Maintenance Policies of A Chemical Plant

8.1 INTRODUCTION

To manufacture a given chemical, a chemical plant is designed. After the plant has come into existence, the problems faced in day-to-day running are twofold. The first is that of monitoring, i.e. identifying snags and removing them. The second one is the maintenance of the plant. In any chemical plant there is always a separate maintenance department with the task of attending to all repair jobs. The important task besides this is to maintain equipment in good running condition for which regular plant shutdowns are ordered and all equipment checked for their efficiency. The latter activity of the maintenance department is very important because it increases the life of the plant. A well maintained equipment lasts longer and intermittent maintenance improves efficiency of the plant as a whole by eliminating all small troubles while it is running. In this context, the importance of a good maintenance policy can hardly be over-emphasized. In the following a few idealized maintenance policies are discussed before taking up the more general ones in the end.

8.2 MAINTENANCE POLICIES

In Chapter 6, it was noted that the configuration of a chemical plant could either be pure series, pure parallel or mixed. These were analyzed when the probability of a given event (the event being the successful operation of equipment) is not time dependent. The time dependent behaviour of an equipment arises because of the ageing process. It is a well known fact that an equipment has a definite life and near its end, the wear-and-tear is high. The changes occurring in the *whole system* are modified by the adopted maintenance policy and for a chemical plant they can be broadly classified as:

1. Irreversible policy, when an equipment is not monitored for its efficiency during its operation and after every period τ it is thoroughly checked.
2. Reversible policy when an equipment is continuously monitored for its efficiency during its operation and any fault is repaired right away.
3. Mixed policy, when major faults are repaired right away, but a general overhaul is done after every time period τ. This is a combination of reversible and irreversible policies and a discussion on this is given after the analysis of the above two policies.

In the reversible policy of the whole chemical plant, the maintenance work is well defined because repairing is started as soon as the fault is detected in any of its equipment. However, in case of irreversible policy, there are the following two possibilities:

1. end-of-the-cycle repair of all failed equipment, and
2. staggered repair of failed equipment.

Quantities of interest for the chemical plant are mean availability and mean fraction of dead time. It is desired to find these from the knowledge of the time-dependent behaviour of individual equipment. Availability is defined to be the fraction of the time the plant is working and its mean value can be found as follows. The behaviour of individual equipment is

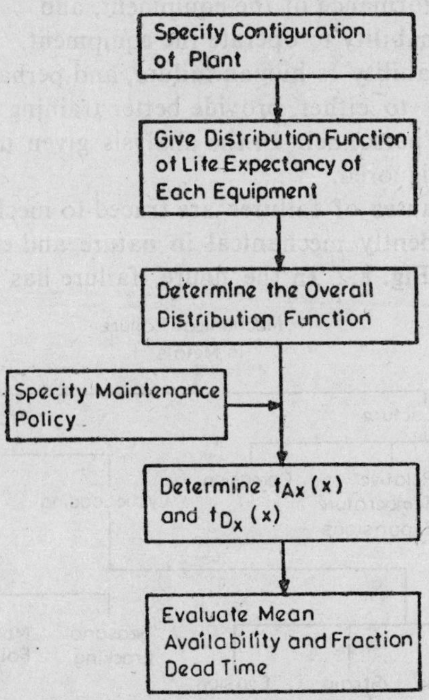

Fig. 8.1 Procedure of finding availability and fractional dead time of a chemical plant

expressed in the form of a distribution function. These are combined to determine the distribution function of life expectancy of the plant. The information, thus generated, is used to find the distribution function, $f_{A_x}(x)$, of the availability A_x and the distribution function $f_{D_x}(x)$ of the fraction dead time, D_x. To determine these functions, one needs to know the type of the maintenance policy. Once these have been specified, to find the mean availability and fraction dead time is a routine affair and is outlined as a flow chart in Fig. 8.1.

In this chapter, a single equipment is considered first and the distribution function of reliability is introduced. The procedure of Fig. 8.1 is then followed to find the mean availability and fraction dead time. After that, pure series, pure parallel and mixed systems are considered to derive similar results. Ultimately they are compared to prove the effect of maintenance on the mean availability of the system. In Sec. 8.3 the reliability of an equipment is discussed first.

8.3 RELIABILITY OF AN EQUIPMENT

When an equipment is installed in a chemical plant, it is required to function as laid down in its detailed design. However, it may or may not perform as desired and the failure in general could be of the following type[1, 2]:

1. breakdown in which the equipment undergoes a mechanical failure,
2. reduced performance of the equipment, and
3. operator's inability to operate the equipment.

The operator's inability is human failure, and perhaps the only solution to this problem is to either provide better training or replace him by a more experienced personnel. In the analysis given in this chapter, this problem has been ignored.

The first two causes of failures are traced to mechanical inadequacy. Breakdown is evidently mechanical in nature and can occur because of reasons shown in Fig. 8.2. In the figure, failure has been subdivided as

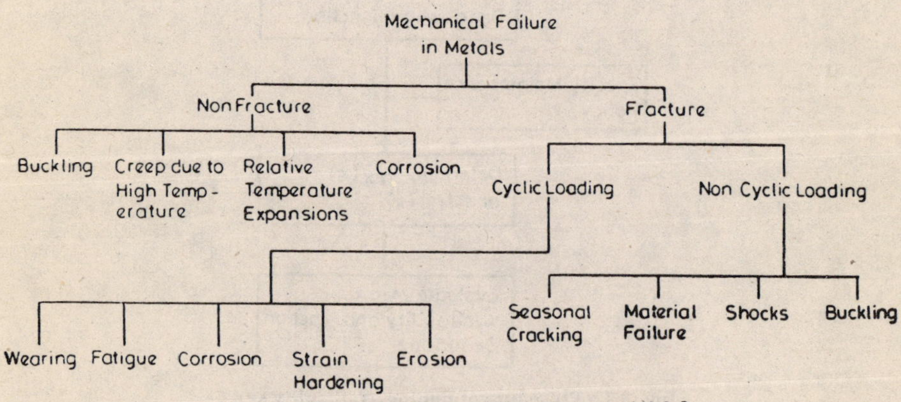

Fig. 8.2 Causes leading to failure in materials[1, 2]

fracture and non-fracture types. Non-fracture type failures normally are caused by changes in dimensions to an extent that the equipment stops working.

Chemical equipment like heat exchangers and distillation columns do not involve any moving parts and therefore can undergo non-fracture type failure. As opposed to this, equipment like pumps and compressors have moving parts and can undergo both types of failures.

Reduced performance of an equipment is caused by defects and they can result from several sources and normally are introduced during the manufacturing process. Figure 8.3 gives some of the sources of defects in an equipment. They can occur either at designing, manufacturing, assembling or the operational level. The mechanical faults can be minimized by strict quality control; however, they can never be completely removed.

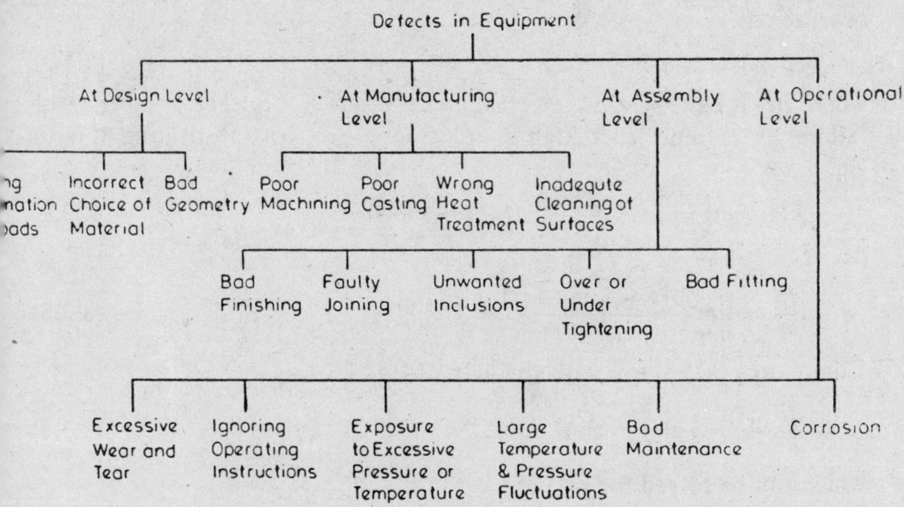

Fig. 8.3 Generation of defects in an equipment[1, 2]

It is clear from the above that sources leading to failures in an equipment are several, and in view of them, to obtain an equipment in working condition at a given time is treated as a stochastic event. The situation here is completely analogous to the survival of an individual in large cities. There are several sources like traffic on road, food poisoning, diseases, etc. which can cause either death or disability, and directly affect the survival of a given person. Since it is difficult to predict when any of these causes will befall him, his life expectancy is treated as stochastic in nature.

In view of the above discussion, the probability of survival of an equipment, $p(t)$, is defined where t is its age counted from the time of its installation. Complementary to this is the probability that the equipment is in a failed state, $\bar{p}(t)$, at time t[3, 4]. Since survival and failure are two mutually exclusive and independent events

$$\bar{p}(t) + p(t) = 1 \tag{8.3.1}$$

It is desired to find $p(t)$ and $\bar{p}(t)$. In order to find it, two times $(t + \delta t)$ and t are considered separated by a small interval δt. The probability $\bar{p}(t + \delta t)$ of an equipment in a failed state at age $(t + \delta t)$ can be written in terms of $\bar{p}(t)$ by noting that

1. the equipment is in the failed state at age t, or
2. the equipment is surviving at age t, but has failed in the time interval δt.

Since these are mutually exclusive events,

$$\bar{p}(t + \delta t) = \bar{p}(t) + p(t)\, \delta\bar{p}(\delta t) \tag{8.3.2}$$

where $\delta\bar{p}(\delta t)$ is the probability of failure in time δt. The probability of event 2 above is found by multiplying this with $p(t)$. $\delta\bar{p}(\delta t)$ can be rewritten as

$$\delta\bar{p}(\delta t) = z(t)\, \delta t + 0(\delta t) \tag{8.3.3}$$

where $z(t)$ is called *hazard function* and is a measure of the frequency of failure. $0(\delta t)$ represents higher order terms in δt. Substituting this in Eq. 8.3.2,

$$\bar{p}(t + \delta t) = \bar{p}(t) + p(t)[z(t)\, \delta t + 0(\delta t)] \tag{8.3.4}$$

or

$$\frac{d\bar{p}}{dt} = \lim_{\delta t \to 0} \frac{\bar{p}(t + \delta t) - \bar{p}(t)}{\delta t} = p(t)\, z(t) \tag{8.3.5}$$

Eliminating $p(t)$ above with the help of Eq. 8.3.1,

$$\frac{d\bar{p}}{dt} + \bar{p}(t)\, z(t) = z(t) \tag{8.3.6}$$

which can be solved for $\bar{p}(t)$ as

$$\bar{p}(t) = 1 + K \exp\left(-\int_0^t z(t)\, dt\right) \tag{8.3.7}$$

K is a constant of integration and can be determined from the initial condition that the equipment is guaranteed to function at time $t = 0$. Or

$$\bar{p}(0) = 0 \tag{8.3.8}$$

Substituting $t = 0$ in Eq. 8.3.8 and using Eq. 8.3.8, one obtains K as -1. $\bar{p}(t)$ is therefore given by

$$\bar{p}(t) = 1 - \exp\left(-\int_0^t z(t)\, dt\right) \tag{8.3.9}$$

and

$$p(t) = \exp\left(-\int_0^t z(t)\, dt\right) \tag{8.3.10}$$

Thus the probability of survival, $p(t)$, of an equipment is seen to be a function of its age, t, and this functionality is governed by the hazard function $z(t)$. Hazard function is a very important parameter, but the vexing problem is how to determine it. One has to resort to experimental means and Fig. 8.4 shows the shape of a general hazard function. It is

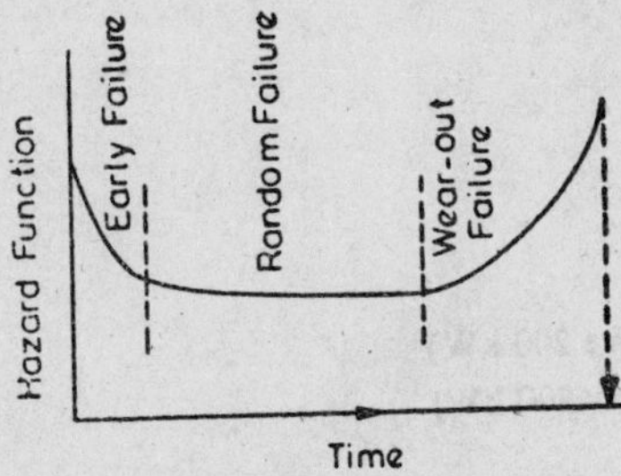

Fig. 8.4 A typical hazard function of an equipment

found to have three distinct regions[1,3,4]. The early failure region is largely attributed to the propagation of minor manufacturing defects leading to complete breakdown. These faults can be generated at the design, installation or commissioning level. As opposed to this, the wear-out region is a result of metallurgical deterioration, looseness of parts due to erosion and sometimes overloading. The more interesting region is the middle one where the equipment fails randomly and it spreads over the major portion of its life time. It may be pointed out that there is no single theoretical distribution function which will describe the entire curve. If the first and third regions are neglected and the hazard function is assumed to be a constant equal to θ, Eqs. 8.3.9 and 8.3.10 can be integrated to obtain

$$\bar{p}(t) = (1 - e^{-\theta t}) \tag{8.3.11a}$$

and

$$p(t) = e^{-\theta t} \tag{8.3.11b}$$

and the corresponding distribution function of life expectancy, $f(t)$, would be

$$f(t) = -\frac{dp}{dt} = \theta\, e^{-\theta t} \tag{8.3.12}$$

If θ is very small and time t is not very large (such that $\theta t \ll 1$), the exponential can be expanded to obtain the following after neglecting higher order terms:

$$\bar{p}(t) \cong \theta t \tag{8.3.13a}$$

and

$$p(t) \cong 1 - \theta t \tag{8.3.13b}$$

Thus it is seen that the hazard function has the meaning of rate of failure. Table 8.1 gives average rates of failure of different equipment of interest for chemical engineers and this is compiled from references 1 and 4. The values of failure rates given in the table depend to a large extent upon the

Table 8.1[1, 4] Failure Rates of Chemical Equipment and Components

S.No.	Items	Failure rate, failure/10^6 h
1.	Boiler pumps	
	6000 H.P.	6.0 faults/year
	7000 H.P.	4.0 faults/year
2.	Motors	
	General	10.0
	Induction (above 200 kW)	10.0
	Induction (below 200 kW)	5.0
	Synchronous	7.0
	Small, general	4.0
	Stepper	5.0
3.	Gas turbines	11.4
4.	Nuclear reactors (with trip facility)	107.0
5.	Leakage through gaskets	0.5
6.	Pipe joints	0.2
7.	Pressure vessels	
	Small	3.0
	Small, high standard	0.3
	Large, high standard	0.1
8.	Relief valves	
	Leakage	2.0
	Blockage	0.5
9.	Hand operated valves	15.0
10.	Ball valves	0.5
11.	Solenoid valves	30.0
12.	Pressure gauges	10.0
13.	Pressure switches	15.0
14.	Bourdon Tube	
	Creep	0.2
	Leakage	0.05
15.	Nozzle and flapper assembly	
	Blockage	6.0
	Breakage	0.2
16.	Fixed orifices	1.0
17.	Variable orifices	5.0
18.	Restrictors	5.0
19.	Pipes	0.2
20.	Thermocouples	0.088

environment. The measurements made in a laboratory normally are found to differ by as much as the factor of four from those made in the actual site of a chemical plant[5] and in this context, the rates given in the table specify the order of magnitude only.

After having derived the probability of survival as a function of age, one is in a position to define reliability of a given equipment. In its most general definition, *reliability* is defined as a measure of performance of an equipment. The measure of performance is obtained when its achievement is compared with the required goal. To illustrate this point further, the flow sheet of Fig. 8.5 is considered in which a pump forces reactant from

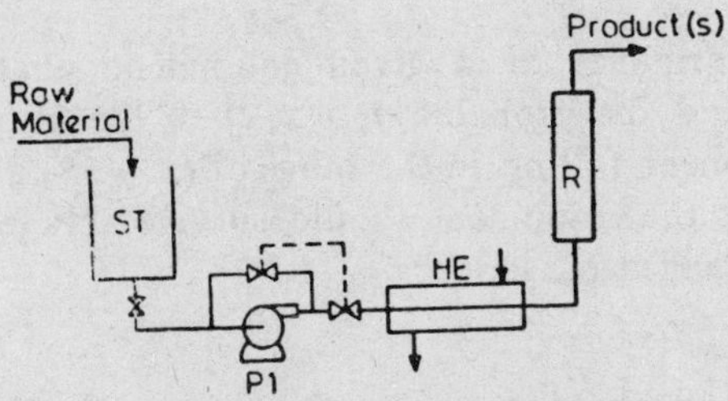

Fig. 8.5 An example flow sheet for calculation of reliability of pump P1

a storage tank into a reactor through a suitable heat exchanger. While designing the reactor, its volume is fixed for a given flow rate so that the desired residence time of reaction is obtained. Consequently, it is necessary that the flow rate of the pump is regulated very closely by the by-pass system installed on it. If Q_{ss} is the desired steady state flow rate and Q is the flow rate developed by the pump before the by-pass system, then the pump is regarded as working satisfactorily as long as

$$Q \geqslant Q_{ss} \tag{8.3.14}$$

This is because for any flow rate more than Q_{ss}, the by-pass system reduces it to the desired level of Q_{ss}. The required goal for the pump is Q_{ss} whereas its achievement is Q and the comparison between the two is set by Eq. 8.3.14.

Acceptability of the pump for the task is the measure of the reliability. It may be pointed out that the required goal depends to a large extent upon the flow sheet. To demonstrate this fact, one considers the situation when the flow rate is controlled by a valve ahead of the pump and not by the by-pass system. The delivered flow rate should then fall within a close limit defined by Q_{min} and Q_{max}, i.e.

$$Q_{min} \leqslant Q \leqslant Q_{max} \tag{8.3.15}$$

Before deriving an explicit function for reliability, R, it may be noted that the performance function of any equipment is rarely deterministic in nature. To demonstrate this for the pump of Fig. 8.5, it is noted that:

(1) Input pump pressure P_i is not equal to one atmosphere but is more by the amount equal to the liquid head in the storage tank ST which is subject to fluctuations.

(2) The pressure drop in equipment ahead may change with time due to depositions in connecting pipelines.

(3) Efficiency of the pump may change with time.

Because these facets cannot be conveniently built in the dynamic equation of the pump, the details of the problem are smeared while writing its performance by a suitable empirical distribution function, $f_Q(Q)$, where Q is the delivered flow rate. In complete analogy with this, a distribution function $f_x(x)$ for the output variable x can be written for any equipment.

Also, the performance of a given equipment changes with time (Eq. 8.3.10). To find the probability, $p(x, t)$ of the output performance of a given equipment falling in the range of $(x + \delta x)$ and x at its age t, one observes that the equipment should survive at age t before it can perform in the specified range, i.e.

$$p(x, t) = p(t)_x f(x)\,dx \tag{8.3.16}$$

where $p(t)$ is the probability of survival given by Eq. 8.3.10. If the successful performance is defined by a lower limit of x_s (Eq. 8.3.14), the reliability, R', of the equipment at its age t is given by

$$R' = p(t) \int_{x_s}^{\infty} f_x(x)\,dx \tag{8.3.17}$$

or

$$R'(t) = p(t)\left[\int_0^{\infty} f_x(x)\,dx - \int_0^{x_s} f_x(x)\,dx\right] \tag{8.3.18}$$

From the definition of distribution function (Chapter 7) one has

$$R'(t) = p(t)[1 - p(x_s)] \tag{8.3.19}$$

where $p(x_s)$ is the probability of occurrence of x_s. Similarly, when the successful performance is defined by a lower and an upper limits, x_l and x_u, the reliability R'' is given by

$$R'' = p(t) \int_{x_l}^{x_u} f_x(x)\,dx$$

$$= p(t)[p(x_u) - p(x_l)] \tag{8.3.20}$$

where $p(x_u)$ and $p(x_l)$ are probabilities of upper and lower limits.

Equations 8.3.19 and 8.3.20 show that the reliability of an equipment depends not only upon constraints of the requirements but also its age. Consequently, to find an average reliability over a time τ, Eq. 8.3.20 is averaged over the time domain

$$R''_{av} = [p(x_u) - p(x_l)]\,\frac{1}{\tau}\int_0^{\tau} p(t)\,dt \tag{8.3.21}$$

Lastly, if there is no requirement constraint like Eqs. 8.3.14 or 8.3.15, the average reliability, R_{av}, is given by

$$R_{av} = \frac{1}{\tau}\int_0^{\tau} p(t)\, dt \equiv p_s \tag{8.3.22}$$

and to determine this experimentally, several equipments (say N) of the same nature must be considered at a given time. All these equipments are run for the same time and the number of them working successfully (say n) are found out. The probability of survival, p_s, in this experiment is given by

$$p_s = \left(\frac{n}{N}\right) \tag{8.3.23}$$

For an equipment of similar design, *installed in the plant*, one can assume it to have the same reliability, i.e.

$$R_{av} = p_s = \left(\frac{n}{N}\right) \tag{8.3.24}$$

After the discussion on reliability of an equipment, its availability and fraction dead time are calculated in Sec. 8.4.

8.4 AVAILABILITY AND FRACTIONAL DEAD TIME OF A SINGLE EQUIPMENT FOR AN IRREVERSIBLE MAINTENANCE POLICY[4]

In a chemical plant, equipment in the main flow sheet is utilized continuously whereas standby equipment, safety devices and some controllers are put to use only when there are failures in the plant. Even though they are used once in a while, it is necessary that they remain in working order *for all time*. It is in view of this, they need to be checked and repaired at regular intervals. In this section a mathematical analysis has been developed for an equipment which undergoes a periodic check up and repairs. The purpose of the following analysis is to develop an expression for the average ratio of working time in a cycle from the failure frequency of the equipment.

There is an equipment which is put to operation at time $t = 0$ and is checked and repaired after every time τ. During this time, it may fail or it may not and in view of it, there are two states, the equipment can take as given below:

1. working state, and
2. failed state.

This has been shown in Fig. 8.6. In the failed state, the failure is shown to occur after time t_s which can have any value between 0 and τ. States of the equipment are discrete in nature

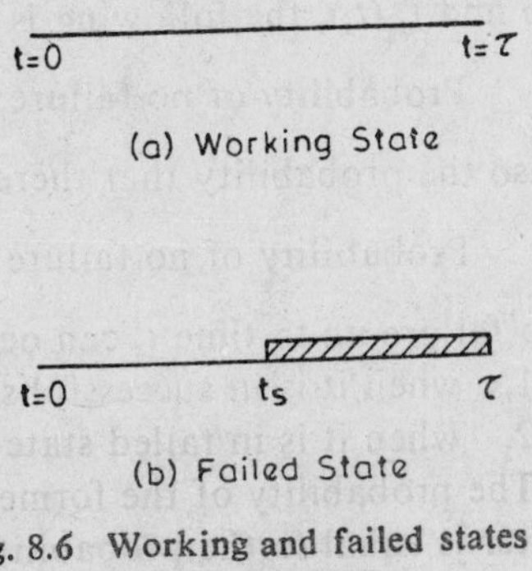

Fig. 8.6 Working and failed states of an equipment

whereas the time of successful operation t_s is continuous and can be represented by a distribution function.

The availability A of the equipment is defined as

$$A = \frac{t_s}{\tau} \tag{8.4.1}$$

and the fraction dead time, D, as

$$D = \frac{\tau - t_s}{\tau} = \frac{t_f}{\tau} \tag{8.4.2}$$

where τ is the time for which the equipment has been run. The probability that it does not fail in time τ (Fig. 8.6a) is given by $[1 - \overline{p}(\tau)]$ whereas the probability that it fails (Fig. 8.6b) is given by $\overline{p}(\tau)$. The availability of the system is found by averaging over these two states as

$$A^* = [1 - \overline{p}(\tau)]\, A_s + \overline{p}(\tau) A_f \tag{8.4.3}$$

whereas A_s and A_f are availabilities in the successful and failed states in Fig. 8.6a and b. Since in the former state, the equipment does not fail, A_s is equal to one. Thus

$$A^* = [1 - \overline{p}(\tau)] + \overline{p}(\tau)\, A_f \tag{8.4.4}$$

Similarly the fraction dead time is averaged over the two states as

$$D^* = [1 - \overline{p}(\tau)]\, D_s + \overline{p}(\tau) D_f \tag{8.4.5}$$

where D_s, D_f and D^* are fractional dead times for the successful state, failed state and the one averaged over both these states respectively. D_s would necessarily be equal to zero for the successful state of Fig. 8.6a. Therefore

$$D^* = \overline{p}(\tau) D_f \tag{8.4.6}$$

From the definition of availability, the distribution function of t_s, $f_{t_s}(t_s)$ is found in terms of the survival of the equipment, $f(t)$, as follows. $f(t)$ is determined by differentiating Eq. 8.3.10 as

$$f(t) = -\frac{dp(t)}{dt} = z(t) \exp\left(-\int z(t)\, dt\right) \tag{8.4.7}$$

To find $f_{t_s}(t_s)$, the following is noted:

$$\text{Probability of no failure in time } \tau = [1 - \overline{p}(\tau)] \equiv p(\tau) \tag{8.4.8}$$

Also the probability that there is no failure till time t_s is given by

$$\text{Probability of no failure in time } t_s = [1 - \overline{p}(t_s)] = p(t_s) \tag{8.4.9}$$

No failure up to time t_s can occur in two ways:

1. when it is in successful state (i.e., no failure up to time τ), and
2. when it is in failed state but the time of failure is more than t_s.

The probability of the former is given by Eq. 8.4.8 whereas that of the latter is equal to the probability that it is in the failed state multiplied by the probability that the time of successful operation t is greater than or equal to t_s.

$$\text{Probability of } t \geqslant t_s = \int_{t_s}^{\tau} f_{t_s}(t_s)\,dt$$
$$= \int_0^{\tau} f_{t_s}(t_s)\,dt - \int_0^{t_s} f_{t_s}(t_s)\,dt$$
$$\equiv [1 - p_{t_s}(t_s)] \tag{8.4.10}$$

where $p_{t_s}(t_s)$ is the probability in the failed state of Fig. 8.6b. Averaging over the two states yields

$$p(t_s) = 1 - \bar{p}(t_s) = [1 - \bar{p}(\tau)] + \bar{p}(\tau)\,[1 - p_{t_s}(t_s)] \tag{8.4.11}$$

On rearrangement, one gets

$$p_{t_s}(t_s) = \frac{\bar{p}(t_s)}{\bar{p}(\tau)} \tag{8.4.12}$$

and on differentiating both sides with respect to t_s, $f_{t_s}(t_s)$ is found as

$$f_{t_s}(t_s) = \frac{1}{p(\tau)}\left[-\frac{d\bar{p}(t_s)}{dt_s}\right] \tag{8.4.13}$$
$$= -\frac{1}{p(\tau)}\left[z(t_s)\exp\left(-\int_0^{t_s} z(t)\,dt\right)\right] \tag{8.4.14}$$

with the help of Eq. 8.4.7 it is simplified to

$$f_{t_s}(t_s) = \frac{-f(t_s)}{\bar{p}(\tau)} \tag{8.4.15}$$

To find the distribution function for the availability $f_{A_f}(A_f)$ for the failed state it is noted that t_s and A_f are similarly distributed because

$$t_s = \tau A_f \tag{8.4.16}$$

Consequently the variable transformation in $p_{t_s}(t_s)$ gives $p_{A_f}(A_f)$ as

$$p_{A_f}(A_f) = p_{t_s}(\tau A_f) \tag{8.4.17}$$

The subscript in the above equation merely denotes the independent variable and this has been done to distinguish it from $f(t_s)$. Further

$$f_{A_f}(A_f) = -\frac{dp_{A_f}(A_f)}{dA_f} \tag{8.4.18}$$

On differentiating Eq. 8.4.13 with respect to A_f

$$\frac{d}{dA_f}\,[p_{A_f}(A_f)] = \frac{\partial}{\partial(t_s/\tau)}\,[p_{t_s}(\tau A_f)] \tag{8.4.19}$$

Or

$$f_{A_f}(A_f) = \tau f_{t_s}(\tau A_f) \tag{8.4.20}$$

Eliminating $f_{t_s}(\tau A_f)$ with the help of Eq. 8.4.15, the distribution function $f_{A_f}(A_f)$ is found in terms of distribution function for the survival of the equipment $f(t_s)$. Then t_s is replaced by (τA_f), i.e.,

$$f_{A_f}(A_f) = -\frac{\tau}{\bar{p}(\tau)} f(\tau A_f) \tag{8.4.21}$$

$f(\tau A_f)$ is the distribution function of the equipment and is assumed to be known. As an example, the exponential distribution is considered. $\bar{p}(\tau)$ is given by

$$\bar{p}(\tau) = (1 - e^{-\theta\tau}) \tag{8.4.22}$$

The distribution function for the survival of the equipment is

$$f(t) = \theta e^{-\theta t} \tag{8.4.23}$$

and with the help of Eqs. 8.4.12 and 8.4.17, $f_{t_s}(t_s)$ and $f_{A_f}(A_f)$ are

$$f_{t_s}(t_s) = \frac{-\theta \exp(-\theta t_s)}{(1 - \exp(-\theta\tau))} \tag{8.4.24a}$$

and

$$f_{A_f}(A_f) = \frac{-\theta\tau \exp\{-\theta\tau A_f)\}}{1 - \exp(-\theta\tau)} \tag{8.4.24b}$$

Information of interest is not $f_{A_f}(A_f)$ but the distribution function for A^*, $f_{A^*}(A^*)$. On rearrangement of Eq. 8.4.4, one finds

$$A_f = \frac{A^* - [1 - \bar{p}(\tau)]}{p(\tau)} \tag{8.4.25}$$

This means that by making suitable transformations in $f_{A_f}(A_f)$, $f_{A^*}(A^*)$ can be found. This is done by extending the result of Eq. 8.4.20 to obtain

$$f_{A^*}(A^*) = \frac{1}{\bar{p}(\tau)} f_{A_f}\left[\frac{A^* - \{1 - \bar{p}(\tau)\}}{\bar{p}(\tau)}\right] \tag{8.4.26}$$

Since the argument of the distribution function can never be negative, therefore the following inequality holds good

$$1 \geqslant A^* \geqslant [1 - \bar{p}(\tau)] \tag{8.4.27}$$

Eliminating $f_{A_f}^{\prime}(A_f)$ in Eq. 8.4.26 using Eq. 8.4.21, $f_{A^*}(A^*)$ is obtained in terms of the distribution function for survival of the equipment as

$$f_{A^*}(A^*) = \frac{-\tau}{[\bar{p}(\tau)} f\left\{\frac{[A^* - \{1 - \bar{p}(\tau)\}]\tau}{\bar{p}(\tau)}\right\} \tag{8.4.28}$$

For exponential distribution of Eq. 8.4.23,

$$\frac{\tau[A^* - \{1 - \bar{p}(\tau)\}]}{\bar{p}(\tau)} = \frac{[A^* - e^{-\theta\tau}]\tau}{(1 - e^{-\theta\tau})} \tag{8.4.29a}$$

and

$$f_{A^*}(A^*) = \frac{-\tau}{(1 - e^{-\theta\tau})^2} \exp\left[-\left\{\frac{\theta\tau[A^* - e^{-\theta\tau}]}{(1 - e^{-\theta\tau})}\right\}\right] \tag{8.4.29b}$$

Mean availability μ_{A^*} can now be found by averaging A^*

$$\mu_{A^*} = \int_{[1-\bar{p}(\tau)]}^{1} \frac{\tau A^*}{[\bar{p}(\tau)]^2} f\left[\frac{[A^* - \{1 - \bar{p}(\tau)\}\tau]}{\bar{p}(\tau)}\right] dA^* \tag{8.4.30}$$

On substituting

$$t = \frac{[A^* - (1 - \bar{p}(\tau)\}]\tau}{\bar{p}(\tau)} \tag{8.4.31a}$$

such that

$$dt = \frac{\tau\, dA^*}{\bar{p}(\tau)} \tag{8.4.31b}$$

the integral of Eq. 8.4.30 becomes

$$\mu_{A^*} = \frac{\tau}{[\bar{p}(\tau)]^2}\int_0^{\tau}\left[\frac{p(\tau)t}{\tau} + 1 - \bar{p}(\tau)\right]\frac{\bar{p}(\tau)}{\tau} f(t)\, dt \tag{8.4.32}$$

or

$$\mu_{A^*} = \frac{1}{\tau}\int_0^{\tau} tf(t)\, dt + \frac{1 - \bar{p}(\tau)}{\bar{p}(\tau)}\int_0^{\tau} f(t)\, dt \tag{8.4.33}$$

This can be further simplified by noting that

$$\int_0^{\tau} f(t)\, dt = \int_0^{\tau} d\bar{p}(t) = \bar{p}(\tau) \tag{8.4.34}$$

and μ_{A^*} is given by

$$\mu_{A^*} = \frac{1}{\tau}\int_0^{\tau} tf(t)\, dt + [1 - \bar{p}(\tau)] \tag{8.4.35}$$

Also integration by part yields

$$\int_0^{\tau} tf(t)\, dt = \Big[t\bar{p}(t)\Big]_0^{\tau} - \int_0^{\tau} \bar{p}(t)\, dt$$

$$= \tau\bar{p}(\tau) - \int_0^{\tau} \bar{p}(t)\, dt \tag{8.4.36}$$

and the average availability becomes

$$\mu_{A^*} = \frac{1}{\tau}\left[\bar{p}(\tau)\tau - \int_0^{\tau} \bar{p}(t)\, dt\right] + [1 - \bar{p}(\tau)]$$

$$= 1 - \frac{1}{\tau}\int_0^{\tau} \bar{p}(t)\, dt \tag{8.4.37}$$

One proceeds in similar fashion to determine the average fractional dead time, μ_{D^*}. $f_{D_f}(D_f)$ is determined first by noting that (See Eq. 8.4.2)

$$t_f = \tau - t_s \tag{8.4.38}$$

Consequently the distribution function $f_{t_f}(t_f)$ for t_f (Eq. 8.4.21) is

$$f_{t_f}(t_f) = f_{t_s}(\tau - t_f) = \frac{f(\tau - t_s)}{\bar{p}(\tau)} \tag{8.4.39}$$

Distribution function, $f_{D_f}(D_f)$ for D_f is similarly given by

$$f_{D_f}(D_f) = [f_{t_f}(D_f\tau)]\tau$$
$$= \frac{\tau f[(\tau - \tau D_f)]}{\bar{p}(\tau)} \tag{8.4.40}$$

The above relation is obtained by substituting (τD_f) in place of t_s in Eq. 8.4.39. The distribution function, $f_{D^*}(D^*)$ is obtained from the relation between D^* and D_f and it is

$$f_{D^*}(D^*) = \frac{f_{D_f}[D^*]\bar{p}(\tau)}{\bar{p}(\tau)} \tag{8.4.41}$$

With the help of Eq. 8.4.40 it becomes

$$f_{D^*}(D^*) = \frac{\tau}{[\bar{p}(\tau)]^2} f\left[\tau - \frac{\tau D^*}{\bar{p}(\tau)}\right] \quad 0 \leqslant D^* \leqslant \bar{p}(\tau) \tag{8.4.42}$$

As an example, for the exponential distribution function of Eq. 8.4.24, these are

$$f_{t_f}(t_f) = \frac{\theta \exp\{-\theta(\tau - t_f)\}}{\{1 - \exp(-\theta\tau)\}^2} \tag{8.4.43a}$$

$$f_{D_f}(D_f) = \frac{\exp\{-\theta(\tau - \tau D_f)\}}{\{1 - \exp(-\theta\tau)\}^2} \tag{8.4.43b}$$

and

$$f_{D^*}(D^*) = \frac{\tau \exp[-\theta\{\tau - \tau D^*/(1 - e^{-\theta\tau})\}]}{(1 - e^{-\theta\tau})^2} \tag{8.4.43c}$$

The average function dead time μ_{D^*} can now be determined as

$$\mu_{D^*} = \int_0^{\bar{p}(\tau)} D^* \frac{\tau}{[\bar{p}(\tau)]^2} f\left[\tau - \frac{\tau D^*}{\bar{p}(\tau)}\right] dD^* \tag{8.4.44}$$

The expression above can be simplified by defining a variable t as

$$t = \tau - \frac{\tau}{\bar{p}(\tau)} D^* \tag{8.4.45}$$

Then

$$D^* = \frac{(\tau - t)}{\tau} \bar{p}(\tau) \tag{8.4.46a}$$

and

$$dD^* = -\frac{\bar{p}(\tau)}{\tau} dt \tag{8.4.46b}$$

Substituting these in Eq. 8.4.44,

$$\mu_{D^*} = -\int_\tau^0 \frac{\tau}{[\bar{p}(\tau)]^2} \frac{(\tau - t)}{\tau} \bar{p}(\tau) f(t) \frac{\bar{p}(\tau)}{\tau} \cdot dt$$
$$= \frac{1}{\tau}\int_0^\tau (\tau - t) f(t)\, dt \tag{8.4.47}$$

Further simplification can be achieved by integrating in parts and noting the defining relation of probability [i.e. $d\bar{p}(t) = f(t)\,dt$]

$$\int_0^{\tau} (\tau - t) f(t)\, dt = \Big[(\tau - t)\bar{p}(t)\Big]_0^{\tau} + \int_0^{\tau} \bar{p}(t)\, dt$$

$$= \int_0^{\tau} \bar{p}(t)\, dt \qquad (8.4.48)$$

Therefore

$$\mu_{D*} = \frac{1}{\tau}\int_0^{\tau} \bar{p}(t)\, dt \qquad (8.4.49)$$

With the help of Eqs. 8.4.37 and 8.4.49 it is seen that

$$\mu_{A*} + \mu_{D*} = 1 \qquad (8.4.50)$$

which is expected from the definitions of the availability and the fractional dead time.

In the analysis presented above it was assumed that the equipment has been operated for a period of time τ and the repair work, if at all required, is done after this time. The time taken for repairs, τ_r, has been assumed to be small compared to τ, or

$$\tau_r \ll \tau \qquad (8.4.51)$$

Now a cyclic operation is considered. If the repair work is assumed to make the equipment as good as new at the end of every cycle, mathematically it will mean that the probability of survival of the equipment at the beginning of every cycle is given by

$$p(i\tau + 0) = 1 \quad \text{for all } i \qquad (8.4.52)$$

where $p(i\tau + t)$ gives the probability at time t after the beginning of ith cycle of operation. Since the differential equation (Eqn. 8.3.6) governing the survival of the equipment and its solution given by Eq. 8.3.7 is valid for any time, the frequency distribution therefore for a given cycle is independent of the past history. This can easily be seen by calculating the constant of integration K in Eq. 8.3.7 using the initial condition given in Eq. 8.4.52. For each cycle, K remains as -1 and the probability of survival becomes

$$\bar{p}(i\tau + t) = \bar{p}(0\cdot\tau + t) = \bar{p}(t) = \left[1 - \exp\left(-\int_0^{t} z(t)\, dt\right)\right] \qquad (8.4.53)$$

Since the average availability μ_{A*}, and the fractional dead time μ_{D*} depend only on $f(t)$, they too become independent of the cycle of operation. Therefore, for the repair work which makes the equipment as good as new at the end of every cycle, only one cycle of the operation needs to be considered for calculating the mean availability.

In actuality, a repair job can never make the equipment as good as new. The initial condition of Eq. 8.4.52 will be true only when the old equipment is replaced by a brand new one at the end of every cycle. But this is never possible for reasons of excessive costs. To take care of the ageing process, a variable t_1 is defined as the life of the equipment. For the ith cycle,

$$t_1 = (i-1)\tau + t \tag{8.4.54}$$

where t is the time variable measured in ith cycle and has a maximum value of τ. In terms of t_1 the survival and failure of the equipment are written as

$$\bar{p}(t_1) = 1 - p(t_1) = 1 + K \exp\left(-\int_0^{t_1} z(t_1)\, dt_1\right) \tag{8.4.55}$$

Because the equipment is only made in working condition through repairing if it has failed during the cycle, it implies that the probability of its being in the failed state at the beginning of every cycle is zero, i.e.

$$\bar{p}[(i-1)\tau] = 0, \quad i = 1, 2, \ldots \tag{8.4.56}$$

The K in Eq. 8.4.55 can now be solved using this condition and it is found to be

$$K = \frac{-1}{\exp\left[-\int_0^{(i-1)\tau} z(t_1)\, dt_1\right]} \qquad i = 1, 2, \ldots \tag{8.4.57}$$

Substituting this in Eq. 8.4.55,

$$\bar{p}(t_1) = 1 - \frac{\exp\left[-\int_0^{t_1} z(t_1)\, dt_1\right]}{\exp\left[-\int_0^{(i-1)\tau} z(t_1)\, dt_1\right]} \tag{8.4.58}$$

Using Eq. 8.3.9, the above relation is simplified to

$$\bar{p}(t_1) = 1 - \frac{1 - \bar{p}_1(t_1)}{1 - \bar{p}_1[(i-1)\tau]} \tag{8.4.59}$$

or

$$\bar{p}(t_1) = \frac{\bar{p}_1(t_1) - \bar{p}_1[(i-1)\tau]}{1 - \bar{p}_1[(i-1)\tau]} \tag{8.4.60}$$

where $\bar{p}_1(t_1)$ is defined by

$$\bar{p}_l(t_1) = \left[1 - \exp\left(-\int_0^{t_1} z(t_1)\, dt_1\right)\right] \tag{8.4.61}$$

In a given cycle of operation, the equipment can have two possible states (working and failed) as shown in Fig. 8.6. To find the distribution function $f_{t_s}(t_s)$ for t_s, one proceeds exactly in the manner as earlier. Probability $\bar{p}(i\tau)$ that there is a failure in ith interval is obtained by substituting $(i\tau)$ in the above equation, i.e.

$$\bar{p}(i\tau) = \frac{\bar{p}_1(i\tau) - \bar{p}_1[(i-1)\tau]}{1 - \bar{p}_1[(i-1)\tau]} \tag{8.4.62}$$

Similarly probability $p(i\tau)$ that there is no failure in ith interval is equal to

$$p(i\tau) = 1 - \bar{p}(i\tau) = \frac{1 - \bar{p}_1(i\tau)}{1 - \bar{p}_1[(i-1)\tau]} \tag{8.4.63}$$

Probability that there is no failure up to t_s in $(i-1)$th cycle is

$$p[t_s + (i-1)\tau] = \frac{1 - \bar{p}_1[t_s + (i-1)\tau]}{1 - \bar{p}_1[(i-1)\tau]} \tag{8.4.64}$$

It is then noted that this situation comprises the following two events:

1. the equipment remains in the successful state for the whole cycle, and
2. the equipment enters in the failed state but the successful operation of the equipment lasts for time more than t_s.

The following relation is rewritten from this exactly as done in Eq. 8.4.12 to get

$$\frac{1 - \bar{p}_1[t_s + (i-1)\tau]}{1 - \bar{p}_1[(i-1)\tau]} = \frac{1 - \bar{p}(i\tau)}{1 - \bar{p}_1[(i-1)\tau]} + \frac{\bar{p}(i\tau) - \bar{p}[(i-1)\tau]}{1 - \bar{p}_1[(i-1)\tau]} \times \{1 - p_{t_s}(t_s)\} \tag{8.4.65}$$

On rearrangement it is found

$$p_{t_s}(t_s) = \frac{\bar{p}_1[t_s + (i-1)\tau]}{\bar{p}_1(i\tau) - \bar{p}_1[(i-1)\tau]} \tag{8.4.66}$$

which on differentiation gives

$$f_{t_s}(t_s) = \frac{f_1[t_s + (i-1)\tau]}{\bar{p}_1(i\tau) - \bar{p}_1[(i-1)\tau]} \tag{8.4.67}$$

For the cyclic process, the definition of availability in the failed state also does not change. However, its averaging over failed and successful states (as done in Eq. 8.4.4) gives a different algebraic expression. A^* for this case is obtained as

$$A^* = \frac{1 - \bar{p}_1(i\tau)}{1 - \bar{p}_1[(i-1)\tau]} + \frac{\bar{p}_1(i\tau) - \bar{p}[(i-1)\tau]}{1 - \bar{p}_1[(i-1)\tau]} A_f \tag{8.4.68}$$

where definitions of A_f and A^* are the same as earlier. From this, A_f is found as

$$A_f = \frac{A^*[1 - \bar{p}_1\{(i-1)\tau\}] - [1 - \bar{p}_1(i\tau)\}}{\bar{p}(i\tau) - \bar{p}_1[(i-1)\tau]} \tag{8.4.69}$$

The distribution function $f_{A_f}(A_f)$ can be found from $f_{t_s}(t_s)$ as

$$f_{A_f}(A_f) = \tau f_{t_s}(\tau A_f)$$

$$= \tau \frac{f_1[\tau A_f + (i-1)\tau]}{\bar{p}_1(i\tau) - \bar{p}_1[(i-1)\tau]} \tag{8.4.70}$$

and the distribution function for A^* is

$$f_{A^*}(A^*) = \frac{[1 - \bar{p}_1\{(i-1)\tau\}]\tau}{\{\bar{p}_1(i\tau) - \bar{p}_1[(i-1)\tau]\}^2} \times f_1\left[\frac{\tau\{A^*[1 - \bar{p}_1[(i-1)\tau]] - [1 - \bar{p}_1(i\tau)]\}}{\bar{p}_1(i\tau) - \bar{p}_1[(i-1)\tau]} + (i-1)\tau\right] \tag{8.4.71}$$

A^* in the distribution function above can have a maximum value of 1 and a minimum value such that A_f does not become negative, i.e.

$$\frac{1 - \bar{p}_1(i\tau)}{1 - \bar{p}_1[(i-1)\tau]} \leqslant A^* \leqslant 1 \tag{8.4.72}$$

Average availability μ_{A^*} can be found by integrating $[A^* f_{A^*}(A^*)\, dA^*]$ over the entire range of A^*, i.e.

$$\mu_{A^*} = \int_{[1-\bar{p}_1(i\tau)]/[1-\bar{p}_1\{(i-1)\tau\}]}^{1} A^* f_{A^*}(A^*)\, dA^* \tag{8.4.73}$$

One defines x as

$$x = \frac{\tau\{A^*(1 - \bar{p}_1[(i-1)\tau]) - [1 - \bar{p}_1(i\tau)]\}}{\bar{p}_1(i\tau) - \bar{p}_1[(i-1)\tau]} + (i-1)\tau \tag{8.4.74}$$

From this A^* is obtained on rearrangement as

$$A^* = \frac{(1/\tau)[x - (i-1)\tau][\bar{p}_1(i\tau) - \bar{p}_1\{(i-1)\tau\}] + [1 - \bar{p}_1(i\tau)]}{1 - \bar{p}_1[(i-1)\tau]} \tag{8.4.75}$$

On differentiation,

$$dA^* = \frac{\bar{p}_1(i\tau) - \bar{p}_1[(i-1)\tau]}{1 - \bar{p}[(i-1)\tau]} \frac{dx}{\tau} \tag{8.4.76}$$

It is noted that the limits of x are $(i-1)\tau$ and $i\tau$ and then Eqs. 8.4.71, 8.4.74, 8.4.75 and 8.4.76 are substituted in Eq. 8.4.73 to obtain

$$\mu_{A^*} = \frac{(1 - \bar{p}_1[(i-1)\tau])\tau}{[\bar{p}_1(i\tau) - \bar{p}_1\{(i-1)\tau\}]^2} \times \int_{(i-1)\tau}^{i\tau} \frac{(1/\tau)[x - (i-1)\tau][\bar{p}_1(i\tau) - \bar{p}_1\{(i-1)\tau\}] + \{1 - \bar{p}_1(i\tau)\}}{1 - \bar{p}_1[(i-1)\tau]} \times f_1(x) \frac{\bar{p}_1(i\tau) - \bar{p}_1[(i-1)\tau]}{1 - \bar{p}[(i-1)\tau]} \frac{dx}{\tau} \tag{8.4.77}$$

Or

$$\mu_{A^*} = \frac{1}{[\bar{p}_1(i\tau) - \bar{p}_1\{(i-1)\tau\}]} \times \int_{(i-1)\tau}^{i\tau} f_1(x) \frac{(1/\tau)[x - (i-1)\tau][\bar{p}_1(i\tau) - \bar{p}_1\{(i-1)\tau\} + \{1 - \bar{p}_1(i\tau)\}}{1 - \bar{p}[(i-1)\tau]} dx$$

$$\equiv \frac{1}{\bar{p}(i\tau) - \bar{p}_1[(i-1)\tau]} \cdot I \tag{8.4.78}$$

The integral I can be further simplified by integrating by parts and using the definition of probability $\bar{p}_1(t_1)$.

$$\begin{aligned} I &= \frac{1-\bar{p}_1(i\tau)}{1-\bar{p}_1[(i-1)\tau]}\int_{(i-1)\tau}^{i\tau} f_1(x)\,dx \\ &\quad + \frac{1}{\tau}\int_{(i-1)\tau}^{i\tau} [x-(i-1)\tau]f_1(x)\,\frac{\bar{p}_1(i\tau)-\bar{p}_1[(i-1)\tau]}{1-\bar{p}_1[(i-1)\tau]}\,dx \\ &= \frac{1-\bar{p}_1(i\tau)}{1-\bar{p}_1[(i-1)\tau]}\,[\bar{p}_1(i\tau)-\bar{p}_1\{(i-1)\tau\}] \\ &\quad + \frac{1}{\tau}\Big\{[x-(i-1)\tau]\bar{p}_1(x)\Big\}_{(i-1)\tau}^{i\tau}\,\frac{\bar{p}_1(i\tau)-\bar{p}_1[(i-1)\tau]}{1-\bar{p}_1[(i-1)\tau]} \\ &\quad - \frac{1}{\tau}\,\frac{\bar{p}_1(i\tau)-\bar{p}_1[(i-1)\tau]}{1-\bar{p}[(i-1)\tau]}\int_{(i-1)\tau}^{i\tau}\bar{p}_1(x)\,dx \end{aligned} \tag{8.4.79}$$

On rearrangement, one gets

$$I = \frac{[\bar{p}_1(i\tau)-\bar{p}_1\{(i-1)\tau\}]}{1-\bar{p}[(i-1)\tau]}\left[1-\bar{p}_1(i\tau)+\bar{p}_1(i\tau)-\frac{1}{\tau}\int_{(i-1)\tau}^{i\tau}\bar{p}_1(x)\,dx\right] \tag{8.4.80}$$

Substituting this in Eq. 8.4.78, μ_{A^*} becomes

$$\mu_{A^*} = \frac{1}{1-\bar{p}_1[(i-1)\tau]}\left[1-\frac{1}{\tau}\int_{(i-1)\tau}^{i\tau}\bar{p}_1(x)\,dx\right] \tag{8.4.81}$$

The $p_1(x)$ above is given by Eq. 8.4.61 which involves the hazard function $z(t)$ of the equipment only. Once the cycle time is specified, mean availability, μ_{A^*}, can be found by integrating $\bar{p}_1(x)$ over the period as given in Eq. 8.4.81.

Average fractional dead time μ_{D^*} can be similarly found. It is first noted that t_f in Eq. 8.4.2 for a given cycle of operation is equal to $(\tau - t_s)$ and $f_{D_f}(D_f)$ is given in terms of $f_{t_s}(t_s)$ by Eq. 8.4.39. It is subsequently noted that D^* is related to D_f by

$$D^* = \frac{\bar{p}_1(i\tau)-\bar{p}_1[(i-1)\tau]}{1-\bar{p}_1[(i-1)\tau]}\,D_f \tag{8.4.82}$$

and $f_{D^*}(D^*)$ would be

$$f_{D^*}(D^*) = \frac{[1-\bar{p}_1\{(i-1)\tau\}]\tau}{[\bar{p}_1(i\tau)-\bar{p}_1\{(i-1)\tau\}]^2}\cdot f_1\left[i\tau - \frac{\tau D^*[1-\bar{p}_1\{(i-1)\tau\}]}{\bar{p}_1(i\tau)-\bar{p}_1[(i-1)\tau]}\right] \tag{8.4.83}$$

Average fractional dead time would be

$$\mu_{D^*} = \frac{1}{1-\bar{p}_1[(i-1)\tau]}\left[\frac{1}{\tau}\int_{(i-1)\tau}^{i\tau}\bar{p}_1(y)\,dy - \bar{p}_1[(i-1)\tau]\right] \tag{8.4.84}$$

where

$$y = i\tau - \frac{\tau D^*[1-\bar{p}_1\{(i-1)\tau\}]}{\bar{p}_1(i\tau)-\bar{p}_1[(i-1)\tau]} \tag{8.4.85}$$

On adding Eqs. 8.4.81 and 8.4.84 one obtains

$$\mu_{D^*} + \mu_{A^*} = \frac{1}{1 - \bar{p}_1[(i - \cdot \,,} [1 - \bar{p}_1\{(i - 1)\tau\}] = 1 \tag{8.4.86}$$

Integrals $\int_{(i-1)\tau}^{i\tau} f_1(x)\,dx$ and $\int_{(i-1)\tau}^{i\tau} f_1(y)\,dy$ are equal even though their arguments are different and this is because the functionality does not change by the change of the variable. Once again, the result of Eq. 8.4.86 is expected because of the definitions of D^* and A^*.

If the equipment is operated for several cycles, one is interested in finding out the average availability $\mu_{A^*,\text{av}}$ over n cycles of operation. This is nothing but the average of availability over individual cycles $\mu_{A^*,i}$, i.e.

$$\mu_{A^*,\text{av}} = \frac{1}{n}\sum_{i=1}^{n} \mu_{A^*,i} \tag{8.4.87}$$

If in the beginning of every cycle, the equipment is made as good as new (i.e. initial condition given by Eq. 8.4.52), availability of all cycles would be identical and this is given by Eq. 8.4.30. However, when the equipment is repaired at the end of each cycle to restore it to the working condition (i.e. when initial condition is given by Eq. 8.4.56), availability in successive operating cycle reduces monotonically because $\bar{p}(t_1)$ is increasing with the age of the equipment. Average fractional dead time $\mu_{D^*,\text{av}}$ for the total of n cycles is found from the information on $\mu_{A^*,\text{av}}$.

$$\mu_{D^*,\text{av}} = (1 - \mu_{A^*,\text{av}}). \tag{8.4.88}$$

The last quantity of interest in cyclic operation of an equipment is the average number of failures μ_n in n cycles. This is obtained by summing the probability of finding the failed state in every cycle. In case when the repair work at the end of the cycle makes the equipment as good as new, the probability of finding the failed state in a given cycle is $\bar{p}(\tau)$ which remains unchanged from cycle to cycle. Therefore

$$\mu_n = \sum_{i=1}^{n} \bar{p}(\tau) = n\bar{p}(\tau) \tag{8.4.89}$$

when the repair work restores the failed equipment to working condition at the end of every cycle, the probability of finding the failed state in cycle i is given by Eq. 8.4.62 and average number of failures is given by

$$\mu_n = \sum_{i=1}^{n} \frac{{}_1(i\tau) - \bar{p}_1[(i-1)\tau]}{1 - \bar{p}_1[(i-1)\tau]} \tag{8.4.90}$$

Lastly one considers the case where the repair time τ_r is not negligible to the period of the cycle. As has been noted, the equipment can be either in the failed state or in working state. If it has failed in time τ, it is repaired in time τ_r. If the equipment does not fail, it is assumed that time τ_r is spent for general overhauling. Time τ_r is treated as a delay time whether the equipment concerned is working or not working. This means that the availability A' averaged over the two states can be written in terms of A^* as

$$A' = \frac{\tau}{(\tau + \tau_r)} A^* \tag{8.4.91}$$

To find distribution function $f_{A'}(A')$ for A', the information on $f_{A^*}(A^*)$ is used. If the repair done at the end of the cycle makes the equipment as good as new, $f_{A^*}(A^*)$ is given by Eq. 8.4.26 and on making the necessary transformation [Eq. 8.4.21), $f_{A'}(A')$ is obtained as

$$\begin{aligned} f_{A'}(A') &= \frac{\partial}{\partial A'} p_{A'}(A') = \frac{\tau + \tau_r}{\tau} \cdot \frac{\partial}{\partial A^*}\left\{p_{A^*}\left[\left(\frac{\tau+\tau_r}{\tau}\right)A'\right]\right\} \\ &= \frac{\tau\left(\frac{\tau+\tau_r}{\tau}\right)}{[\bar{p}(\tau)]^2} f\left[\frac{\tau\left(\frac{\tau+\tau_r}{\tau}\right)A' - \tau\{1-\bar{p}(\tau)\}}{\bar{p}(\tau)}\right] \end{aligned} \tag{8.4.92}$$

Or

$$f_{A'}(A') = \frac{(\tau + \tau_r)}{[\bar{p}(\tau)]^2} f\left[\frac{(\tau+\tau_r)A' - \tau\{1-\bar{p}(\tau)\}}{\bar{p}(\tau)}\right] \tag{8.4.93}$$

Since A^* can have values between $[1 - \bar{p}(\tau)]$ and 1, A' from its definition will have the range

$$\frac{[1-\bar{p}(\tau)]\tau}{(\tau+\tau_r)} \leqslant A' \leqslant \frac{\tau}{(\tau+\tau_r)} \tag{8.4.94}$$

To find average availability $\mu_{A'}$, $[A' f_A(A')\, dA']$ is integrated between its limits and after proceeding similarly as earlier, it is found as

$$\mu_{A'} = \frac{\tau}{(\tau+\tau_r)} - \frac{1}{(\tau+\tau_r)} \int_0^\tau \bar{p}(t)\, dt \tag{8.4.95}$$

Comparing the above equation with Eq. 8.4.37,

$$\mu_{A'} = \frac{\tau}{\tau+\tau_r} \mu_{A^*} \tag{8.4.96}$$

Fractional dead time D' is similarly defined for this case as

$$D' = \frac{\tau}{\tau+\tau_r} D^* \tag{8.4.97}$$

and $f_{D'}(D')$ is found by similar transformations as

$$f_{D'}(D') = \frac{(\tau+\tau_r)}{[\bar{p}(\tau)]^2} f\left[\frac{\tau\bar{p}(\tau) + \tau_r - D'(\tau+\tau_r)}{\bar{p}(\tau)}\right] \tag{8.4.98}$$

Average fractional dead time $\mu_{D'}$ is found to be

$$\mu_{D'} = \frac{\tau_r}{(\tau+\tau_r)} + \frac{\tau}{\tau+\tau_r} \mu_{D^*} \tag{8.4.99}$$

In this section, relations for mean availability and fractional dead time of a given equipment have been derived. The study of this section reveals that there are two very important informations:

1. distribution function of survival of the equipment, and
2. time τ of a single cycle.

The time τ determines the frequency of maintenance work carried out and it can be seen to be a very important parameter. Intuitively it is felt that the shorter this time τ, one would be quicker to discover faults (or failure) in the equipment, if any, this way increasing its availability. Since availability is a direct measure of reliability of the equipment, it is, therefore, expected that shorter time τ would mean larger reliability.

When the hazard function is constant, the distribution function is exponential given by Eq. 8.3.12. Probability of survival is obtained by integrating it and is substituted in Eq. 8.4.37 to obtain μ_{A*}

$$\mu_{A*} = 1 - \frac{1}{\tau}\int_0^{\theta\tau} (1 - e^{-\theta t})\, dt \tag{8.4.100}$$

where the equipment is assumed to be repaired to as good as new at the end of every cycle. On integration,

$$\mu_{A*} = \frac{1 - e^{-\theta\tau}}{(\theta\tau)} \tag{8.4.101}$$

Similarly the average fractional dead time is found as

$$\mu_{D*} = 1 - \frac{1 - e^{-\theta\tau}}{\theta\tau} \tag{8.4.102}$$

μ_{A*} and μ_{D*} are calculated in Table 8.2 for different values of $(\theta\tau)$. It is found that μ_{A*} monotonically decreases whereas μ_{D*} increases as $[\theta\tau]$ increases in value, which is in conformation to the intuitive argument stated earlier.

Table 8.2 Mean Availability and Fractional Load Time of an Equipment with Exponential Distribution Function for Different Cycle Time τ

$\theta\tau$	$e^{-\theta\tau}$	μ_{A*}	μ_{D*}
0.00	1.000	1.000	0.000
0.25	0.779	0.884	0.116
0.50	0.607	0.786	0.214
1.00	0.368	0.632	0.368
2.00	0.136	0.432	0.568
3.00	0.050	0.317	0.683
4.00	0.018	0.266	0.734

For $\theta\tau$ equal to zero, it is discovered that both numerator $(1 - e^{-\theta\tau})$ and denominator in Eq. 8.4.101 simultaneously go to zero. Therefore to find the limiting value of μ_{A*}, L'Hospital rule is applied

$$\mu_{A*} = \lim_{\theta\tau\to 0} \frac{\dfrac{d}{d(\theta\tau)}(1 - e^{-\theta\tau})}{\dfrac{d}{d(\theta\tau)}(\theta\tau)} = \lim_{\tau\to 0} \frac{e^{-\theta\tau}}{1} = 1 \tag{8.4.103}$$

which is 1 for $\theta\tau$ equal to zero. For nonzero θ, $\tau = 0$ means that the equipment is monitored continuously and as seen above, it gives the maximum mean availability.

Situation arising from continuous monitoring leads to the concept of reversible maintenance policy which is discussed next.

8.5 AVAILABILITY AND FRACTIONAL DEAD TIME OF A SINGLE EQUIPMENT FOR A REVERSIBLE MAINTENANCE POLICY[4]

In the previous section, it was stated that standby equipment, safety devices and some controllers need to be checked and repaired periodically to ensure that they remain in working condition. However, this strategy cannot be adopted for main train equipment of the chemical plant. This is because if these fail, it would mean the entire plant has to be shut down. In view of this, any delay in repairing leads to loss of production which is not desirable.

Reversible policy of maintenance of an equipment refers to the operation of the equipment in which it is allowed to operate as long as it does not fail and on its failure, the repair work is ordered immediately. It may be appreciated that this is not precisely the same as the irreversible policy with $\tau = 0$. The similarity lies only ir the fact that the equipment is continuously monitored for any failure or fault. In reversible policy, the equipment is allowed to work as long as it does not fail which is not the same as in the irreversible policy where it is stopped and checked after time τ. In this section, availability of an equipment has been derived in terms of its hazard function. As stated earlier, the repair work could be of two kinds: one that restores the failed equipment to working condition and the second one which makes it as good as new. The expressions for availability for these two cases are found to be different except when distribution functions for repair and survival of the equipment are exponential in nature in which case they are the same.

8.5.1 When the Repair-Work Restores the Equipment to As Good As New Condition

Figure 8.7 shows the operation of a new equipment installed in a plant. Life of the equipment is denoted by t_l whereas u_1, u_2, u_3 etc. denote the time when the failure occurs. Immediately after it occurs, the repair work starts and it is assumed that it takes time t_1, t_2, t_3 etc. to make the equipment as good as new. As earlier, the equipment has two possible states, working and failed state as shown in Fig. 8.6. Since repair starts immediately after the failure, the duration in which it remains in the failed state is determined by the speed of the repair work. The total cycle time τ is the sum of the operation time (u_1, u_2 etc.) and the repair time (t_1, t_2 etc.) and unlike irreversible state, it is found to vary from cycle to cycle. The cycle time is stochastic in nature because it is impossible to know the failure time deterministically. The repair time depends upon the nature of the failure as well as the efficiency of the workers and is also stochastic in nature.

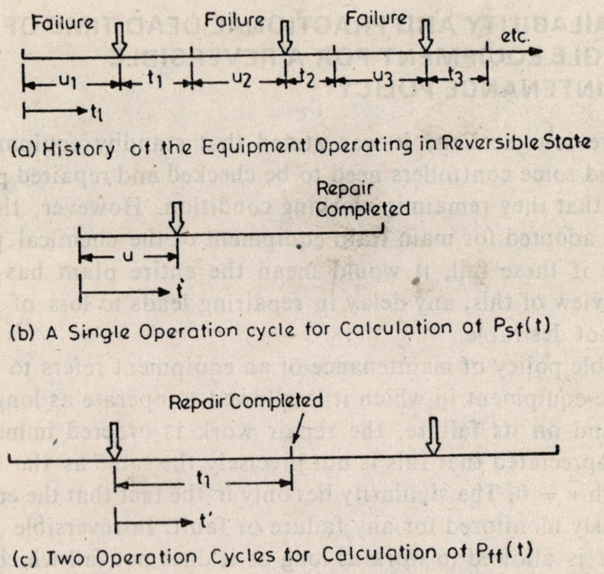

Fig. 8.7 Operation of an equipment with reversible maintenance policy

The fact that the equipment becomes as good as new in the beginning of every cycle, its behaviour in one cycle, on an average, is expected to be the same as in any other. As a result of this, one considers any one of these cycles shown in Fig. 8.7b. Time t denotes the time from the completion of the repair and not the life of the equipment. To solve this problem mathematically, the following probabilities and distribution functions are defined:

$p_{sf}(t)$: Probability of the equipment in failed state at time t starting from successful state at $t = 0$ (8.5.1a)

$p_{ff}(t)$: Probability of the equipment in failed state at time t starting from failed state at $t = 0$ (8.5.1b)

$f_{t_f}(t_f)$: Distribution function of failure time t_f (8.5.1c)

$f_{t_r}(t_r)$: Distribution function of repair time t_r (8.5.1d)

To find $p_{sf}(t)$ for the reversible policy for an equipment, it is noted that the equipment will be in the failed state at time t provided it fails at time u, $0 \leqslant u \leqslant t$, and the repair work does not complete in the remainder of the time $(t - u)$. The probability of failure at u is found from its distribution function and is given by $f_{t_f}(u)$. Since the equipment is new at $t = 0$, $p_{sf}(0)$ would necessarily be zero and in view of this

$$p_{sf}(t) = p_{sf}(t) - p_{sf}(0) = \int_0^t p_{ff}(t-u) f_{t_f}(u)\, du \tag{8.5.2}$$

where according to the definition, $p_{ff}(t-u)$ gives probability that the equipment remains in the failed state for time $(t-u)$ starting from the failed state at $t = u$. This is a convolution integral involving $p_{ff}(t)$ which is yet to be determined. The best way of representing a convolution integral is to write it in Laplace transform which is defined for a function $g(x)$ as follows:

$$L[g(x)] \equiv G(s) = \int_0^\infty e^{-sx} g(x)\, dx; \; s < 1 \tag{8.5.3}$$

where $L[g(x)]$ stands for the Laplace transform of function $g(x)$. Since the integration has been performed on x, the resultant function $G(s)$ is going to be a function of the dummy variable, s, only. The inverse of the Laplace transform can be found to get back $g(x)$

$$g(x) = L^{-1}[G(s)] \tag{8.5.4}$$

The integral of Eq. 8.5.3 has been solved for almost all standard functions and Laplace transforms and their inverse defined by Eq. 8.5.4 can be found in any mathematical handbook[4,6]. For some commonly occurring function, they are given in Table 8.3.

One now considers the Laplace transform of a convolution integral $h(x)$ defined as

$$h(x) = \int_0^x g_1(u) g_2(x-u)\, du \tag{8.5.5}$$

Its Laplace transform would be

$$\begin{aligned} H(s) = L[h(x)] &= \int_{x=0}^\infty e^{-sx}\left[\int_0^x g_1(u) g_2(x-u)\, du\right] dx \\ &= \int_{x=0}^\infty \int_{u=0}^x g_1(u) g_2(x-u) e^{-sx}\, du\, dx \end{aligned} \tag{8.5.6}$$

In the above equation, integration is performed with respect to u first and then x. Since the functions to be integrated are continuous, the integration can be performed with respect to x first and then u as long as the limits of integration are adjusted properly. While integrating with respect to x first, it is noted that it cannot be less than u whereas its upper limit is ∞ whereas u can vary from 0 to ∞, i.e.

$$H(s) = \int_{u=0}^\infty \int_{x=u}^\infty g_1(u) g_2(x-u) e^{-sx}\, dx\, du \tag{8.5.7}$$

Since $g_1(u)$ is not a function of x, on rearrangement it becomes

$$H(s) = \int_{u=0}^\infty g_1(u)\, e^{-su}\left[\int_u^\infty g_2(x-u) e^{-s(x-u)}\, dx\right] du \tag{8.5.8}$$

Substituting

$$y = (x-u) \tag{8.5.9}$$

Table 8.3[4,5] Laplace Transform of Some of Important Functions

S. No.	$G(s) = Lg(x)$	$g(x)$
1.	1	$\delta(x)$
2.	$1/s$	$U(x)$
3.	a/s	$aU(x)$
4.	$1/s^2$	x
5.	$1/s^n \; (n > -1)$	$x^{n-1}/\Gamma(n)$
6.	$1/(s+a)$	e^{-ax}
7.	$\dfrac{1}{(s+a)(s+b)} \; (a \neq b)$	$\dfrac{1}{(b-a)}[e^{-ax} - e^{-bx}]$
8.	$\dfrac{1}{(s+a)(s+b)(s+c)} \; (a \neq b \neq c)$	$\dfrac{(c-b)e^{-ax} + (a-c)e^{-bx} + (b-a)e^{-cx}}{(a-b)(b-c)(c-a)}$
9.	$\dfrac{s}{(s+a)(s+b)(s+c)} \; (a \neq b \neq c)$	$\dfrac{a(b-c)e^{-ax} + b(c-a)e^{-bx} + c(a-b)e^{-cx}}{(a-b)(b-c)(c-a)}$
10.	$\log \dfrac{s+a}{s+b}$	$\dfrac{1}{x}(e^{-bx} - e^{-ax})$
11.	e^{-as}	$\delta(x-a)$
12.	$\dfrac{W}{s^2 + W^2}$	$\sin Wx$
13.	$\dfrac{s}{W^2 + s^2}$	$\cos Wx$
14.	$\lim\limits_{x \to \infty} sG(s)$	$\lim\limits_{x \to 0} g(x)$
15.	$\lim\limits_{s \to 0} sG(s)$	$\lim\limits_{x \to \infty} g(x)$
16.	$aF(s)$	$af(x)$
17.	$G_1(s) + G_2(s)$	$g_1(x) + g_2(x)$
18.	$e^{-as}G(s)$	$g(x-a)$
19.	$G(s-a)$	$e^{ax}g(x)$
20.	$aG(as)$	$g(x/a)$
21.	$\dfrac{d^nG(s)}{ds^n}$	$(-1)^n x^n g(x)$
22.	$\int_s^\infty G(s)\,ds$	$g(x)/x$
23.	$\left[G^{-n}(s)\right]_s^\infty$	$\dfrac{g(x)}{x^n}$
24.	$s^nG(s) - s^{n-1}(0+) - s^{n-2}g(0+)$	$\dfrac{d^ng(x)}{dx^n}$
25.	$\dfrac{\partial}{\partial a} G(s, a)$	$\dfrac{\partial}{\partial a}[g(x, a)]$
26.	$\dfrac{G(s)}{s} + \dfrac{g^{-1}(0+)}{s}$	$\int g(x)\,dx$
27.	$\dfrac{G(s)}{s^2} + \dfrac{g^{-1}(0+)}{s^2} + \dfrac{g^{-2}(0+)}{s}$	$\int\left[\int g(x)\,dx\right]dx$
28.	$G_1(s)G_2(s)$	$\int_0^x g_1(x-u)g_2(u)\,du$

Explanations

$\delta(x)$: Dirac Delta function

$U(x)$: Step function

in Eq. 8.5.8, one gets

$$H(s) = \int_{u=0}^{\infty} g_1(u)e^{-su}\,du \int_{y=0}^{\infty} g_2(y)e^{-sy}\,dy = G_1(s)G_2(s) \qquad (8.5.10)$$

where $G_1(s)$ and $G_2(s)$ are Laplace transforms of functions $g_1(x)$ and $g_2(x)$ respectively. With this knowledge, the Laplace transform of both sides of Eq. 8.5.2 is taken to obtain

$$P_{sf}(s) = P_{ff}(s)F_{t_f}(s) \qquad (8.5.11)$$

where $P_{sf}(s)$, $P_{ff}(s)$ and $F_{t_f}(s)$ are the Laplace transform of functions $p_{sf}(t)$, $p_{ff}(t)$ and $f_{t_f}(u)$ respectively.

To calculate $p_{ff}(t)$, the analysis must extend from one failed state to another and as a result of these two operation cycles are shown in Fig. 8.7c. The time origin is fixed at t' from the first failure. The equipment will be in the failed state if the repair started at $t' = 0$ is not completed at the time t' and the probability of this event, $[1 - p_{t_r}(t')]$, is given by

$$[1 - p_{t_r}(t')] = 1 - \int_0^{t'} f_{t_r}(x)\,dx \qquad (8.5.12)$$

The equipment will also be in the failed state if after the repair work completed at time v, it fails in the remainder of the time $(t' - v)$. Probability of the completion of the repair work at time v is given by $f_{t_r}(v)\,dv$. Since the equipment is in successful state at time v, the probability of finding it in the failed state at t' is given by $p_{sf}(t')$.

$$p_f(t') = \int_{v=0}^{t'} p_{sf}(t' - v)f_{t_r}(v)\,dv \qquad (8.5.13)$$

Since both these events contribute with equal likelihood to finding the equipment in failed state at t', therefore $p_{ff}(t')$ is sum of Eq. 8.5.12 and 8.5.13, i.e.

$$p_{ff}(t') = [1 - p_{t_r}(t') + \int_0^{t'} p_{sf}(t' - v)f_{t_r}(v)\,dv \qquad (8.5.14)$$

The Laplace transform of the above equation is taken to obtain

$$p_{ff}(s) = \frac{1}{s} - P_{t_r}(s) + P_{sf}(s)F_{t_r}(s) \qquad (8.5.15)$$

where $p_{ff}(s)$, $p_{t_r}(s)$ and $F_{t_r}(s)$ are the Laplace transforms of functions $p_{ff}(t)$, $p_{t_r}(t)$ and $f_{t_r}(t')$ respectively. $P_{ff}(s)$ is then eliminated from Eq. 8.5.11 with the help of Eq. 8.5.15 to get

$$P_{sf}(s) = \frac{1}{s}[1 - sP_{t_r}(s) + sP_{sf}(s)F_{t_r}(s)]F_{t_f}(s) \qquad (8.5.16)$$

Solving for $P_{sf}(s)$

$$P_{sf} = \frac{1}{s}\,\frac{F_{t_f}(s)[1 - sP_{t_r}(s)]}{1 - F_{t_r}(s)F_{t_f}(s)} \qquad (8.5.17)$$

Because $p_{t_r}(t)$ and $f_{t_r}(t)$ are related through Eq. 8.5.12, a relation must exist between $P_{t_r}(s)$ and $F_{t_r}(s)$ and to find it, the differentiation of a Laplace transform function $G(s)$ is examined below:

$$G(s) = L[g(x)] = \int_0^\infty g(x)e^{-sx}\, dx \tag{8.5.18}$$

Laplace transform of $\frac{d}{dx} g(x)$ is given by

$$L\left[\frac{d}{dx} g(x)\right] = \int_0^\infty \frac{dg(x)}{dx}\, e^{-sx}\, dx \tag{8.5.19}$$

Integration of Eq. 8.5.18 by parts leads to

$$\begin{aligned} L[g(x)] &= \left[\frac{g(x)e^{-sx}}{s}\right]_0^\infty + \frac{1}{s}\int_0^\infty \frac{dg}{dx}\, e^{-sx}\, dx \\ &= \frac{g(0)}{s} + \frac{1}{s}\, L\left[\frac{d}{dx} g(x)\right] \end{aligned} \tag{8.5.20}$$

Since integration is the reverse of differentiation, using the above relation on Eq. 8.5.12, one gets

$$P_{t_r}(s) = \frac{1}{s}\, F_{t_r}(s) \tag{8.5.21}$$

and substituting this in Eq. 8.5.17,

$$P_{sf}(s) = \frac{1}{s}\, \frac{F_{t_f}(s)[1 - F_{t_r}(s)]}{1 - F_{t_r}(s)F_{t_f}(s)} \tag{8.5.22}$$

Lastly to find the probability, $p_{sf}(t)$ of the equipment in successful state at time t, it is noted that it can be either in the working or in the failed state. Therefore at a given time

$$p_{ss}(t) + p_{sf}(t) = 1 \tag{8.5.23}$$

Taking Laplace transform of both the sides above

$$P_{ss}(s) + P_{sf}(s) = \frac{1}{s} \tag{8.5.24}$$

and with help of Eq. 8.5.22, one obtains

$$P_{ss}(s) = \left(\frac{1}{s} - P_{sf}(s)\right) = \frac{1 - F_{t_f}(s)}{s[1 - F_{t_f}(s)F_{t_r}(s)]} \tag{8.5.25}$$

As an example, exponential failure and repair time distributions are considered

$$f_{t_f}(t_f) = \theta \exp(-\theta t_f) \tag{8.5.26a}$$

and

$$f_{t_r}(t_r) = \frac{1}{\tau} \exp(-t_r/\tau) \tag{8.5.26b}$$

Their Laplace transforms are found with the help of Table 8.3.

$$F_{t_f}(s) = \frac{\theta}{(s + \theta)} \tag{8.5.27a}$$

$$F_{t_r}(s) = \frac{1}{1 + \tau s} \tag{8.5.27b}$$

Substituting these in Eq. 8.5.22

$$P_{sf}(s) = \frac{\theta}{s[s + (\theta + 1/\tau)]} \tag{8.5.28}$$

and $P_{sf}(t)$ is found by inverting the Laplace transform,

$$P_{sf}(t) = \frac{\theta\tau}{1 + \theta\tau}[1 - e^{-t(\theta + 1/\tau)}] \tag{8.5.29}$$

For more complex distribution functions, finding the inverse to get $p_{sf}(t)$ is not an easy task. It is in view of this, one presents the following asymptotic result. This is possible because $L[f(x)]$ can be expanded into a series involving moments of x, M_k, where $f(x)$ is its distribution function. By its definition

$$F(s) = L[f(x)] = \int_0^\infty e^{-sx} f(x)\, dx \tag{8.5.30}$$

Expanding e^{-sx} into an infinite series

$$F(s) = \int_0^\infty \left[1 - sx + \frac{(sx)^2}{2!} - \frac{(sx)^3}{3!} + \ldots\right] f(x)\, dx$$

$$= 1 - sM_1 + \frac{s^2}{2!} M_2 - \frac{s^3}{3!} + \ldots \tag{8.5.31}$$

where

$$M_k = \int_0^\infty x^k f(x)\, dx \tag{8.5.32}$$

For large times, terms containing s^2 and higher are neglected as first approximation

$$F(s) \simeq 1 - sM_1 + 0(s^2) \tag{8.5.33}$$

where $0(s^2)$ represents terms of order s^2 and higher, which is expected to be small. Similar expansions are written for $F_{t_f}(s)$ and $F_{t_r}(s)$ and substituted in Eq. 8.5.25

$$P_{ss}(s) = \frac{1 - [1 - sM_{t_f} + 0(s^2)]}{s[1 - \{1 - sM_{t_f} + 0(s^2)\}\{1 - sM_{t_r} + 0(s^2)\}]} \tag{8.5.34}$$

where M_{t_f} and M_{t_r} are first moments (or average) of the time of failure t_f, and the time of repair t_r. If $0(s^2)$ is dropped

$$P_{ss}(s) \simeq \frac{1}{s} \frac{M_{t_f}}{(M_{t_f} + M_{t_r})} \tag{8.5.35}$$

Similarly

$$P_{sf} \simeq \frac{1}{s} \frac{M_{t_r}}{M_{t_f} + M_{t_r}} \tag{8.5.36}$$

Noting the following from Table 8.3

$$\lim_{t\to\infty} f(t) = \lim_{s\to 0} sF(s) \tag{8.5.37}$$

one gets

$$p_{s\infty} \equiv \lim_{t\to\infty} p_{ss}(t) = \frac{M_{t_f}}{M_{t_r} + M_{t_f}} \tag{8.5.38a}$$

and

$$p_{f\infty} \equiv \lim_{t\to\infty} p_{sf}(t) = \frac{M_{t_r}}{(M_{t_r} + M_{t_f})} \tag{8.5.38b}$$

The $p_{f\infty}$ is substituted in Eqs. 8.4.37 and 8.4.48 to get

$$\mu_{A*} = \frac{M_{t_f}}{(M_{t_r} + M_{t_f})} \tag{8.5.39a}$$

$$\mu_{D*} = \frac{M_{t_r}}{(M_{t_r} + M_{t_f})} \tag{8.5.39b}$$

8.5.2 When Repair-Work Restores the Equipment to Working Condition

When repair done on an equipment does not make it as good as new, it is not possible to analyze a single cycle. This is because condition of the equipment at the time of consideration depends upon the degree of wear-and-tear in previous cycles which is equivalent to saying that it has a memory.

For the analysis, the entire life of the equipment prior to the operation cycle must be considered. Figure 8.8 shows the entire cycle in which t_l

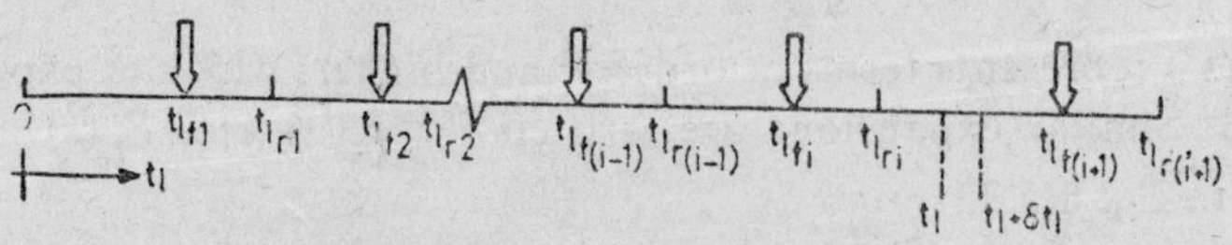

Fig. 8.8 Operation of an equipment with reversible maintenance with repair restoring it to the working state

denotes the life of the equipment and t_{lr_i} and t_{lf_i} are times of ith repair and failure respectively measured from $t_1 = 0$. One considers its operation between the time span t_l and $(t_l + \delta t_l)$ and would like to find if the equipment is in successful state or in the failed state. To do so, the following two probabilities are defined:

$p_f(t_l)$: probability of finding the equipment in failed state at time t_1, and (8.5.40a)

$p_s(t_l)$: probability of finding the equipment in successful state at time t_1 (8.5.40b)

From definition of these probabilities, at a given time t_l

$$p_s(t_l) + p_f(t_l) = 1 \tag{8.5.41}$$

It is now desired to find $p_f(t_l)$ and to find it, attention is focussed upon a small time span of δt_l as shown in Fig. 8.8. It is noted that at t_l the equipment can be either in the failed state or in successful state. It shall be in the failed state at $(t_l + \delta t_l)$ if

1. it is in failed state at t_l and does not get repaired in time δt_l
2. it is in successful state at t_l but it fails during time δt_l.

Or

$$p_f(t_l + \delta t_l) = p_s(t_l)\ \delta p_f(\delta t_l) + p_f(t_l)\ \delta p_r(\delta t_l) \tag{8.5.42}$$

where $\delta p_f(\delta t_l)$ is the transition probability of entering in the failed state in time δt_l and $\delta p_r(\delta t_l)$ is the probability of a repair work remaining incomplete in the time. Transition functions $z(t_0)$ and $z_{t_r}(t_r)$ are defined such that

$$\delta p_f(\delta t_l) = z(t_{0i})\ \delta t_{0l} = z(t_{0i})\ \delta t_l \tag{8.5.43a}$$

and

$$\begin{aligned} p_r(\delta t_l) &= 1 - z_{t_r}(t_{r_i})\ \delta t_{r_i} \\ &= 1 - z_{t_r}(t_{r_i})\ \delta t_l \end{aligned} \tag{8.5.43b}$$

where t_{0l} is the time of operation measured from the completion of the previous repair and t_{r_i} is the time measured from the previous failure i.e.

$$t_{0l} = (t_l - t_{lr_i}) \tag{8.5.44a}$$

and

$$t_{r_i} = (t_l - t_{lf_i}) \tag{8.5.44b}$$

In Eq. 8.5.43a, t_{0i} appears because the wear-and-tear of the equipment occurs only during its operation after it has been set in working condition after repairs. Similarly because in reversible maintenance policy, repair work is started only after the failure has occurred, z_{t_r} should be equal to zero for repair time $t_{r_i} < 0$. Also $z(t_{0l})$ has the same meaning as the hazard function discussed earlier. Substituting Eqs. 8.5.43 and 8.5.44 in Eq. 8.5.42,

$$p_f(t_f + \delta t_l) = [1 - z(t_l - t_{lr_i})\ \delta t_l]p_f(t_f) + p_f(t_l)z_{t_r}\ \delta t_l \tag{8.5.45}$$

Rearranging the above equation and in the limit of $\delta t_l \rightarrow 0$, one obtains

$$\frac{dp_f}{dt_l} = \lim_{\Delta t_l \rightarrow 0} \frac{p_f(t_l + \delta t_l) - p_f(t_l)}{\delta t_l} = p_s(t_l)z(t_{0l}) - z_{t_r}(t_{r_i})p_f(t_l) \tag{8.5.46}$$

Eliminating $p_s(t_l)$ using Eq. 8.5.41

$$\frac{dp_f(t_l)}{dt_l} + [z_{t_r}(t_{r_i}) + z(t_{0i})]p_f(t_l) = z(t_{0i}) \tag{8.5.47}$$

This is a first order linear differential equation and can be solved for $p_f(t_l)$ as

$$p_f(t_l) = V(t_l)\left[K + \int_0^{t_l} z(t_{0i}) \exp\left(+\int_0^t [z_{t_r}(t_{r_i}) + z(t_{0i})]\, dt\right) dt\right] \tag{8.5.48}$$

where

$$V(t_l) = \exp\left(-\int_0^{t_l} [z_{t_r}(t_{r_i}) + z(t_{0i})]\, dt\right] \tag{8.5.49}$$

K is the constant of integration which can be determined from the initial condition that it is new at $t_l = 0$, i.e.

$$p_s(0) = 1 \tag{8.5.50a}$$

and

$$p_f(0) = 0 \tag{8.5.50b}$$

This gives

$$K = 0$$

and $p_f(t_l)$ is given by

$$p_f(t_l) = V(t_l) \int_0^{t_l} z(t_{0i}) \exp\left[+\int_0^t [z_{t_r}(t_{r_i}) + z(t_{0i})]\, dt_l\right] dt_l \tag{8.5.51}$$

As an example one considers the ith cycle of operation starting with a new equipment. If the time to repair and time to failure both follow an exponential distribution $z(t_r)$ and $z(t_0)$ both would be constant. One assumes these to be

$$z_{t_r}(t_{r_i}) = (1/\tau) \tag{8.5.52a}$$

$$z(t_{0i}) = \theta \tag{8.5.52b}$$

which are independent of t_{ri} and t_{0i}. In view of this independence, Eq. 8.5.51 can be integrated without any problem. $V(t_l)$ for any cycle i is given by

$$V(t_l) = \exp\left(-\int_0^{t_l} [\theta + 1/\tau]\, dt\right) = \exp\left[-\left(\frac{\theta\tau + 1}{\tau}\right)t_l\right] \tag{8.5.53}$$

and $p_f(t_f)$ is

$$p_f(t_l) = \exp\left[-\left(\frac{\theta\tau + 1}{\tau}\right)t_l\right]\int_0^{t_l} \exp\left[\left(\frac{\theta\tau + 1}{\tau}\right)t_l\right] dt_l \tag{8.5.54}$$

or

$$p_f(t_l) = \frac{\theta\tau}{1 + \theta\tau} \exp\left[-\left(\frac{\theta\tau + 1}{\tau}\right)t_l\right]\left[\exp\left(\frac{\theta\tau + 1}{\tau}\, t\right) - 1\right] \tag{8.5.55}$$

Or

$$p_f(t_l) = \frac{\theta\tau}{1+\theta\tau}\left[1 - \exp\left\{-\left(\frac{\theta\tau+1}{\tau}\right)t_l\right\}\right] \tag{8.5.56}$$

On comparison of this equation with Eq. 8.5.29, they are found to be same in functional relation. However $p_f(t_l)$ and $p_{sf}(t)$ differ slightly in their meanings. $p_f(t_l)$ is the probability of finding the equipment in failed state even though the equipment may have failed and been repaired several times from the beginning of its operation, whereas $p_{sf}(t)$ is the probability of its *first* failure. Lastly it needs to be pointed out that Eq. 8.5.56 is independent of the cycle of operation because $z(t_{0i})$ and $z_{t_r}(t_{ri})$ are time independent.

In Secs. 8.4 and 8.5 maintenance of a single equipment has been considered. Since a plant consists of several equipment, one considers next their combination and examines availability and fractional dead time.

8.6 AVAILABILITY AND FRACTIONAL DEAD TIME OF A PURE SERIES SYSTEM

In Chapter 6 it was noted that there could exist three possible configurations in a chemical plant

1. pure series,
2. pure parallel, and
3. mixed.

In this section, the pure series configuration having n components is discussed. The system as a whole could be operated either with irreversible or with reversible maintenance policy depending upon how the failures are monitored. Exactly as done for a single equipment, the repair work can make all components of the system as good as new or merely restore it (or them) to the working condition. In the former case, only one failure cycle must be considered, whereas in the latter, the whole life of the system must be analyzed. The mathematical procedure for calculating the availability and fractional dead time, however, remains unaltered from that for single equipment presented earlier.

Instead of analyzing an n-component system, a system consisting of *two* components only is examined because the results derived for this can be easily generalized. The system is assumed to consist of two equipments A and B in series. Because each operates independent of the other, the system as a whole can be in the working state only when both A and B are working. Therefore the following probabilities are defined:

$\bar{p}_A$: probability that A is in the failed state (8.6.1a)

$\bar{p}_B$: probability that B is in the failed state (8.6.1b)

and

$\bar{p}_S$: probability that the system is in the failed state (8.6.1c)

$\bar{p}_S$ is related to $\bar{p}_A$ and $\bar{p}_B$ by

$$(1 - \bar{p}_S) = (1 - \bar{p}_B)(1 - \bar{p}_A) \tag{8.6.2}$$

where $(1 - \bar{p}_S)$, $(1 - \bar{p}_A)$ and $(1 - \bar{p}_B)$ are probabilities that the system as a whole, components A and B respectively are in the working condition. On simplification of the equation above,

$$\bar{p}_S = \bar{p}_A + \bar{p}_B - \bar{p}_A\bar{p}_B \tag{8.6.3}$$

The system as a whole can either be in the failed state or in the working condition, and therefore its availability and fractional dead time can be calculated exactly as earlier. To find the mean availability μ_{A^*}, $\bar{p}_S$ is substituted in Eq. 8.4.37 to obtain

$$\mu_{A_S^*} = 1 - \frac{1}{\tau}\int_0^\tau \bar{p}_A(t)\,dt - \frac{1}{\tau}\int_0^\tau \bar{p}_B(t)\,dt + \frac{1}{\tau}\int_0^\tau \bar{p}_A(t)\bar{p}_B(t)\,dt \tag{8.6.4}$$

Thus it is seen that the mean availability of the combination of the equipment is not equal to the product of mean availabilities of individual equipment μ_A and μ_B, i.e.

$$\mu_{A_S^*} \neq \mu_A\mu_B \tag{8.6.5}$$

In view of the above result for a system consisting of n components in series, one is therefore left with no choice but to evaluate the mean availability by integrating the general equation (Eq. 8.4.37). If the probability of survival of the ith component in an n-component system is written as (see Eq. 8.3.10)

$$p_i(t) = \exp\left\{-\int_0^t z_i(t)\,dt\right\} \tag{8.6.6}$$

The probability of survival $p_S(t)$ of the system would be

$$p_S(t) = \prod_{i=1}^{n} p_i(t) = \exp\left\{-\int_0^t \left[\sum_{i=1}^{n} z_i(t)\right] dt\right\} \tag{8.6.7}$$

Since the system can either be in the failed or in the working state, the probability of the system in the failed state $\bar{p}_S(t)$ is given by

$$\bar{p}_S(t) = 1 - p_S(t) = 1 - \exp\left\{-\int_0^t \left[\sum_{i=1}^{n} z_i(t)\right] dt\right\} \tag{8.6.8}$$

To find the mean availability of the system, information on $z_i(t)$ must be known. If it is assumed as constant equal to θ_i, the above equation becomes

$$\bar{p}_S(t) = 1 - \exp\left\{-\left(\sum_{i=1}^{n} \theta_i\right)t\right\} \tag{8.6.9}$$

For small e_i (or small time), the exponential can be expanded to obtain

$$\bar{p}_S(t) \simeq (\sum \theta_i)t \tag{8.6.10}$$

and the mean availability μ_{A_S} of the system is obtained as

$$\mu_{A_S} = 1 - \frac{(\sum \theta_i)}{\tau} \int_0^\tau t \, dt$$
$$= 1 - (\sum \theta_i)(\tau/2) \tag{8.6.11}$$

Another quantity which is of great interest is the average time of first failure, μ_f, and is given by

$$\mu_f = \int_0^\infty \exp\left\{-\int_0^t (\sum z_i(t)) \, dt\right\} dt \tag{8.6.12}$$

For constant $z_i(t)$, it is

$$\mu_f = \int_0^\infty \exp\{-t(\sum \theta_i)\} \, dt$$
$$= \frac{1}{\sum_{i=1}^{n} \theta_i} \tag{8.6.13}$$

Because θ_i is a positive constant for all values of i, it is seen that μ_f becomes smaller and smaller as the total number of components, n, in the system increases. Smaller μ_f implies that it takes a shorter time for the system to fail and this is also expected intuitively because of the following reason. It may be recalled that the system is in the working condition *only* when all its components are working. Since each component can fail independent of each other, the system can attain the failed state with greater ease as the number of components increases.

Now one considers several cycles of operation of the system consisting of equipment in series. In the irreversible policy of maintenance, the repair work will be now shown to play a major role. In a system with more than one component, for this policy, the following variations could exist:

1. repair makes the failed component as good as new,
2. repair restores the failed component to the working condition,
3. all components of the system are checked at the end of the cycle period, and/or
4. all components are checked exactly after time τ but their times of testing are staggered.

The first two types are a measure of the efficiency of the repair work and their effect on the performance of the system is exactly the same as that for a single equipment. If the repair makes the total system as good as new, it is necessary to analyze only one time cycle. When it merely restores it to the working condition, the system has a memory and it is not possible to analyze a single cycle in isolation.

The latter two types 3 and 4 above are the variation on the irreversible maintenance policy. In case 3 all equipment are checked at the end of the period sequentially. If it takes τ_r time to check one, the total idle time is $n\tau_r$ and according to Eq. 8.4.91, the availability A'_S would be

$$A_S = \frac{\tau}{\tau + n\tau_r} A_S^* \tag{8.6.14}$$

The mean availability $\mu_{A'_S}$ of the system would be

$$\mu_{A'_S} = \frac{\tau}{\tau + n\tau_r} \mu_{A_S^*} \tag{8.6.15}$$

However, in case of staggered testing and repairing, the situation is entirely different because at a given time of operation, different component in the system have different life times. As a result, probability of failure of different components are different which has to be properly accounted for in calculating its mean availability and fractional dead time. Staggered testing may or may not be better compared to non-staggered one, and the superiority of one over the other depends upon the functional relation of the probability of failure.

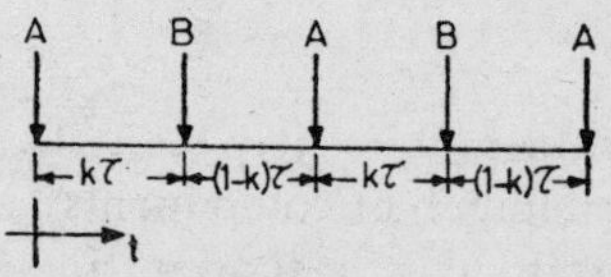

Fig. 8.9 An operation cycle of a system consisting of two equipments. Arrows show the time of repair

As earlier, one first considers the system consisting of only two equipment, A and B, in series. Figure 8.9 shows two cycles of operation of the system. Each equipment is checked and repaired after the cycle time τ, however, their times of repair are separated by $k\tau$ where k is a constant less than one. In the analysis given here it is assumed that the repair work makes the equipment as good as new. In a given cycle, there are two intervals and availability in both these intervals would be different. Supposing they are $\mu_{A_{S_1}}$ and $\mu_{A_{S_2}} \cdot \mu_{A_{S_1}}$ can be found from Eq. 8.6.4 to be

$$\mu_{A_{S_1}} = 1 - \frac{1}{k\tau}\int_0^{k\tau} \bar{p}_A(t)\, dt - \frac{1}{k\tau}\int_0^{k\tau} \bar{p}_B[t + (1-k)\tau]\, dt + \frac{1}{k\tau}\int_0^{k\tau} \bar{p}_A(t)\bar{p}_B[t + (1-k)\tau]\, dt \tag{8.6.16}$$

The probability $\bar{p}_B$ is evaluated at time $[t + (1-k)\tau]$ because the equipment B has been working for this duration of time. Similarly in the second interval, B starts as new at time $k\tau$ and therefore

$$\mu_{A_{S_2}} = 1 - \frac{1}{(1-k)\tau}\int_{k\tau}^{\tau} \bar{p}_A(t)\, dt - \frac{1}{(1-k)\tau}\int_{k\tau}^{\tau} \bar{p}_B(t - k\tau)\, dt + \frac{1}{(1-k)\tau}\int_{k\tau}^{\tau} \bar{p}_A(t)\bar{p}_B(t - k\tau)\, dt \tag{8.6.17}$$

Substituting in the above equation

$$t' = t - k\tau \tag{8.6.18}$$

one has

$$\mu_{A_{S2}} = 1 - \frac{1}{(1-k)\tau}\int_0^{(1-k)\tau} \bar{p}_A(t' + k\tau)\, dt' - \frac{1}{(1-k)\tau}\int_0^{(1-k)\tau} \bar{p}_B(t')\, dt' + \frac{1}{(1-k)\tau}\int_0^{(1-k)\tau} \bar{p}_A(t' + k\tau)\bar{p}_B(t')\, dt' \tag{8.6.19}$$

and the system has the average availability μ_{A_S} given by

$$\mu_{A_S} = k\mu_{A_{S1}} + (1-k)\mu_{A_{S2}} \tag{8.6.20}$$

and the mean fractional dead time μ_{D_S} by

$$\mu_{D_S} = 1 - \mu_{A_S} \tag{8.6.21}$$

As an example, the exponential probability of failure of equipment A and B is considered

$$\bar{p}_A = 1 - \exp(-\theta_a t) \tag{8.6.22a}$$

$$\bar{p}_B = 1 - \exp(-\theta_b t) \tag{8.6.22b}$$

Substituting these in Eqs. 8.6.16 and 8.6.19, μ_{D_S} is obtained as

$$\mu_{D_S} = 1 - \frac{1}{(\theta_a + \theta_b)\tau}[\{1 - \exp(-\theta_a\tau)\}\exp(-(1-k)\theta_b\tau) + \{1 - \exp(-\theta_b\tau)\}\exp(-k\theta_a\tau)] \tag{8.6.23}$$

For small θ_a and θ_b, it is

$$\mu_{D_S} = \frac{(\theta_a + \theta_b)}{2} \tag{8.6.24}$$

which is the same as that for two elements in series with the repair done at the end of the period. Thus it is seen that the system's availability and fractional dead time remain unchanged if probability is given by a linear function. However, for exponential function it indeed does, as seen from Table 8.4. The table has been generated by substituting $\theta_a\tau = \theta_b\tau = 1$ in Eq. 8.6.23. For these values Eq. 8.6.23 becomes

$$\mu_{D_S} = 1 - \tfrac{1}{2}[1 - e^{-1}][e^{-k} + e^{-(1-k)}] \tag{8.6.25}$$

Table 8.4 Calculation of Availability and Fractional Dead Time for $\theta_a\tau = \theta_b\tau = 1$ for Different Values of k

k	0	1/4	1/2	3/4	1
μ_{D_S}	0.57	0.60	0.62	0.60	0.57
μ_{A_S}	0.43	0.40	0.38 ↑ Minimum	0.40	0.43

The case of $k = 0$ (or equal to 1) corresponds to the situation when A and B are checked and repaired at the same time, i.e. non-staggered maintenance policy. Values of k other than these belong to the staggered policy and the mean fractional dead time for this is found to be higher than that for the non-staggered one. Thus all the maintenance work should be done at the end of the period when the probability of failure of equipment is exponential.

Now a pure series system undergoing *reversible maintenance policy* is considered. As an example a two-element system is analyzed first. To find the probability of system in successful state at time t, it is noted that

$$p_S(t) = p_A(t)p_B(t) \tag{8.6.26}$$

where $p_A(t)$ and $p_B(t)$ are probabilities of finding its components A and B in the working condition. These are given by Eq. 8.5.25 as

$$p_A(t) = L^{-1}\left[\frac{1 - F_{t_{fA}}(s)}{s[1 - F_{t_{fA}}(s)F_{t_{rA}}(s)]}\right] \tag{8.6.27a}$$

and

$$p_B(t) = L^{-1}\left[\frac{1 - F_{t_{fB}}(s)}{s[1 - F_{t_{fB}}F_{t_{rB}}(s)]}\right] \tag{8.6.27b}$$

Substituting these in Eq. 8.6.26, one gets

$$p_S(t) = L^{-1}\left[\frac{1 - F_{t_{fA}}}{s[1 - F_{t_{fA}}(s)F_{t_{rA}}(s)]}\right]L^{-1}\left[\frac{1 - F_{t_{fB}}(s)}{s[1 - F_{t_{fB}}(s)F_{t_{rB}}(s)]}\right] \tag{8.6.28}$$

The long time behaviour can be easily found using the limiting relation of Table 8.3 to obtain the mean availability of the system μ_{A_S} as

$$\mu_{A_S} = \frac{M_{t_{fA}}}{M_{t_{rA}} + M_{t_{fA}}} \cdot \frac{M_{t_{fB}}}{M_{t_{rB}} + M_{t_{fB}}}$$

$$= \mu_{Aa}\mu_{Ab} \tag{8.6.29}$$

The mean fractional dead time μ_{D_S} would consequently be

$$\mu_{D_S} = 1 - \mu_{A_S} = 1 - \mu_{Aa}\mu_{Bb} \tag{8.6.30}$$

These results can be easily generalized to n-component series system for mean availability and fractional dead time as

$$\mu_{A_S} = \prod_{i=1}^{n} \mu_{A_i} \tag{8.6.31a}$$

and

$$\mu_{D_S} = 1 - \prod_{i=1}^{n} \mu_{A_i} \tag{8.6.31b}$$

In this section, an n-component system in pure series has been analyzed for mean availability and fractional dead time. It was noted that a composite system could undergo irreversible or reversible maintenance policy. In the former, either staggered or non-staggered testing can be had. It was shown that for linear probability functions, both these policies were equivalent, otherwise they were not. For reversible change of state, the mean availability of the system is the product of the mean availability of individual components.

8.7 AVAILABILITY AND FRACTIONAL DEAD TIME OF A PURE PARALLEL SYSTEM

A system consisting of equipment in parallel differs from the one with equipment in series in that in the former all its components must fail for the system to be in the failed state. If $\bar{p}_s(t)$ is the probability of finding the system in the failed state, then

$$\bar{p}_S(t) = \prod_{i=1}^{n} \bar{p}_i(t) \tag{8.7.1}$$

where $\bar{p}_i(t)$ is the probability of finding its ith component in the failed state. As earlier, irreversible or reversible maintenance policy can be had for the system; in the following the irreversible policy is considered first. Equation 8.4.37 for the mean availability and Eq. 8.4.49 for mean fractional dead time hold good for parallel systems and one substitutes the expression for $\bar{p}_S(t)$ (Eq. 8.7.1) to obtain the result. As an example, a two component parallel system is considered whose component probabilities of failure are exponential

$$\bar{p}_A = 1 - \exp(-\theta_a t) \tag{8.7.2a}$$

and

$$\bar{p}_B = 1 - \exp(-\theta_b t) \tag{8.7.2b}$$

According to Eq. 8.7.1 $\bar{p}_S(t)$ is

$$\begin{aligned} \bar{p}_S(t) &= (1 - \exp(-\theta_a t))(1 - \exp(-\theta_b t)) \\ &= 1 - \exp(-\theta_a t) - \exp(-\theta_b t) + \exp\{-(\theta_a + \theta_b)t\} \end{aligned} \tag{8.7.3}$$

The plant availability would therefore be

$$\begin{aligned} \mu_{A_s} &= 1 - \frac{1}{\tau}\int_0^{\tau} [1 - \exp(-\theta_a t) - \exp(-\theta_b t) + \exp\{-(\theta_a + \theta_b)t\}]\, dt \\ &= \frac{1}{(\theta_a + \theta_b)}[1 - \exp\{-(\theta_a + \theta_b)\tau\}] - \frac{1}{\theta_a \tau}[1 - \exp(-\theta_a \tau)] \\ &\quad - \frac{1}{\theta_b \tau}[1 - \exp(-\theta_b \tau)] \end{aligned} \tag{8.7.4}$$

The average time of first failure, μ_f, would be given by

$$\begin{aligned} \mu_f &= \int_0^{\infty} p_S(t)\, dt \\ &= \int_0^{\infty} [1 - \{1 - \exp(-\theta_a t)\}\{1 - \exp(-\theta_b t)\}]\, dt \end{aligned} \tag{8.7.5}$$

which on integration gives

$$\mu_f = \frac{1}{\theta_a} + \frac{1}{\theta_b} - \frac{1}{(\theta_a + \theta_b)} \tag{8.7.6}$$

If probability of failure of a single component is a linear with equal hazard function, θ, in an n-component parallel system

$$\bar{p}_i(t) = \theta t \tag{8.7.7}$$

and

$$\bar{p}_S(t) = \theta^n t^n \tag{8.7.8}$$

The mean availability and fractional dead time are

$$\mu_{A_S} = 1 - \frac{1}{\tau}\int_0^\tau \theta^n t^n \, dt$$

$$= 1 - \frac{(\theta\tau)^n}{(n+1)} \tag{8.7.9}$$

and

$$\mu_{D_S} = \frac{(\theta\tau)^n}{(n+1)} \tag{8.7.10}$$

Table 8.5 gives values of μ_{D_S} for different $(\theta\tau)$ and n. For a given $\theta\tau$, as n increases, μ_{D_S} is found to reduce drastically. This means that the mean availability increases to a value almost equal to one.

Table 8.5 Calculation of Average Fractional Dead Time as a Function $\theta\tau$ and n

	μ_{D_S} for different $\theta\tau$		
n	$\theta\tau = 0.5$	$\theta\tau = 0.1$	$\theta\tau = 0.01$
0	1	1	1
1	2.5×10^{-1}	5×10^{-2}	5.0×10^{-3}
2	8.0×10^{-2}	3.3×10^{-3}	3.3×10^{-5}
3	4.1×10^{-2}	2.5×10^{-4}	2.5×10^{-7}
4	1.3×10^{-2}	2.0×10^{-5}	2.0×10^{-9}
5	5.0×10^{-3}	1.7×10^{-6}	1.7×10^{-11}

In the irreversible policy, as pointed out earlier, there are two kinds of maintenance policies: staggered and non-staggered. In the latter, all equipment are checked and repaired after the time period τ. If the checking is sequential and if it takes τ_r time to check a single component, the total idle time would be $n\tau_r$ for the system. The availability of the system is then given by

$$A' = \frac{\tau}{\tau + n\tau_r} A^* \tag{8.7.11}$$

and one can carry out similar corrections as done for a single equipment to find the mean availability. In the staggered testing and repairing, all components are assumed to be operated for time τ, but in a given cycle, the repairing is done at equal space of time τ_l as shown in Fig. 8.10. If

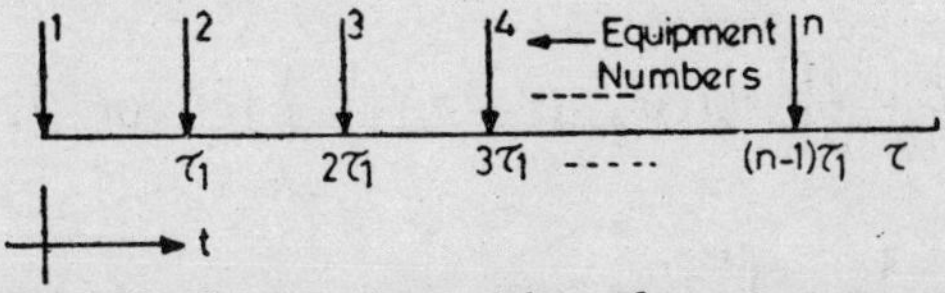

Fig. 8.10 Staggered repairing of an n-component system in parallel. Arrows show when the equipment was repaired

the system consists of n-components, the total operation consists of n such periods each of duration τ_1. The mean fractional dead time of the system would be given by

$$\mu_{D_S} = \frac{1}{\tau_1}\int_0^{\tau_1} \bar{p}_1(t)\bar{p}_2[t + (n-1)\tau_1]\ldots\bar{p}_n[t+\tau_1]\,dt$$
$$+ \frac{1}{\tau_1}\int_{\tau_1}^{2\tau_1} \bar{p}_1(t)\bar{p}_2[t-\tau_1]\ldots\bar{p}_n[t+2\tau_1]\,dt$$
$$+ \frac{1}{\tau_1}\int_{2\tau_1}^{3\tau_1} \bar{p}_1(t)\bar{p}_2[t-\tau_1]\bar{p}_2[t-2\tau_1]\ldots\bar{p}_n[t+3\tau_1]\,dt + \ldots \quad (8.7.12)$$

In the relation above, it is assumed that it takes a much shorter time to test and repair compared to time τ.

Since it is a parallel system, there is yet another maintenance policy possible: without stopping the system, one checks and repairs the components of the system sequentially at the end of cycle time τ. This is possible because at a given time, only one of the components needs to work for the successful operation of the system. As will be seen now, this is a better maintenance policy. It is assumed that each element takes τ_r time to repair and the first element is repaired between time τ and $(\tau + \tau_r)$, second between $(\tau + \tau_r)$ and $(\tau + 2\tau_r)$ and so on till the last element which is repaired between time $[\tau + (n-1)\tau_r]$ and $(\tau + n\tau_r)$. The total operation cycle in this case is really spread over time $(\tau + n\tau_r)$ and therefore in Eq. 8.4.49 one should use $(\tau + n\tau_r)$ instead of τ. Further, between time 0 to τ, one has n-components working, between time τ and $(\tau + \tau_r)$ there are only $(n-1)$ components working because the first one is being repaired and so on. In view of this, the probability of failure $\bar{p}_S(t)$ of the system is given by

$$\bar{p}_S(t) = \begin{cases} \bar{p}_n(t)\bar{p}_{n-1}(t+\tau_r)\ldots\bar{p}_2[t+(n-2)\tau_r]\bar{p}_1[t+(n-1)\tau_r] & \text{for } 0 < t < \tau \\ \bar{p}_n(t)\bar{p}_{n-1}(t+\tau_r)\ldots\bar{p}_2[t+(n-2)\tau_r] & \text{for } \tau < t < (\tau+\tau_r) \\ \bar{p}_n(t)\bar{p}_{n-1}(t+\tau_r)\ldots\bar{p}_3[t+(n-3)\tau_r]\bar{p}_1[t-(\tau+\tau_r)] & \text{for } (\tau+\tau_r) < t < [\tau+2\tau_r] \\ \vdots & \vdots \\ \bar{p}_{n-1}[t-\{\tau+(n-1)\tau_r\}]\ldots\bar{p}_2[t-(\tau+2\tau_r)] \times \bar{p}_1[t-(\tau+\tau_r)] & \text{for } [\tau+(n-1)\tau_r] < t < [\tau+n\tau_r] \end{cases} \quad (8.7.13)$$

Substituting these in Eq. 8.4.49, the average fractional dead time is obtained as

$$\mu_{D_S} = \frac{1}{(\tau + n\tau_r)} \int_0^{\tau} \bar{p}_n(t)\bar{p}_{n-1}(t + \tau_r) \ldots \bar{p}_2[t + (n-2)\tau_r] \times \bar{p}_1[t + (n-1)\tau_r]\, dt$$
$$+ \frac{1}{(\tau + n\tau_r)} \int_{\tau}^{\tau+\tau_r} \bar{p}_n(t)\bar{p}_{n-1}(t + \tau_r) \ldots \bar{p}_2[t + (n-2)\tau_r]\, dt$$
$$+ \frac{1}{(\tau + n\tau_r)} \int_{\tau+\tau_r}^{\tau+2\tau_r} \bar{p}_n(t)\bar{p}_{n-1}(t + \tau_r\} \ldots \bar{p}_3[t + (n-3)\tau_r] \times \bar{p}_1[t - (\tau + \tau_r)]\, dt + \ldots \tag{8.7.14}$$

As an example one considers a linear probability function

$$\bar{p}_j(t) = \theta t \tag{8.7.15}$$

and on substitution in Eq. 8.7.14 the mean availability is found to be

$$\mu_{D_S} = \frac{1}{(\tau + n\tau_r)} \left[\int_0^{\tau} \theta^n t(t + \tau_r)(t + 2\tau_r) \ldots [t + (n-1)\tau_r]\, dt + \int_{\tau}^{\tau+\tau_r} \theta^{n-1} t(t + \tau_r) \ldots [t + (n-2)\tau_r]\, dt + \ldots \right] \tag{8.7.16}$$

The integration can be performed term by term by expanding the polynomial. When $n\tau_r \ll \tau$, the above equation becomes

$$\mu_{D_S} = \frac{1}{\tau} \int_0^{\tau} \theta^n t^n\, dt + \frac{1}{\tau} \int_{\tau}^{\tau+\tau_r} \theta^{n-1} t^{n-1}\, dt + \int_{\tau+\tau_r}^{\tau+2\tau_r} \theta^{n-1} t^{n-2} [t - (\tau + \tau_r)\, dt + \ldots \tag{8.7.17}$$

The first two terms in the above equation are significant. Neglecting others, one has

$$\mu_{D_S} = \frac{\theta^n \tau^n}{(n+1)} + \theta^{n-1}\tau^{n-2}\tau_r \tag{8.7.18}$$

For $\tau_r \to 0$, this becomes the same as Eq. 8.7.10.

Finally the reversible policy of maintenance is analyzed for the parallel system. For individual equipment, the probability of failure is given either by Eq. 8.5.22 or 8.5.51 depending upon the nature of the repair work. For the system, however, the probability of failure is given by

$$\bar{p}_S(t) = \mathop{\pi}_{i=1}^{n} \bar{p}_i(t) \tag{8.7.19}$$

Assuming the case of complete renewal (which is equivalent to saying that the repair work makes the equipment as good as new), Eq. 8.5.22 is substituted above to obtain

$$\bar{p}_S(t) = \mathop{\pi}_{i=1}^{n} L^{-1} \left[\frac{1}{s} \frac{F_{t_{fi}}(s)[1 - F_{t_{ri}}(s)]}{1 - F_{t_{ri}}(s) F_{t_{fi}}(s)} \right] \tag{8.7.20}$$

The behaviour of the system for long times can easily be found out. The asymptotic value of fractional dead time of the system would be given by

$$\mu_{D_S} = \prod_{i=1}^{n} \mu_{D_i} = \prod_{i=1}^{n} \frac{M_{t_{ri}}}{M_{t_{ri}} + M_{t_{fi}}} \tag{8.7.21}$$

where $M_{t_{ri}}$ and $M_{t_{fi}}$ have been defined earlier. The mean availability of the system would therefore be

$$\mu_{A_S} = 1 - \mu_{D_S} = 1 - \prod_{i=1}^{n} \frac{M_{t_{ri}}}{M_{t_{ri}} + M_{t_{fi}}} \tag{8.7.22}$$

8.8 MEAN AVAILABILITY OF MIXED CONFIGURATIONS: VOTING SYSTEM

In the last two sections, series and parallel configurations were considered which have the lowest and the highest availabilities for a given number of components in the system. The availability for any configuration other than these would lie between these limits. As an example, for a nuclear reactor whose temperature has to be controlled with great precision, it would be best to have more than one thermocouple arranged in parallel. However, in practice, it is found that the reactor tends to have several "spurious" shutdowns[3, 4]. The larger the number of sensors, the larger the number of such shutdowns and to suppress these, voting system of sensors are used. This is described below.

A system consisting of n independent components are considered. The voting system of m-out-of-n $(m < n)$ is defined to be in the successful state when any m components are in the working condition. Figure 8.11

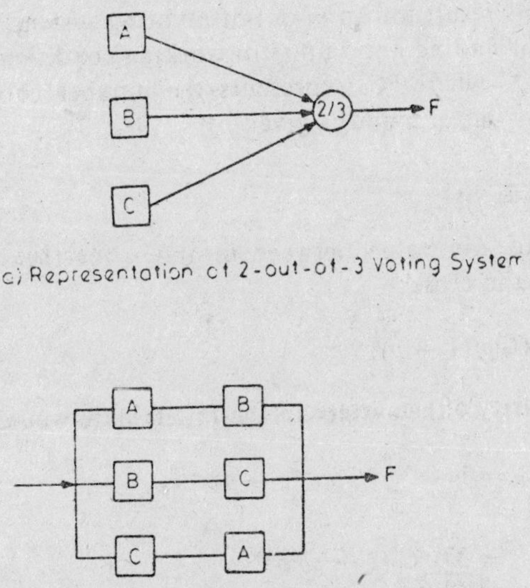

(b) Logical Arrangement of the System

Fig. 8.11 Representation of 2-out-of-3 voting system

shows a 2-out-of-3 voting system as an example. From Fig. 8.11b, one can write its Boolean function F as

$$F = AB + BC + CA \tag{8.8.1}$$

and it is seen that F is true when any two of the variables A, B and C are true. It may be recalled that functions AB, BC and CA are not mutually exclusive, even though variables A, B and C are, which can be confirmed through the mapping method described in Chapter 6. Therefore this is written in the non-overlapping form as

$$\begin{aligned} F &= AB(C + \bar{C}) + BC(A + \bar{A}) + CA(B + \bar{B}) \\ &= ABC + AB\bar{C} + ABC + \bar{A}BC + ABC + A\bar{B}C \end{aligned} \tag{8.8.2}$$

After removing redundancies, function F reduces to

$$F = ABC + AB\bar{C} + A\bar{B}C + \bar{A}BC \tag{8.8.3}$$

If p_a, p_b and p_c are probabilities of having components A, B and C in successful states, then the probability of the system, in the successful state, p_S, would be

$$\begin{aligned} p_S &= p_a p_b p_c + p_a p_b(1 - p_c) + p_a(1 - p_b)p_c + (1 - p_a)p_b p_c \\ &= p_a p_b + p_b p_c + p_c p_a - 2p_a p_b p_c \end{aligned} \tag{8.8.4}$$

If all these probabilities are equal

$$p_a = p_b = p_c = p \text{ (say)}, \tag{8.8.5}$$

Equation 8.8.4 becomes

$$p_S = 3p^2 - 2p^3 \tag{8.8.6}$$

To generalize this result for an m-out-of-n voting system, it is noted that the probability of finding any j units in working condition out of n would be ${}^nC_j p^j(1 - p)^{n-j}$ where ${}^nC_j$ represents the number of possible combinations of j units out of n and is given by

$${}^nC_j = \frac{n!}{j!\,(n - j)!} \tag{8.8.7}$$

Since j can take any value between m and n for the system to be in successful state, therefore

$$p_S = \sum_{j=m}^{n} {}^nC_j\, p^j(1 - p)^{n-j} \tag{8.8.8}$$

and the probability of the system in the failed state would be

$$\begin{aligned} \bar{p}_S &= 1 - p_S = 1 - \sum_{m}^{n} {}^nC_j\, p^j(1 - p)^{n-j} \\ &= \sum_{j=r}^{n} {}^nC_j(1 - p)^j p^{n-j} \end{aligned} \tag{8.8.9}$$

where

$$r = (n - m + 1) \tag{8.8.10}$$

When the probability of success, p, is close to one, it is possible to approximate $\bar{p}_S$ as

$$\bar{p}_S \cong {}^nC_r(\bar{p})^r \tag{8.8.11}$$

where

$$\bar{p} = (1 - p) \tag{8.8.12}$$

One first considers the system undergoing irreversible maintenance policy and assumes a complete renewal after the period of the cycle τ. If all the repairs are done at the end of the duration and it takes negligible time to repair, Eqs. 8.4.37 and 8.4.49 are used for mean availability and fractional dead time. If a linear probability function is assumed

$$\bar{p}_i = \theta t \tag{8.8.13}$$

the $\bar{p}_S$ for *m*-out-of-*n* voting system would be given by

$$\bar{p}_S = {}^nC_r\theta^r t^r \tag{8.8.14}$$

Substituting this in Eq. 8.4.49,

$$\mu_{D_S} = \frac{{}^nC_r\theta^r}{\tau}\int_0^\tau t^r\,dt$$

$$= {}^nC_r\frac{(\theta\tau)^r}{(r+1)} \tag{8.8.15}$$

Since $(\theta\tau)$ is less than one, on comparison of the above equation with Eq. 8.7.10 it is found that it has a higher fractional dead time compared to that for pure parallel system.

Next, the maintenance policy, in which repair is done for individual components sequentially after the cycle period τ, is analyzed. As an example, in the analysis of the 2-out-of-3 voting system, it is assumed that the first component is repaired at time τ, the second component at $(\tau + \tau_r)$ and the third component at $(\tau + 2\tau_r)$. During the regular operation (i.e. between time 0 to τ), the three elements have different lifetimes which must be accounted for in calculating the fractional dead time of the system. For 2/3 voting system (symbolic representation of the two-out-of-three voting system), the probability of successful operation is given by Eq. 8.8.4 and the probability of finding the system in the failed state, $\bar{p}_S$ would be found by negating it

$$\bar{p}_S = \bar{p}_a\bar{p}_b + \bar{p}_b\bar{p}_c + \bar{p}_c\bar{p}_a - 2\bar{p}_a\bar{p}_b\bar{p}_c \tag{8.8.16}$$

Even though all components have the same function for probability of failure (Eq. 8.8.13), yet Eq. 8.8.15 cannot be used due to the fact that at any given time, their lifetimes are different. In other words, $p_S(t)$ between time 0 to τ is given by

$$\begin{aligned}\bar{p}_S(t) = {} & \bar{p}(t)\bar{p}(t+\tau_r) + \bar{p}(t+\tau_r)\bar{p}(t+2\tau_r) \\ & + \bar{p}(t)\bar{p}(t+2\tau_r) - 2\bar{p}(t)\bar{p}(t+\tau_r)\bar{p}(t+2\tau_r)\end{aligned} \tag{8.8.17}$$

Substituting Eq. 8.8.13 above, one gets

$$\begin{aligned}\bar{p}_S &= \theta^2 t(t+\tau_r) + \theta^2(t+\tau_r)(t+2\tau_r) \\ &\quad + \theta^2 t(t+2\tau_r) - 2\theta^3 t(t+\tau_r)(t+2\tau_r) \\ &= \theta^2[3t^2 + 6t\tau_r + 2\tau_r^2] + 2\theta^3[t^3 + 3t^2\tau_r + 2\tau_r^2 t]\end{aligned} \tag{8.8.18}$$

The time period of the cycle above is $(\tau + 3\tau_r)$. If τ_r is much smaller than τ, one considers the 0 to τ period and neglects the period between τ to $(\tau + 3\tau_r)$ as a first approximation to obtain the mean fraction dead time as

$$\begin{aligned}\mu_{D_S} &\simeq \frac{1}{\tau + 3\tau_r}\int_0^\tau \bar{p}_S(t)\,dt \\ &= \frac{1}{(\tau + 3\tau_r)}\left[(\tau^3 + 3\tau^2\tau_r + 2\tau_r^2) + 2\theta^3\left(\frac{\tau^4}{4} + \tau^3\tau_r + \tau_r^2\tau^2\right)\right]\end{aligned} \tag{8.8.19}$$

When an m/n voting system undergoes a *reversible change* of state, the procedure of analysis remains identical. Since the performance of all equipment is mutually exclusive probability p_i of an ith equipment in successful state would therefore be

$$p_i = L^{-1}\left[\frac{F_{t_{fi}}(s)[1 - F_{t_{ri}}(s)]}{s[1 - F_{t_{ri}}(s)F_{t_{fi}}(s)]}\right] \tag{8.8.20}$$

and the probability of being in the failed state would be

$$\bar{p}_i = L^{-1}\left[\frac{1 - F_{t_{fi}}(s)}{s[1 - F_{t_{fi}}(s)F_{t_{ri}}(s)]}\right] \tag{8.8.21}$$

A Boolean function for the system is next developed in terms of the mutually exclusive events as was done for 2/3 voting system in Eq. 8.8.3. After that, Eqs. 8.8.20 and 8.8.21 are substituted in the probability relation developed from it. One now considers a special case when all components of the system are identical. On proper integration of Eq. 8.8.9, the mean availability and fractional dead time are found. Even though this problem can be solved for specified distribution functions for the system's failure and repair time, the algebra for a general system is complex. However, the asymptotic result for large times can be easily found. It is noted that for a given equipment for large times, the availability of a single equipment would be

$$\lim_{t\to\infty} \mu_{A_i} = \lim_{t\to\infty} p_i = \frac{M_{t_f,i}}{M_{t_r,i} + M_{t_f,i}} \tag{8.8.22a}$$

and

$$\lim_{t\to\infty} \mu_{D_i} = \lim_{t\to\infty} \bar{p}_i = \frac{M_{t_r,i}}{M_{t_r,i} + M_{t_f,i}} \tag{8.8.22b}$$

The mean availability of the system is obtained by substituting these in

Eq. 8.8.8

$$\mu_{A_S} = \lim_{t\to\infty} p_S = \sum_{j=m}^{n} {}^nC_j \mu_A^j (1-\mu_A)^{n-j} \tag{8.8.23}$$

and the mean fractional dead time is given by

$$\mu_{D_S} = 1 - \mu_{A_S} = \lim_{t\to\infty} \bar{p}_S$$

$$= \sum_{j=r}^{n} {}^nC_j \mu_D^j \mu_A^{n-j} \tag{8.8.24}$$

In this section a more complex configuration of voting system has been analyzed for different maintenance policies. It is noted that the information required for calculating the mean availability and fractional dead time is the probability of failure. If the fault tree analysis (Chapter 6) of a system is known, the Boolean function for it can be written from which the system's probability of failure is found. Information developed in this chapter is next applied to a practical example of a high-pressure-trip system for the protection of a pressure vessel.

8.9 DESIGN OF HIGH-PRESSURE-TRIP SYSTEM OF A FACTIONATOR[7]

As discussed in Chapter 5, all distillation columns are equipped with safety valves which release the content of the column on development of excessive pressure. The same can be attained through an arrangement which cuts off all inputs to the column, which checks the pressure rise immediately. This arrangement is called the trip-system. In a reactor where there is a possibility of runaway reaction, the trip system cuts the raw material inputs whereas in a fractionator with excessive pressure buildup, it cuts off the feed along with the heat input into it.

Before taking up a specific design problem, a trip system is compared with safety valves[8]. Table 8.1 gives the failure rate for a safety valve under blockage as 0.5 faults per 10^6 hours. The probability of failure is approximated as

$$\bar{p} = \theta t \tag{8.9.1}$$

where θ for the safety valve,

$$\theta = 0.5 \times 10^{-6} \frac{\text{faults}}{\text{h}} \frac{24 \times 365\ \text{h}}{\text{yr}}$$

$$= 0.0044\ \text{faults/year} \tag{8.9.2}$$

The fractional dead time μ_D is therefore given by

$$\mu_D = \frac{1}{\tau} \int_0^\tau \theta t\, dt = (\theta\tau/2) \tag{8.9.3}$$

If it is decided to inspect the valve yearly, τ will be equal to one year and

$$\mu_D = \frac{0.0044 \times 1}{2} = 0.0022 \tag{8.9.4}$$

One assumes that the column gets overpressurized once in five years or 0.2 times/year. Since every occurrence of overpressure does not lead to a hazardous situation, the hazard rate h_S is given by

$$\begin{aligned} h_S &= (\text{rate of over-pressure occurrence})\mu_D \\ &= 0.2 \times 0.0022 \\ &= 4.4 \times 10^{-4} \text{ faults/year} \\ &= 1 \text{ in } 2270 \text{ years (on an average)} \end{aligned} \tag{8.9.5}$$

If the safety valve is inspected once in two years it is seen from Eqs. 8.9.4 and 8.9.5, that the hazard rate, on an average, is doubled or 1 in 1135 years. If it is done once in 10 years, it becomes 1 in 227 years, and in this way, the importance of the maintenance work can be seen.

The trip system is extremely important for those chemicals which cannot be permitted in atmosphere under any circumstances, like for a nuclear reactor. This is because the trip system automatically cuts off the incoming raw materials to control the runaway situation. In spite of its great advantage, trip systems have higher chances of failure. However, it is possible to tinker with the system's configuration so as to match the frequency of hazard with that of the safety valve as shown below.

The simplest structure of the trip system consists of a pressure switch connected to a motor valve. A motor valve is a valve operated by a motor, and is placed on the pipeline which is desired to be isolated at the time of emergency. A system of this nature has about one fault in $1\frac{1}{2}$ years,[8] or

$$\theta = \frac{1}{3/2} = 0.67 \text{ per year} \tag{8.9.6}$$

If a regular check of the trip system is made once every week (i.e. $\tau = 1$ week), the mean fractional dead time for this would be

$$\begin{aligned} \mu_D = \tfrac{1}{2}\theta\tau &= \frac{0.67 \times 7}{2 \times 365} \\ &= 6.45 \times 10^{-3} \end{aligned} \tag{8.9.7}$$

If the occurrence of the rate of over-pressure is the same as 0.2 per year, the hazard rate h_t for this system would be

$$h_t = 0.2 \times 6.45 \times 10^{-3} = 1.29 \times 10^{-3} \text{ per year} \tag{8.9.8}$$

This would amount to having one accident in 783 years. If daily check up of the system is made, it is seen that the fractional dead time is reduced by a factor of 7 and have, on an average, one accident in 5470 years. This is better than a safety valve but the daily check is not always possible. It is possible to improve upon the rate of occurrence of accidents by installing two such trip systems in parallel even when $\tau = 1$ week. The fractional dead time of the parallel system can be found using Eq. 8.7.10 ($\tau = 1$ week) as

$$\mu_{D_2} = \frac{1}{3}\left(\frac{0.67 \times 7}{365}\right)^2 = 5.5 \times 10^{-5} \tag{8.9.9}$$

If the rate of pressure rise is the same as 0.2 per year, the hazard rate, h_{t2}, would be

$$\begin{aligned} h_{t2} &= 0.2 \times 5.5 \times 10^{-5} \\ &= 1.1 \times 10^{-5} \text{ faults/year} \end{aligned} \tag{8.9.10}$$

This would mean the occurrence of one fault in 91,000 years when weekly checks are made.

Now one considers a more difficult problem of controlling pressure in the fractionator of Fig. 8.12. In the column; the feed is heated by hot water

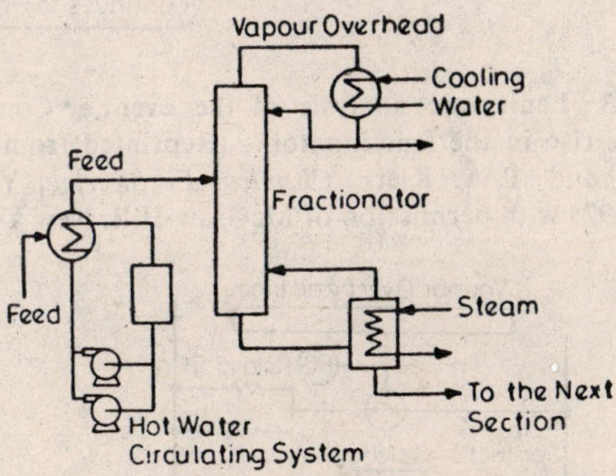

Fig. 8.12 A fractionator with its accessories.[7] (Reprinted from H. G. Lawley and T. A. Kletz, *Chemical Engineering*, Vol. 81, May 1975 with the permission of McGraw-Hill, New York.)

whereas the reboiler gets its heat through steam as shown. The head event of pressure rise in the column is analyzed. Figure 8.13 gives the fault tree analysis of the situation and from this, it is found that the primary events giving rise to this situation are the supply of uncontrolled heat to the reboiler and heating water to the feed. In view of this, a trip system is installed on both these as shown in Fig. 8.14.

In Fig. 8.14, the pressure switch, which is the sensing unit, is installed on the exit vapour line. It gives the trip signal in case of an excessive pressure buid-up. The electrical signal thus generated acts upon the pump contractor as well as a solenoid valve which opens the air supply to the pneumatic valve installed on the steam supply line to the reboiler. The study of the trip system given in the figure reveals that it has three subsections:

1. trip initiator section,

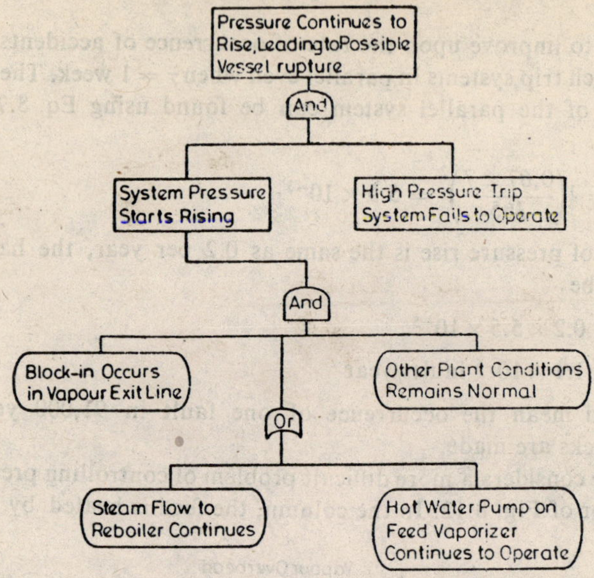

Fig. 8.13 Fault tree analysis of the event, "Continued pressure rise in the fractionator". (Reprinted from H. G. Lawley and T. A. Kletz, *Chemical Engineering*, Vol. 81, May 1975 with permission of McGraw-Hill, New York.)

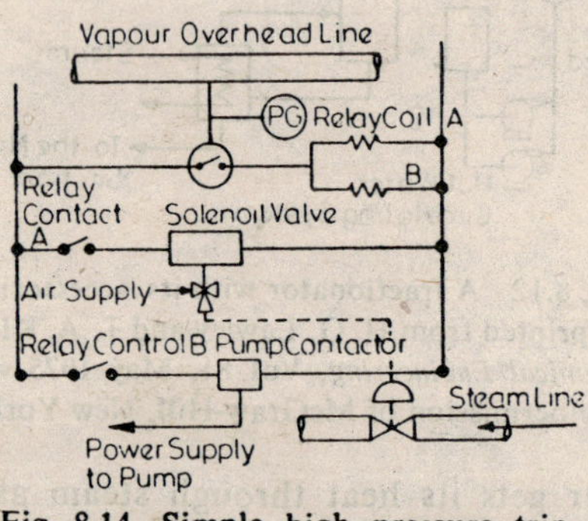

Fig. 8.14 Simple high pressure trip system. (Reprinted from H. G. Lawley and T. A. Kletz, *Chemical Engineering*, May 1975 with permission of McGraw-Hill, New York.)

2. steam shutoff section, and
3. pump shutoff section.

Each of these sections can fail individually for various reasons and the rates of failure due to these are given in Table 8.6. For example, in the trip initiator section, the pressure switch is directly connected to the vapour overhead line through an impulse line and the pressure switch would fail to detect the pressure rise in the fractionator when this line is

Table 8.6[7] Assumed Equipment Fault Rates

Type of Faults	Total Rate Faults/Year θ_i	Fail-Danger Faults/Year	Fail-Safe Faults/Year
Trip Initiator			
1. Impulse-line blockage	0.03	0.03	0.00
2. Impulse-line leakage	0.06	0.06	0.00
3. Pressure switch (contacts open to give trip on rising pressure)	0.13	0.10	0.03
4. Cable fractured or severed	0.03	0.00	0.03
5. Loss of electrical supply	0.05	0.00	0.05
6. Total for single initiator ($\Sigma\ \theta_i$)	0.30	0.19	0.11
Steam Shutoff System			
7. Relay coil (de-energize to trip)	0.05	0.00	0.05
8. Relay contact	0.02	0.01	0.01
9. Relay terminals and wires	0.01	0.00	0.01
10. Solenoid valve (de-energize to trip)	0.30	0.10	0.20
11. Loss of power supply to solenoid valve circuit	0.05	0.00	0.05
12. Trip valve (closes on air failure)	0.25	0.10	0.15
13. Blocked or crushed air line	0.01	0.01	0.00
14. Fractured or holed air line	0.01	0.00	0.01
15. Loss of air supply	0.05	0.00	0.05
16. *Total* for single shut off system ($\Sigma_i\ \theta_i$)	0.75	0.22	0.53
Pump Shutoff System			
17. Relay coil, contact, terminals and wires (as above)	0.08	0.01	0.07
18. Total for single shutoff system ($\Sigma_i\ \theta_i$)	0.08	0.01	0.07

(Reprinted from H. G. Lawley and T. A. Kletz, *Chemical Engineering*, May 1975 with permission of McGraw-Hill Publishing Co., New York.)

blocked *or* it is leaking, causing the loss of pressure at the switch. The pressure switch can also fail to give the electric signal due to its own fault *or* it may not be connected properly to the cable transferring the signal, and so on.

In Table 8.6, fault rates in terms of number of faults per year are given against each primary event that leads the specific section to the state of failure. Since not all failures lead to dangerous conditions, the total failure rates have been further classified in the next two columns as fail danger and fail safe situations. Because one is interested in finding out the occurrence of hazardous situations, therefore the fail-safe rates are discounted from the total failure rates and the values of θ_i given in the 'fail-danger' column are used.

The components leading to failure of any section have all been arranged in pure series. Therefore if the probability of a component in successful state is exponential, the probability of the whole section in the failed state is given by Eq. 8.6.9. When $(\sum \theta_i)$ is small compared to 1, it can be approximated by a linear function given in Eq. 8.9.10 and in 6th, 16th and 18th rows of Table 4.6, $(\sum \theta_i)$ have been calculated for each section.

Now one proposes the following maintenance policy for the trip system. After a regular interval of time τ, the trip initiator system is isolated and thoroughly examined. *At the same time* the remainder of the sections are also examined, so that after the examination is over, the system starts *as good as new.* If it takes about one hour to finish this job (or 1×10^{-4} year), it represents the repair time τ_r. The probability of failure of the system, $\bar{p}_S$, according to Eq. 8.6.10, is

$$\bar{p}_S(t) = (0.19 + 0.22 + 0.01)t$$

$$= 0.42t \tag{8.9.11}$$

When $\tau_r = 0$, the fractional dead time of the system, $\mu_{D_S^*}$, is given by

$$\mu_{D_S^*} = \frac{1}{\tau}\int_0^\tau \bar{p}_S(t)\,dt = \tfrac{1}{2}0.42\tau$$

$$= 0.21\tau \tag{8.9.12}$$

When this equation is corrected for the dead time arising due to τ_r, Eq. 8.4.99 is used, i.e.

$$\mu_{D_S} = \frac{\tau_r}{\tau + \tau_r} + \frac{\tau}{\tau + \tau_r}\mu_{D_S^*} \tag{8.9.13}$$

If $\tau_r \ll \tau$, the above relation can be approximated as

$$\mu_{D_S} = \frac{\tau_r}{\tau} + \mu_{D_S^*} \tag{8.9.14}$$

Or, the mean fractional dead time for the trip system would be

$$\mu_{D_S} = \frac{1 \times 10^{-4}}{\tau} + 0.21\tau \tag{8.9.15}$$

From above, it is seen that there is a value of τ for which μ_{D_S} undergoes a minimum and is obtained by setting $\frac{\partial}{\partial \tau}\mu_{D_S}$ equal to zero as follows:

$$\frac{\partial \mu_{D_S}}{\partial \tau} = -\frac{1 \times 10^{-4}}{\tau^2} + 0.21 = 0 \tag{8.9.16}$$

or

$$\tau_{opt} = 10^{-2}\sqrt{\frac{1}{0.21}} = 2.18 \times 10^{-2} \text{ years}$$

$$\simeq 8 \text{ days} \tag{8.9.17}$$

Substituting this in Eq. 8.9.15, the minimum mean fractional dead time is calculated as

$$\mu_{D_S} = 0.21 \times 2.18 \times 10^{-2} + \frac{1 \times 10^{-4}}{2.18 \times 10^{-2}}$$

$$= 0.00916 \qquad (8.9.18)$$

If it is desired that the mean fractional dead time of the trip system is better than 0.001, the simple trip system of Fig. 8.14 will not work because of its higher value of μ_{D_S}. To meet the higher reliability required, one suggests the one-out-of-two cross connected configuration of the trip system given in Fig. 8.15a. In this, there are two pressure switches and any *one* of them is independently capable of tripping the power supply to the circulating pump as well as shutting off any one of the two steam line valves. A careful analysis of the figure reveals that there are two trip initiators in parallel followed by a two steam shut off section in parallel followed by two pump shut off sections in parallel. These three sections, however, remain in series as shown in Fig. 8.15b.

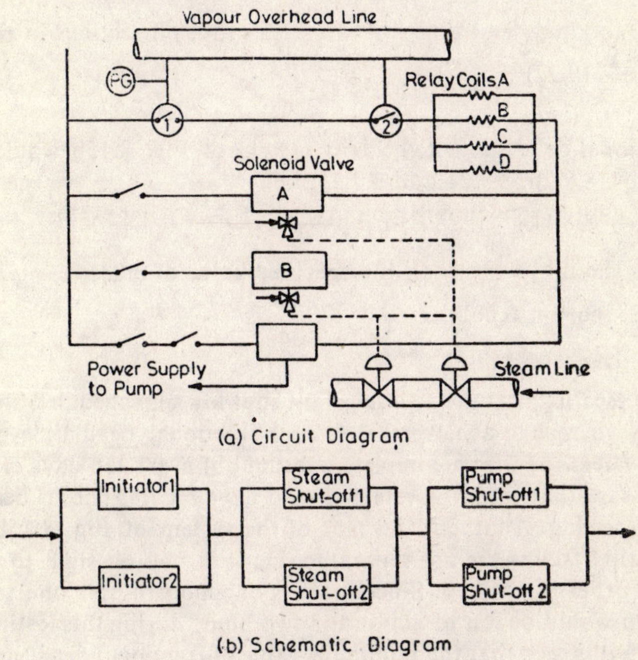

Fig. 8.15 One-out-of-two cross connected high pressure trip systems. (Reprinted from H. G. Lawley and T. A. Kletz, *Chemical Engineering*, May 1975 with permission of McGraw-Hill, New York.)

With the trip system of Fig. 8.15 installed, it is also proposed to have general plant shutdown after every 16 weeks when everything is thoroughly checked for the general overhaul of the plant. The question that needs to be answered is whether one can do this with the trip system proposed and at the same time meet the requirement of the average fractional dead time of 0.001. To find this out, it is assumed that the general plant shutdown occurs after time τ (it would be checked later that $\tau = 16$ weeks).

For a parallel system with two components, the average fractional dead time is obtained by substituting $n = 2$ in Eq. 8.7.10, or

$$\mu_{D_S} = \frac{(\theta\tau)^2}{3} \tag{8.9.19}$$

Using this relation, the fractional dead time, $\mu_{D,I}$, for the initiator section is calculated as

$$\mu_{D,I} = \frac{(0.19)^2\tau^2}{3} = 0.012033\tau^2 \tag{8.9.20}$$

Similarly, the average fractional dead time, $\mu_{D,P}$ for the pump shut off section is

$$\mu_{D,P} = \frac{(0.01)^2\tau^2}{3} = 3.33\times10^{-5}\tau^2 \tag{8.9.21}$$

The probability of failure, $\bar{p}_S$, for steam shut off section is given by

$$\bar{p}_S = 0.22t \tag{8.9.22}$$

and the fractional dead time for two steam shut off sections in parallel is

$$\mu_{D,S} = \frac{(0.22)^2\tau^2}{3} = 0.0161\tau^2 \tag{8.9.23}$$

The fractional dead time of the trip system of Fig. 8.15b would be the sum of Eqs. 8.9.20, 8.9.21 and 8.9.23, i.e.

$$\mu_{D_S} = (0.012033 + 0.0161 + 0.000033)\tau^2 = 0.028166\tau^2 \tag{8.9.24}$$

Since this should be less than the specified value of 0.001,

$$0.028166\tau^2 < 0.001$$

$$\tau < 1.42 \text{ weeks} \tag{8.9.25}$$

It may be recalled that the general plant shutdown is scheduled after every 16 weeks, whereas to achieve the desired fractional dead time of 0.001, the trip system requires a complete overhaul of every ten days. Therefore it is necessary that the trip system should have regular checks besides the shutdown testing. Another drawback of the system of Fig. 8.15b is that the two trip systems are not separable, i.e. it is not possible to test one while the other is working. Since it takes one hour to test one trip unit, the system would be out of action for two hours during this testing. From Eq. 8.9.13 it is seen that the minimum value of fractional dead time would occur when the system is guaranteed to work during its operation and its value is obtained as (τ_r/τ) by substituting $\mu_{D_S^*} = 0$. Therefore for the required fractional dead time of 0.001, in a year (=8760 hours) one can spend *almost* 0.001×8760 or 8.76 hours in on-line testing or in all, 4 testings during the plant operation. Thus too many on-line tests also cannot be made to meet the requirement of the fractional dead time of 0.001.

Before proceeding any further, it would be desirable to discuss the difference between on-line testing and the shutdown testing. The system involves a valve on the steam line, and to test a valve two things must be found:

1. does the valve respond to the air pressure, and
2. does the valve leak when it is tightly shut.

The former can occur when the valve is seized, the spring is damaged or the actuator vent is blocked whereas the latter can occur when the valve plug is blocked or damaged or the valve seat is corroded. It is possible to check the response of the valve to the air pressure during the on-line testing whereas the latter is not possible because one cannot afford to stop the steam flow during the plant operation. This can be done *only* during the shutdown testing. The total valve fail-danger rate is assumed to be 0.10 faults/year out of which 0.07 faults/year is due to failure of valve to respond to the air pressure and 0.03 faults/year due to its inability to shut completely.

The inadequacy of the configuration of Fig. 8.15 for the trip system has already been pointed out earlier and it is in view of this, that its structure is improved in Fig. 8.16. It consists of two pressure switches,

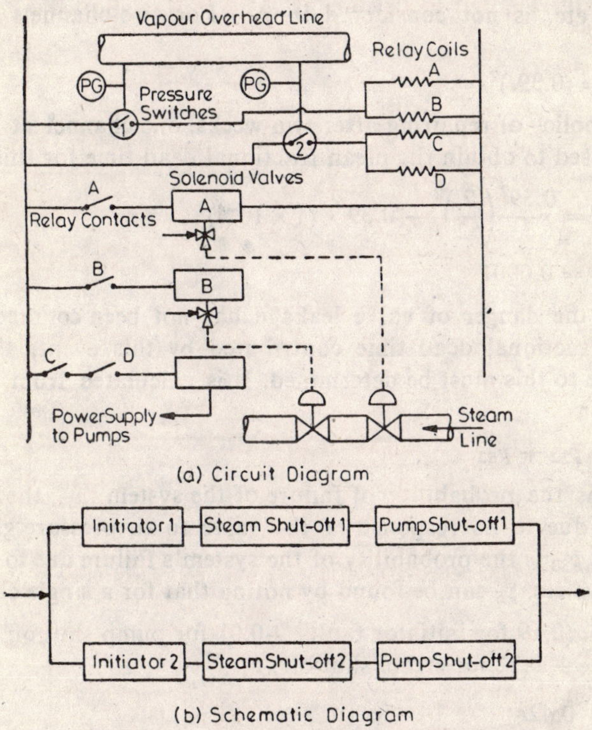

Fig. 8.16 Parallel configuration of the trip system. (Reprinted from H. G. Lawley and T. A. Kletz, *Chemical Engineering*, May 1975, with permission of McGraw-Hill, New York.)

two solenoid valves, two relay contacts for pump shut offs and two steam valves arranged in pure parallel configuration shown in Fig. 8.16b.

For the trip system of Fig. 8.16b, the following maintenance policy is adopted. There would be a general plant shutdown after every 16 weeks. During this time, the fractionator is inoperative and in view of this, it would not contribute to the fractional dead time of the trip system. This is because there is no danger of over-pressure during this time, which means $(\sum\theta)$ for this period would strictly be equal to zero. During the regular operation of the plant, there is an on-line testing after every two weeks. During this testing, one examines for faults in

1. trip initiator section,
2. pump shut-off sections, and
3. response of the valve towards air pressure.

The probability of failure associated with these events for a *single channel* is

$$\bar{p}_{S1} = [0.19 \text{ for initiator faults} + 0.01 \text{ for pump shut off} + (0.22 - 0.03) \text{ for steam shut off}]\ t \quad (8.9.26)$$

Above 0.03 has been subtracted because the failure of the valve due to corrosion etc. is not considered in $\bar{p}_{S1}$. For two-channels in parallel, therefore

$$\bar{p}_{S2} = (0.39t)^2 \quad (8.9.27)$$

With the policy of repairing after two weeks, one channel at a time Eq. 8.7.18 is used to obtain the mean fractional dead time for this as

$$\mu_{D_{S1}} = \frac{0.39^2}{3}\left(\frac{2}{52}\right)^2 + 0.39 \times (1 \times 10^{-4}) = 0.0001 \quad (8.9.28)$$

However, the danger of valve leakage has not been counted above. To find the fractional dead time contributed by this event, the system's failure due to this must be determined. It is calculated from the following relation

$$\bar{p}_S = \bar{p}_{S2} + \bar{p}_{S3} \quad (8.9.29)$$

where $\bar{p}_S$ is the probability of failure of the system, $\bar{p}_{S2}$, the probability of failure due to no response of the valve to air pressure given by Eq. 8.9.27 and $\bar{p}_{S3}$ is the probability of the system's failure due to the leakage of the valve etc. $\bar{p}_S$ can be found by noting that for a single channel

$$\bar{p}_{S4} = (0.19 \text{ for initiator faults} + 0.01 \text{ for pump shut off} + 0.22 \text{ for steam shut off})\ t = 0.42t \quad (8.9.30)$$

and

$$\bar{p}_S = (\bar{p}_{S4})^2 = (0.42)^2 t^2 \quad (8.9.31)$$

Therefore $\bar{p}_{S3}$ in Eq. 8.9.29 is given by

$$\bar{p}_{S3} = (0.42^2 - 0.39^2)t^2 = 2.43 \times 10^{-2} t^2 \quad (8.9.32)$$

and the mean fractional dead time for the system for this event would be

$$\mu_{D_{S2}} = \frac{(2.43 \times 10^{-2})^2}{3}\,\tau^2 = 0.0081\tau^2 \tag{8.9.33}$$

For $\tau = 16$ weeks or (16/52) years, it is equal to

$$\mu_{D_{S2}} = 0.0081\left(\frac{16}{52}\right)^2 = 0.00077 \tag{8.9.34}$$

Therefore the mean fractional dead time, μ_{D_S}, of the system would be

$$\mu_{D_S} = \mu_{D_{S1}} + \mu_{D_{S2}} = 0.00087 \tag{8.9.35}$$

This is less than the required fractional dead time 0.001 and therefore the design of the trip system of Fig. 8.16 is satisfactory *provided* there is an on-line testing every two weeks and there is a plant shutdown testing after every 16 weeks.

In this section an example of designing a control system has been considered in which principles developed in previous sections of this chapter were applied. It can be clearly seen that the mean plant availability and the fractional dead time are extremely important parameters and they are greatly affected by the configuration of the system. Depending upon the reliability of the system required, the configuration is adjusted to meet the specification of the desired value as done in the example.

8.10 OPTIMAL MIXED MAINTENANCE POLICY[9, 10]

In previous sections of this chapter two idealized maintenance policies, irreversible and reversible, were considered for systems of different configurations. In the analysis presented, it was assumed that a system can either be in working or failed condition. The effects of different maintenance policies were then examined on the performance of system. These analyses, however, tended to be idealized in nature because in reality, for a chemical plant, a combination of irreversible as well as reversible policies occur simultaneously. If there is a chemical plant in which any of the main-train equipment breaks down, one has no choice but to isolate and repair it. This way the system undergoes a reversible maintenance. Along with this, all equipment slowly deteriorates due to usage; and to reduce the maintenance cost due to breakdowns, plant shutdowns are produced regularly for general overhaul and repairs. In this way, it is undergoing an irreversible maintenance.

Any given equipment of a chemical plant can be modelled to be in any of the following three states:

(1) good,
(2) deteriorated, and
(3) failed.

This is to be contrasted with the earlier analysis where the equipment could either be in working or in failed condition. The plant shutdown is scheduled at regular intervals so as to repair 'deteriorated' equipment whereas failed equipment must be repaired immediately. If an equipment is in 'good' condition at one instant of time, it has finite probability at a *subsequent instant* to be classified either 'deteriorated' or 'failed'. However, if it is in a 'deteriorated' condition, it can make a transition into 'failed' condition only. Because one immediately repairs failed equipment, the only transition that it can make is to the 'good' condition. As a result, one defines transition probability t_{ij} (these have the same sense as z_i discussed earlier) from the initial condition i to the final condition j and arranges them in the following matrix form.

$$\begin{array}{ll} & \text{Final condition } (j) \longrightarrow \quad 1 \quad\; 2 \quad\; 3 \\ & \text{Initial condition } (i) \\ T = & \qquad\qquad \downarrow \qquad\quad \begin{array}{c} 1 \\ 2 \\ 3 \end{array} \begin{bmatrix} t_{11} & t_{12} & t_{13} \\ 0 & t_{22} & t_{23} \\ 1 & 0 & 0 \end{bmatrix} \end{array} \tag{8.10.1}$$

These transition probabilities are assumed to be independent of time. With this knowledge it is designed to find the probabilities of the equipment being in 'good', 'deteriorated' and 'failed' conditions at any other given time. To do this, the total time elapsed is divided into equal intervals and these probabilities of finding the system in a given state are constant in any given interval. The following discrete probabilities are defined:

$p_1^{(n)}$: probability of finding the equipment in 'good' condition in the nth interval (8.10.2a)

$p_2^{(n)}$: probability of finding the equipment in 'deteriorated' condition in the nth interval (8.10.2b)

$p_3^{(n)}$: probability of finding the equipment in 'failed' condition in the nth interval (8.10.2c)

It is noted that an equipment can be in the 'good' condition in the $(n + 1)$th interval provided it was in the 'good' condition in nth state and it remains so till the next interval or it was in the failed state in the nth interval and was repaired in the last interval. The probability of the former is $p_1^{(n)}t_{11}$ whereas that of the latter is $p_3^{(n)}$. Therefore, $p_1^{(n+1)}$ is given by

$$p_1^{(n+1)} = t_{11}p_1^{(n)} + p_3^{(n)} \tag{8.10.3}$$

One similarly obtains

$$p_2^{(n+1)} = t_1p_1^{(n)} + t_{22}p_2^{(n)} \tag{8.10.4}$$

and

$$p_3^{(n+1)} = t_{13}p_1^{(n)} + t_{23}p_2^{(n)} \tag{8.10.5}$$

These equations can be rewritten in the matrix form

$$[p_1^{(n+1)} \quad p_2^{(n+1)} \quad p_3^{(n+1)}] = [p_1^{(n)} \quad p_2^{(n)} \quad p_3^{(n)}]\begin{bmatrix} t_{11} & t_{12} & t_{13} \\ 0 & t_{22} & t_{23} \\ 1 & 0 & 0 \end{bmatrix} \tag{8.10.6}$$

or

$$\mathbf{P}^{(n+1)} = \mathbf{P}^{(n)}\mathbf{T}; \; n = 0, 1, 2, \ldots \tag{8.10.7}$$

where

$$\mathbf{P}^{(n+1)} = [p_1^{(n+1)} \quad p_2^{(n+1)} \quad p_3^{(n+1)}] \tag{8.10.8a}$$

and

$$\mathbf{P}^{(n)} = [p_1^{(n)} \quad p_2^{(n)} \quad p_3^{(n)}] \tag{8.10.8b}$$

The matrix **T** is already defined by Eq. 8.10.1. For different values of n, Eq. 8.10.7 represents a set of recurring equations. Before giving the final solution, the special nature of Eq. 8.10.7 may be pointed out first: the probability value of any interval depends *only* upon their values in the previous interval and no other. Any process which has this special characteristic is called a *Markov Chain.*

If it is given that the equipment starts as new, i.e. $P^{(0)} = [1, 0, 0]$, these recurring equations can be solved in terms of initial condition

$$\mathbf{P}^{(n+1)} = \mathbf{P}^{(0)}\mathbf{T}^{n+1} \tag{8.10.9}$$

where $\mathbf{T}^{n+1}$ means that the matrix **T** has been multiplied $(n + 1)$ times. If there are N_i intervals, then at the end of the interval the probability of finding the equipment in deteriorated condition is equal to $p_2^{(N_i)}$ whereas the expected number of failures, μ_{N_i} in the duration would be given by

$$\mu_{N_i} = \sum_{j=1}^{N_i} p_3^{(j)} \tag{8.10.10}$$

Similarly the expected number of preventive repairs μ_{R_i} in N_i intervals is equal to

$$\mu_{R_i} = \sum_{j=1}^{N_i} p_2^{(j)} \tag{8.10.11}$$

Thus, μ_{N_i} and μ_{R_i} can be easily calculated once the transition matrix **T** is specified. The procedure outlined above does not specify the incremental time interval, but for the calculation the shorter it is the better would be the results for μ_{N_i} and μ_{R_i}. However, this should be large enough such that if the equipment is in the failed state, it can be diagnosed and repaired within the time interval. As an example, the transition matrix, **T**, is taken as

$$\mathbf{T} = \begin{bmatrix} 0.90 & 0.09 & 0.01 \\ 0 & 0.90 & 0.10 \\ 1 & 0 & 0 \end{bmatrix} \tag{8.10.12}$$

μ_{N_i} and μ_{R_i} are calculated on computer and are given in Fig. 8.17 as a function of inspection times, τ.

After having calculated μ_{N_i} and μ_{R_i}, the stage is now set to find the optimal maintenance policy. One assumes the regular plant shutdown scheduled after every time τ. During this shutdown, not all equipment may be tested and repaired. It is assumed that an equipment is tested and repaired in $(k_i\tau)$ time where k_i could take integer values 1, 2, 3 etc. If $k_i = 2$ and is the same for all equipment, it would mean that half of the equipment is tested in every shutdown overhaul.

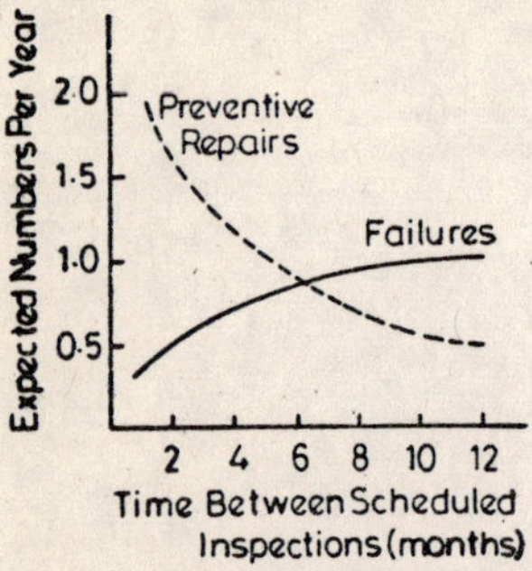

Fig 8.17 Expected number of repairs and failures as a function of time of failure. (Reprinted from Reference 9 with permission of John Wiley & Sons Inc., New York.)

During the plant shutdown, two things are done:

1. all equipment is inspected, and
2. deteriorated equipment is repaired.

Consequently, there is a labour cost involved in inspecting them. After locating the deteriorated equipment, it is repaired and associated with those where there is a labour charge. Also, if the number of maintenance men employed is small, the inspection and the repair work would progress slowly, that way increasing the plant shutdown time. In view of this, the cost of inspection C_{1i} of ith equipment would be

$$C_{1i} = \frac{W_{1i}C_D}{M} \tag{8.10.13}$$

and the cost of preventive repair, C_{2i}, of ith equipment is

$$C_{2i} = \frac{W_{2i}C_D}{M} \tag{8.10.14}$$

where M is the number of repair men employed, C_D the hourly cost of the system downtime, W_{1i}, the average man-hour for preventive inspection of ith equipment and W_{2i}, the average man-hour for its preventive repair. The equipment can also fail while the plant is running. If it takes D man-hours in locating the failed equipment and repairing it, the cost associated with this failure C_{3i} would be equal to

$$C_{3i} = DC_D \tag{8.10.15}$$

If in the period of $(k_i\tau)$, ith equipment fails μ_{N_i} times and requires μ_{R_i} preventive repairs per year, then the yearly cost would be

$$C_i = \frac{C_{1i}}{(k_i\tau)} + C_{2i}\mu_{R_i} + C_{3i}\mu_{N_i} \qquad (8.10.1\ldots)$$

Supposing it costs C_0 amount of money (fixed cost) in carrying out the overhaul, then the total cost C_T of the shutdown per unit time would be

$$C_T = \left(\frac{C_0}{\tau}\right) + MC_m + \sum_i C_i \tag{8.10.17}$$

where C_m is the yearly wages of single maintenance man and $(\sum_i C_i)$ is the total cost of repairing all equipment of the chemical plant. To find out the best maintenance policy, one finds the optimal values of k_i, τ and M and this is done as follows. Values of τ and M are assumed and k_i is allowed to vary in Eq. 8.10.16. The value of k_i which gives a minimum C_i is found and if there is no minimum one assumes it to be equal to one. Then τ and M are varied to find minimum C_T.

To demonstrate this procedure, a chemical plant with 100 identical equipments is considered and each is assumed to have the same transition matrix given by Eq. 8.10.12. Their expected number of failures and preventive repairs are given by Fig. 8.17. Other relevant information required in Eqs. 8.10.13 to 8.10.17 are given as follows:[9]

$$W_1 = 0.2 \text{ man-hours}; \; W_2 = 1 \text{ man-hour}$$

$$D = 4 \text{ hours}; \; C_0 = 5000; \; C_m = 10{,}000; \; C_D = 500 \tag{8.10.18}$$

The first step is to assume a value of M and τ and it is taken as two workers and one month in the first trial. Then the cost factors are

$$C_1 = \frac{500 \times 0.2}{2} = 50 \tag{8.10.19a}$$

$$C_2 = \frac{500 \times 1}{2} = 250 \tag{8.10.19b}$$

Next k_i is varied to find minimum C_i. When $k_i = 2$, a given equipment is inspected after every two months. The values of μ_{N_i} and μ_{R_i} are read from Fig. 8.17 for two months and C_i calculated. Results are given in Table 8.7 which shows a minimum for $k_i = 2$. However, when $\tau = 1$ and $M = 4$, as k_i increases, there is no minimum in C_i with respect to k_i; consequently $k_i = 1$.

After determining the best value of k_i, C_T is calculated from Eq. 8.10.17. For example, for the case of $k_i = 2$, $\tau = 1$ month and $M = 2$, $C_i = 1700$. Since there are 100 identical units in the system, $(\sum C_i)$ is equal to 1,70,000. The periodic shutdown costs 5000 units of money and there are 12 shutdowns in a year, the annual cost would therefore be 5000×12 or 60,000 money units. The manpower cost is equal to $10{,}000 \times 2$ or 20,000 money units per year. Therefore C_T is calculated as 2,50,000 money units per year. Table 8.8 gives costs C_T for different values of M and τ and it is found that the optimal maintenance policy would have $k_i = 1$, $M = 4$ men and $\tau = 2$ months.

Table 8.7 Minimization of C_1 with respect to k_i

$\tau = 1$ month, $M = 2$ $\quad C_1 = 50, C_2 = 150$

k_i	μ_{R_i}	μ_{N_i}	P.M.	P.R.	F	C_i
1	1.95	0.34	50 × 12	488	680	1768
2	1.60	0.50	50 × 6	400	1000	1700 ← min
4	1.20	0.75	50 × 3	300	1500	1950
6	0.85	0.80	50 × 2	210	1600	1910
12	0.49	~1	50 × 1	125	2000	2175

$\tau = 1$ month, $M = 4$ $\quad C_1 = 25, C_2 = 125$

k_i	μ_{R_i}	μ_{N_i}	P.M.	P.R.	F	C_i
1	1.95	0.34	25 × 12	244	680	1224
2	1.60	0.50	25 × 6	200	1000	1350
4	1.20	0.75	25 × 3	150	1500	1725
6	0.85	0.80	25 × 2	105	1600	1755
12	0.49	~1	25 × 1	63	2000	2088

P.M. = Preventive maintenance charges
P.R. = Preventive repair charges
F = Fixed charges

Table 8.8[0] Calculation of C_T for Different Values of M in Thousand of Money Units

Time between inspection (Months) ↓ \ Number of maintenance men →	1	2	4	8	12
1	286	250	222	236	255
2	267	218	|203|	226	260
4	252	219	217	246	283
8	246	232	241	275	313
12	247	241	253	289	328

(Reprinted from Ref. 8 with the permission of John Wiley & Sons Inc., New York.)

In this section, an optimal maintenance policy was found where equipment is not only inspected and repaired at regular intervals but along with that, the failed equipment is brought back to working condition also. The example solved in the section assumes that all equipments are identical; however, the method can be easily extended to an ensemble of non-identical equipments.

8.11 CONCLUSION

In this chapter, the reliability in maintenance policy of a chemical plant has been discussed. It was shown that the plant availability is a direct measure of the former which is greatly affected by the maintenance work. In this chapter this has been examined for different configurations of controllers and equipment. The plant configuration relates the probability of failure of the system to the probabilities of failure of its components. Once this is known, one can find out the average plant availability and fractional dead time for a given maintenance policy. In the end these concepts have been applied to the design of controllers for a distillation column. In chemical engineering literature, there exist few examples of this kind[11–14] and the work of Henley et al is noteworthy where the operating policies of storage tanks to improve the plant availability is discussed.[12] Interested readers are referred to the original publication.

REFERENCES

1. Collacot, R. A., *Mechanical Fault Diagnosis and Condition Monitoring*, 1st edn, Chapman and Hall, London, 1977.
2. Brownell, L. E. and Young, E. H., *Process Equipment Design*, 1st edn, Wiley, N.Y., 1959.
3. Bazovsky, I., *Reliability Theory and Practice*, Prentice-Hall, N.J., 1961.
4. Green, A. E. and Bourne, A. J., *Reliability Technology*, 1st edn, Wiley, Interscience, 1972.
5. Anyakora, S. N., Lees, F. P. and Engel, G. F. M., *Chem. Engr.*, Nov. 1971.
6. Abromowitz, M. and Stegun, I. A., *Handbook of Mathematical Functions*, 1st edn, Dover, N.Y., 1968.
7. Lawley, H. G. and Kletz, T. A., *Chem. Eng.*, **81**, May 12, 1975.
8. Kletz, T. A., *Chem. Processing*, p. 66, Sep. 1974.
9. Goldman, A. S. and Slattery, T. B., *Maintainability: A Major Element of Systems Effectiveness*, 1st edn, John Wiley, New York, 1964.
10. von Alven, W. H. (ed.), *Reliability Engineering*, Prentice-Hall, New Jersey, 1964.
11. Lawley, H. G., *C.E.P.*, 70 (4), 45, 1974.
12. Henley, E. J. and Hoshino, H., *I. & E. C. Fundam.*, **16**, 439, 1977.
13. Cherry, D. H., Grogan, J. C., Holmes, W. A. and Perris, F. A., *C.E.P.*, **74**(1), 55, 1978.
14. Kumamoto, H. and Henley, E. J., *A.I.Ch.E.J.*, **25**, 108, 1979.

PROBLEMS

8.1 Information on the reliability of different chemical equipment of interest cannot be found in literature and is not given in Table 8.1. However it can be built from basic information given in it.

Consider a filter press which would be regarded in non-working condition, if

(a) the slurry passes through but there is no proper separation; or

(b) the slurry does not pass through it.

Prepare a fault tree for each in terms of failure of basic components and then write the reliability of these in terms of primary events. What more information do you require ?

8.2 Consider conditions of failure of a distillation column. It can develop following faults:

(a) over-pressure or under-pressure;

(b) plate temperature going beyond the permissible range; and

(c) disbalance in liquid and vapour flow rates.

Part (a) has been considered in the text (Fig. 8.13). Do a similar analysis for (b) and (c). Find the probability of the head event in terms of the primary event probabilities.

8.3 Do an analysis similar to that in P. 8.2 for a heat exchanger. First find out under what conditions it would fail and then treating them as the head event, develop a fault tree. Next find the reliability of an exchanger.

8.4 Find the relation between the reliability and the availability of a system.

8.5 Supposing the failure of an equipment is expressed by the Poisson distribution function, find its availability under irreversible maintenance policy. Express the function in a series expansion and integrate term by term.

8.6 Repeat your calculations for the Poisson distribution function for reversible maintenance policy.

8.7 Consider equipments A and B in series with A having a standby. Find the availability of the system when an irreversible maintenance policy is kept. Derive results when the equipments are repaired to 'as good as new' condition and when they are restored to working condition only.

8.8 Solve P. 8.7 when reversible maintenance policy is maintained.

8.9 Consider the control design problem of the ethylene oxide reactor of Chapter 6. Supposing there is an exponential decay in the performance of a single controller. How do you solve for the optimal value of N under:

(a) irreversible checking of individual controller after fixed time τ when they are repaired to 'as good as new';

(b) reversible checking when they are repaired to 'as good as new'.

8.10 Consider a storage tank in which there is one upstream (US) and there is one downstream (DS) as shown in below.

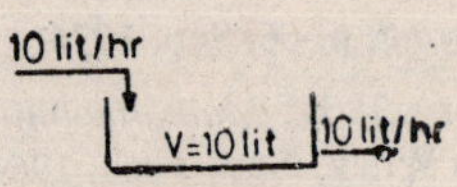

Fig. P8.10

If the volume of the tank is 10, and US fails, the downstream DS would fail after one hour. Write this fact in an equation relating the failure of DS to the failure of US. It is seen that a storage tank acts as a pure time delay system.

Supposing the US is getting repaired with exponential probability, adjust the equation above accordingly.

8.11 Supposing there are two equipment with an interment storage tank as shown below.

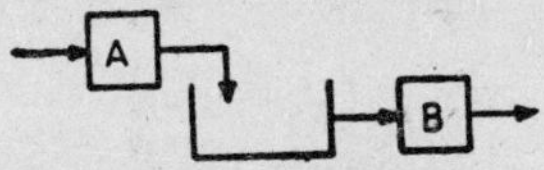

Fig. P8.11

If output from B is regarded as successful state, show successful and failed states diagrammatically (as done in Fig. 8.6). Next determine the availability of the system for irreversible maintenance. Also show that there is an improvement in availability. Assume the equipment are repaired to as-good-as-new conditions.

8.12 Repeat the derivation of P8.11 if the system is operated with reversible policy.

8.13 Consider a storage tank with any amount of equipment upstream and downstream as shown below.

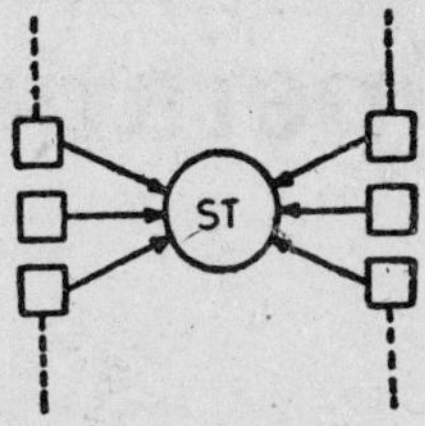

Fig. P8.13

Analyse this configuration for the system with irreversible policy.

8.14 Analyse the configuration of Fig. P8.13 for it to be operated with reversible policy.

8.15 In view of P. 8.11 to 8.14, suggest an optimal plant flow sheet for an acetaldehyde plant.

9

Estimation of Chemical Equipment and Plant Costs and Direct Operating Labour

9.1 INTRODUCTION

In Chapters 1 to 3, it was explained how the steady state design of a chemical plant is carried out. Subsequent chapters explained what more must be done to this design to make it realistic, and after the plant has come into existence what it takes to run it efficiently. However, in these discussions, the economic factors of the overall plant design were ignored. It may be recalled that producing a chemical is an economic venture and in order to carry it out, capital is required at different stages of the chemical plant. As it is desired to maximize the capital return per unit capital invested, a design can never be complete without a thorough analysis of all economic factors.

The total financial liability of a chemical plant can be divided into fixed and recurring costs. The fixed cost approximately includes the cost of land, building and equipment and it is spent once during the installation of the plant. The recurring cost consists of all expenses which are (essentially) proportional to the amount of product processed, e.g., raw materials, direct labour, utility, etc. In this chapter, a method of estimating cost of chemical equipment has been presented which can be used to find the money requirement for its installation. Another piece of information which is of great interest in this context is the direct labour requirement. It will be shown in the next chapter that for a given chemical to be produced, the fixed capital and the direct operating labour required decide the total capacity of the plant and in view of this, they represent important information. In the following, the cost estimation of equipment is considered first.

9.2 COST OF CHEMICAL EQUIPMENT

Chemical equipment is varied in design and size and Table 9.1 gives their classification[1–3]. Since the price depends heavily upon the design as well as size, the variation in their costs is found to be phenomenal. Some of the factors affecting the cost besides the design are:

1. size of the equipment;
2. material of construction;
3. operating conditions;
4. the country where the equipment is fabricated; and
5. the inflation rate.

It may be pointed out that the best way of obtaining the cost is to obtain direct quotations from manufacturers for the specific equipment, which is indeed done while finalizing the design. However, to be able to do this, the complete engineering details must be known and this is possible only when the design is in its final stages. In the early stages of design, while evaluating different alternatives, only an-order-of-magnitude

Table 9.1 Kinds of Chemical Equipment[1–3]

Sl.No.	Equipment	Types
1.	*Pumps*	Centrifugal, canned motor, magnetic, acid egg, sump, diffuser type, axial-flow, turbine, positive displacement, piston, plunger, diaphragm, gear, screw, jet pumps, etc.
2.	*Compressors*	Centrifugal fans, turbo-blower, axial and rotary compressors, etc.
3.	*Mixers*	Turbine blower for gases, agitated vessels for liquids, scheibel columns and ball, pebble, hammer, impact, cage, attrition mills for solids, etc.
4.	*Heat Exchangers*	Cocurrent, countercurrent, multipass shell-and-tube exchangers, evaporators, condensers
5.	*Separators*	See Table 3.6 of Chapter 3
6.	*Furnaces*	Incinerators, boilers, evaporators, direct and indirect heaters
7.	*Material handling*	Conveyors, elevators, pallets, fork trucks, pipelines
8.	*Reactors*	Stirred tank reactors, tubular reactors, pressure vessels
9.	*Storage tanks*	API conical vessels with floating head
10.	*Process Plant Drivers*	Electric motors, steam and gas turbines, reciprocating engines using steam, internal combustion engines

estimation is required where one can afford to be less accurate. Between these limits of accuracies of cost estimation, there are several other estimates (described in Table 9.2) which are required at different stages

Table 9.2[2,4,5] Different Kinds of Cost Estimates during the Plant Design

Sl.No.	Types of estimate	Accuracy	Minimum information required
1.	Order of magnitude	±50%	Product capacity, location
2.	Study (rough approximation, factored)	±30%	Sketches of major equipment and main buildings
3.	Budget (scope, preliminary economics)	±20%	Preliminary sizes of major equipment and main buildings
4.	Project control (refined approximation)	±10%	Equipment, building data sheet
5.	Firm (final, detailed, definitive)	±5%	Complete engineering design, bill of quantities of supplier's quotations for major items

(Reprinted from Ref. 2 with permission of Dr. O. P. Kharbanda, O. P. Kharbanda Associates, Bombay.)

of design. The following correlations give only an order-of-magnitude estimate.

9.3 COST ESTIMATION

9.3.1 Cost of Chemical Equipment

The total cost of an equipment consists of the cost of the material, fabrication, transportation to the site and installation, i.e.

$$C'_T = c_M M_t + c_F T_F + C_{tr} + C_I \tag{9.3.1}$$

where C'_T is the total cost, M_t is the total amount of material required, T_F is the total time of fabrication, C_{tr} is the cost of transportation and C_I is the installation charge. The constants c_M and c_F are the cost per unit weight of the material and the cost of the labour (including factory overheads) respectively. The normal procedure of accounting transportation and installation charges is to take it as some fraction of C_T, defined by the sum of the cost of material and fabrication charges only:

$$C_T = c_M M_t + c_F T_F \tag{9.3.2}$$

To estimate T_F is not a simple task as this requires a careful analysis of all the fabrication steps. As an example, in Table 9.3, a sequence of mechanical operation for a heat exchanger is given. The time required for each step is estimated and by adding them all, T_F is determined.

Calculation of T_F is a time-consuming process, but some general characteristics of the cost, C_T can be easily derived. T_F in general depends upon the total amount of material and the amount of fabrication work to be done on it, or,

$$T_F = f(M_t, W) \tag{9.3.3}$$

Table 9.3 Sequence of Fabrication for a Heat Exchanger

Step	Activity	Nature of work
1	Layout	(a) To mark the sheet for the shell
		(b) To mark the sheet for the heads
		(c) To mark baffles
		(d) To mark end-sheets
		(e) To mark flanges
		(f) To mark separators
		(g) To mark vents and drains
2	Cutting	(a) Cut sheet for shell and heads
		(b) Cut plates for baffles
		(c) Cut flanges
		(d) Cut baffle tie rods
		(e) Cut tubes
		(f) Cut plates for separator
3	Drilling	(a) Drill holes in tube sheets and baffles
		(b) Drill holes in flanges
		(c) Drill holes in the sheet for the shell for nozzles and vents
4	Rolling	(a) Roll sheet to make shell
		(b) Roll sheet to make head
5	Welding	(a) Weld the shell ends
		(b) Weld flanges
		(c) Weld separator
		(d) Weld nozzles and vents
6	Assembly	(a) Insert tubes through baffles and tube sheets
		(b) Insert the tube set through the shell
		(c) Close the exchanger heads

where W is some measure of the density of fabrication work, to be contrasted with the total amount of work done to produce the equipment. For a specific equipment, W is expected to be approximately independent of its size and in view of this, Eq. 9.3.2 becomes

$$C_T \simeq c_M M_t + c_F f(M_t) \tag{9.3.4}$$

The functional dependence of T_F on W above has been deleted because of its constancy. Thus, the best way of correlating the cost of a fabricated equipment is its weight. In the literature, however, this is rarely done and all cost correlations are presented with respect to a variable, measuring its capacity. For example, the capacity variable for a distillation column is the volume of the column V and for a heat exchanger it is its heat transfer area A. Inadequacies of these variables as the basis can hardly be overemphasized.

A distillation column, as shown in Fig. 9.1, is a tall vessel consisting of several sections joined together by suitable flanges. Normally its one section consists of one actual plate which is done to facilitate an easy cleaning of the column. In view of this, the increase in volume by the change of height is an easy matter which can be achieved by adding a suitable number of more sections to the column. Consequently, the change in the cost of the column is expected to be proportional to the height change. However, the change in volume by the change in the diameter of the column is a different matter because this requires a uniform change in diameter of all its sections. Similarly in a heat exchanger, the change in heat transfer area can be obtained by increasing its diameter as well as its length. Both these are not equivalent which can be seen by examining Table 9.3.

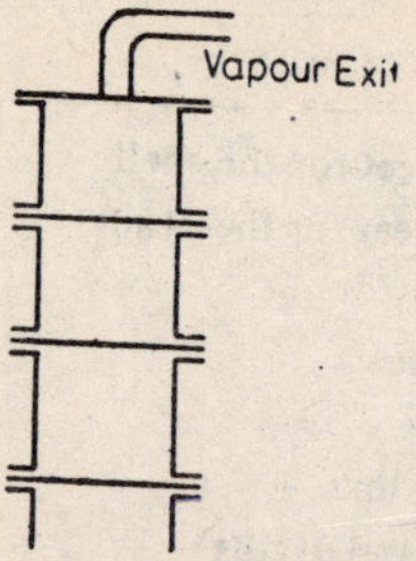

Fig. 9.1 Schematic representation of modular units constituting distillation column

In cases where cost data are correlated in terms of the capacity variable, V_c, one first determines the relation between the amount of material M_t and V_c and the result is substituted in Eq. 9.3.4 to obtain the functionality between C_T and V_c. This is never done for estimating costs of chemical equipment and one simply uses the following empirical relation[2, 7-9]:

$$\frac{C_{T1}}{C_{T2}} = \left[\frac{V_{c1}}{V_{c2}}\right]^n \tag{9.3.5}$$

where V_{c1} and V_{c2} are two capacities and C_{T1} and C_{T2} are their costs for any given *type* of the equipment. It can be clearly seen that this equation is an approximation and it is only in view of this, the exponent n is found to vary slightly with capacity[8]. The value of n normally lies around 0.6, however there is no theoretical constraint on its value and at times it is greater than 1 also.

In Figs. 9.2 to 9.7, the costs of chemical equipment have been given as nomographs and they have been prepared using Eq. 9.3.5 as the basis. These nomographs can be used with the help of Tables 9.4 to 9.12 as follows. One first searches the equipment in these tables and finds the point and the figure number corresponding to it that governs its cost. Under the column of 'size/capacity' of the table, the capacity variable is specified and using the appropriate capacity variable the cost is found from the nomograph. If necessary, this is corrected according to footnotes of these tables. As an example, it is desired to find the cost of a catalytic reformer furnace at pressure 34.01 atm and heat duty 105.51×10^6 kJ/h. Since the catalytic reformer is a furnace, the point number corresponding to furnace in Table 9.4 is 2 in the nomograph of Fig. 9.2. According

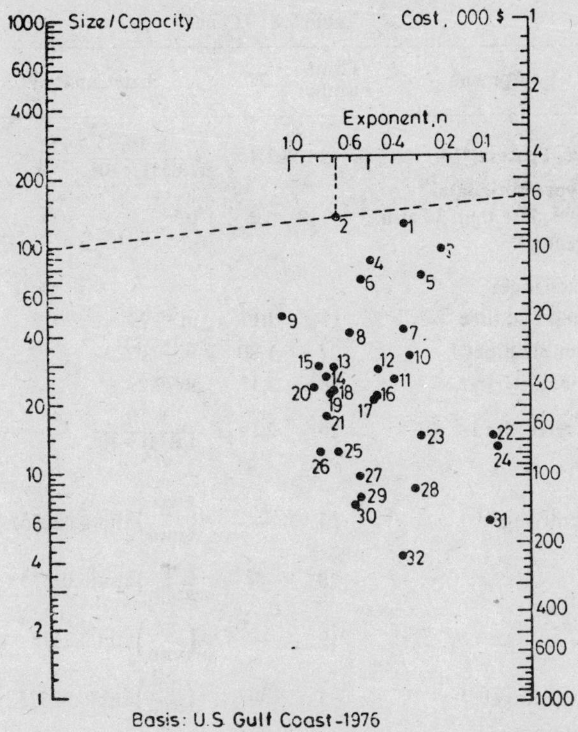

Fig. 9.2 Cost of equipment given in Table 9.4. (Reprinted from Ref. 2 with permission of Dr. O. P. Kharbanda, O. P. Kharbanda & Associates, Bombay.)

Table 9.4[2] Equipment Appearing in the Nomograph of Fig. 9.2

Sl. No.	Equipment	Point number	M	Size/Capacity	Range
1.	Compressor units				
	(a) Centrifugal	30	2.15	kW/74.57	2–100
	(b) Reciprocating	29	2.15	kW/74.57	2–100
	(c) Direct acting	27	2.15	kW/74.57	1–15
2.	Driver, pump and compressor				
	(a) Gas turbine	21	3.30	kW/7.46	10–1000
	(b) Gas engine	14	—	kW/7.46	10–1000
	(c) Motor drive	9	—	kW/7.46	6–1000
	(d) Gear assembly	1	—	kW/7.46	20–1000
3.	Driver turbine				
	(a) Condensing	16	—	kW/7.46	20–1000
	(b) Non-condensing	11	—	kW/7.46	25–800
4.	Driver, auxiliary				
	(a) Gas turbine	7	—	kW/74.57	1–100
	(b) Gas engine	23	—	kW/74.57	1–100
	(c) Turbine condensing	10	—	kW/74.57	1–100
	(d) Turbine non-condensing	5	—	kW/74.57	1–100

(*Contd.*)

Table 9.4 (*Contd.*)

Sl. No.	Equipment	Point number	M	Size/Capacity	Range
5.	Furnace, Process(*) (Box, type, horizontal CS, tubes, less than 34 atm., field erected)	2	2.4	$\dfrac{\text{k J/h}}{1.0551\times10^6}$	20–500
6.	Heat exchanger				
	(a) Shell and tube	13	3.17	$m^2/9.29$	2–50
	(b) Double pipe(*)	22	1.80	$m^2/0.0929$	2–50
	(c) Air cooler(*)	18	2.17	$m^2/9.29$	3–70
	(d) Heater, fired	26	2.19	$\dfrac{\text{k J/h}}{1.0551\times10^6}$	2–30
7.	Pumps				
	(a) Centrifugal(*)	24	—	$\left(\dfrac{m^3}{min}\right)$atm/0.0002575	1–200
		28	—	$\left(\dfrac{m^3}{min}\right)$atm/0.02575	2–100
		19	—	$\left(\dfrac{m^3}{min}\right)$atm/0.2575	10–400
	(b) Centrifugal(*) (in line)	3	—	$\left(\dfrac{m^3}{min}\right)$atm/0.02575	7–50
		4	1.75	$\left(\dfrac{m^3}{min}\right)$atm/0.02575	50–400
	(c) Chemical injection (*) variable speed	6	2.83	$\left(\dfrac{m^3}{min}\right)$atm/0.0002575	40–600
	Fixed speed	8	2.83	$\left(\dfrac{m^3}{min}\right)$atm/0.002575	4–200
	(d) Reciprocating(*)	31	—	$\left(\dfrac{m^3}{min}\right)$atm/0.0002575	40–60
	upto 69 atm*	32	—	$\left(\dfrac{m^3}{min}\right)$atm/0.02575	2–100
		25	—	$\left(\dfrac{m^3}{min}\right)$atm/0.2575	10–400
	(69–341 atm)	17	—	$\left(\dfrac{m^3}{min}\right)$atm/0.2575	10–200
	(e) Turbine(*)	12	1.80	$\left(\dfrac{m^3}{min}\right)$atm/0.2575	1–500
	(f) Vertical axial	15	2.05	$\left(\dfrac{m^3}{min}\right)$atm/3.785	1–300
	(g) Mixed	20	1.70	$\left(\dfrac{m^3}{min}\right)$atm/3.785	1–200

**Notes*:

(1) For process furnaces (Sl. No. 5), multiply the cost found by 100.

(2) For double pipe heat exchanger (Sl. No. 6b), divide the cost found from the nomograph by 100.

(*Contd.*)

(3) For air coolers (Sl. No. 6c), divide calculated area by 15.5 (fin tube factor) before entering the nomogram.

(4) For centrifugal pumps found from points 28 and 19 (Sl. No. 7a), divide the found costs by 100 and 10 respectively. For motor driven pumps, $M = 3.30$.

(5) For centrifugal (in-line) pumps (Sl. No. 7b), divide the cost found by 10.

(6) For chemical injection pumps (Sl. No. 7c) divide the cost found by 10.

(7) For reciprocating pumps up to the pressure of 69 atm (Sl. No. 7d), divide the cost found by 100. However for equipment with $\left(\frac{(m^3/min)atm}{0.2575}\right)$ between 10 and 400 atm divide it by 10.

(8) For turbines (Sl. No. 7e), divide the cost by 10.

(9) All costs must be corrected for changes in design d, choice of the material of construction m and pressure p and these parameters are given in Table 9.11.

(10) To find the cost of fully installed unit, the nomograph reading is multiplied by M.

(Reprinted from Ref. 2 with permission of Dr. O. P. Kharbanda, O. P. Kharbanda Associates, Bombay.)

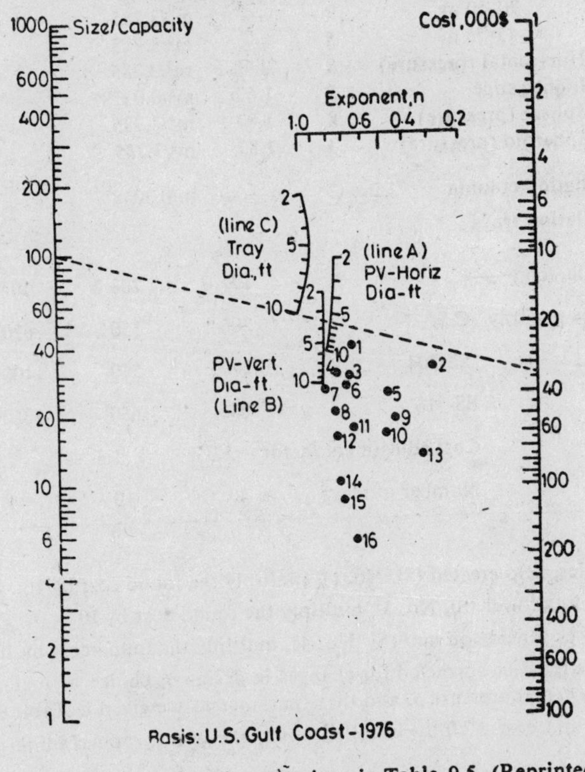

Fig. 9.3 Cost of equipment given in Table 9.5. (Reprinted from Ref. 2 with permission of Dr. O. P. Kharbanda, O. P. Kharbanda and Associates, Bombay.)

Table 9.5[a] Equipment Appearing in the Nomograph of Fig. 9.3

Sl. No.	Equipment	Point number	M	Size/Capacity	Range
1.	Boiler, field erected*	7	1.8	kg/h/453.59	50–1000
2.	Boiler package	12	1.8	kg/h/453.59	20–700
3.	Cooling tower*	1	1.7	$\left(\frac{m^3}{min}\right)/0.3785$	10–600
4.	Power generation*	4	1.46	kVA/100	7–200
5.	Power emergency				
	Gas engine	16	1.46	kVA/100	2–80
	Diesel	15	1.46	kVA/100	2–80
6.	Process vessel				
	Horizontal	Line A	3.05	m/0.3048	Length 4–100
	Vertical	Line B	4.16	m/0.3048	Height 2–10
7.	Refrigeration (Mech.)	14	1.53	Tons	7–300
8.	Tanks				
	(a) API conical	13	1.63	$m^3/0.3785$	3–500
	(b) API conical	11	1.94	$m^3/37.85$	3–400
	(c) Elevated steel				
	height 15–25 m	10	—	$m^3/3.785$	20–800
	30–50 m	9	—	$m^3/3.785$	20–800
	45–75 m	5	—	$m^3/3.785$	20–800
	(d) Horizontal (pressure)	6	2.08	$m^3/3.785$	2–80
	(e) Light gauge	2	1.63	$m^3/0.03785$	2–200
	(f) Sphere (pressure)	8	1.87	$m^3/3.785$	10–300
	(g) Spheroid (pressure)	3	1.87	$m^3/3.785$	10–300
9.	Distillation column	Line C	—	m/0.3048	Height 6–400
10.	Distillation trays				Dia. 2–10

Outer dia. (cm) ⟶		101.6	203.2	304.8	457.2
Cost in $ per tray ↓	C.S.	200	350	620	1550
	SS-304	270	540	1100	3000
	SS-316	340	650	1350	3700
	Cost adjustment factor	1.0	1.5	1.9	2.4
	Number of trays	40	10	4	1

**Notes*

(1) For boiler, field erected (Sl. No. 1), multiply the found cost by 10.

(2) For cooling tower (Sl. No. 3), multiply the found cost by 10.

(3) For power generation unit (Sl. No. 4), multiply the found cost by 100.

(4) All costs must be corrected for changes in design d, choice of material of construction m and pressure p, and these parameters are given in Table 9.12.

(5) To find the cost of fully installed units, the nomograph reading is multiplied by M.

(Reprinted from Ref. 2 with permission of Dr. O. P. Kharbanda, O. P. Kharbanda Associates, Bombay.)

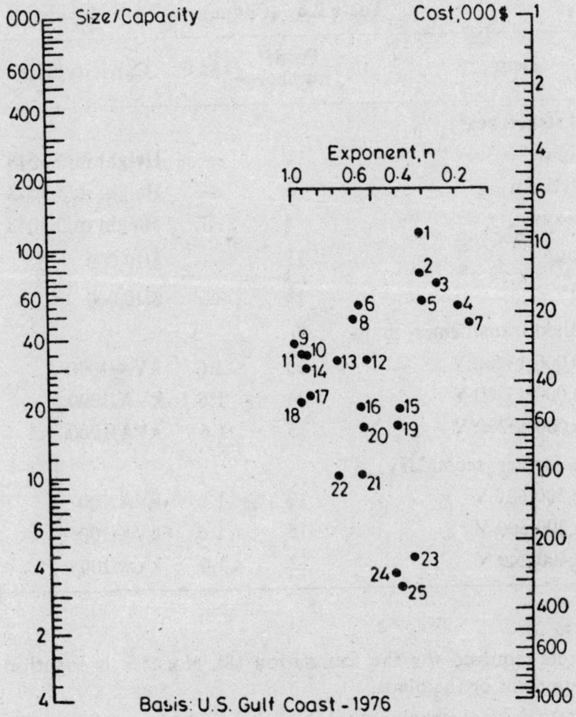

Fig. 9.4 Cost of equipment given in Table 9.6. (Reprinted from Ref. 2 with permission of Dr. O. P. Kharbanda, O. P. Kharbanda and Associates, Bombay.)

Table 9.6[2] Special and Miscellaneous Equipment Appearing in the Nomograph of Fig. 9.4

Sl.No.	Item	Point number	*M*	Capacity/Size	Range
1.	Concrete*	10	—	\$/10,000	1–10,000
2.	Concrete structures*				
	Light	3	—	m/0.3048	10–200
	Medium	4	—	m/0.3048	10–200
	Heavy	7	—	m/0.3048	10–200
3.	Ductwork*				
	Stainless steel	21	—	m^2/0.0929	1–100
	Galvanized sheets	20	—	m^2/0.0929	1–100
	Aluminium	16	—	m^2/0.0929	1–100
4.	Electrical*	14	—	\$/10,000	1–1000
5.	Hoppers, industrial*	22	1.08	m^3/28.317	1–70
6.	Instruments	17	—	\$/10,000	1–1000
7.	Insulation*	9	—	\$/10,000	1–1000
8.	Panel-graphic	13	—	Linear m/0.3048	2–100
	Miniature	8	—	Linear m/0.3048	2–100
	Large instrument	6	—	Linear m/0.3048	2–100

(Contd.)

Table 9.6 (*Contd.*)

Sl.No.	Item	Point number	*M*	Capacity/Size	Range
9.	Steel structures*				
	Light	11	—	Height m/0.3048	10–300
	Medium	2	—	Height m/0.3048	10–300
	Heavy	5	—	Height m/0.3048	10–300
10.	Piping*	11	—	\$/10,000	2–1000
11.	Steel*	18	—	\$/10,000	1–1000
12.	Oil filled transformer, main				
	70,000/13200 V	25	1.6	kVA/1000	2–50
	40,000/13200 V	24	1.6	kVA/1000	2–50
	25,000|13200 V	23	1.6	kVA/1000	2–50
13.	Transformer, secondary				
	12,300/600 V	19	1.6	kVA/100	2–60
	4,200/600 V	15	1.6	kVA/100	2–60
	2,400/240 V	12	1.6	kVA/100	2–15

*Notes**

(1) The concrete required for the foundation (Sl. No. 1) a is function of the cost of the equipment or the plant.

(2) For concrete or steel structures (Sl. Nos. 2 and 9), the total cost, C_T is given by

$$C_T = (\text{Installation factor}) \times (\text{Area}) \times \text{Height}$$

In the nomograph, installation factor, I.F., are reported as a function of height.

For concrete and steel structures, $\text{I.F.} = \dfrac{\text{Nomograph reading}}{10{,}000}$

(3) For ductwork (Sl. No. 3), the nomograph reading is found for the specified cross-section.

$$\text{Cost of ductwork per unit length, \$} = \frac{\text{Nomograph reading}}{1{,}000}$$

(4) Electrical (Sl. No. 4) cost is correlated with the total equipment cost.

(5) Industrial hopper (Sl. No. 5) cost is found by dividing the nomograph reading by 100.

(6) Insulation (Sl. No. 7) cost is correlated with the total equipment cost. To find the cost, the nomograph reading is divided by 10. If it is a hot insulation, the cost is,

$$\text{Cost} = 0.75\left(\frac{\text{Nomograph reading}}{10}\right)$$

(7) Piping (Sl. No. 10) cost is correlated with total material + equipment cost. To find the cost the nomograph reading is multiplied with 10.

(8) To find the steel cost (Sl. No. 11), the nomograph reading is divided by 10. The cost is correlated with total material + equipment cost.

(9) To find the cost of fully installed units, the nomograph reading is multiplied by *M*.

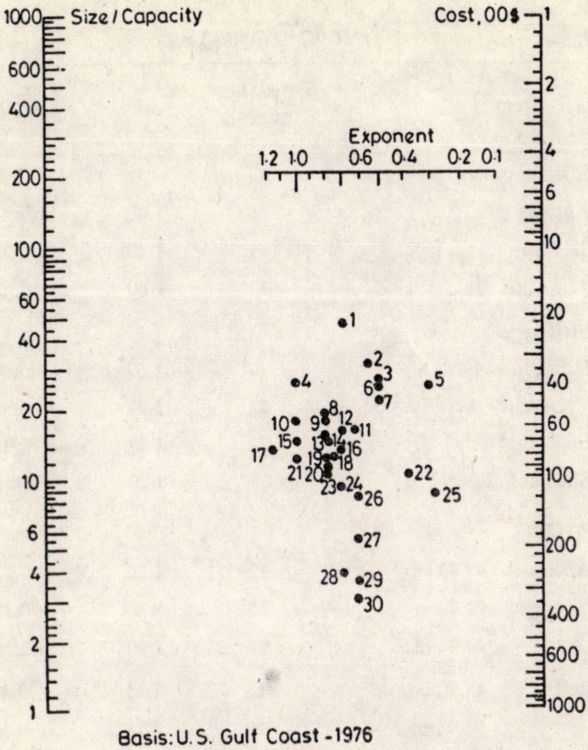

Fig. 9.5 Cost of equipment given in Table 9.7. (Reprinted from Ref. 2 with permission of Dr. O. P. Kharbanda, O. P. Kharbanda and Associates, Bombay.)

Table 9.7[2] Special and Miscellaneous Equipment Appearing in the Nomograph of Fig. 9.5

Sl.No.	Item	Point number	*M*	Size/Capacity
1.	Agitators			
	Propeller	3	1.12	kW/0.7457
	Turbine	5	1.12	kW/0.7457
2.	Air compressors			
	9.5 atm	25	1.25	$\left(\frac{m^3}{min}\right)/0.0283$
3.	Air dryer	2	1.25	$\left(\frac{m^3}{min}\right)/0.0283$
4.	Ball mill	18	1.10	M Tons/day/1.016
5.	Blender	23	1.12	$m^3/0.0283$
6.	Blower/fan*	1	1.10	$\left(\frac{m^3}{min}\right)/0.0283$

(*Contd.*)

Table 9.7 (*Contd.*)

Sl.No.	Item		Point number	*M*	Size/Capacity
7.	Industrial boiler				
	2 atm		6	1.40	kg/h/0.4536
	21.5 atm		7	1.40	kg/h/0.4536
	42.0 atm		11	1.40	kg/h/0.4536
8.	Centrifuges				
	Horizontal basket		17	1.57	diameter, cm/2.54
	Vertical basket		21	1.57	diameter, cm/2.54
	Bird solid bowl		22	1.27	kW/0.7457
	Sharples super		28	1.22	kW/0.7457
9.	Conveyors				
	Apron	45.70 cm	12	1.20	m/0.3048
		91.4 cm	16	1.20	m/0.3048
		274.3 cm	24	1.20	m/0.3048
	Belt	45.70 cm	14	1.25	m/0.3048
		91.4 cm	19	1.25	m/0.3048
		274.3 cm	20	1.25	m/0.3048
	Vibrating	30.5 cm	4	1.15	m/0.3048
		45.70 cm	10	1.15	m/0.3048
		66.00 cm	15	1.15	m/0.3048
	Screw	22.9 cm	8	1.10	m/0.3048
		30.5 cm	9	1.10	m/0.3048
		40.6 cm	13	1.10	m/0.3048
10.	Cranes				
	Span	3.05 cm	26	—	MT/1.016
		9.15 m	27	—	MT/1.016
		15.24 m	29	—	MV/1.016
		30.48 m	30	—	MT/1.016

**Note*:

(1) For blower/fans (Sl. No. 6), divide the nomograph reading by 10 to find the cost of the equipment.

(2) To find the cost of fully installed unit, the nomograph reading is multiplied by *M*.

(Reprinted from Ref. 2 with permission of Dr. O. P. Kharbanda, O. P. Kharbanda Associates, Bombay.)

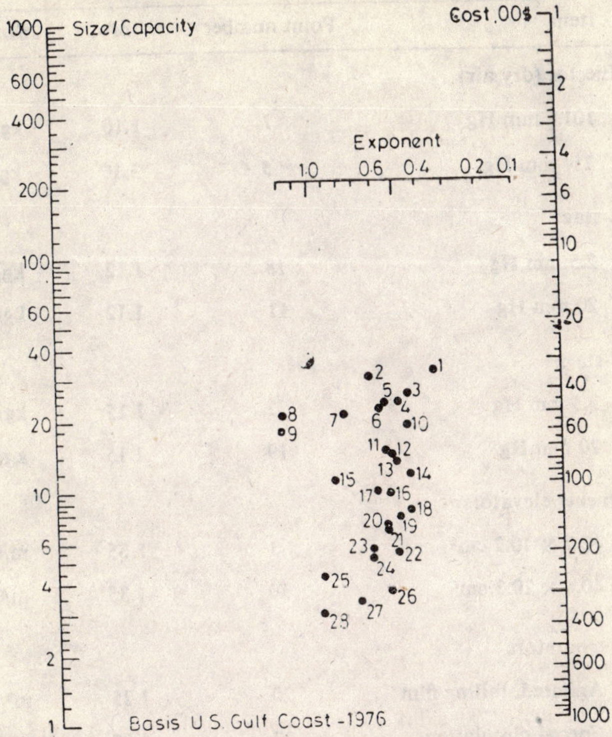

Fig. 9.6 Cost of equipment given in Table 9.8. (Reprinted from Ref. 2 with permission of Dr. O. P. Kharbanda, O. P. Kharbanda Associates, Bombay.)

Table 9.8[2] Special and Miscellaneous Equipment Appearing in the Nomograph of Fig. 9.6

Sl.No.	Item	Point number	*M*	Size/Capacity
1.	Crushers			
	Cone	15	1.08	MT/h/1.016
	Gyratory	8	1.08	MT/h/1.016
	Jaw	9	1.08	MT/h/1.016
	Pulverizer	1	1.10	kg/h/0.4536
2.	Crystallizer			
	Growth	24	1.12	MT/day/1.016
	Forced circulation	26	1.28	MT/day/1.016
	Forced batch	2	1.60	MT/day/1.016
3.	Dryer			
	Pan	17	1.25	m²/0.0929
	Rotary vacuum	21	1.25	m²/0.0929
	Drum	23	1.60	m²/0.0929

(*Contd.*)

Table 9.8 (*Contd.*)

Sl.No.	Item	Point number	*M*	Size/Capacity
4.	Ejector (dry air)			
	101.6 mm Hg	7	1.10	kg/h/0.4536
	254 mm Hg	5	1.10	kg/h/0.4536
	4-stage			
	2.5 mm Hg	18	1.12	kg/h/0.4536
	20 mm Hg	11	1.12	kg/h/0.4536
	5-stage			
	2.5 mm Hg	22	1.15	kg/h/0.4536
	20 mm Hg	19	1.15	kg/h/0.4536
5.	Bucket elevators			
	12.7 × 10.2 cm^2	3	1.35	m/0.3048
	40.6 × 20.3 cm^2	10	1.35	m/0.3048
6.	Evaporators			
	Agitated, falling film	20	1.25	m^2/0.0929
	Forced circulation	27	1.90	m^2/0.0929
	Vertical tube	16	1.90	m^2/0.0929
	Horizontal tube	12	1.90	m^2/0.0929
	Jacketed vessel	13	1.74	m^2/0.0929
7.	Filters			
	Plates and press filter	6	1.80	m^2/0.0929
	Pressure leaf			
	Wet	4	1.65	m^2/0.0929
	Dry	14	1.65	m^2/0.0929
8.	Hydraulic press			
	7.8 atm	25	1.78	m^2/0.0929
	69.0 atm	28	1.78	m^2/0.0929

Note: To find the cost of fully installed unit, the nomograph readings are multiplied by *M*.

(Reprinted from Ref. 2 with permission of Dr. O. P. Kharbanda, O. P. Kharbanda Associates, Bombay.)

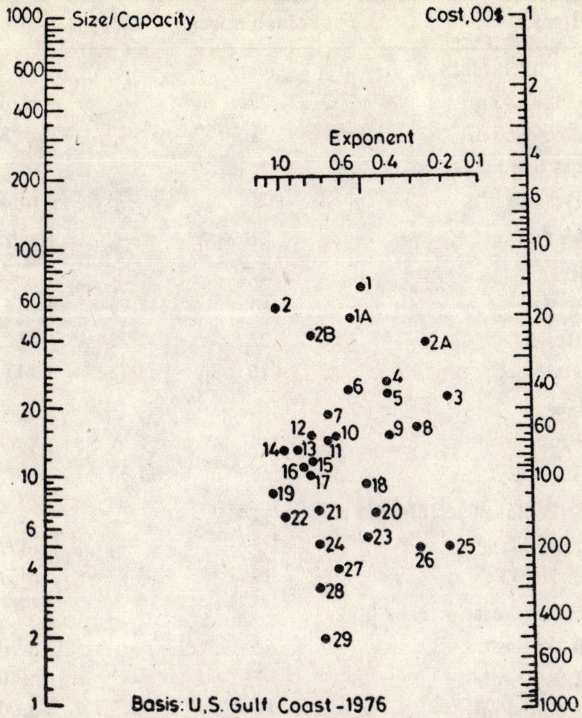

Fig. 9.7 Cost of equipment given in Table 9.9. (Reprinted from Ref. 2 with permission of Dr. O. P. Kharbanda, O. P. Kharbanda and Associates, Bombay.)

Table 9.9[2] Special and Miscellaneous Equipments Appearing in the Nomograph of Fig. 9.7

Sl.No.	Item	Point number	*M*	Size/Capacity
1.	Autoclave	5	—	m^3/0.0283
2.	Electric boiler			
	0.75–2.60 kW	4	—	kW/0.7457
	3.00–11.20 kW	7	—	kW/0.7457
3.	Water cooler	22	—	kJ/h/10551
4.	Pulp dewaterer			
	25–400 MT/h	17	—	MT/h/1.016
	37.3–596.5 kW	12	—	kW/0.7457
5.	Distillation column*	1A	—	cm^3/16381 (Dia) × (Height) × (Thickness)
6.	Thin film evaporator	18	—	m^2/0.0929
7.	Wiped film evaporator			
	30.5 cm	26	—	m^2/0.0929
	91.4 cm	23	—	m^2/0.0929

(Contd.)

Table 9.9 (*Contd.*)

Sl.No.	Item	Point number	*M*	Size/Capacity
8.	Glass lined tanks			
	0.38–3.80 m^3	2A	—	m^3/0.003785
	3.80–9.50 m^3	1	—	m^3/0.3785
	Glass lined jacketed			
	Reactor	20	—	m^3/0.003785
9.	LPG storage*	2B	—	m^3/3.785
10.	Mill			
	Ball	10	1.78	MT/h/1.016
	Roller	27	1.78	MT/h/1.016
	Hammer	16	1.78	MT/h/1.016
11.	Mixer			
	Propeller	6	1.38	kW/0.7457
	Turbine	8	1.38	kW/0.7457
12.	Horizontal mixer/blender ribbon type			
	1–2 MT	9	—	MT/1.016
	3–51 MT	19	—	MT/1.016
13.	Pressure, rotator turba-film			
	0.3–1.9 m^2	25	—	m^2/0.0929
	1.9–7.4 m^2	1	—	m^2/0.0929
	7.4–20.4 m^2	13	—	m^2/0.0929
14.	Refrigeration units			
	−6.7°C	21	1.38	Tons
	−29°C	24	1.38	Tons
	−40°C	28	1.38	Tons
	−80°C	29	1.38	Tons
15.	Scrubber, FRC			
	57–113 m^3/min	3	—	$\left(\frac{m^3}{min}\right)/28.317$
	141–991 m^2/min	11	—	$\left(\frac{m^3}{min}\right)/28.317$
16.	Separator-clarifier	14	—	$\left(\frac{m^3}{min}\right)/0.003785$
17.	Spray dryer, 2.70–6.1 m*	2	—	m/0.3048
18.	Tank, FRP	15	—	m^3/3.785

**Notes*:

(1) For Sl. Nos. 5 and 9, multiply the nomograph reading by 10 to find the cost.

(2) For Sl. No. 17, multiply the nomograph reading by 1000 to find the cost.

(3) To find the cost of fully installed unit, the nomograph reading is multiplied by *M*.

(Reprinted from Ref. 2 with permission of Dr. O. P. Kharbanda, O. P. Kharbanda Associates, Bombay.)

Table 9.10 Equations Governing Multiplying Factors for Different Equipment

Sl.No.	Equipment	Multiplying factors	
1.	Furnace	$m_{F,\,Fur} = (d + p + m)$ when	$d, p, m \neq 0$
		$= (1 + m)$	$d = p = 0$
		$= (1 + p)$	$d = m = 0$
		$= (1 + p + m)$	$d = 0$
		$= 1$	$d = p = m = 0$
2	Cooler	$m_{F,\,Cool} = (d + p + m)$	$d, p, m \neq 0$
		$= 1 + m$	$d = p = 0$
		$= 1 + d$	$p = m = 0$
		$= p$	$d = m = 0$
		$= (1 + m + d)$	$p = 0$
		$= 1$	$d = p = m = 0$
3.	Shell and tube heat exchanger	$m_{F,\,HX} = (d + p_s + p_t)m$	$d, p_s, p_t, m \neq 0$
		$= (1 + p_s + p_t)m$	$d = 0$
		$= (d + p_s)m$	$p_t = 0$
		$= (d + p_t)m$	$p_s = 0$
		$= dm$	$p_s = p_t = 0$
		$= (d + p_s + p_t)$	$m = 0$
		$= 1 + p_s + p_t$	$d = m = 0$
		$= 1 + p_s$	$p_t = d = m = 0$
		$= 1 + p_t$	$p_s = d = m = 0$
		$= 1$	$d = p_s = p_t = m = 0$
4.	Pumps	$m_{F,\,P} = mp$	$p, m \neq 0$
		$= p$	$m = 0$
		$= m$	$p = 0$
		$= 1$	$m = p = 0$
5.	Boilers	$m_{F,\,B} = (p + d)$	$p, d \neq 0$
		$= p$	$d = 0$
		$= (1 + d)$	$p = 0$
		$= 1$	$p = d = 0$
6.	Process vessels	$m_{F,\,Pro} = pm$	$p, m \neq 0$
		$= p$	$m = 0$
		$= m$	$p = 0$
		$= 1$	$m = p = 0$
7.	Refrigeration	$m_{F,\,Ref} = d$	
		$= 1$	when $d = 0$
8.	Storage tanks	(a) API conical (Sl.No. 8a of Table 9.5) and Light gauge tank (Sl.No. 8e)	
		$m_{F,\,T_1} = m$	
		$= 1$	$m = 0$

(*Contd.*)

Table 9.10 (*Contd.*)

Sl.No.	Equipment	Multiplying factors	
		(b) API conical (Sl.No. 8b of Table 9.5)	
		$m_{F, T_2} = d + m$	$d, m = 0$
		$= d$	$m = 0$
		$= (1 + m)$	$d = 0$
		$= 1$	$d = m = 0$
		(c) Horizontal tank (Sl.No. 8c of Table 9.5) and Sphere (Sl.No. 8f)	
		$m_{F, T_3} = p$	$p \neq 0$
		$= 1$	$p = 0$
9.	Distillation columns	$m_{F, Dist.} = (d_s + d_{ty} + m)$	$d_s, d_{ty}, m \neq 0$
		$= (1 + d_{ty})$	$d_s, m = 0$
		$= (d_s + m)$	$d_{ty} = 0$
		$= 1$	$d_s = d_{ty} = m = 0$
		d_s = parameter to account for change in tray spacing	
		d_{ty} = parameter to account for change in tray type	

Table 9.11[a] Correction of Costs of Equipment of Table 9.4 for Changes in Design, Material of Construction and Pressure

Sl.No.	Equipment		Correction factor
1.	Multiplier for design change (d)		
	(a) Furnace		d
	Pyrolysis		1.10
	Reformer		1.35
	(b) Shell and tube exchanger		
	Kettle reboiler		1.34
	U-tube		0.84
	Fixed tube		0.79
	(c) Air cooler		
	Tube length	4.88 m	0
		6.10 m	0.03
		7.32 m	0.10
		9.14 m	0.15
	(d) Heater, fired		
	Dowtherm		1.33
2.	Multiplier m for the change in material of construction		
	(a) Furnace		m
	Cr-Mo		0.25
	SS		0.35
	(b) Heat exchanger, shell/tube		
	CS/ADM		1.20
	Mo/Mo		1.90
	CS/Mo		1.60
	SS/SS		3.00

(*Contd.*)

Table 9.11 *(Contd.)*

Sl.No.	Equipment	Correction factor
	CS/SS	2.25
	Ti/Ti	13.00
	CS/Ti	7.00
	Monel/Monel	3.75
	(c) Air cooler	
	Al	0.50
	SS-316	2.00
	(d) Heater fired	
	Cr/Mo	0.44
	SS	0.75
	(e) Centrifugal	
	CI	1.00
	CS	1.38
	SS	2.03
	(f) Centrifugal (in-line)	
	CS	1.28
	SS	1.64
	(g) Chemical injection	
	CS	1.25
	SS	1.95
	(h) Reciprocating	
	CI	1.00
	CS	1.82
	SS	2.25
3.	Multiplier p for pressure change	

Equipment	Multiplier p for pressure in atm 34.0	68.0	204.0	10.2	102.0
(a) For process furnace	0.04	0.10	0.36	—	—
(b) For pyrolysis furnace	0.05	0.15	0.42	—	—
(c) For reformer furnace	0.08	0.20	0.60	—	—
(d) For kettle reboiler					
Shell	0.16	0.24	0.45	—	—
Tube	0.02	0.05	0.09	—	—
(e) For floating head exchanger					
Shell	0.05	0.13	0.35	—	—
Tube	0.04	0.06	0.10	—	—
(f) For U-tube exchanger					
Shell	0.06	0.15	0.35	—	—
Tube	0.04	0.06	0.11	—	—
(g) For double pipe exchanger	—	~0.10	1.0	—	0.3
(h) For fired heater (tube press)	—	0.14	—	~0.11	—
For Dowtherm	—	0.11	—	~0.11	0.17
(i) For air cooler	1.10	1.15	—	—	—
(j) Centrifugal pumps (suction pressure)	1.62	2.12	—	1.0	—
(k) Centrifugal (in line)	1.48	1.92	—	—	—
(l) Chemical injection pumps	1.37	1.79	—	—	—
(m) Reciprocating pumps	1.32	1.53	—	1.00	—

Table 9.12[2] Factors for Changes in Design, Materials of Construction and Pressures of Operation in Equipment of Table 9.5

Sl.No.	Equipment	Correction factor
1.	Changes in design, *d*	
	(a) Boiler, field erected	
	Super heat temp. 100°C	0.09
	200°C	0.18
	300°C	0.22
	400°C	0.28
	(b) Boiler, package	
	Super heat temp. 200°C	0.10
	250°C	0.18
	300°C	0.24
	(c) Cooling tower	
	Cooling range of 8.33°C	1.0
	Cooling range of 11.11°C	1.58
	Cooling range of 13.89°C	1.97
	(d) Refrigeration (Mech.)	
	Evaporation temp. of 40°C	1.00
	Evaporation temp. of 30°C	1.50
	Evaporation temp. of 0°C	1.90
	Evaporation temp. of −20°C	2.20
	Evaporation temp. of −40°C	3.50
	(e) A.P.I. conical	
	Floating roof	1.09
	Lifter roof	1.26
	(f) Distillation columns	
	Tray spacing 60.96 cm	1.0
	Tray spacing 45 72 cm	1.3
	Tray spacing 30.48 cm	1.2
	Tray type	
	Sieve	0.0
	Trough/valve	0.3
	Bubble cap	1.6
	Koch kascade	3.2
2.	Changes in the materials of construction, *m*	
	(a) Process vessels	
	SS 304 cladding	2.30
	SS 304 solid	3.50
	SS 316 cladding	2.60
	SS 316 solid	4.25
	Monel cladding	4.50
	Monel solid	9.75
	Titanium cladding	4.89
	Titanium solid	10.60

(Contd.)

Sl.No.	Equipment	Correction factor
	(b) API conical tanks	
	Aluminium as the material	1.46
	Rubber lining	1.52
	Lead lining	1.58
	Glass lining (max. size of 37850 lit)	4.24
	Stainless steel	3.25
	(c) API conical	
	Stainless steel material	2.35
	Rubber lining	0.49
	Rubber lining	0.58
	(d) Light gauge tank	
	Aluminium material	1.46
	Rubber lining	1.52
	Lead lining	1.58
	Glass lining (max. size of 37850 lit)	4.24
	Stainless steel lining	3.25
	(e) Distillation columns	
	Stainless steel	1.5
	Monel	3.5
3.	Change in pressure, *p*	
	(a) Boiler, field erected	
	Pressure, 34.02 atm	1.05
	47.06 atm	1.18
	68 atm	1.35
	204.0 atm	1.58
	(b) Boiler, package	
	Pressure, 20.4 atm	1.08
	27.2 atm	1.31
	40.8 atm	1.74
	(c) Process vessels, horizontal and vertical	
	Pressure, 7.14 atm	1.25
	40.82 atm	3.00
	71.40 atm	4.20
	136.00 atm	6.50
	272.00 atm	11.25
	340.0 atm	13.75
	(d) Tanks (horizontal) under pressure (Sl.No. 8d of Table 9.5)	
	Pressure range	
	10.20–13.60 atm	1.18
	13.60–17.01 atm	1.38
	(e) Tanks, spherical	
	Pressure, 3.40 atm	1.08
	6.80 atm	1.26
	13.60 atm	1.52

to the table, the size/capacity axis is in kJ/(1.0551×10^6) and therefore point 2 is aligned with 100 on this scale and one finds the nomograph reading as 6000. When going through the footnote of Table 9.4 it is found that the cost of the furnace would be 100 times this value, or \$ 6,00,000. The exponent n can also be found from this nomograph by drawing a vertical line on the 'exponent' axis from the point number. For the furnace, n is read to be 0.68 as shown in the nomograph.

To give another example, the cost of a distillation column with diameter of 304.8 cm and height 30.48 cm is desired to be found. The column cost would be \$ 34,000 as read from the nomograph of Fig. 9.4. If the tray spacing is taken as 60.96 cm, it shall have 50 plates. Table 9.5 gives the cost of a single carbon steel (CS) tray of 304.8 cm diameter as \$ 620. Therefore the total cost C_T is given by

$$C_T = (34{,}000 + 620 \times 50) = 65{,}000 \tag{9.3.6}$$

This represents the ex-factory price of the column. To find the cost of an installed unit on the plant site, this must be multiplied by factor M given in Tables 9.4 to 9.9, against every equipment. For example, for furnace it is 2.14 and the cost of the installed furnace (discussed above) would be $2.14 \times 600{,}000$ or \$ 1,284.000.

In the previous section, the cost of equipment was shown to depend upon design, operating conditions and the material of construction. In the following these have been examined and shown how they are accounted for.

9.3.2 Effect of Variation of Design, Material of Construction and the Operating Pressure on the Cost of Chemical Equipment

When accounting for variation of design, material of construction and operating pressure, experience shows that the exponent n in Eq. 9.3.5 does not change significantly. This would imply that these variations merely shift the line horizontally on nomographs without changing its slope. As a result, they show up as a multiplying factor m_F in equipment cost. It is desired to relate this to different changes in the equipment.

One defines the following:

$$d = \text{factor to account for the change in design} \tag{9.3.7a}$$

$$m = \text{factor to account for the change in material of construction} \tag{9.3.7b}$$

$$p = \text{factor to account for the change in the operating pressure} \tag{9.3.7c}$$

There is no theoretical guideline for the relation between m_F and d, p and m. The empirical correlations for different equipments are summarized in Table 9.10 and the numerical values of these parameters are given in Tables 9.11 and 9.12.

To give an example, the problem of the catalytic reformer furnace discussed earlier is reconsidered for which it is decided to have stainless steel tubes operating at 34.01 atm. From Table 9.11, it is found that

$$d = 1.35 \tag{9.3.8a}$$

$$p = 0.08 \tag{9.3.8b}$$

$$m = 0.35 \tag{9.3.8c}$$

and $m_{F,\,Fur}$ is equal to 1.78. The cost of this furnace would then be (600,000) × 1.78 or \$ 10,68,000 and the cost of an installed one would be (600,000 × 2.14) × 1.78 or \$ 22,90,000.

As another example, it is desired to find the cost of a kettle reboiler with specifications as below:

Surface area = 278.70 m^2

Material shell—Stainless steel; Pressure shell = 68.03 atm

Material tube—Stainless steel; Pressure tube = 136.05 atm

Kettle reboiler is a shell and tube heat exchanger and as a result, from Table 9.4 the appropriate nomograph is found to be Fig. 9.2 and point 13 corresponds to the equipment. The cost of the equipment is read as \$ 34,000. Values of d, p_s, p_t and m are read from Table 9.11 as

$$d = 1.34$$

$$p_s = 0.24$$

$$p_t = 0.09$$

$$m \text{ for SS/SS} = 3.00$$

Therefore $m_{F,\,HX}$, according to Table 9.10, is 5.10 and the cost of an uninstalled kettle reboiler is found to be 34,000 × 5.10 or \$ 1,74,000 whereas that of an installed one as 1,74,000 × 3.17 or \$ 5,52,000.

Thus it is seen that the cost of an equipment can be found in dollars if the design is known. Since these nomographs have been prepared in terms of costs at Gulf Coast, they must be corrected for any other country. Further, there is continued inflation due to which there is fall in values of currencies. These two factors have been known to affect equipment cost and are accounted for by the location index and inflation index discussed below.

9.4 LOCATION AND INFLATION INDICES FOR EQUIPMENTS AND CHEMICAL PLANTS

Location index, LI, is defined as the ratio of the cost of an equipment at a given place to that in the USA

$$\text{LI} = \frac{\text{cost at a place}}{\text{cost in USA}} \tag{9.4.1}$$

and it is assumed to be independent of the nature of the equipment. Since a chemical plant is nothing but a congregation of equipment it is expected that the location index remains the same for the plant as a whole. Table 9.13 gives a cost comparison on an international level that exists in

Table 9.13[1, 2] Relative Costs of Equipment and Plants in Different Countries

Sl.No.	Items	France	Japan	Italy	U.K.	U.S.A.	W. Germany
1.	Equipment						
	Tank	95	80	90	95	100	95
	Pump	80	85	85	90	100	85
	Heat exchanger	100	110	90	96	100	90
2.	Material						
	Piping	90	90	90	85	100	90
	Electrical	95	80	85	110	100	95
3.	Labour						
	Skilled	23	17	35	17	100	28
	Others	19	15	30	16	75	26

(Reprinted from Ref. 2 with permission of Dr. O. P. Kharbanda, O. P. Kharbanda Associates, Bombay.)

literature[1, 2, 10, 11] from which different equipments are found to have different trends of variation. In view of it, the average location index for a country is found by using suitable weighting factors. The location indices are given in Table 9.14 and are the average values for different

Table 9.14[2] Location Index of Different Countries with the U.S.A. Cost as the Basis

Country	Location index
U.K.	96
France	94
India	90
Italy	94
Japan	64
West Germany	99
U.S.A.	100

(Reprinted from Ref. 2 with permission of Dr. O. P. Kharbanda, O. P. Kharbanda Associates, Bombay.)

countries. To account for the change in equipment cost with time, an empirical parameter called inflation index is used. There are several kinds of inflation indices depending upon how they have been calculated (e.g.,

ENR index, Nelson index, Marshall and Stevens index; Chemical Engineering Equipment index, USA indices and EPE index). Table 9.15 gives the recommended inflation index for different countries which would

Table 9.15[a] Recommended Inflation Indices for Different Countries
(Base: 1970 = 100)

Year	U.K.	France	India	Italy	Japan	U.S.A.	West Germany
1965	76	66	80	60	50	81	78
1966	78	69	85	68	62	83	81
1967	82	72	90	70	69	87	81
1968	85	77	91	76	77	91	84
1969	93	87	93	91	91	97	94
1970	100	100	100	100	100	100	100
1971	120	115	104	105	106	106	114
1972	129	131	110	111	111	110	125
1973	140	155	128	123	124	116	138
1974	177	183	165	146	154	134	160
1975	225	205	175	171	172	146	172
1976	270	238	161	200	188	154	186
1977	289	262	175	236	204	169	190
1978	328	295	179	265	211	184	199
1979*	355	314	215	282	228	195	212
1980*	385	334	237	310	246	206	225
1981*	416	355	261	341	256	220	239
1982*	450	377	288	375	277	234	254
1983*	487	400	317	414	300	250	270

*Calculated on the basis of *assumed* annual inflation rates for France, U.S.A. and Germany as 6%, for U.K. and Japan as 8% and for India and Italy as 10%.
(Reprinted from Ref. 2 with permission of Dr. O. P. Kharbanda, O. P. Kharbanda Associates, Bombay.)

be used here.[a] To find the cost for any future (or past) year, the following relation is used:

$$\frac{\text{Cost in year 1}}{\text{Cost in year 2}} = \frac{\text{Inflation index in year 1}}{\text{Inflation index in year 2}} \tag{9.4.2}$$

This method of finding future cost gives satisfactory results as long as the year difference does not exceed 10.

9.5 COST DATA OF SMALL EQUIPMENT IN INDIA

From the method outlined in Sec. 9.4, it is possible to calculate the cost of an equipment in India. Supposing there is a shell and tube exchanger having a surface area of 92.9 m^2. From Fig. 9.2, its cost would be 16,000 dollars at the U.S. Gulf Coast in 1976. Since the location index for India is 90, its cost in India would be ($16000 \cdot \frac{90}{100}$) or 14,400 dollars. To correct this cost for 1979, the inflation indices from Table 9.15 are found for these two years as 161 and 215 and the value in the year 1978 as ($14{,}400 \cdot \frac{215}{161}$) or 19,152 dollars. Since the international price of a dollar is about 10 rupees, the cost of the exchanger in 1978 is 1,91,520 rupees and this value just about maches with the actual cost of a shell and tube exchanger in India[12]. However, in the case of small fabricated equipment (storage tanks, heat exchangers, reactors and distillation columns) discrepancies creep in because of the following reasons.

In India there are several small fabrication shops with a holding no more than a couple of machines and a small shed in which they carry out their work. These machines have been used for several years and are kept in running condition through good maintenance. Because of the large number of years of working, their book value is zero and therefore c_F in Eq. 9.3.2 consists of mainly

1. direct operating labour charges, and
2. maintenance charges

and in view of this, c_F is small. In a large modernized plant, as opposed to this, there are several more complex fabrication machines which are continually displaced to keep up with technological innovations. They use sophisticated mechanical testing procedures to produce reliable chemical equipment. Unlike small fabricators, they have large establishment charges and due to the depreciated costs of these machines, c_F swells up to a very large value. As has been discussed in Sec. 9.4, it is only because of the predominance of the latter term, the empirical Eq. 9.3.5 is found generally to hold good. When c_F is small, which is the case for small fabrication companies, the prices of equipment are quoted in terms of the weight of the material required. Indeed there are several companies in India which quote prices in terms of the weight of steel required. Table 9.16 summarizes average numbers for different equipment in this

Table 9.16[13] Average Cost Numbers for Fabricated Equipment in India for the Year 1974-75

Sl.No.	Item	Cost trend		
1.	Material	Estimated ex-factory price in Rs. per kg of the material required		
		MS	SS	Aluminium
	(a) Storage tanks	10	75	55
	(b) Atmospheric process vessels	12	80.25	60
	(c) Reactors, towers	15	100	70
2.	Heat Exchanger	Rs. per m^2 area		
		MS		SS
		2566		8,000–10,000

(*Contd.*)

Sl.No.	Item	Cost trend
3.		Approximate cost of linings, Rs. per m^2 of the surface 3 mm thick
	(a) Rubber	210
	(b) Lead lining	675
	(c) Lead bonding	800
	(d) FRP lining	350
	(e) PVC bonding	300
	(f) PVC lining	230
	(g) Fine brick lining	400
4.	Utility costs	
	(a) Steam based on coal	Rs. 50 per MT
	(b) Water	Re. 0.60 per 1000 l
	(c) Power	Re. 0.22 per kWh
5.	Cost of chemicals	They can be obtained from *Chemical Weekly*, Sevak Publications, Bombay, India.

country[13]. One should be cautious in using these numbers because they are limited due to the following:

1. equipment do not require high technology of fabrication;
2. they are small in capacity; and
3. they are not as reliable as the ones produced by big fabricators.

After having determined the fixed cost of the plant, another important quantity which must be estimated is the direct operating labour required. It will be seen from Chapter 10 that this affects the throughput of the chemical plant considerably.

9.6 ESTIMATION OF DIRECT OPERATING LABOUR FOR A CHEMICAL PLANT[8, 14–18]

It was pointed out in Chapter 4 that running of a chemical plant consists of sets of activities which must be coordinated. If an activity cannot be performed by a controller, it has to be done by men. People required for the production of the chemical are defined to constitute the direct operating labour which must be differentiated from the clerical staff, maintenance and sales personnel.

To find the operating labour of a plant precisely is a hard task because it requires a complete specification of the plant layout, controllers and description of all the work to be done and the order of carrying them out. The easiest way of finding the labour force is to study a similar existing plant and extrapolate results for the plant to be designed; however this is not always possible. In the absence of information on existing plants, a complete analysis of the activity must be made.

As an example, one considers the chemical plant of Fig. 9.8. The total process involves three raw materials, A, B and C. Two liquid materials,

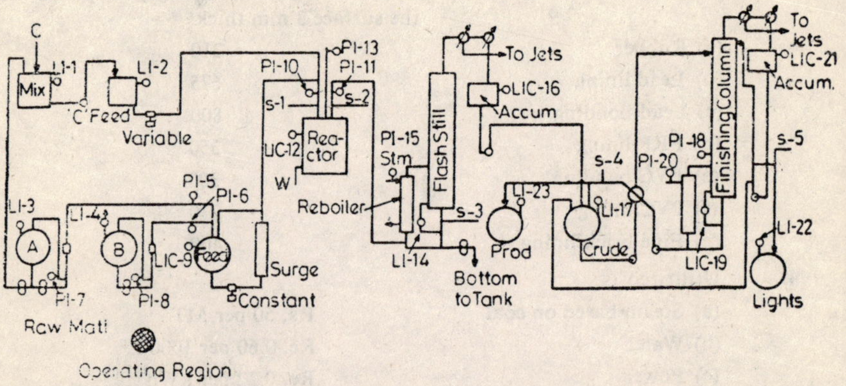

Fig. 9.8 An arbitrary flow sheet for calculation of labour force. (Reprinted from Ref. 15 with permission of the American Institute of Chemical Engineers, New York.)

A and B, are mixed in a feed tank and the mixture is withdrawn through a constant displacement pump into the reactor. Raw material A is withdrawn separately to dissolve solid C and the solution is prepared batchwise. This is subsequently transferred to a feed tank from which it is withdrawn through a variable displacement pump into the reactor. The variable pump ensures the reactants at a known ratio of feeds into the reactor. The crude product is continuously withdrawn and fed into the flash column. The bottom is rejected whereas the top product is condensed and redistilled in the finishing column. The lights from the flash column and the desired product, collected at the bottom, are suitably stored in tanks as shown in the figure. For smooth functioning of the chemical plant, pressure indicators (PI), level indicators (LI) and level indicator controllers (LIC) are suitably placed in the flow sheet. The reactor, being a very important equipment, consists of pressure indicator, level indicator controller and a feed ratio recorder controller (FRRC).

The kind of work personnel in a chemical plant perform has to be contrasted with that of a mechanical assembly or packaging industry. In the latter types, there are very well defined but repetitive tasks. Once the person is oriented, the nature of the job does not require any decision making. In contrast to this, in a chemical plant, an operator has to use his past experience, mental skill of analysis and process know-how to be effective.

Any given task can always be broken into basic unit tasks and these are considerably different from what one encounters in mechanical industries. Some of the unit tasks which are normally performed in a chemical plant are given in Table 9.17 and the corresponding standard times required for them. The latter is defined as the time taken to carry out a unit task in an actual plant operation. These times depend conside-

Table 9.17[15] Standard Time Required for Different Unit Tasks

Sl.No.	Unit task	Standard minutes	Per unit
1.	Walk	0.458	100 ft
2.	Walk downstairs	0.100	10 stairs
3.	Walk upstairs	1.122	10 stairs
4.	Open or close elevator gate	0.044	each time
5.	Check, PIC, PRC or FRC	0.088	check
6.	Check FRRC	0.110	check
7.	Check sight glass line or level indicator	0.099	check
8.	Check manometer	0.154	check
9.	Check PI or TR	0.044	check
10.	Get or aside data board	0.050	each time
11.	Record instrument reading	0.099	recording
12.	Check pump	0.100	check
13.	Check stuck control valve	0.047	check
14.	Adjust variable flow pump	2.830	each time
15.	Adjust control instrument	0.330	each time
16.	Adjust pump packing	0.091	each time
17.	Attention to filling feed tank	1.330	filling
18.	Attention to filling mix tank	2.140	filling
19.	Attention to drain sludge mix tank	0.430	draining
20.	Attention to pumping out of flash column	2,500	pump out
21.	Arrange by-pass valves	0.330	each time
22.	Open or close plug cock	0.091	each time
23.	Open or close hand valve (upto 6.4 cm)	0.143	each time
24.	Open or close sample line valve	0.057	each time
25.	Open or close cabinet	0.062	each time
26.	Pick up sample carrier	0.055	pick up
27.	Remove graduate from carrier	0.066	each time
28.	Flush graduate	0.089	each time
29.	Draw sample	0.078	each time
30.	Set full graduate in carrier	0.088	each time
31.	Get or away wrench	0.100	each time
32.	Tunk valve	0.158	each time
33.	Dismantle air line to control valve	0.240	each time
34.	Air shock control valve	2.269	each time
35.	Unbolt screen from feed line	1.420	each time

(*Contd*)

Table 9.17 (*Contd.*)

S.No.	Unit task	Standard minutes	Per Unit
36.	Clean screen	1.100	each time
37.	Replace screen in line	1.030	each time
38.	Wipe off wrench	0.078	each time
39.	Pick up or replace grease gun	0.071	each time
40.	Grease plug cock	0.250	plug cock
41.	Wipe off grease gun	0.094	each time
42.	On or off switch	0.033	each time
43.	Switch filling tank	0.750	each time
44.	Remove recorder chart from box	0.140	chart
45.	Note date and code number on chart	0.222	chart
46.	Open case, remove old chart	0.160	chart
47.	Position new chart, close case	0.490	chart
48.	Set old charts in rack	0.100	each time
49.	Get or aside ink bottle	0.050	each time
50.	Ink instrument pen	0.189	pen
51.	Notify foreman of operational difficulties	2.600	each time
52.	Get or aside bag cutter (sheath)	0.072	each time
53.	Cut bag open	0.196	bag
54.	Dump 50 lb bag into mix tank	0.350	bag
55.	Make titration	2.050	each time
56.	Take specific gravity	2.740	each time
57.	Run flash column distillation analysis	3.980	each time
58.	Run finishing column distillation analysis	5.060	each time
59.	General area check—open area	0.121	check
60.	General area check—normal area	0.242	check
61.	General area check—congested area	0.418	check
62.	Check consistency	1.210	each time

(Reprinted from Ref. 15 with permission of the American Institute of Chemical Engineers, New York.)

rably upon the working environment and in view of this, they change from place to place and time to time. The measurements given in the table are for a specific company (Dow Chemical Company, Michigan, U.S.A.) and are used for the flow sheet example of Fig. 9.8[15].

Controllers placed on different equipment will normally be separated spatially and for convenience of operation, they are displayed on panel boards. In Fig. 9.9, controllers have been divided on two panel boards, one for the combination of the reactor and flash column and the other for the finishing column. To find the direct operating labour, the complete

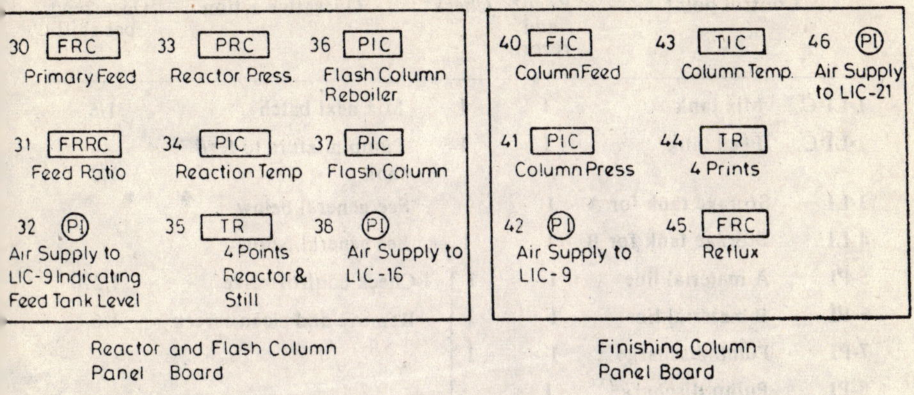

Fig. 9.9 Panel boards for the flow sheet of Fig. 9.8

work task of the plant is broken down as follows. It consists of checking different indicators and correspondingly adjusting the necessary controllers. There exists no guideline to decide upon the frequency of these checks in a shift, except for the fact that the variables to which the overall performance is sensitive are controlled closely. Table 9.18 gives the complete analysis of the frequency of checks and corresponding actions that are required to be done.

After dividing the total work into its elemental tasks, it is an easy matter to find the labour force. One simply assigns these to individual workers and subsequently confirms through the standard times for different unit tasks (Table 9.17) that he is occupied during the entire duration of the shift. Before this is done it is necessary to evolve a systematic time schedule of the work. This is done as follows.

It is noted from Table 9.18 that there are several items of work which have to be repeated 32 times in an 8-hour shift. This would mean that they have to be carried out at every 15 minutes and they are lumped together into a single set of work. A stepwise breakdown of this set of work would be to (1) walk from the operating region to the panel board for the reactor and the flash column, (2) check FRRC-31, (3) check FRC-30, PRC-33, PIC-34, PIC-36 and PIC-37, (4) check PI-32 and PI-38, (5) check four temperature points of TR-35, (6) walk from the panel board to flash still, (7) check LI-14, (8) walk to the finishing column panel board, (9) check FIC-40, PIC-41, TIC-43 and FRC-45, (10) check PI-42 and PI-46, (11) check four temperatures of TR-44, and (12) walk from the panel to the operating region.

The shift is assumed to start at 8 A.M., and end at 4 P.M. Consequently the 15-minute round starts from 8.15 A.M. onwards. The checks and measurements to be done only 8 times in the shift (after every hour starting from 8 A.M.) can be combined with the regular 15-minute checks.

Table 9.18[15] Analysis of Checks and Actions—Check Frequency Per Shift

Control point		Read* and record	Check*	Operative action	Occurrence per shift
1-LI-C	Mix tank	1	1	Mix next batch	1/6
2-LI-C	Feed tank	1	1	Pump mixture to feed tank	1/3
3-LI	Storage tank for A	1	1	See general below	
4-LI	Storage tank for B	1	1	See general below	
5-PI	A material line	1	1 }	Check control valve	1/270
6-PI	B material line	1	1 }	Remove and clean screen	1/6
7-PI	Pump discharge	1	1 }		
8-PI	Pump discharge	1	1 }		
30-FRC	Primary feed flow	4	28	General adjustment	1/3
				Check control valve	1/90
31-FRRC	Secondary feed ratio flow	4	28	Adjust feed ratio specific gravity by sampling as s-2	1/9
				Check control valve	1/90
9-LIC	Feed tank	1	1	See 32	
32-PI	Output air on 9	4	28	Check control valve	1/90
33-PRC	Reactor outlet	4	28	Check control valve	1/3
10-PI	Reactor feed line	4	4 }		
11-PI	Reactor outlet	4	4 }	Reactor wash out	remote
12-LIC	Reactor jacket water	1	1		
13-PI	Reactor Steam vent	—	1/270	See 34	
34-PIC	Reactor jacket water vapour	4	28 }	Check control valve	1/270
35-TR	4 points	4	28 }	In conjunction with others	
14-LI	Flash still reboiler	4	28	General adjustment	16
15-PI	Steam to flash	—	1/270	Draw off fluid from bottom	2
36-PIC	Steam to flash still reboiler	4	28	Check control valve	1/270
37-PIC	Flash still	4	28	Check control valve	1/270
16-LIC	Flash still accumulator	1	1	See 38	
38-PI	Output air on 16	4	28	Check control valve	1/270
17-LI	Crude tank	1	1	See 40	
40-FIC	Finishing column feed	4	28	General adjustment	1/3
18-PI	Finishing column bottom	4	4	See 41	

(Contd.)

	Control point	Read* and record	Check*	Operative action	Occurrence per shift
41-PIC	Finishing column top	4	28	Check control valve	1/270
19-LIC	Finishing column reboiler	4	4	See 42	
42-PI	Output air on 19	4	28	Check control valve	1/270
20-PI	Steam to finishing column	—	1/270	See 43	
44-TR	4 points in column	4	28	General adjustment	1/3
43-TIC	Finishing column, steam to reboiler	4	28	Check control valve	1/270
45-FRC	Finishing column reflux	4	28	General adjustment Check control valve	1/3 1/270
21-LIC	Finishing column accumulator	1	1	See 46	
46-PI	Output air on 21	4	28	Check control valve	1/270
22-LI	Lights tank	1	1	Switch to alternate tank	1/9
23-LI	Finished product tank	1	1	Switch to alternate tank	1/9
s-1	Reactor feed		4	Adjust variable pump	1/90
s-2	Reactor discharge		4	-do-	1/3
s-3	Flash still bottom		4	Adjust at the instrument	
s-4	Finished product		4	Adjust at the instrument	
s-5	Lights in finishing column		4	Adjust at the instrument	

General and related duties

Notify foreman of trouble, operation problems etc.	1/3
General area check	4
Receive and exchange shift information	1
Grease plugcock	2
Adjust pump packing	1/3
Change charts	1/3

*Total number of checks and recording is sum of the two columns; e.g., LI-1 has to be checked and recorded twice in a shift etc.

(Reprinted from Ref. 15 with permission of the American Institute of Chemical Engineers, New York.)

The round of measurements to be done only twice in the shift can be staggered uniformly for the reason of distributing the total work load. Table 9.19 gives the complete work schedule.

One analyzes the total standard time taken in a single round after every fifteen minutes, which is given in Table 9.20. It is noted that the

Table 9.19[15] Complete Schedule of Checkings to be done for the Chemical Plant of Fig. 9.8

No.	Occurrence/ shift	Occurrence time	Control points covered
1.	20	8.15 A.M. onwards after every 15 min.	Explained in text and also in Table 9.20
2.	4	8 A.M., 10 A.M., 12 o'clock, 2 P.M.	No. 1 + recording in No. 1 + check and record 10, 11, 18, 19*
3.	4	9 A.M., 11 A.M., 1 P.M., 3 P.M.	No. 1 + check 10, 11, 18, 10 + take samples s-1 to s-5 + general area check
4. (a)	1	8.30 A.M.	No. 1 + check 5–8, tank levels, 9 and 12†
(b)	1	12.30 A.M.	Same as 4a + recording
5. (a)	1	10.30 A.M.	No. 1 + check 16, 21‡
(b)	1	2.30 A.M.	Same as 5a + recordings

*10, 11, 18 and 19 stand for controllers PI-10, PI-11, PI-18 and LIC-19.

†Checking of tank levels means monitoring of LI-1, LI-2, LI-3, LI-4, LI-17, LI-22, and LI-23.

9 and 12 stand for controllers LIC-9 and LIC-12.

‡16 and 21 stand for controllers LIC-16 and LIC-21.

(Reprinted from Ref. 15 with permission of the American Institute of Chemical Engineers, New York.)

Table 9.20 Standard Times Taken for Composite Events

Sl. No.	Steps	Standard time	Factor	Time taken
1.	Time taken for completion of task 1 of Table 9.19			
	(a) Walk a distance from operating region to Reactor Panel Board (assumed 15 ft.)	0.458	15/100	0.069
	(b) Check 31	0.110	1	0.110
	(c) Check 30, 33, 34, 36, 37	0.088	5	0.440
	(d) Check 32, 38	0.044	2	0.088
	(e) Check 35 (three points)	0.044	3	0.132
	(f) Walk from Reactor Panel Board to flash still (assumed 12 ft.)	0.458	12/100	0.055
	(g) Check 14 (site glass)	0.099	1	0.099
	(h) Walk from flash still to Finishing Column panel board (assumed 33 ft.)	0.458	33/100	0.151
	(i) Check 40, 41, 43 and 45	0.088	4	0.352
	(j) Check 42, 46	0.044	2	0.088
	(k) Check 44 (four points)	0.044	4	0.176
	(l) Walk from panel to the operating region (assumed 5 ft.)	0.458	5/100	0.023
		Total		1.783

(Contd)

Sl. No.	Steps	Standard time	Factor	Time taken
2.	Pump C-Mixture to Feed Tank			
	(a) Walk from operation region to stairway (assumed 15 ft.)	0.458	15/100	0.069
	(b) Down 15 steps	0.100	15/10	0.150
	(c) Walk to C-Mix tank (96 ft.)	0.458	96/100	0.440
	(d) Open plug cock of C-Mix tank	0.091	1	0.091
	(e) Walk to C-Feed tank (assumed 75 ft.)	0.458	75/100	0.344
	(f) Open plug cock of C-feed tank	0.091	1	0.091
	(g) Switch on to start pump	0.033	1	0.033
	(h) Attention to transfer	1.330	1	1.330
	(i) Switch off to stop pump	0.033	1	0.033
	(j) Close plug cock of C-feed tank	0.091	1	0.091
	(k) Walk to C-mix tank (75 ft.)	0.458	75/100	0.344
	(l) Close plug cock of C-mix tank	0.091	1	0.091
	(m) Walk to stairway (96 ft.)	0.458	96/100	0.440
	(n) Climb 15 steps	0.122	15/10	0.183
	(o) Walk to the operating region (15 ft.)	0.458	15/100	0.069
		Total		3.799
3.	Check control valve			
	(a) Walk from instrument to valve (16 ft. on an average)	0.458	16/100	0.073
	(b) Check the position of the valve	0.047	1	0.047
	(c) Walk to the operating region (10 ft.)	0.458	10/100	0.046
	(d) Get wrench from rack	0.100	1	0.100
	(e) Walk to valve (10 ft.)	0.458	10/100	0.046
	(f) Dismantle union of air to diaphragm	0.240	1	0.240
	(g) Air-shock the valve	0.269	1	0.269
	(h) Assemble union	0.171	1	0.171
	(i) Walk to bench (10 ft.)	0.458	10/100	0.046
	(j) Replace wrench	0.100	1	0.100
	(k) Go back to the next job (6 ft. on an average)	0.458	6/100	0.026
		Total		1.166
4.	Cleaning of screens of raw material tanks A and B			
	(a) Get a wrench from rack	0.100	1	0.100
	(b) Walk from the operating region to stairway (15 ft.)	0.458	15/100	0.069
	(c) Down 15 steps	0.100	15/10	0.150
	(d) Walk to the raw material tanks (50 ft.)	0.458	50/100	0.229
	(e) Close the valve (Two)	0.158	2	0.316
	(f) Unbolt the screen from feed line (Two)	1.420	2	2.840
	(g) Clean the screen (Two)	1.100	2	2.200
	(h) Replace screen in line (Two)	1.030	2	2.060
	(i) Open valve (Two)	0.158	2	0.316
	(j) Switch filling tank on (Two)	0.750	2	1.500
	(k) Walk to stairways (50 ft.)	0.458	50/100	0.229
	(l) Climb 15 steps	0.100	15/10	0.150
	(m) Walk to the operating region (15 ft.)	0.458	15/100	0.069
		Total		10.228

(*Contd.*)

Table 9.20 (*Contd.*)

Sl. No.	Steps	Standard time	Factor	Time taken
5.	Mixing of a batch of C			
	(a) Walk from operating region to stairway (15 ft.)	0.458	15/100	0.069
	(b) Down 15 steps	0.100	15/10	0.150
	(c) Walk to C-Mix tank (96 ft.)	0.458	96/100	0.440
	(d) Get a bag cutter	0.072	1	0.072
	(e) Cut ten bags open	0.196	10	1.960
	(f) Put the bag cutter aside	0.072	1	0.072
	(g) Dump ten 50 lb bags into the mixing tank	0.350	10	3.500
	(h) Close the outgoing plug cock of the tank	0.091	1	0.091
	(i) Open the inlet plug cock of the tank	0.091	1	0.091
	(j) Walk to tank A (75 ft.)	0.458	75/100	0.344
	(k) Open the plug cock of tank A	0.091	1	0.091
	(l) Walk back to tank C (75 ft.)	0.458	75/100	0.344
	(m) Switch on the pump	0.033	1	0.033
	(n) Attention to filling mix tank	2.140	1	2.140
	(o) Switch off the pump	0.033	1	0.033
	(p) Close plug cock of tank C	0.091	1	0.091
	(q) Walk to tank A (75 ft.)	0.458	75/100	0.344
	(r) Close plug cock of the tank	0.091	1	0.091
	(s) Walk to stairways (100 ft.)	0.458	100/100	0.458
	(t) Climb stairs (15)	0.122	15/10	0.183
	(u) Walk to the operation region (15 ft.)	0.458	15/100	0.069
		Total		10.666
6.	Draw flash still bottom			
	(a) Walk to the flash still (50 ft.) from the operating region	0.458	50/100	0.229
	(b) Open the plug cock	0.091	1	0.091
	(c) Attention to pumping of flash still fluid	2.500	1	2.500
	(d) Close the valve	0.091	1	0.091
	(e) Walk back to the operating region (50 ft.)	0.458	50/100	0.229
		Total		3.140
7.	Time taken for Task 2 of Table 9.19			
	(a) Time for task 1	(See 1)	1	1.783
	(b) Recording of 22 instruments	0.099	22	2.178
	(c) Check and record PI-10, PI-11 and PI-18	(0.044+0.099)	3	0.429
	(d) Check and record LIC-19	(0.099+0.099)	1	0.198
	(e) Extra distance to be walked (about 50 ft.)	0.458	50/100	0.229
		Total		4.817

(*Contd.*)

Sl. No.	Steps	Standard time	Factor	Total time
8.	Time taken for task 3 of Table 9.19			
	(a) Time for task 1	1.783	1	1.783
	(b) Check 10, 11 and 18	0.044	3	0.132
	(c) Check 19	0.099	1	0.099
	(d) Extra distance to be walked in checking 10, 11, 18 and 19 (50 ft.)	0.458	50/100	0.229
	(e) Extra distance to be walked in going to sample points (100 ft.)	0.458	100/100	0.458
	(f) Pick up sample carriers (one)	0.055	1	0.055
	(g) Remove graduates from carriers (five)	0.066	5	0.330
	(h) Flush graduates (five)	0.089	5	0.445
	(i) Draw samples (five)	0.078	5	0.390
	(j) Set full graduates in the carrier (five)	0.088	5	0.440
	(k) Bring back samples for testing (100 ft.)	0.458	100/100	0.458
	(l) General area check (normal)	0.242	1	0.242
		Total		5.061
9.	Time taken for task 4(a) and 4(b) of Table 9.19			
	(a) Check task 1	1.783	1	1.783
	(b) Go to stairways (15 ft.)	0.458	15/100	0.069
	(c) Go down 15 stairs	0.100	15/10	0.150
	(d) Walk to raw material tanks (50 ft.)	0.458	50/100	0.229
	(e) Check 1, 2, 3 and 4	0.099	4	0.396
	(f) Check LIC-9	0.088	1	0.088
	(g) Walk to product tanks (100 ft.)	0.458	100/100	0.458
	(h) Check 17, 22 and 23	0.099	3	0.297
	(i) Walk back to stairs (50 ft.)	0.458	50/100	0.229
	(j) Climb up stairs (15)	0.122	15/10	0.183
	(k) Walk to the reactor (15 ft.)	0.458	15/100	0.069
	(l) Check LIC-12	0.088	1	0.088
	(m) Walk to the original point (30 ft.)	0.458	30/100	0.138
		Total time for task 4(a)		4.177
	(n) Time for recording 9 instruments	0.099	9	0.891
		Total time for task 4(b)		5.068
10.	Time taken for task 5(a) and 5(b) of Table 9.19			
	(a) Time for task 1	1.783	1	1.783
	(b) Extra distance to be walked to check 16 and 21 (about 200 ft.)	0.458	200/100	0.916
	(c) Check LIC-16 and 21	0.088	2	0.176
		Time for task 5(a)		2.875
	(d) Time for recording two instruments	0.099	2	0.198
		Time for task 5(b)		3.073

(*Contd.*)

Table 9.20 (*Contd.*)

Sl. No.	Steps	Standard time	Factor	Total time
11.	Time for changing single chart			
	(a) Remove a new recorder chart from box	0.140	1	0.140
	(b) Note date and code number on it	0.222	1	0.222
	(c) Open case and remove old chart	0.160	1	0.160
	(d) Position new chart, close case	0.490	1	0.490
	(e) Get an ink bottle	0.050	1	0.050
	(f) Ink the instrument	0.189	1	0.189
	(g) Put the bottle aside	0.050	1	0.050
	(h) Set the old chart in rack	0.100	1	0.100
		Total		1.401
	There are twelve recorders in all and therefore total time required for changing charts would be 16.812 min.			
12.	Time taken for greasing plug cocks			
	(a) Pick up the grease gun	0.071	1	0.071
	(b) Grease the plug cock	0.250	1	0.250
	(c) Wipe off the grease gun	0.094	1	0.094
	(d) Replace the grease gun	0.071	1	0.071
	(e) Extra distance to be walked because of this task (50 ft.)	0.458	50/100	0.229
		Total	0	0.715
13.	Time taken for adjusting packings of nine pumps			
	(a) Extra distance to be walked for checking pumps (150 ft.)	0.458	150/100	0.687
	(b) Check the pump (9 in number)	0.100	9	0.900
	(c) Adjust the packing	0.091	9	0.819
		Total		2.406

act of checking control valve, pumping of C mixture to C-FEED tank, etc. are not single unit tasks but are a combination of several and these are analyzed in this table by breaking them into individual unit tasks. With the help of Table 9.17, the total time required for each of these is found. In Table 9.21, all the activities of the chemical plant are given with the frequencies of their occurrences and time taken in a given single shift. From these, the total standard time of direct operating labour is determined.

The basic purpose of the above example is to examine the effect of the plant capacity upon the labour force requirements. To determine this, the nature of work done by operators is analysed and is found to consist of:

1. checking and attention to instruments;
2. operating adjustment;
3. recording of readings;

4. sampling and testing;
5. work duties; and
6. travel movements.

Table 9.21 Calculation of Total Standard Time Required in a Shift

Sl. No.	Tasks	Standard time	Actual occurrence per shift	Standard time, min/shift
1.	Task 1 of Table 9.19	1.783	20	35.660
2.	Task 2	4.817	4	19.268
3.	Task 3	5.061	4	20.244
4.	Task 4(a)	4.177	1	4.177
5.	Task 4(b)	5.068	1	5.068
6.	Task 5(a)	2.875	1	2.875
7.	Task 5(b)	3.073	1	3.073
8.	Titration (s-1, s-2 and s-3)	2.050	3×4	24.600
9.	Specific gravity (s-2)	2.740	4	10.960
10.	Consistency (s-3)	1.210	4	4.840
11.	Distillation analysis of s-4	3.980	4	15.920
12.	Distillation analysis of s-5	5.060	4	20.240
13.	Mix batch C	10.666	1/6	1.778
14.	Pump C-mixture to feed tank	3.799	1/3	1.266
15.	Check control valve	1.666	10/270	0.062
16.	Check control valve	1.666	3/90	0.056
17.	Check control valve	1.666	1/3	0.555
18.	Remove and clean screen	10.228	1/6	1.703
19.	Adjust FRC-30	0.330	1/3	0.110
20.	Adjust FRRC-31	0.330	1/9	0.037
21.	Adjust PIC-36	0.330	16	5.280
22.	Draw flash still bottom	3.140	2	6.280
23.	Adjust FIC-40	0.330	1/3	0.110
24.	Adjust TIC-43	0.330	1/3	0.110
25.	Adjust FRC-45	0.330	1/3	0.110
26.	Switch tanks-22	0.750	1/9	0.083
27.	Switch tanks-23	0.750	1/9	0.083
28.	Adjust variable flow pump	2.830	1/90	0.031
29.	Adjust variable flow pump	2.830	1/3	0.943
30.	Notify foreman	2.600	1/3	0.867
31.	Grease plug cock	0.715	2	1.430
32.	Adjust pump packing	2.407	1/3	0.802
33.	Change charts	16.812	1/3	5.604
34.	Exchange shift information	2.600	1	2.600
		Total		196.825
		10 per cent allowance		19.683

Total direct labour time required = 216.508 min.

As the plant size increases, the number of controllers per equipment remains essentially unchanged but the distance of separation between different equipment is altered due to their increased dimensions. In view of this, as the plant capacity increases, the time taken for completing travel movements is mainly affected. However, this increase cannot be unbounded and experience shows that it has an upper limit. One considers a situation when one person is assigned to every major equipment and all these people can communicate with each other. Whatever tests have to be carried out are done at the site itself and results communicated to appropriate workers for further action. Since checking and attention to instruments, operating adjustments, recording, sampling and testing are little affected by the capacity of the plant and travel movements have been completely eliminated by having one worker per major equipment, it is therefore expected that this represents the upper limit upon the direct labour.

Instead of carrying out detailed calculations as above, normally a rough value of the labour force is required while designing a new plant. Correlations have been developed and given in Fig. 9.10. The figure gives

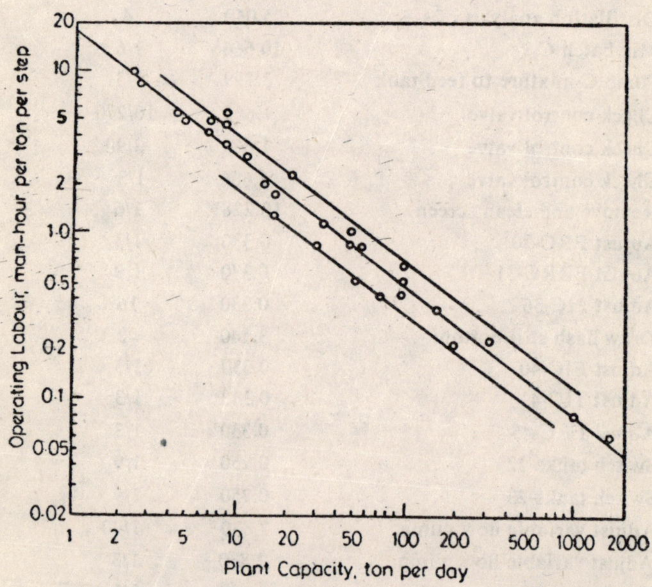

Fig. 9.10 Correlation for finding the operating labour for specified plant capacity. (Reprinted from H. E. Wessel, *Chemical Engineering*, July 1956, with permission of McGraw-Hill, New York.)

three lines: the upper line is meant for batch processes, the middle line, for normally encountered plants whereas the lowermost line, for highly instrumented continuous processes. The procedure of finding the operating labour is to find the number of major steps in obtaining the final product.

For example, in the flow sheet of Fig. 9.8, the number of steps are two: chemical reaction and separation. From the knowledge of the plant capacity the operating labour per step is found from Fig. 9.10.

The analytical equation that Fig. 9.10 represents is given by[8]

$$\frac{MH}{V_c} = t\,\frac{\text{(Number of steps)}}{V_c^{0.75}} \tag{9.6.1}$$

where MH is the man-hour of the operating labour required, V_c is the plant capacity and t is a constant for a given type of the process. The constant t is equal to 23 for batch processes, 17 for continuous (or semi-continuous) processes and 10 for highly instrumented ones. The above equation can be rearranged to

$$MH = KV_c^{0.25} \tag{9.6.2}$$

where

$$K = t\text{ (number of steps)} \tag{9.6.3}$$

and is independent of the plant capacity. Equation 9.6.2 gives the effect of plant capacity upon the labour force required.

In this section, a rigorous method of finding the operating labour force has been given and to calculate it, complete information on plant layout must be available. It is further noted that the man-hour requirements cannot increase unbounded with the plant size. In the early stages of designing when the plant layout is not decided, the empirical relation given in Fig. 9.10 can be used. After the discussion of equipment cost and labour estimation, the overall plant design is now considered.

9.7 OVERALL CHEMICAL PLANT DESIGN[4, 5, 19–23]

As explained in Chapter 4, there are design variables associated with every equipment. The steady state designing of a chemical plant consists of finding these variables for all the equipment in the plant. The total cost, C_t, incurred in producing the product depends to a great extent upon the choice of these variables and the best design is the one which minimizes it, i.e.,

$$\text{Minimize } C_t(x_1, x_2, \ldots, x_n) \tag{9.7.1}$$

where x_1 to x_n are the design variables. However, they cannot take all possible values because the decision upon them is normally constrained by safety considerations, raw material availability, management restrictions, handling problems, mass and energy balance equations, market conditions, etc. These constraints are written as

$$G_i(x_1, x_2, \ldots, x_n)\begin{bmatrix}\le\\=\\\ge\end{bmatrix}b_i \quad i = 1, 2, \ldots, m \tag{9.7.2}$$

The optimality problem consists of finding the best values of x_1 to x_n and the procedure of solving them can be had in any standard textbook of optimization[24].

The aim of this section is to point out that the design of a chemical plant becomes extremely complex because these variables are stochastic in nature. Also along with this, because the plant is dynamic in nature, one must specify a range upon these variables, within which they must be controlled. Table 9.22 gives some typical ranges for different variables. Parameters like density, friction factor, specific heats, heat transfer coefficients, rate constants etc. are also used in equipment design. These are called technological coefficients and a list is given in Table 9.23.

Table 9.22 Typical Control Range of Some Design Variables[21]

Sl. No.	Design variable	Range
1.	Temperature	±2°C at 250°C
2.	Pressure	±0.5%
3.	Differential pressure	±1%
4.	Liquid level	±2%
5.	Flow	±2 to 5%
6.	Tubing I.D.	±0.4%
7.	Electrical power	±1%
8.	Force load or weight	±0.25 to 1%
9.	Current and voltage	±0.25 to 1%

(Reprinted with permission from J. Weisman and A. G. Holtzman, *I. & E. C. Proc. Des. Dev.*, **11**, 386, 1972. Copyright by the American Chemical Society.)

Table 9.23 Examples of Variables and Constraints in Chemical Plant Design[21]

Typical design variables	Typical technological coefficients	Typical constraints
Pressure Temperature Flow	Reaction rate constant Density Friction factor	Rate of heat transfer to be above given value
Chemical composition Diameter	Viscosity Specific heat	Maximum temperature shall be below given value
Wall thickness Volume	Flow Diameter	Wall thickness adequate for operating conditions
Concentration Heat flux	Heat transfer coefficient Pressure loss Thermal conductivity Relative volatility Vapour pressure Volume	Output meets required specifications etc.

(Reprinted with permission from J. Weisman and A. G. Holtzman, *I. & E. C. Proc. Des. Dev.*, **11**, 386, 1972. Copyright by the American Chemical Society.)

These are experimentally determined quantities and are subject to experimental errors. These uncertainties must also be accounted for.

In view of the discussion above, there exists an uncertainty in the production level of the plant. If the range of the production level is specified, it would be of great interest to a designer, to specify the ranges of control upon the design variables and in this section, a method has been given to calculate them.

Since the constraints of Eq. 9.7.2 involve technological coefficients, y_1 to y_r, this equation is rewritten as

$$G_i(x_1, x_2, \ldots, x_n, y_1, \ldots, y_r) \leqslant b_i, \quad i = 1, 2, \ldots, m \tag{9.7.3}$$

The 'greater than or equal to' sign has been dropped from Eq. 9.7.2 for the sake of convenience and the mathematical development given here is independent of this.

The constraint of Eq. 9.7.3 physically means that a margin on design variables and technological coefficients has been placed. Because x_1 to x_n and y_1 to y_r are continuous random variables, G_i is a random variable too and can take any value. If the constraint is generally satisfied, then the probability of $G_i > b_i$ would be very low (if not zero). Mathematically this is written as

$$p_i \equiv p[G_i(x_1, x_2, \ldots, x_n, y_1, y_2, \ldots, y_r) > b_i] \leqslant K_i \qquad i = 1, 2, \ldots, m \tag{9.7.4}$$

where K_i is some small number and p_i is the probability of the occurrence of G_i greater than the design margin b_i.

A violation of constraints of Eq. 9.7.4 implies some kind of a physical failure either immediately or in the near future. In Table 9.24, four types of failure are considered and their short- and long-term repercussions to the plant owner as well as the designer are given. If a chemical plant does not operate at the desired level, there is not only a fall in the plant output but a loss of reputation (and therefore the future sale) of the designer and so on. Table 9.24 is not general but for a given situation, it is always possible to list different modes of losses when the constraints are violated. In view of this, with every violated constraint, an expected value of loss, $E(C_i)$ can be associated. The expected value of C_t, $E(C_t)$, consists of not only the expected values of fixed and operating costs of the plant but also of failure costs. This is written as

$$E(C_t) = \sum_h E(C_h) + \sum_{i=1}^{m} p_i E(C_i) \tag{9.7.5}$$

The expected value of loss, $E(C_i)$, is assumed to be constant in the analysis given here. The probability of failure p_i is obtained from Eq. 9.7.4 and the term $\sum_i p_i E(C_i)$ represents the total penalty associated with the violation of different constraints. The problem of chemical plant design now consists of minimizing $E(C_t)$ and the approach adopted in finding the solution is called *chance constrained programming*. A few characte-

Table 9.24[21] Failure Cost Enumeration

Sl. No.	Failure type	Costs: To chemical plant owner	Costs: To designer
1.	Inability to meet required performance	(a) Direct losses due to lost sales, poor efficiency, incr[illegible] material consumption etc. (b) Costs of repairs and additions required to meet performance specifications	(a) Contract penalties (b) Cost of repairs which are required to be done by designer to restore performance (c) Losses due to reduction in future sales
2.	Short useful lifetime	(a) Costs of replacement and repair of that portion of the system which failed (b) Costs of replacement and repair of other portions of the system damaged due to failure (c) Costs of lost sales, loss of goodwill during downtime required for repair (d) Consequential effect on other systems due to nonavailability	(a) Warranty costs for repair of replacement (b) Loss due to reduction in future sales
3.	Poor system availability	(a) Costs of lost sales and goodwill during downtime required for repair (b) Costs of repairs and/or replacement parts (c) Consequential effects on other systems due to nonavailability	(a) Warranty costs (b) Losses due to reduction in future sales
4.	Disastrous failure	(a) Costs of repair or replacement of that portion of the system which fails (b) Costs of repair, replacement or compensation for consequential damage and loss of insurance coverage (c) Costs of lost sales, loss of goodwill during downtime required for repairs	(a) Warranty costs (b) Repair, redesign, and replacement cost undertaken by designer to minimize loss in goodwill (c) Losses due to reduction in future sales

(Reprinted with permission from J. Weisman and A. G. Holtzman, *I. & E. C. Proc. Des. Dev.*, **11**, 386, 1972. Copyright by the American Chemical Society.)

ristics of this minimization problem may be pointed out here. *All* the constraint equations of Eq. 9.7.3 have been built in the probability of failure, p_i, and finding minimum $E(C_t)$ is an unconstrained optimization problem. K_i in Eq. 9.7.4 are merely arbitrary constants and at the global minimum solution, they have unique values which can be found.

In order to find p_i, it is important to know the nature of distribution functions of design variables, x_i, and technological coefficients, y_i. In the analysis presented here, they are assumed to be normally distributed. The distribution of G_i is approximated as normal and its expected value, $E[G_i(x_1, x_2, \ldots)]$, and its variance $\sigma^2[G_i(x_1, x_2, \ldots)]$ are found as follows. The Taylor's expansion of G_i is carried out and the first term is retained only to obtain

$$G_i(x_1, x_2, \ldots, y_1, y_2) = G_i(x_{10}, x_{20}, \ldots, y_{10}, y_{20}, \ldots) + \sum_j \left(\frac{\partial G_i}{\partial x_j}\right)_{x_{j0}} (x_j - x_{j0}) + \sum_j \left(\frac{\partial G_i}{\partial y_j}\right)_{y_{j0}} (y_j - y_{j0}) \tag{9.7.6}$$

where x_{i0} and y_{i0} are mean values of variables x_i and y_i for all permitted values of i. The expected value of G_i is obtained as

$$E[G_i(x_1, x_2, \ldots, y_1, y_2, \ldots)] = G_i(x_{10}, x_{20}, \ldots, y_{10}, y_{20}, \ldots) = G_i[E(x_1), E(x_2), \ldots, E(y_1), E(y_2) \ldots] \tag{9.7.7}$$

The latter two summation terms of Eq. 9.7.6 do not contribute to $E[G_i]$ because the expected value of x_j and y_j are x_{j0} and y_{j0} and they vanish. Similarly the variance is found as

$$\sigma^2(G_i) = \sigma^2[G_i(x_1, x_2, \ldots, y_1, y_2, \ldots)] = \sum_k \left\{\frac{\partial G_i}{\partial x_k}\right\}^2 \sigma^2(x_k) + \sum \left(\frac{\partial G_i}{\partial y_j}\right)^2 \sigma^2(y_j) \tag{9.7.8}$$

One defines z as

$$z = \frac{b_i - E(G_i)}{\sigma(G_i)} \tag{9.7.9}$$

and finds p_i in terms of it as

$$p_i[G_i \geqslant b_i] = \int_z^{\infty} \frac{1}{(2\pi)^{1/2}} \exp(-z^2/2)\, dz \tag{9.7.10}$$

The procedure of finding the optimal solution consists of assuming values of design variables, calculating $E(G_i)$ and $\sigma(G_i)$ from Eqs. 9.7.7 and 9.7.8 and finding p_i from Eq. 9.7.10. Values of p_i for all i are substituted in Eq. 9.7.5 to find $E(C_t)$ and it is calculated for all possible choices of design variables. After tabulation of the result, the values of x_1 to x_m which give minimum $E(C_t)$ are found. It may be evident that these would be less than their values when design variables are assumed to be deterministic in nature and the difference between them represents the safety margins of these variables in the process design.

To solve any real problem in reasonable time where uncertainties are involved in design variables and technological coefficients, one is forced to take recourse to a high speed computer (IBM 360 or any equivalent one). A number of programming techniques have been used in obtaining the solution to such kinds of problems[20-22] and in the following, the one used by Weisman, et al. is described. Equation 9.7.5 is rewritten as

$$E(C_t) = \sum_h E(C_h) + K \sum_i p_i E(C_i) \tag{9.7.11}$$

where K is some positive and small arbitrary constant. Initially K assumes some very small value. A search is made for the optimum design variables. The value of K is increased by a certain factor (it is suggested to be 8) and the subsequent search starts from the optimum values at the previous value of K. This search continues until the optimum values are obtained for $K = 1$. The reason for this elaborate search is to make sure that the global minimum values are obtained instead of wrongly concluding calculations at some local minimum.

9.8 CONCLUSION

In this chapter the cost estimation of equipment and the operating labour involved in chemical plant design is described in view of the fact that they constitute a major share of the total cost. The cost estimation procedure suggested here is unique because of its international nature. Even though costs have been reported in dollars, they can be conveniently converted in any other currency.

If all design variables are deterministic in nature the overall chemical plant design is nothing but an optimization problem with constraints. Unfortunately uncertainties exist not only in these but also in technological coefficients which are used for calculating them. It is in view of this that safety margins are built into these variables while designing a chemical plant. To find optimum values of these margins, the penalty function approach has been suggested here which converts this task into an unconstrained optimization problem. In Chapter 10, the industrial economics involved in chemical plant design are discussed.

REFERENCES

1. Perry, J. H. and Chilton, C. H. (ed.), *Chemical Engineer's Handbook*, 5th edn., McGraw-Hill, New York, 1973.
2. Kharbanda, O. P., *Process Plant and Equipment Cost Estimation*, 1st edn., Sevak Publication, Bombay, 1977.
3. Foust, A. S., Wenzel, L. A., Clump, C. W., Mous, L. and Anderson, L. B., *Principles of Unit Operations*, 1st edn., Wiley, 1960.
4. Garcia-Borras, T., *Hydrocarbon Proc.*, 137, Dec. 1976.

5. Garcia-Borras, T., *Hydrocarbon Proc.*, 171, Jan. 1977.
6. McNeill, T. F., and Clark, D. S., *Cost Estimating and Contract Pricing*, 1st edn., Elsevier, New York, 1966.
7. Peter, M. S. and Timmerhous, K. D., *Plant Design and Economics For Chemical Engineers*, 2nd edn., McGraw-Hill, New York, 1968.
8. Jelen, F. C., *Cost and Optimization Engineering*, 1st edn., McGraw-Hill, New York, 1970.
9. Chandrasekharan, K. D., *Bibliography on Cost Estimation of Chemical Plants, Processes and Equipment*, Chemical Engineering Education Development Centre, Indian Institute of Technology, Madras, 1974.
10. Gallaghar, G. J. T., *Chem. Eng.*, **76**, 196, June 2, 1969.
11. Johnson, R. J., *Chem. Eng.*, **76**, 146, March 10, 1969.
12. Tipnis, U. V., Personal Communication, Engineers India Ltd., New Delhi.
13. Sharma, M. M., Personal Communication, University Department of Chemical Technology, University of Bombay.
14. Gropper, F., *C.E.P.*, **53** (10), 464, 1957.
15. Haines, T. B., *C.E.P.*, **53** (11), 556, 1957.
16. Wobus, R. S., *C.E.P.*, **53** (12), 581, 1957.
17. Wessel, H. E., *Chem. Eng.*, p. 209, July, 1952.
18. O'Connel, F. P., *Chem. Eng.*, p. 150, Feb. 19, 1962.
19. Tckamatsu, L., Hashimoto, I. and Ohno, H., *I. & E. C. Proc. Des. Dev.*, **9**, 368, 1970.
20. Gottfried, B. S., Bruggink, P. R. and Harwood, E. R., *I. & E. C. Proc. Des. Dev.*, **9**, 581, 1970.
21. Weisman, J. and Holzman, A. G., *I. & E. C. Proc. Des. Dev.*, **11**, 386, 1972.
22. Weisman, J., Pullido, H. and Khanna, A., *I. & E. C. Proc. Des. Dev.*, **14**, 51, 1975.
23. Nishida, N., Liu, Y. A. and Ichikawa, A., *A.I.Ch. E.J.*, **22**, 539, 1976.
24. Denn, M. M., *Optimization By Variational Methods*, McGraw-Hill, New York, 1969.

PROBLEMS

9.1 Consider a plant manufacturing 1000 kg/day diethyl ether from ethanol. Suggest a plant layout, define the work task and carry out the labour force analysis.

9.2 Determine the cost of the installed plant of P. 9.1. Also determine the cost using Table 9.16 and find the difference. Discover the source of difference.

9.3 Consider P. 1.8. Write down the total work task and determine the total standard time required. Find the total cost of the plant.

9.4 Do the same as above for P. 1.12.

10

Industrial Economics in Chemical Plant Design

10.1 INTRODUCTION

Like any other economic venture, the installation of a chemical plant has the basic purpose of making profits. In view of this, the study of industrial economics is inseparable from chemical plant design. In this chapter, the total plant cost has been shown to consist of fixed and running expenses and these are determined by systematically accounting for all expenses. This information will be shown to be extremely useful in deciding between different alternatives. Also for a given chemical with a specified process route, it fixes the minimum throughput of the plant.

The major complication that arises in the analysis above is that all the cash flow does not occur at the same time and because there is a time value of money, it is necessary that a proper discounting be carried out in obtaining the total cost. The time value of money and the method of such discounting is discussed first.

10.2 TIME VALUE OF MONEY AND RISKS IN INVESTMENTS

If an entrepreneur has a certain amount of money, he can either keep it locked in the house, deposit it in the bank to earn interest or install a factory to obtain profits through the sale of products. In the latter two situations, the money grows, and as a result, the two cash receipts separated by time are not equivalent. A more subtle way by which the change in the value of money can occur is through inflation or depression. Inflation gives rise to loss of purchasing power per unit money. Depression is the reverse of this and arises when a glut in the market is created through supply being more than the consumption.

When a money is deposited in the bank, it grows at an interest fixed by the bank. The situation becomes more complex when money is invested in a chemical plant and the growth of capital occurs through profit earned by the sale of product(s). Since profit to a large extent depends upon varying market forces, it is therefore stochastic in nature. In view of this, the rate of return (or the interest on the invested capital) is characterized by an average value. When this is done, the investment in a chemical plant compares with the deposition of money in a bank except for the fact that the rate of return in the former is normally much higher (thus justifying the investment).

Inflation, which is found in reality to always exist, indicates the loss of value of money. In other words, the same thing costs more with time. This loss is continuous in nature and is a very difficult phenomena to quantify. One of the simplest methods of quantifying it is to assume a certain inflation rate, completely analogous to the rate of return or the interest rate discussed earlier. This, however, differs from bank interest in one fashion: the interest earned on money deposited in a bank is discrete in nature whereas the loss of value of money through inflation is continuous.

Investment in a chemical plant differs from investment of money in bank or the phenomena of inflation in that in the former there is a risk element associated with it. Like any other investment, a chemical plant can also be insured (insurance against accident) and the value of insurance premium can serve as an indicator of the element of risk involved. Table 10.1 gives insurance rates for different types of projects.

One considers a project which yields 100 units of money at the end of the year with 50% probability of success in comparison with another project profiting 1000 money units with 10% probability of success. They are assumed to have the same level of investment. Even though the

Table 10.1[1] Typical Insurance Rates for Different Projects

Sl.No.	Type of project	Insurance premium per unit investment
1.	*High risk*: Projects involving considerable novelty or based on unproven sales data, products or raw material	0.20–1.00 or more
2.	*Fair risk*: Projects outside the present field of activity, or novel projects that have not been thoroughly investigated	0.10–0.20
3.	*Average risk*: Projects in present field of operations but with some novel features or indefinite market information	0.05–0.10
4.	*Good risk*: Expansion of existing operations in a known market	0.01–0.05
5.	*Excellent risk*: Cost reduction in operating processes in stable environment	0.00–0.01

(Reprinted from Ref. 1 with permission of John Wiley & Sons, Inc., New York.)

basic goal of any venture is to maximize profit, the former is most likely to be preferred because of its lower risk of failure. Thus a venture yielding ten times the financial gain may not necessarily be ten times better. As a matter of fact each incremental wealth is usually valued less highly than the previous one[2] and this introduces nonlinearity in the utility of money. The plant design under this situation is complex mathematically and is not considered here. In the following, it is assumed that with every investment, it is possible to associate a rate of return exactly as one can do with any money deposited in a bank. This assumption is justified only when the non-linearity of money and the stochastic nature of returns are ignored.

10.3 DISCOUNTING

The comparison of two ventures with cash flows at different times cannot be compared because of the time value of money. They can be compared only when a proper correction is made for the time factor. This is called *discounting* and is done as follows.

The growth of money is assumed to occur discretely at a constant rate of return i at the end of every year. If there is sum of money P invested at time $t = 0$, its values, $S_1, S_2, \ldots$, at the end of the first, second year, etc. are given by

$$S_1 = P(1 + i) \tag{10.3.1a}$$

$$S_2 = S_1(1 + i) \tag{10.3.1b}$$

$$S_3 = S_2(1 + i) \tag{10.3.1c}$$

or in general

$$S_n = S_{n-1}(1 + i) \tag{10.3.1d}$$

where S_n is the value of money P at the end of the nth year. Eliminating S_{n-1} from the above equation, one obtains

$$S_n = P(1 + i)^n \tag{10.3.2}$$

This equation shows the change in value of money with time and is the same as the compound-interest formula, but with different interpretations of S_n and P. S_n is the *future worth* after n years of an investment made *now*. Conversely P is the *present worth* of an investment which is going to be made after n years from now.

Whenever two projects of *equal life time* with cash receipts and expenditure occurring at different times are required to be compared, the present worths of all these are found first. The total present worth of one is subsequently compared with the total present worth of the other. To demonstrate the working procedure, the *example* of Table 10.2 is considered. There are two projects with a life time of 8 years with fixed costs of 100,000 and 150,000 money units. After these plants are put to stream,

Table 10.2 Comparison of Two Projects

Sl. No.	Item	Cash flow of project A	Cash flow of project B
1.	Life of the chemical plant*	7 years	6 years
2.	Time taken for complete installation*	1 year	2 years
3.	Cost of equipment, building and land C_F	100,000	150,000
4.	Repair costs every year C_R	10,000	5,000
5.	Net profit resulting from sale of chemicals P_r	60,000 for first four years and then 20,000	80,000
6.	Duration of regular shut-down overhaul	1 year (for first four years)	4 years
7.	Expenses of overhauling C_{ov}	5,000	5,000
8.	Penalty of polluting environment C_P	5,000 at the end of 2nd year	—
9.	Bank interest rate i	10%	10%

*After adding the time taken for installation, total life times of these projects are equal.

the net profit from these are 60,000 and 80,000 due to difference in cost of the chemical manufactured. However, the profit of the former goes down to 20,000 after four years of operation due to inefficiencies.

It is best to show the cash flows on a cost diagram as done in Fig. 10.1.

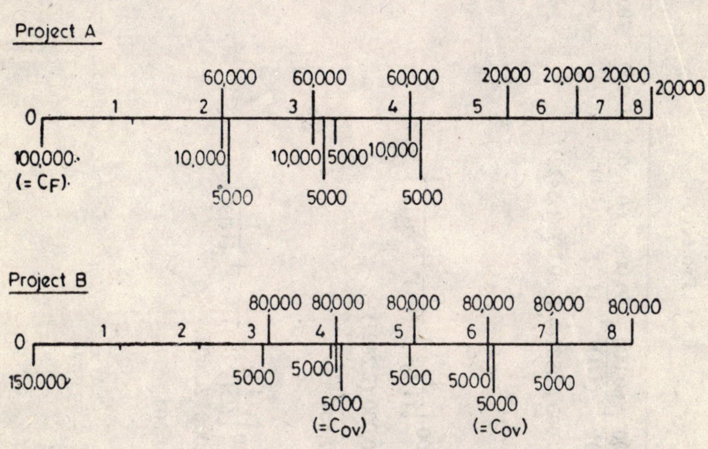

Fig. 10.1 Cost diagram of projects A and B of equal life given in Table 10.2

It is nothing but the time coordinate on which cash receipts are shown above the axis and expenses, below it.[3] The present worth of each cash item is found by rearranging Eq. 10.3.2 as

$$P = \frac{S_n}{(1+i)^n} \tag{10.3.3}$$

and results are given for both the projects in Table 10.3. Total gain for

Table 10.3 Calculation of Present Worth for Different Cash Flows of Table 10.2

Sl. No.	Item	Project A	Project B
1.	Discounting net profit	$\frac{60{,}000}{(1.1)^2}+\frac{60{,}000}{(1.1)^3}+\frac{60{,}000}{(1.1)^4}+\frac{60{,}000}{(1.1)^5}+\frac{20{,}000}{(1.1)^6}+\frac{20{,}000}{(1.1)^7}+\frac{20{,}000}{(1.1)^8}$ $= 60{,}000\ (2.8816) + 20{,}000(1.5422)$ $= 2{,}03{,}780$	$80{,}000\left[\frac{1}{(1.1)^3}+\frac{1}{(1.1)^4}+\frac{1}{(1.1)^5}+\frac{1}{(1.1)^6}+\frac{1}{(1.1)^7}+\frac{1}{(1.1)^8}\right]$ $= 80{,}000[3.5994]$ $= 2{,}87{,}952$
2.	Discounting of repair cost	$10{,}000\left[\frac{1}{(1.1)^2}+\frac{1}{(1.1)^3}+\frac{1}{(1.1)^4}\right]$ $= 10{,}000\ (2.2607)$ $= 22{,}607$	$5{,}000\left[\frac{1}{(1.1)^3}+\frac{1}{(1.1)^4}+\frac{1}{(1.1)^5}+\frac{1}{(1.1)^6}+\frac{1}{(1.1)^7}\right]$ $= 5{,}000(3.1330)$ $= 15{,}665$
3.	Discounting cost of shut-down overhaul	$5{,}000\left[\frac{1}{(1.1)^2}+\frac{1}{(1.1)^3}+\frac{1}{(1.1)^4}\right]$ $= 5{,}000\ (2.2607)$ $= 11{,}304$	$5{,}000\left[\frac{1}{(1.1)^4}+\frac{1}{(1.1)^6}\right]$ $= 5{,}000\ (1.2475)$ $= 6{,}238$
4.	Discounting of penalty	$\frac{5{,}000}{(1.1)^3} = 3{,}757$	—
5.	Net profit	$(1) - (2) - (3) - (4) - C_F$ $= 66{,}112$	$(1) - (2) - (3) - C_F$ $= 1{,}16{,}049$

projects A and B are found by summing individual present worths; and on comparison, project B is found to be better. In this example, discounting has been done against the bank interest rate, which means that all transactions have been made through bank and the cash in the bank earns interest. In reality, this is normally the case.

There are several instances by which growing money can be put to an advantage. Supposing there is an entrepreneur with 100,000 money units whereas a chemical plant requires 200,000 money units of investment. He can deposit his money in the bank and allow it to grow to double its value. The time taken for this to occur can be easily calculated from Eq. 10.3.3 by noting that $P = 100{,}000$ and $S_n = 200{,}000$. If the interest rate, i, is 10% on substitution

$$(1.10)^n = (S_n/P) = 2 \qquad (10.3.4)$$

from which n is found to lie between seven and eight years.

Normally a bank loan is taken when a plant is to be installed. Interest is charged from the day the loan is sanctioned up to the day it is completely paid. If the money borrowed is 100 units and it takes five years to repay the loan, the money paid to the bank, S_5, can be calculated as

$$S_5 = 100\,(1 + 0.10)^5 = 161.05 \qquad (10.3.5)$$

Thus, the rental paid to the bank on the loan would be $(S_5 - P)$ or 61.05 money units!

10.4 COST EQUIVALENCE

The discussion above explains the importance of the time factor in any investment. In general, the duration must be specified with every investment. Two investments of the same length of time are equal only when their present worths are equal. In this section, two investments with different lengths of duration are compared.

Two schemes A and B with their life times n_A and n_B are considered for comparison. Each of them is assumed to have different cash flows; however, for each scheme, it is possible to reduce them into their present worth values P_{A,n_A} and P_{B,n_B} as shown in Fig. 10.2a. If $n_A = n_B$, one can conveniently evaluate the merit of one over the other, but when they are different, the comparison of present worth leads to incorrect conclusions.

To compare two schemes of different durations, their life times can be artificially matched as follows. For scheme A, a similar scheme is reinstalled after every n_A years at the cost of P_{A,n_A} (equal to the present worth) and this is done n_B-times, this way making the total project life as $(n_A n_B)$ years. One does the same to scheme B n_A-times to make their total lives equal and the resultant cost diagrams are given in Fig. 10.2b.

The present worths of the fictitious schemes of Fig. 10.2b, P_A^* and P_B^*,

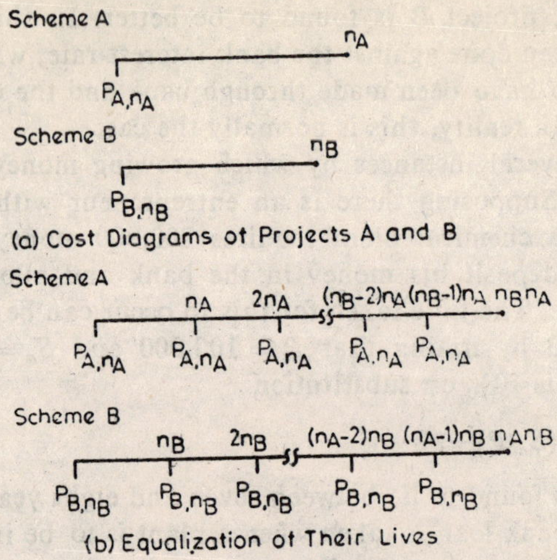

Fig. 10.2 The equalization of two costs of different life-span

can be determined using the procedure outlined in the previous section and they are

$$P_A^* = P_{A,n_A}\left[\frac{1}{(1+i)^{n_A}} + \frac{1}{(1+i)^{2n_A}} + \ldots\right]$$

$$\equiv P_{A,n_A} \sum_{j=1}^{(n_B-1)} \frac{1}{(1+i)^{jn_A}} \tag{10.4.1}$$

and

$$P_B^* = P_{B,n_B}\left[\frac{1}{(1+i)^{n_B}} + \frac{1}{(1+i)^{2n_B}} + \ldots\right]$$

$$\equiv P_{B,n_B} \sum_{j=1}^{(n_A-1)} \frac{1}{(1+i)^{n_B j}} \tag{10.4.2}$$

These are geometric series with common ratios given by $1/(1+i)^{n_A}$ and $1/(1+i)^{n_B}$ and can be summed up to obtain

$$P_A^* = P_{A,n_A} \frac{1 - \dfrac{1}{1+i)^{n_A n_B}}}{1 - \dfrac{1}{(1+i)^{n_A}}}$$

$$= P_{A,n_A} \frac{[(1+i)^{n_A n_B} - 1](1+i)^{n_A}}{(1+i)^{n_A n_B}[(1+i)^{n_A} - 1]} \tag{10.4.3}$$

and

$$P_B^* = P_{B,n_B} \frac{1 - \dfrac{1}{(1+i)^{n_A n_B}}}{1 - \dfrac{1}{(1+i)^{n_B}}}$$

$$= P_{B,n_B} \frac{[(1+i)^{n_A n_B} - 1](1+i)_{n_B}}{(1+i)^{n_A n_B}[(1+i)^{n_B} - 1]} \tag{10.4.4}$$

If these two schemes are financially identical,

$$P_A^* = P_B^* \tag{10.4.5}$$

or

$$\frac{P_{A,n_A}[(1+i)^{n_A}]}{(1+i)^{n_A} - 1} = \frac{P_{B,n_B}[(1+i)^{n_B}]}{[(1+i)^{n_B} - 1]} \tag{10.4.6}$$

On rearrangement,

$$P_{A,n_A} = P_{B,n_B} \frac{[(1+i)^{n_A} - 1]/(1+i)^{n_A}}{[(1+i)^{n_B} - 1]/(1+i)^{n_B}} \tag{10.4.7}$$

The above equation is very important because it converts a given scheme of one life-span into a completely equivalent scheme of *different* life-span. This method of artificial change in the life-span of a given scheme facilitates cost comparison with any other. Two special situations of $n_A = 1$ and $n_A = \infty$ in Eq. 10.4.7 give rise to concepts of unacost and capitalized cost discussed below.

10.5 UNACOST AND CAPITALIZED COST[3,4]

A given amount of money P borrowed from the bank *now* can be repaid in equal instalments U at the end of each year. This instalment is called unacost. The cost diagram of this situation is shown in Fig. 10.3. The

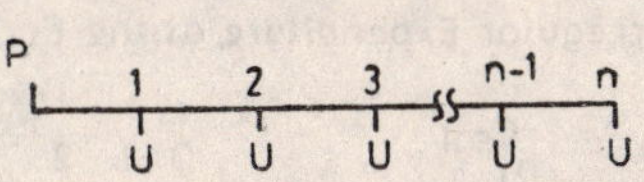

Fig. 10.3 Cost diagram of an irregular expenditure at the end of x years

loan would be considered completely paid only when the present worth of all these payments equals the amount of the money borrowed.

Or

$$P = \sum_{j=1}^{n} \frac{U}{(1+i)^j} \tag{10.5.1}$$

where n is the total number of instalments. The geometric series with common ratio $1/(1+i)$ above can be summed to obtain

$$P = U\frac{\frac{1}{(1+i)}\left[1-\frac{1}{(1+i)^n}\right]}{\left[1-\frac{1}{(1+i)}\right]} = U\frac{(1+i)^n-1}{i(1+i)^n} \tag{10.5.2}$$

and the unacost is found for a given present worth P as

$$U = P\left[\frac{i(1+i)^n}{(1+i)^n-1}\right] \tag{10.5.3}$$

One now re-examines Eq. 10.4.7 after substituting $n_A = 1$ in it. $P_{A,1}$ is given by

$$\begin{aligned} P_{A,1} &= P_{B,n_B}\frac{[(1+i)-1]/(1+i)}{[(1+i)^{n_B}-1]/(1+i)^{n_B}} \\ &= \frac{P_{B,n_B}}{(1+i)}, \frac{i(1+i)^{n_B}}{(1+i)^{n_B}-1} \end{aligned} \tag{10.5.4}$$

It is further noted that $P_{A,1}$ is the payment in the beginning of the year and the equivalent payment S_1 at the end of the year would be

$$S_1 = P_{A,1}(1+i) \tag{10.5.5}$$

Or

$$S_1 = P_{B,n_B}\frac{i(1+i)^{n_B}}{(1+i)^{n_B}-1} \tag{10.5.6}$$

which is found to be the same as unacost U. Thus it is seen that two schemes of different cash flows and life-spans can be compared by finding their unacosts.

Unacost earlier was defined in a very limited sense. This can be found for any money, whether a payment or a receipt, occurring at any time. A few situations of practical significance are given in the form of cost diagrams and their unacosts can be determined. In Fig. 10.4a, there is an

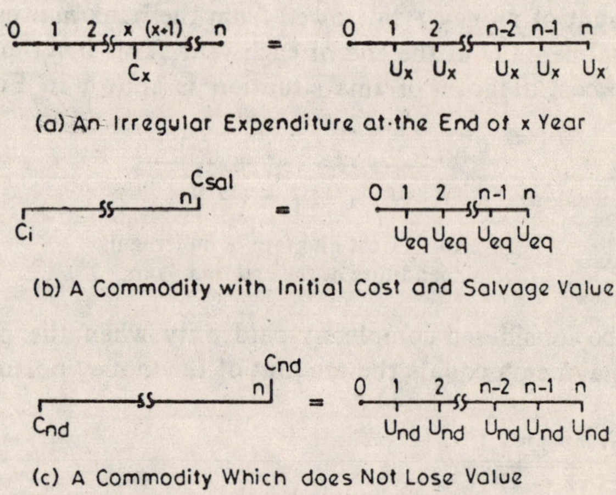

Fig. 10.4 Calculation of unacost of a non-depreciable money

irregular expenditure at the end of the xth year. Its present worth P_x is

$$P_x = \frac{C_x}{(1+i)^x} \tag{10.5.7}$$

and its unacost U_x is obtained by substituting this in Eq. 10.5.3 as

$$U_x = \frac{C_x}{(1+i)^x} \frac{i(1+i)^n}{(1+i)^n - 1} \tag{10.5.8}$$

Normally after the use of any equipment for n years, it has some salvage value, C_{sal}. This situation is represented in Fig. 10.4b where C_i is its initial cost. The present worth P_{eq} for this cost diagram would be

$$P_{eq} = C_i - \frac{C_{sal}}{(1+i)^n} \tag{10.5.9}$$

and on substituting this in Eq. 10.5.3, equivalent unacost U_{eq} is found as

$$U_{eq} = C_i \frac{i(1+i)^n}{(1+i)^n - 1} - C_{sal} \frac{i}{(1+i)^n - 1} \tag{10.5.10}$$

There are several commodities, for example land, which do not lose value with time (on the contrary, land normally has a tendency to appreciate). This would mean that the cost price can be recovered whenever it is sold in future. This is depicted in the cost diagram of Fig. 10.4c and its present worth, P_{nd}, is found to be

$$P_{nd} = C_{nd} - \frac{C_{nd}}{(1+i)^n} = \frac{(1+i)^n - 1}{(1+i)^n} C_{nd} \tag{10.5.11}$$

Its unacost U_{nd} is given by

$$U_{nd} = \frac{(1+i)^n - 1}{(1+i)^n} C_{nd} \frac{i(1+i)^n}{(1+i)^n - 1} = iC_{nd} \tag{10.5.12}$$

Next it is noted that a chemical plant consists of several equipment each having different life-spans, initial costs and salvage values. In this case, the unacosts of all equipment are determined individually and the unacost of the whole plant U_P is the sum of all these, i.e.

$$U_p = \sum_{j=1}^{p} U_j \tag{10.5.13}$$

where p is the total number of equipments in the plant.

If $n_A \to \infty$ in Eq. 10.4.7, it is noted that

$$\lim_{n_A \to \infty} \left[\frac{(1+i)^n - 1}{(1+i)^n}\right] = \lim_{n_A \to \infty} \left[1 - \frac{1}{(1+i)^n}\right] = 1 \tag{10.5.14}$$

and $P_{A\infty}$ becomes

$$K_A \equiv P_{A\infty} = P_{B,n_B} \frac{(1+i)^{n_B}}{(1+i)^{n_B} - 1} \tag{10.5.15}$$

On comparison with Eq. 10.5.3,

$$U_A = iK_A \tag{10.5.16}$$

K_A is called *capitalized cost* and the relation above has been shown in the cost diagram of Fig. 10.5. Physically this means that on deposition of an

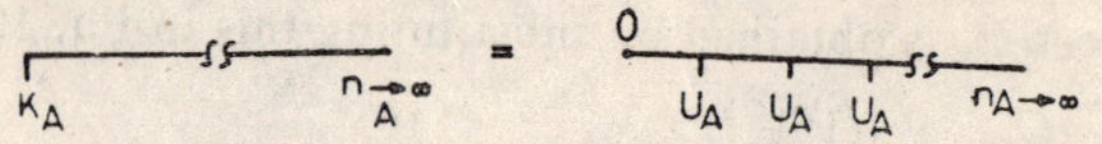

Fig. 10.5 Relation between capitalized cost and unacost

amount K_A in a bank now, a sum of U_A can be withdrawn at the end of every year *forever*. For a chemical plant having p equipment, the capitalized cost K_p for it would be the sum of individual ones, i.e.

$$K_p = \sum_{j=1}^{p} K_j \tag{10.5.17}$$

Table 10.4 gives the capitalized cost for different situations.

The method of unacost and capitalized cost both can be used for cost comparison of any two schemes and is demonstrated through an *example*. Oxalic acid is prepared from the reaction of concentrated nitric acid upon molasses and in this process nitric oxide gas is liberated. The feasibility of the process depends to a large extent upon the recovery of nitric oxide and it is done in an absorber operating at 70°C. The absorption column can either be packed with 7 cm (cylindrical ceramic) raschig rings or 7 cm pieces of polyethylene pipes of diameter 7 cm. Ceramic rings are unaffected at the conditions of the column and last forever; however, polyethylene pipe pieces start fragmenting after some time and need to be replaced every year. It is desired to decide upon the two alternatives.

One assumes that 10,000 pieces of ceramic rings (or plastic pieces) are required to fill the column. If raschig rings are available for 10 money units per piece, the total cost would be 100,000 money units and because they last forever, it would be the same as the capitalized costs. If a polyethylene pipe is available for 17.5 money units per metre, the total cost would then be 12,500 units (assuming 14 pieces per metre). Since this has to be replaced every year, this represents the unacost of this scheme. If the bank interest rate is 10%, its capitalized cost would be 1,25,000 units and it would be economical to use raschig rings. However, if the bank interest rate is 15%, the capitalized cost would be 83,333 and therefore it would be advantageous to fill the column with polyethylene pipe pieces and replace them every year. Thus the choice of a specific process is decided by the availability of money.

To demonstrate the use of industrial economics in deciding between alternatives, another *example* is given below[3]. A new storage tank is assumed to be available for 10,000 money units which lasts for 5 years. It is required to find out whether a new tank should be purchased or the old one be repaired. The old tank is assumed to cost 1000 units for repair for 1 more year of service, 2000 money units for one more year of service after this, 3000 units for another year of service after the second year and

Table 10.4[3] Mathematical Relations Between P, U and K

Sl. No.	Cost Description	Present worth	Unacost	Capitalized cost
1.	P, initial cost for n years	P	$P\dfrac{i(1+i)^n}{(1+i)^n-1}$	$P\dfrac{(1+i)^n}{(1+i)^n-1}$
2.	U, uniform end of the year payment	$U\dfrac{(1+i)^n-1}{i(1+i)^n}$	U	U/i
3.	C_x, irregular expenditure at the end of xth year	$\dfrac{C_x}{(1+i)^x}$	$\dfrac{C_x}{(1+i)^x}\dfrac{i(1+i)^n}{(1+i)^n-1}$	$\dfrac{C_x}{(1+i)^x}\dfrac{(1+i)^n}{(1+i)^n-1}$
4.	Initial cost C_i and salvage value C_{sa1} at the end of nth year	$(C_i - C_{sa1}) + C_{sa1}\dfrac{(1+i)^n-1}{(1+i)^n}$	$(C_i - C_{sa1})\dfrac{i(1+i)^n}{(1+i)^n-1} + iC_{sa1}$	$(C_i - C_{sa1})\dfrac{(1+i)^n}{(1+i)^n-1} + C_{sa1}$
5.	C_{nd}, non-depreciable cost	$C_{nd}\dfrac{(1+i)^n-1}{(1+i)}$	iC_{nd}	C_{nd}

(Reprinted from F. C. Jelen, *Cost and Optimization Engineering*, 1970 with permission of McGraw-Hill, New York.)

so on. If the bank interest rate is 10% per year, the unacost, U_{new}, for the new tank is given by

$$U_{new} = 10{,}000 \frac{i(1+i)^5}{(1+i)^5 - 1} = 2638 \text{ money units} \qquad 10.5.18)$$

If instead of buying a new tank, it is decided to repair the old one for one more year of service, the U_{repl} of the expense of repairing is given by

$$U_{repl} = 1000 \frac{i(1+i)}{(1+i) - 1} = 1100 \qquad (10.5.19)$$

which is lower than U_{new}. Therefore one should repair instead of buying a new tank. After the first year is over, a similar decision has to be made—either to buy a new tank or repair the old one for 2000 money units for one more year of service. The unacost, $U_{rep\,2}$, is given by

$$U_{rep\,2} = 2000 \frac{i(1+i)}{(1+i) - 1} = 2200 \qquad (10.5.20)$$

and on comparison it is still found economical to repair. After the end of two years, $U_{rep\,3}$ is found as

$$U_{rep\,3} = 3000(1.1) = 3300 \qquad (10.5.21)$$

and it is found advantageous to buy a new tank. It may be added here that after the end of the first year, the choice to be made is independent of previous repair expenses and it is in view of this that $U_{rep\,1}$ and $U_{rep\,2}$ were not added before comparing with U_{new}. If it takes 2000 money units every year to repair an old tank, it is seen that a new tank should never be purchased.

In this section, unacosts and capitalized costs have been discussed in connection with comparison of different economic schemes. One now analyzes the nature of expenses incurred in running a chemical plant.

10.6 ANALYSIS OF COSTS IN A CHEMICAL PLANT

The total money required for running a chemical plant is called capital investment and it has the following two elements:

(1) fixed plant cost, and
(2) working cost.

Fixed plant cost is the capital required to provide all facilities needed for the plant. This is the money spent only once in the life-span of the project; as opposed to this, the working cost is the money required to meet recurring expenses.

Fixed costs have several components and are given in Table 10.5. Percentages given against each item are normally applicable for large plants and are approximate values. Recurring expenses are given in Table 10.6. Even among these, some are proportional to the total plant capacity and some are not. The former is called variable cost. It is

Table 10.5[3, 11] Components of Fixed Plant Costs

Sl.No.3	Item	Typical percentage	Average percentage for small plants of item 1
1.	Direct process cost		
	(a) Principal equipment items	15–40	
	(b) Installation of equipment	6–16	
	(c) Process piping	2–25	
2.	Other direct costs		
	(a) Instruments	1–8	7
	(b) Electrical	1–5	5
	(c) Buildings, process	0.5–15	15
	(d) Land	—	2
	(e) Contractor's fee	—	10
	(f) Utilities (installed)	3–20	
	(g) General services	0.5–10	
	(h) Buildings, general	4–14	
	(i) Receiving, shipping, storage	2–12	
3.	Indirect costs		
	(a) Engineering, overheads, etc	8–16	19
	(b) Contingencies	9–22	20

Table 10.6[3] Components of Annual Cost of a Chemical Plant

Sl.No.	Components
Annual Variable Cost (60-70%)	
1.	Raw materials, make up solvents, catalysts, etc.
2.	Direct labour (operators). Add 25% for pensions, vacation etc.
3.	Process services (steam, power, water, refrigeration)
4.	Maintenance (labour and materials), about 4% to 6% of fixed plant investment
5.	Miscellaneous supplies and other at about 0.5% of fixed plant investment
6.	Direct supervision (foreman) about 10% of direct labour
7.	Laboratory charges (process control)
8.	Royalty (charges for using some other company's process)
9.	Packing and storage charges
10.	Credit for by-products
11.	Spoilage and other losses (items 9 and 11 can be estimated at 5% of total labour plus services).

(*Contd.*)

Table 10.6 (*Contd.*)

Sl.No.	Components
Annual Fixed Costs (30–40%)	
1.	Indirect plant costs, or plant burden (15%)
	(a) Investment costs, 5%
	(i) Depreciation about 10% of fixed plant cost
	(ii) Taxes, 2% of fixed plant cost
	(iii) Insurance, 1% of fixed plant cost
	(iv) Financial costs (interest on inventory plant equipment, total capital where allowed)
	(v) Other assessments, penalties etc., for certain items of inventory
	(b) Overhead, 10%
	(i) Technical (engineering)
	(ii) Non-technical (office force, plant protection etc.)
	(iii) Supplies (those not chargeable to direct costs)
	(iv) Rent (included where equipment, buildings, land or services rented)
	(v) Others
2.	Management expenses, 5%
	(a) Executives
	(b) Legal
	(c) Research (technical research and market research)
3.	Selling or distribution expense, 10 to 20% (depends upon market forces)
	(a) Cost of selling
	(b) Delivery and warehouse costs
	(c) Technical service, 3 to 5%

(Reprinted from F. C. Jelen, *Cost and Optimization Engineering*, 1970 with permission of McGraw-Hill, New York.)

possible to include the fixed plant cost in the non-varying portion of the working cost through depreciation, which is discussed below.

10.7 DEPRECIATION

Depreciation is a means by which a fixed cost is distributed over a given period of years. The depreciated sum is treated as an expense for the specific year and in Table 10.6, this is shown as an annual fixed cost. Since an expense is tax deductible; the schedule of allocation of depreciation cannot be arbitrary and must meet the government regulation. Before discussing methods of depreciation the following terms are defined.

If an equipment has an initial cost C_i and can be sold for C_{sal} (salvage value) after n years of use, the depreciable first cost C_d is defined as

$$C_d = (C_i - C_{sal}) \tag{10.7.1}$$

The amount of money depreciated in mth year, D_m, is some fraction, D_{fm}, of this amount. Or

$$D_m = C_d D_{fm} \tag{10.7.2}$$

The book value of the asset in the mth year, B_m, is obtained by subtracting the accumulated depreciated money from C_i, i.e.

$$B_{m'} = C_i - C_d \sum_{m=1}^{m=m'} D_{fm} \tag{10.7.3}$$

Normally there are five accepted ways of depreciation as given in Secs. 10.7.1 to 10.7.5.

10.7.1 Straight-line Depreciation (SLD)

In this method, the depreciable costs are uniformly allocated over the years, i.e.

$$D_{f1} = D_{f2} = \ldots = D_{fn} = \frac{1}{n} \tag{10.7.4}$$

where n is the total number of years. The book value at the end of the mth year can be found to be

$$B_m = C_i - C_d \sum^{m} \left(\frac{1}{n}\right) = C_i - C_d\left(\frac{m}{n}\right)$$

$$= C_{sal} + C_d\left(1 - \frac{m}{n}\right) \tag{10.7.5}$$

In this expense allocation, the time value of money is not considered. When $D = C_d/n$ is the yearly depreciated expense, the cost diagram for this situation would appear as in Fig. 10.6 and the present worth P_{SLD}

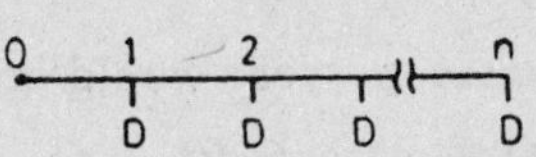

Fig. 10.6 Cost diagram of straight-line depreciation

of this is given by

$$P_{SLD} = \frac{C_d}{n}\left[\frac{1}{(1+i)} + \frac{1}{(1+i)^2} + \ldots + \frac{1}{(1+i)^n}\right]$$

$$= \frac{C_d}{n} \frac{(1+i)^n - 1}{i(1+i)^n} \tag{10.7.6}$$

10.7.2 Sinking-fund Depreciation (SF)

In this method a constant amount of money is annually deposited in the bank such that after the life-span of the venture (n years) one gets the amount C_d back. This implies that

$$S_n = C_d \tag{10.7.7}$$

and the present worth of this P_{SF} would be

$$P_{SF} = \frac{C_d}{(1+i)^n} \tag{10.7.8}$$

The amount of money deposited at the end of every year R_{SF} is found from Eq. 10.5.3 as

$$R_{SF} = C_d \frac{i}{(1+i)^n - 1} \tag{10.7.9}$$

The cost diagram of sinking fund depreciation is shown in Fig. 10.7.

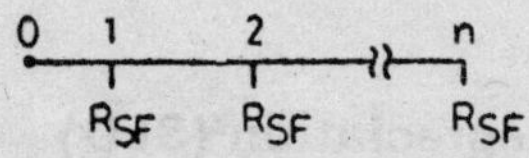

Fig. 10.7 Cost diagram of sinking fund depreciation

10.7.3 Declining-balance Depreciation (DB)

In this method of depreciation, a constant fraction of the book value at the beginning of the year is depreciated for that period. If f is the fraction, the amount depreciated D_1 for the first year is

$$D_1 = C_i f \tag{10.7.10}$$

The book value is then found as

$$B_1 = C_i - C_i f = C_i(1-f) \tag{10.7.11}$$

The amount D_2 depreciated in the second year is given by

$$D_2 = B_1 f = C_i(1-f)f \tag{10.7.12}$$

The book value B_2 for the second year would be

$$B_2 = B_1 - D_2 = C_i(1-f) - C_i f(1-f)$$
$$= C_i(1-f)^2 \tag{10.7.13}$$

In general

$$D_m = C_i(1-f)^{m-1} f \tag{10.7.14}$$

and

$$B_m = C_i(1-f)^m \tag{10.7.15}$$

Since the book value of the investment after n years is the same as the salvage value, one has

$$B_n = C_i(1-f)^n = C_{sal} \tag{10.7.16}$$

and from this, the fraction f is found as

$$f = 1 - (C_{sal}/C_i)^{1/n} \tag{10.7.17}$$

This method of depreciation cannot be used when the salvage value is zero.

10.7.4 Double Declining-balance Depreciation

This method is the same as the earlier one, except for the fact that the value of f is specified as

$$f = 2/n \tag{10.7.18}$$

in which case the book value and the depreciated sum at the end of the mth year is given as

$$B_m = C_i\left(1 - \frac{2}{n}\right)^m \tag{10.7.19}$$

and

$$D_m = C_i\left(1 - \frac{2}{n}\right)^{m-1}\left(\frac{2}{n}\right) \tag{10.7.20}$$

In this case, however, the book value after n years would not be equal to the salvage value, C_{sal}.

10.7.5 Sum-of-digit Depreciation (SD)

In this method, the fraction F_{SD} declined in the mth year for a project of life-span of n years is specified as

$$F_{SD} = \frac{n - m + 1}{\sum_{i=1}^{n} i} = \frac{2(n - m + 1)}{2(n + 1)} \tag{10.7.21}$$

If $n = 10$, the denominator is 110 and F_{SD} for the first year is 20/55, for the second year, 18/110 and so on. Thus it is seen that a higher fraction is depreciated initially while this reduces as the time progresses.

10.7.6 Tax Benefits with Depreciation

It has already been pointed out earlier that it is advantageous to distribute an expense over a period of time. The benefit results in the form of tax savings as follows. It is to be recalled that taxes are levied on profits only. Any expense is tax deductible and it is in view of this, if a depreciated sum ($C_d D_{fm}$) is shown as an expense for the mth year, it should be tax-deductible. If there is an item available for an initial cost of C_i having a salvage value of C_{sal} after n years, the depreciable cost C_d would be ($C_i - C_{sal}$) and the cost diagram is given in Fig. 10.8. In the figure, t is the taxation rate.

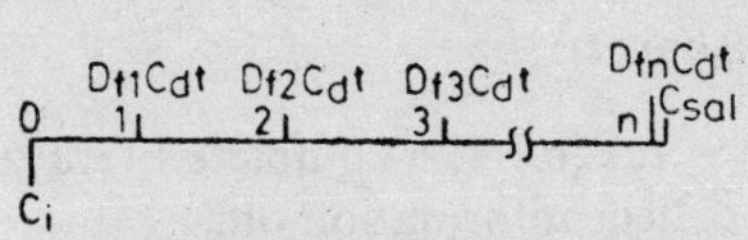

Fig. 10.8 Cost diagram and equipment with a salvage value

One of the immediate effects of tax is to reduce the net interest rate to the entrepreneur. Since the interest earned on a capital is a profit, it is taxable. The rate of return r after taxes would be

$$r = i(1 - t) \tag{10.7.22}$$

The present worth P for the cost diagram of Fig. 10.8 can be easily calculated as

$$P = C_i - C_d\, t\left[\frac{D_{f1}}{(1+r)} + \frac{D_{f2}}{(1+r)^2} + \ldots + \frac{D_{fn}}{(1+r)^n}\right] - \frac{C_{sal}}{(1+r)^n} \tag{10.7.23}$$

or

$$P = C_d(1 - t\psi) + C_{sal}\,\frac{(1+r)^n - 1}{(1+r)^n} \tag{10.7.24}$$

where

$$\psi = \sum_{j=1}^{n} \frac{D_{fj}}{(1+r)^j} \tag{10.7.25}$$

For any given depreciation method, D_{fj} above is known and the parameter ψ can be evaluated for given n. The present worth can subsequently be found from Eq. 10.7.24.

To show the nature of costs involved in a chemical plant, an acetaldehyde plant design is now given as an example. There are two types of information to be found in any plant design:

(1) break even capacity, and
(2) optimal capacity,

and these can be found through a cost analysis as shown in Sec. 10.8.

10.8 PRELIMINARY DESIGN OF AN ACETALDEHYDE PLANT

The central feature of a plant design lies in deciding an optimal flow sheet for the process and this is done using information developed in the previous chapters of this book. Once this is completed, an economic analysis is carried out to decide the plant throughput. However, this can be done only after mass and energy balance have been made. Subsequently individual equipment can be designed using the model equations governing its operations. The steps involved in the overall design are shown schematically in Fig. 10.9. The following design of an acetaldehyde plant is taken up as an example.

10.8.1 Literature Search

Before a plant design is taken up, a complete literature search is carried out. The search is to collect information on:

(1) different routes for producing the desired product,
(2) information on rate and equilibrium of chemical reactions involved,
(3) equilibrium data of relevant mixtures,
(4) physical properties of all relevant systems.

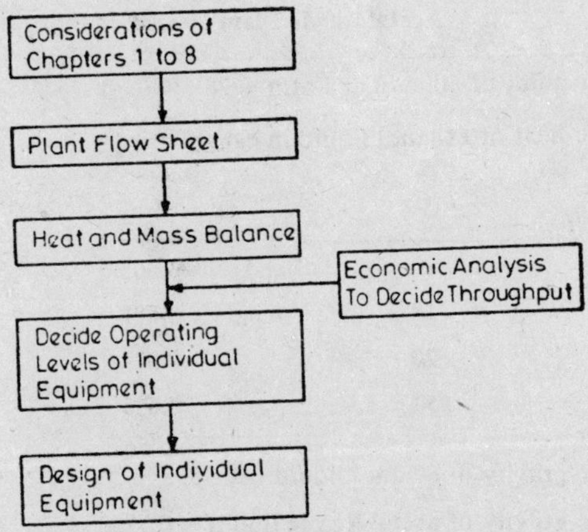

Fig. 10.9 Steps involved in a chemical plant design

The literature search on acetaldehyde yields data tabulated in Table 10.7. The processes of this product have been discussed in references 5, 6 and 7 and the one selected here is based upon ethanol as the raw material. The reaction to be carried out is the dehydrogenation in presence of a Cu-Co-Cr_2O_3 supported catalyst between the temperature range of 280–350°C. The raw material is vaporized and the vapours, so generated, are heated in a heat exchanger to the reaction temperature by hot product stream. The product stream is cooled to -10°C and in doing it, all unreacted ethanol and acetaldehyde are condensed. The outgoing gaseous stream, containing hydrogen mainly, is scrubbed with dilute alcohol (alcohol + water) to remove uncondensed products and the undissolved gas. The remaining pure hydrogen (~98% pure) is burnt in stack.

The kinetic study of dehydrogenation of ethanol to acetaldehyde has been reported in references 7 and 10. The temperature range studied is 225° to 285°C with the reactor pressure varying between 1 to 10 atm. The rate controlling mechanism was found to be the surface reaction step without dissociation. Rate constants have been measured and reported in the reference. Relevant physical properties have been found from literature and reported in Table 10.7.

Figure 10.10 shows the flow sheet of the process in which ethanol is vaporized in V_p and heated to the reactor temperature in exchanger HX1. The heated vapours are passed through the converter. The product stream is first cooled in HX1 and then in condensers C_1 and C_2 to 10°C using water and liquid ammonia respectively. This condenses most of the unreacted ethanol and the acetaldehyde formed in reactor R. The escaping gas, which is almost pure hydrogen, is further scrubbed by ethanol to remove all the traces of the product. The liquid stream, consisting of mainly ethanol and acetaldehyde, is distilled in column DC1 to get acetaldehyde.

Table 10.7[7–10] **Physical Properties and Equilibrium Data Required for Acetaldehyde Plant Design**

1. Boiling point of ethanol at 1 atm = 78.4°C
2. Specific heat of ethanol liquid in cal/g°C

T°C	C_p
−20	0.505
10	0.532
20	0.579
50	0.635

3. Specific gravity of ethanol liquid = 0.7893
4. Specific gravity of acetaldehyde liquid = 0.7780
5. Latent heat of ethanol = 200.6 cal/g
6. Specific heat of ethanol vapour

T°C	C_p, cal/g-°C
90	0.432
140	0.463
150	0.471
270	0.549

7. Latent heat of acetaldehyde = 139.5 cal/g
8. Specific heat of acetaldehyde liquid at 10°C = 0.522 cal/g-°C
9. Specific heat of acetaldehyde vapour, cal/g-°C

T°C	C_p
	0.36
25	0.336
140	0.426
270	0.528

10. Specific heat of hydrogen gas cal/g-°C

T°C	C_p
10	3.381
140	3.462
270	4.47

11. Latent heat of pure ammonia at −33°C = 590 cal/g

(Contd.)

12. Vapour-liquid equilibrium data for ethanol-acetaldehyde mixture

(a) Total pressure $P = 699$ mm Hg

Boiling point °C	Mole % ethanol in vapour	Mole % ethanol in liquid
20.1	0	0
25.7	0	20
29.5	0	30
33.7	0	40
41.5	4.1	55
47.4	12.3	65
58.1	31.8	80
65.8	50.5	90
76.2	100	100

(b) $P = 398$ mm Hg.

Boiling point °C	Mole % ethanol in vapour	Mole % ethanol in liquid
5.8	0	0
11.5	0	20
21.2	0	40
30.5	6.4	55
36.8	16.7	65
43.5	29.0	75
57.3	49.2	85
62.8	100	100

(c) $P = 77$ mm Hg

Boiling Point °C	Mole % ethanol in vapour	Mole % ethanol in liquid
−23.9	0	0
−15.2	0	35
0	0	50
4.6	92	60
17.0	22.4	70
22.7	33.0	80
28.0	50.0	90
31.2	78.8	95
34.3	100	100

(Contd.)

Table 10.7 (*Contd.*)

13. Vapour pressure data of pure acetaldehyde

Vapour pressure, (mm Hg)	Boiling Temp., T°C	Vapour pressure, (mm Hg)	Boiling Temp., T°C
1	−81.5	60	−31.4
5	−65.1	100	−22.6
10	−56.8	200	−10.0
20	−47.8	400	+4.9
40	−37.8	760	20.2

14. The rate equation of the dehydrogenation reaction

$$\frac{W}{F} = \frac{-x - 2\ln(1 - x)}{k_1(P/RT)}$$

where

W = weight of the catalyst

F = flow rate in kg/h

x = final conversion

(P/RT) = density of the gas at the temperature

k_1 = 3 L/g-h at 310°C

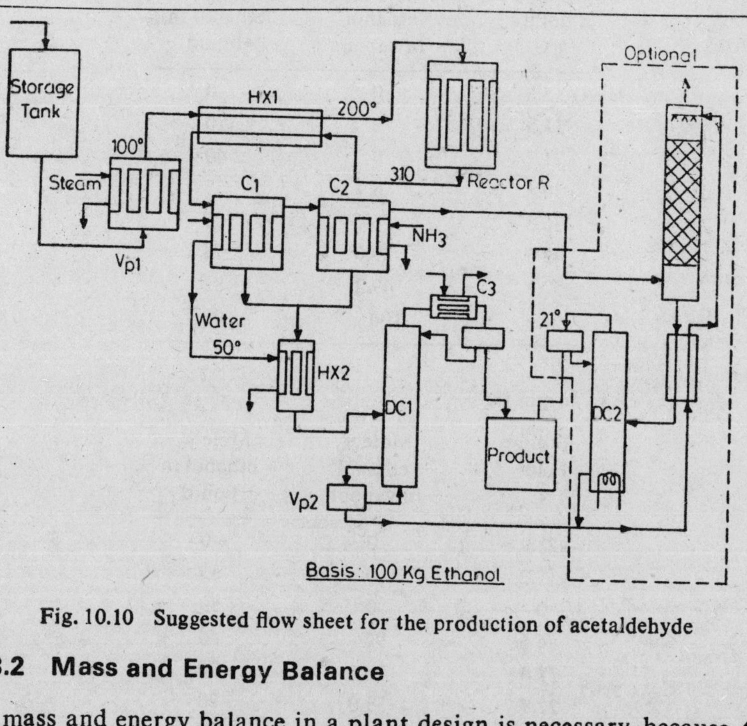

Fig. 10.10 Suggested flow sheet for the production of acetaldehyde

10.8.2 Mass and Energy Balance

The mass and energy balance in a plant design is necessary because this fixes the relative flow rates of different streams and temperatures in the flow sheet. Calculations for the acetaldehyde plant have been carried out for 100 kg *pure ethanol feed* as the basis.

In the *vaporizer*, ethanol liquid enters in at an average temperature of 30°C and leaves as superheated vapour at 100°C. The heating fluid is assumed to be saturated steam here and to provide sufficient temperature gradient, it is taken to be at about 3 atmospheric pressure. At this pressure it condenses at 132.9°C and because process steams are normally available at this pressure, this is not a serious assumption. The amount of steam required $\dot{M}_{stm}$ per hour can be calculated from energy balance as follows

$$\dot{M}_{stm}\Delta H_{stm} = 100(78.4 - 30)0.635 + 200.6 + (100 - 78.4)0.432$$

$$= 24066 \text{ kcal/h} \quad (10.8.1)$$

From steam table, ΔH_{stm} to be 517.6 kcal/kg at atm.
Therefore

$$\dot{M}_{stm} = \frac{24066}{517.6} = 46.5 \text{ kg/h} \quad (10.8.2)$$

The *heat exchanger* HX1 is only for heat recovery. It is assumed that ethanol vapour is heated upto 200°C by the product stream at temperature 310°C. As will be shown later, the product stream consists of 89.9 kg acetaldehyde, 6 kg ethanol and 4.1 kg hydrogen. If T is the outlet temperature of the product, one has the following from heat balance:

$$(89.9 \times 0.528 + 6 \times 0.549 + 4.1 \times 3.47)(310 - T)$$

$$= 100 \times 0.471 \times (200 - 100) \quad (10.8.3)$$

$$\therefore \quad T = 237.53°\text{C} \quad (10.8.4)$$

The reaction occurring in the *reactor* is the following:

$$C_2H_5OH \rightarrow CH_3CHO + H_2 \quad (10.8.5)$$

which is endothermic in nature. The optimum reaction temperature has been reported to be 310°C at which one obtains a conversion of 94%[10]. The occurrence of minor secondary reaction of ethanol to ethyl acetate has been neglected here. At this temperature, the conversion of ethanol is 94% which means that $94 \times \frac{44}{46}$ or 89.9 kg of acetaldehyde and $94 \times \frac{2}{46}$ or 4.10 kg of hydrogen are formed. If it is decided to use saturated steam at 133 atm. (steam at this pressure condenses at 335.5°C), it means that the reactor should be a pressure vessel. This proposition is rejected because of high costs and it is decided to use saturated vapours of dowtherm condensing at 350°C for heating purposes. Its latent heat is 56.5 kcal/kg and the heat of reaction for ethanol dehydrogenation is 332.64 kcal/kg of ethanol[8]. From this, the rate of dowtherm flow $\dot{M}_d$ can be found as

$$56.5\dot{M}_d = 100 \times 0.549\ (310 - 200) + 332.64 \times 94$$

$$= 37307.16$$

or $$\dot{M}_d = 660.30 \text{ kg/h} \quad (10.8.6)$$

In *condensor* C1 it is decided to use cooling water which is at 30°C. The outlet temperature of cooling water is not allowed to go above 50°C, because above this temperature, there is the problem of vaporization. Normally the approach temperature difference is about 10°C, therefore the product can at best be cooled to 40°C. At this temperature, the product stream would be a two-phase mixture and the mixture composition can be found from vapour-liquid equilibrium. Since complete information on equilibrium is not available, one uses the information available at the pressure of 699 mm Hg as a first approximation. From Table 10.7, it is found that at this temperature 4.1 mole % ethanol is in vapour and 55 mole % ethanol in liquid phase. To find out the amount of liquid condensed, it is assumed that there are m_l moles of liquid consisting of ethanol (abbreviated as E) and acetaldehyde (abbreviated as A) and m_v moles of vapour consisting of E and A. The mole balance on liquid and vapour phase for E and A are

$$0.55m_l + 0.041m_v = 0.13 \tag{10.8.7}$$

$$0.45m_l + 0.959m_v = 2.04 \tag{10.8.8}$$

In m_v, the amount of hydrogen is at present discounted. From these m_l and m_v are calculated to be 0.084 and 2.091. The composition of liquid phase would be 2.125 kg (or 0.046 moles) of E and 1.663 kg (or 0.038 moles) of A. Similarly the composition of vapour would be 88.22 kg (2.005 moles) of E, 3.94 kg (0.086 moles) of A and 4.1 (2.05 moles) of hydrogen. To find the amount of cooling water required, the heat balance yields

$$\begin{aligned}\text{Heat given out by hydrogen} &= 4.1 \times 3.462(237.5 - 40)\\ &= 2803.4 \text{ kcal}\end{aligned} \tag{10.8.9a}$$

$$\begin{aligned}\text{Heat given out by acetaldehyde} &= 89.9 \times 0.426(23.5 - 40)\\ &\quad + 1\,663 \times 139.5\\ &= 7795.7 \text{ kcal}\end{aligned} \tag{10.8.9b}$$

$$\begin{aligned}\text{Heat given out by alcohol} &= 6 \times 0.463(237.5 - 40)\\ &\quad + 2.125 \times 200.6\\ &= 975.4 \text{ kcal}\end{aligned} \tag{10.8.9c}$$

The total heat exchanged is 11574.5 kcal/h and if $\dot{M}_w$ is the cooling water flow rate,

$$\dot{M}_w \times 1 \times (50 - 30) = 11574.5 \tag{10.8.10}$$

Or

$$\dot{M}_w = 578.7 \text{ kg/h} \tag{10.8.11}$$

Results are shown in Fig. 10.11a.

In *condensor* C2, it is desired to condense all ethanol and acetaldehyde. If the working pressure is 1 atm from the equilibrium data it is seen that for temperatures below 30°C, there is going to be no ethanol in vapour phase and acetaldehyde would exert its vapour pressure at that temperature. If it is desired to achieve about 90% recovery of A, accor-

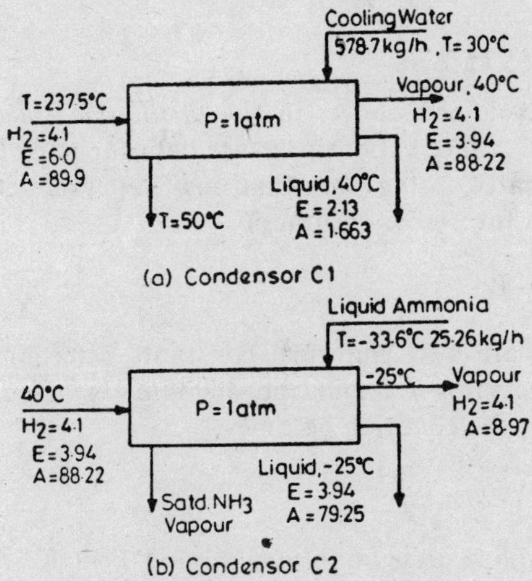

Fig. 10.11 Mass and energy balance on condensers C1 and C2

ding to Table 10.7, the outlet temperature of the product stream should be about −25°C. This is because at −22.6°C its vapour pressure is 100 mm Hg and the vapour phase will consist of 13.15 mole %. In view of this, the cooling fluid chosen is saturated liquid ammonia at 1 atm, at which it boils at −33.6°C. The coolant flow rate can be found out from heat balance after noting that the liquid consists of 1,663 kg of ethanol and 79.25 kg (= 0.9 × 89.9 − 1.663) of acetaldehyde. Also,

$$\text{Heat given by } H_2 = 4.1 \times 3.381(40+25) = 901.0 \text{ kcal/h} \tag{10.8.12a}$$

$$\text{Heat given by A} = 88.24 \times 0.36(40 + 25) + 79.25 \times 139.5$$
$$= 13120.20 \text{ kcal/h} \tag{10.8.12b}$$

$$\text{Heat given by E} = 3.94 \times 0.36(40 + 25) + 3.94 \times 200.6$$
$$= 882.56 \text{ kcal/h} \tag{10.8.12c}$$

Thus the total heat exchanged is 14903.8 kcal/h and noting that the latent heat of ammonia is 590 kcal/kg, the amount of coolant required would be 25.26 kg/h as shown in Fig. 10.11b.

The preheater to the distillation column is necessary because the feed plate will be completely chilled if the feed is not heated. The water stream from condensor Cl is available at 50°C and is used in the preheater. If the maximum approach temperature difference is 10°C, the product stream can at best be heated to 40°C. The distillation column pressure is chosen to be 1588 mm Hg so that pure acetaldehyde is obtained as liquid product at 40°C. In view of this, the stream coming out of the preheater is liquid and making the heat balance, one obtains

$$\dot{M}_w(50 - 30) = (80.91 \times 0.522 + 6 \times 0.532)(40 + 25) \quad (10.8.13a)$$

or

$$\dot{M}_w = 147.65 \text{ kg/h} \quad (10.8.13b)$$

As has been explained above, in the *distillation column* acetaldehyde condenses at 40°C. Since vapour pressure data for pure gas is not available, it is estimated using the Clausius-Clapeyron equation (or the Antoine equation for higher accuracy)

$$\ln P = \frac{A}{T} + B \quad (10.8.14)$$

where A and B are two empirical constants determined from boiling point data at pressures 760 mm and 400 mm Hg given in Table 10.7. For acetaldehyde, the equation becomes

$$\ln P = 17.5780 - \frac{3209}{T} \quad (10.8.15)$$

where P is the vapour pressure in mm Hg and T in K. At 40°C, vapour pressure is found to be 1519 mm Hg. Mass balance upon the column is made assuming 98% acetaldehyde recovery to obtain the total overhead flow rate of 82.33 kg/h and at the bottom as 4.58 kg/h as shown in Fig. 10.12. To find out the heat duty of the total condenser and partial reboilers, the reflux ratio must be known. If it is taken as 3, the vapour

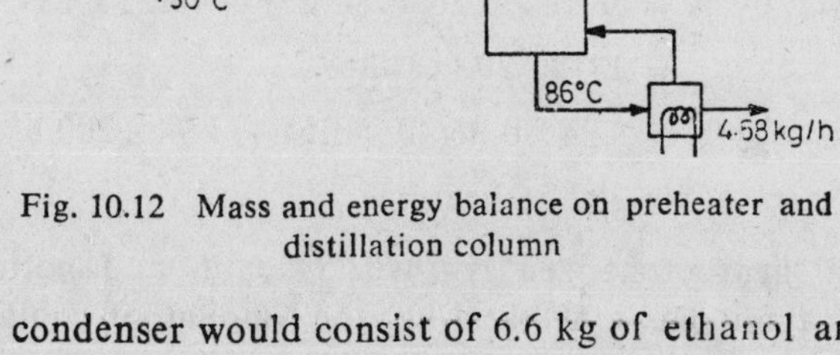

Fig. 10.12 Mass and energy balance on preheater and distillation column

going to the condenser would consist of 6.6 kg of ethanol and 322.76 kg of acetaldehyde. Consequently the heat load of the condenser is $(6.6 \times 200.6 + 322.76 \times 139.5)$ or 46348.98 kcal/h. Similarly the heat duty of the partial reboiler is 46491 kcal/h.

Now the *absorber* is considered through which the acetaldehyde and hydrogen mixture coming from condenser C2 is passed. The column operates at 1 atm and average temperature T_{ab} (assumed). Figure 10.13 shows the schematic diagram of a countercurrent absorption tower with numerical values given in it. It is assumed that there is 95 mole %

recovery of acetaldehyde. If the column is long enough such that the outgoing streams are in equilibrium (hydrogen is regarded as inert) the

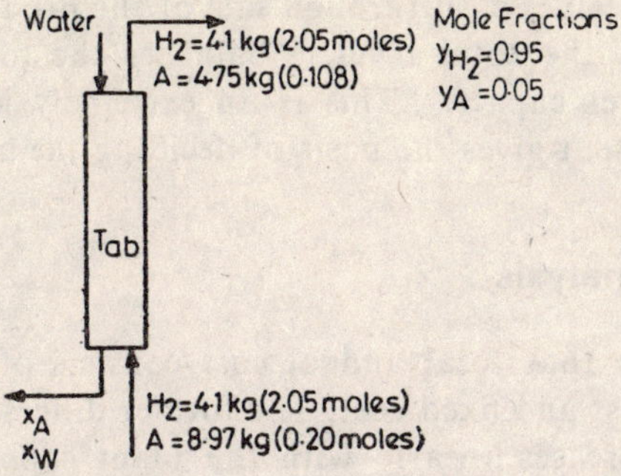

Fig. 10.13 Mass and energy balance on absorber

concentration of acetaldehyde can be calculated in the outgoing water stream. The value, thus calculated, is the upper limit and in reality lower values will be obtained. Assuming Roult's law holds good for the mixture, one has

$$p_A x_A + p_w x_w = P(1 - Y_{H_2}) \tag{10.8.16}$$

Hydrogen above is acting as an inert gas and is not present in the liquid phase. T is the total pressure whereas p_A and p_w are vapour pressures of acetaldehyde and water respectively at the column temperature P_{ab}. If the column temperature is given to be 15°C, p_A can be found from Eq. 10.8.15 as 615.6 mm Hg. Vapour pressure of water at 15°C is 12.8 mm and from this information one finds

$$x_A = \frac{P(1 - Y_{H_2}) - p_w}{p_A - p_w}$$

$$= \frac{760(1 - 0.95) - 12.8}{615.6 - 12.8} = 4.18 \times 10^{-2} \tag{10.8.17}$$

This amounts to 9.64% weight of acetaldehyde in water solution which is a very small concentration of acetaldehyde. Since this is to be stripped from the solution in a suitable distillation column, it may not be economical to recover A from the stream coming out of the condenser C2. In view of this, it is decided to burn it in stack. It may be added here that a decision of this nature can be made only after overall economic evaluation and elimination of the absorber system from the flow sheet of Fig. 10.10 is a first order approximation which should be subsequently modified.

After the mass and energy balance has been made, the minimum throughput of the chemical plant must be fixed. This is obtained by balancing the total revenue with total expenditure and this analysis is called the break-even analysis as discussed in Sec. 10.9.

10.9 BREAK-EVEN ANALYSIS OF A CHEMICAL PLANT

In running a chemical plant, there are different costs involved listed in Table 10.6. One receives cash through sale of the product and the plant capacity at which the total revenue balances the total annual cost is called the break-even capacity. This is an extremely important piece of information because it gives the basis of deciding the actual plant size.

10.9.1 Linear Analysis

Table 10.6 reveals that total annual cost consists of two components, namely variable cost and fixed cost. The former differs from the latter in that it (almost) increases linearly with the plant capacity. Examination of the table indicates that components of fixed costs are less sensitive to plant size even though the total layout of expenses are affected by the size of the project.

As a first approximation it is assumed that the total variable cost C_v is proportional to the plant throughput Q

$$C_v = VQ \tag{10.9.1}$$

where V is a constant independent of Q and C_v and has the dimension of cost per year. Similarly the annual fixed cost, C_F is assumed to be independent of the plant size. If C_T is the total cost per year, one has

$$C_T = VQ + C_F - SQ \tag{10.9.2}$$

where S is the sale price of product produced. From the definition of the break-even point, the total cost C_T is equal to zero at this capacity and therefore one obtains

$$Q_{B.E.} = \frac{C_F}{(V - S)} \tag{10.9.3}$$

where $Q_{B.E.}$ is the break-even capacity and is shown in Fig. 10.14 in an economic-production chart.

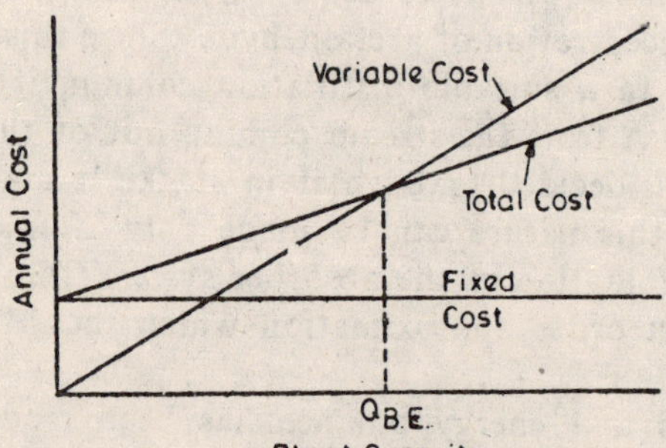

Fig. 10.14 Economic-production chart giving the break-even capacity

10.9.2 Analysis for a Chemical Plant

The linear analysis presented above is extremely idealized and does not yield any meaningful result. The procedure for finding the break-even point remains the same as earlier except for the fact that the annual fixed and variable costs are related in a more complex way. This is modelled as follows.

In the analysis presented here, the annual variable costs are assumed to consist of mainly

1. raw material cost,
2. direct operating labour cost, and
3. process services (steam, power, water, refrigeration, etc.)

From the mass and energy balance, it is found that raw material and utility requirements are directly proportional to the plant capacity. Contrary to this, as has been discussed in the previous chapter, the direct operating labour charges are proportional to $Q^{0.25}$. In view of this, the annual variable charges C_{VP} of the chemical plant would be

$$C_{VP} = (c_R Q + c_U Q + c_L Q^{0.25}) \tag{10.9.4}$$

where c_R, c_U and c_L are constants of proportionality. It is thus seen that C_{VP} is a non-linear expression.

The fixed annual cost of a chemical plant is found by first determining the total fixed cost of the entire plant. This is next depreciated (assuming linear) over a given period to determine the annual cost. Since other cost items constituting the indirect plant cost (Table 10.6) are dependent upon the total plant cost, in the absence of any information, it is taken proportional to the depreciated cost. If the cost involved with the selling of products (for example, expenses of advertisement, etc.) is ignored for preliminary calculation purposes, C_F is given by

$$C_F = \frac{\alpha}{n} \sum_{i=1}^{N} c_i (CV_i)^{b_i} \tag{10.9.5}$$

where the cost of ith equipment C_i has been written as

$$C_i = c_i (CV_i)^{b_i} \tag{10.9.6}$$

(CV_i) stands for the capacity variable for the equipment, n is the number of years over which the plant has been depreciated and α is the constant of proportionality. If S is the sale price of the product, the total cost C_T is given by

$$C_T = (c_R Q + c_U + c_L Q^{0.25}) + \frac{\alpha}{n} \sum_{i=1}^{N} c_i (CV_i)^{b_i} - SQ \tag{10.9.7}$$

The break-even capacity $Q_{B.E.}$ is found by setting C_T equal to zero and is given by

$$\frac{\alpha}{n} \sum c_i (CV_i)^{b_i} + (c_R Q_{B.E.} + c_U Q_{B.E.} + c_L Q_{B.E.}^{0.25}) = S Q_{B.E.} \tag{10.9.8}$$

and can be obtained approximately by the technique demonstrated through the example of the acetaldehyde plant.

10.10 MINIMUM THROUGHPUT OF AN ACETALDEHYDE PLANT

An examination of Eq. 10.9.8 would reveal that one requires the complete design of the whole plant before $Q_{B.E.}$ can be solved. Therefore the rigorous trial-and-error procedure would consist of following steps:

1. A plant capacity Q is assumed.
2. From the mass and energy balance the operating level of individual equipment is determined.
3. The cost of individual equipment is determined through the correlations of Chapter 9.

The costs are substituted in Eq. 10.9.8 to check if it is satisfied.

To reduce the amount of calculations, capacity variables CV_i must be estimated based upon the mass and energy balance carried out in the earlier section. To do so, values of the technological coefficients must be guessed and the first approximations of these are given in Table 10.8

10.10.1 Cost of 100 kg/h Acetaldehyde Plant

For the example of the acetaldehyde plant of Sec. 10.8 and as a first trial it is assumed that $Q = 100$ *kg/h of acetaldehyde*. From the flow sheet it is found that two storage tanks are required. One for the raw material ethanol is assumed to be big enough to hold one month's supply. This would mean that it has a capacity of 72000 kg or at the ambient condition 89.93 m^3. If it is a 15.24 m elevated steel tank, its cost would be $ 70,000 (point 10 of Fig. 9.3). The second storage tank is required for the product acetaldehyde. It is assumed to have one week's capacity or a volume of 17.28 m^3. The tank is an API conical type with floating head. Using point 13 of Fig. 9.3, its cost is found to be $ 110,000 and including in the factor for change in design and installation, its cost becomes $ 195437 ($m_F = 1.09$, $M = 1.63$).

The chemical plant consists of three condensers C1, C2 and C3, two preheaters, HX1 and HX2, one vaporizer, V_{p1}, and one reboiler V_{p2}. The heat load of C1 is 11574.5 kcal/h and log mean temperature difference, ΔT_{C1} is 60.61°C (Fig. 10.11a). From Table 10.8, the overall heat transfer coefficient, U_{C1}, is taken as 48.8 kcal/hr-m^2-°C. Similarly ΔT_{C2} and ΔT_{C3} are 34.48°C and 24.7°C and for $U = 48.8$ kcal/h-m^2-°C their areas A_{C2} and A_{C3} are 8.85 and 38.43 m^2. Because A_{C1} is small, it is decided to employ a double pipe exchanger for this purpose whereas for C2 and C3 one uses a shell and tube exchanger. Points 22 and 13 of Fig. 9.2 are used to find their costs 750, 2900 and 9000 dollars. M for double pipe exchanger is 1.80 whereas for shell-and-tube heat exchanger, it is 3.17. Therefore the cost of installed units is 1350, 9193 and 28,530 dollars. One proceeds similarly to

Table 10.8[12, 13] A Quick Procedure for the Determination of Capacity Variables of Different Equipments

Sl. No.	Description	Quick formula	Application and comments
1.	Cost indices		Given in Chapter 9. Also see Ref. 11
2.	Depreciation	Use straight line with 10% of fixed capital per year	Order of magnitude estimate Equipment lasts for 10 years with zero salvage value
3.	Energy balance	Negligible sensible heat of steam etc.	
4.	Mass balance	Assume equilibrium for two phase mixtures	
5.	HE cooling water	30–50°C	
6.	Approach temperatures	10°C	
7.	Condenser area	$Q = UA\, \Delta T_{lm}$	Q is energy, exchanges A is heat transfer area, ΔT_{lm} is log mean temperature difference
8.	Overall heat transfer coefficients	48.8 kcal/h-m²-°C	Air/vapour
		244 -do-	Water/oil
		48.8 -do-	Vapour/water
		244–488 -do-	Oil/oil
		195.2–732 -do-	Steam/oil
9.	Heat transfer coefficients, *h*	48.8 kcal/h-m²-°C	Gas
		488 -do-	Liquid
		2440 -do-	Boiling liquid
		4880 -do-	Condensing liquid
10.	Horsepower efficiencies	90%	Electric motors
		40–80%	Pumps
		60%	Compressors
11.	Labour	Use nomogram of Chapter 9 or use one person per major equipment	
12.	Reactors	5 h^{-1} space velocity	Used only when no other information available
13.	Reflux ratios	1–4	
14.	Average flooding velocity for vapours in distillation columns	1.5–2 m/s	
15.	Specific heats of liquid	Same as water	
16.	Vessel sizing	1 : 1 to 3 : 1 as the ratio of length : diameter	
17.	Raw material storage tank	Volume calculated assuming one month raw material requirements	
18.	Product storage tank	About one week capacity	

find areas of HX1 and HX2 as 0.77 and 0.41 m^2 (U_{HX1}=48.8 kcal/h-m^2-°C, ΔT_{HX1} = 125.14°C, U_{HX2} = 244, ΔT_{HX2} = 29.41°C) and treating them as double pipe exchangers, the costs of installed units are 1170.00 and 1116.00 dollars respectively. Reboiler and vaporizers are treated as agitated falling film evaporators with their overall heat transfer coefficients as 244 kcal/h-m^2-°C and for specified temperature and heat duties, one finds the surface area requirements as 3.00 m^2 and 13.60 m^2. Using point 20 of Fig. 9.6, their costs (installed units) are found to be 42,500 and 95,000 dollars.

To find the cost of a *distillation column* (Table 9.11 and point 1A of Fig. 9.7), its height H, diameter D and the shell thickness t must be specified. Shell thickness can be taken as 0.635 cm, whereas the height H of the column is independent of the throughput. From the information on relative volatility, the number of equilibrium stages is found and from the knowledge of plate efficiency the actual number of plates is determined. These have been carried out in Appendix 10.1 where it is found that 9 plates are required for the column. If tray spacing is assumed to be 60.96 cm, the height H would be 5.49 m. The diameter is calculated from the knowledge of reflux ratio, R. For R = 3 (Table 10.8) the mass flow rate of vapour at the top of the column would be 82.33×4 or 319.32 kg/h. Assuming density of the vapour as 2.40 kg/m^3 and allowable vapour velocity as 1.524 m/s (Table 10.8), the column diameter is found to be 15.85 cm. Consequently the factor (DHt) is (15.84×549×0.635) or 5522.06 cm^3. Examination of Table 9.11 reveals that this is beyond the prescribed range of the nomograph and in view of this, the cost is extrapolated from the following equation

$$(\text{cost})_{\text{Dist}} = 26896.70\left(\frac{DHt}{16387}\right)^{0.58} \tag{10.10.1}$$

This equation has been determined from the nomograph (Fig. 9.7) itself and yields the cost of the column as \$ 14312.25.

To calculate the cost of the *reactor*, it is taken as a fired heater (point 26, Fig. 9.2) with dowtherm as heating fluid. The reactor heat duty is known from energy balance as 3.95×10^4 k J/h and the cost can be obtained only through extrapolation from

$$(\text{cost})_{\text{R}} = 22468\ (\text{mBTU})^{0.82} \tag{10.10.2}$$

The above equation has once again been found from the nomograph and yields the reactor cost as \$ 3295.3.

Including spares and standbys, about 20 *liquid pumps* with (m^3/min) (atm) as 0.0011 are assumed to be required and from point 24 in Fig. 9.2 the cost of one pump is found as \$ 7.0. The total amount of steam required (in vaporizer V_{p1} and reboiler V_{p2}) is 132.95 kg/h and the cost of an *installed boiler* is determined from point 6 of Fig. 9.5 as \$ 16,100. Similarly the total *refrigeration* required is 61252.78 kcal/h or 20.16 Tons (1 Ton = 12660 k J/h). From point 28 of Fig. 9.7, the cost of an installed refrigeration unit is determined as \$ 1,65,600.

Table 10.9 gives the cost of different equipment of the plant and the total cost of equipment is found to be \$ 6,52,834. Piping cost is determined from point 11 of Fig. 9.4 and thus the fixed cost of a 100 kg/h acetaldehyde plant is found as \$ 7,10,834.

Table 10.9 Total Cost of Installed Equipments of a 100 kg/h Acetaldehyde Plant

Sl.No.	Equipment	Cost in Dollars
1.	Storage tanks	
	Ethanol storage	70,000
	Acetaldehyde storage	1,95,437
2.	Double pipe exchangers	
	HX1	1,170
	HX2	1,116
	C1	1,350
3.	Shell and tube exchangers	
	C2	9,193
	C3	28,530
4.	Evaporators	
	V_{p1}	42,500
	V_{p2}	95,000
5.	Reactor	3,295
6.	Distillation column	14,312
7.	Pump (Twenty in number)	14,000
8.	Steam Boiler	16,100
9.	Refrigeration unit	1,65,600
	Total	6,57,603
10.	Piping cost (point 11, Fig. 9.4)	58,000
	Total	7,15,603

10.10.3 Variation of Plant Fixed Cost with Capacity

Kharbanda correlates the cost of an acetaldehyde plant using the following relation:[11]

$$C = K(T/1000)^{0.63} \tag{10.10.3}$$

where C is the fixed cost, T is its capacity in Tons/day and K is a constant. K can be evaluated from the information on 100 kg/h plant and is

$$K = \frac{7,15,603}{\left(\frac{24 \times 100}{1000}\right)^{0.63}} = 4,12,229 \tag{10.10.4}$$

which gives the correlation equation as

$$C = 4,12,229\ (0.024Q)^{0.63} \tag{10.10.5}$$

10.10.4 Calculation of Break-even Capacity

It has already been noted in Chapter 9 that if one person is allotted per major equipment, the size of the labour force would become independent of the plant throughput. If the plant is divided in 10 regions and a single worker is allotted to it, the labour cost (at a rate of \$ 500 per person) to the company would then be \$ 5000 per month. Assuming 25 working days in a month, the cost would be \$ 8.33 per hour. If the fixed cost is depreciated in 10 years, the depreciated cost per hour D, would be

$$D = \frac{4,12,229 \times (0.024Q)^{0.63}}{300 \times 10 \times 24}$$

$$= 5.73 \times (0.024Q)^{0.63} \qquad (10.10.6)$$

The cost of acetaldehyde and ethanol in the market is found to be 1.10 \$/kg and 0.14 \$/kg respectively[11] and making the cost balance, one gets

$$5.73(0.024Q_{BE})^{0.63} + 8.33 = (1.10 - 0.14)Q_{BE}$$

$$= 0.96Q_{BE} \qquad (10.10.7)$$

From this Q_{BE} is determined as 11.30 kg/h by trial and error.

10.10.5 Calculation of Actual Break-even Point

It may be recalled that Q_{BE} calculated above does not account for several expenses which actually occur. In view of this, the analysis above is refined as follows. As a second trial, the plant capacity is taken as 25 kg/h of acetaldehyde. Table 10.10 gives the total fixed capital required. It also gives the working capital requirements. In Table 10.11 the annual cost has been calculated after neglecting the working capital requirements because it is expected to get renewed every month. In this table, the item 'direct supervision' has been removed and put under the section 'overheads'.

Total revenue obtained every year through the sale of acetaldehyde would be \$ 2,29,824 which just about matches with the total annual cost. Therefore the actual break-even capacity is 25 kg/h *of acetaldehyde*.

The actual size of the plant should be greater than this capacity (i.e. 25 kg/h) which is an economic necessity. Examination of Eq. 10.9.7 reveals that the profit (which is equal to $-C_T$) is a monotonically increasing function of Q and in view of it, higher and higher profits would be obtained as Q is increased. In Appendix 10.1 detailed design of reactor has been carried out for 25 kg/h of acetaldehyde plant. Design of heat exchangers for this flow rate (not given here) using the method of Kern[12] reveals that they have low efficiencies of heat transfer. It is felt that the smallest heat exchanger available is big for the task.

The process flow sheet is then synthesized using the method outlined in Chapter 4. The resultant flow sheet is given in Fig. 10.15. It may be noted

Table 10.10 Total Fixed and Working Capital Requirements for 25 kg/h Acetaldehyde Plant Processing 30.5 kg/h of Ethanol

Sl.No.	Head of Expenditure	Requirement in dollars
1.	Fixed Capital	
	(a) Principal equipment, piping and their installation	3,35,630
	(b) Instruments and controls (@ 7% of 1)	23,494
	(c) Electrical (@ 5% of 1)	16,781
	(d) Building (@ 15% of 1)	50,344
	(e) Land (@ 2% of 1)	6,712
	(f) Engineering and supervision (@ 19% of 1)	63,770
	(g) Construction expenses (@ 20% of 1)	67,126
	(h) Contractor's fee (@ 10% of 1)	33,563
	(i) Contingency (@ 20% of 1)	67,126
	Total	6,64,546
2.	Working Capital	
	(a) One month stock of ethanol (0.14 dollars/kg)	3,074.40
	(b) One week stock of acetaldehyde	4,788.00
	(c) One month trade credit	19,152.00
	(d) One month salary	5,000.00
	Total	32,014.40

Table 10.11 Calculation of Annual Cost of Acetaldehyde Plant Processing 30.5 kg/h Ethanol

Annual Variable Cost in Dollars		Annual Fixed Cost in Dollars	
1. Raw material	36,893	1. Indirect plant cost	
2. Direct labour	60,000	(a) Depreciation, 10%	66,455
3. Power required*		(b) Taxes, 2%	13,291
(a) Steam	1,542	(c) Insurance 1%	6,646
(b) Refrigeration†	19,152	2. Overheads	
(c) Water	10.39	Engineering and supervision	12,000
(d) Dowtherm heating	177.14	3. Management	12,000
4. Maintenance (4% of depreciated cost)	2,658		

Total Annual Cost = 2,30,824.53 dollars

*Costs of Utility are[11]

(1) Power	2 cents/kWh
(2) Steam	4.40 dollars/Ton
(3) Cooling water	20 cents/1000 gal
(4) Plant fuel	1.80 dollars/m Btu

†Assumed that COP for the refrigeration system is 2 and the system is run by electricity.

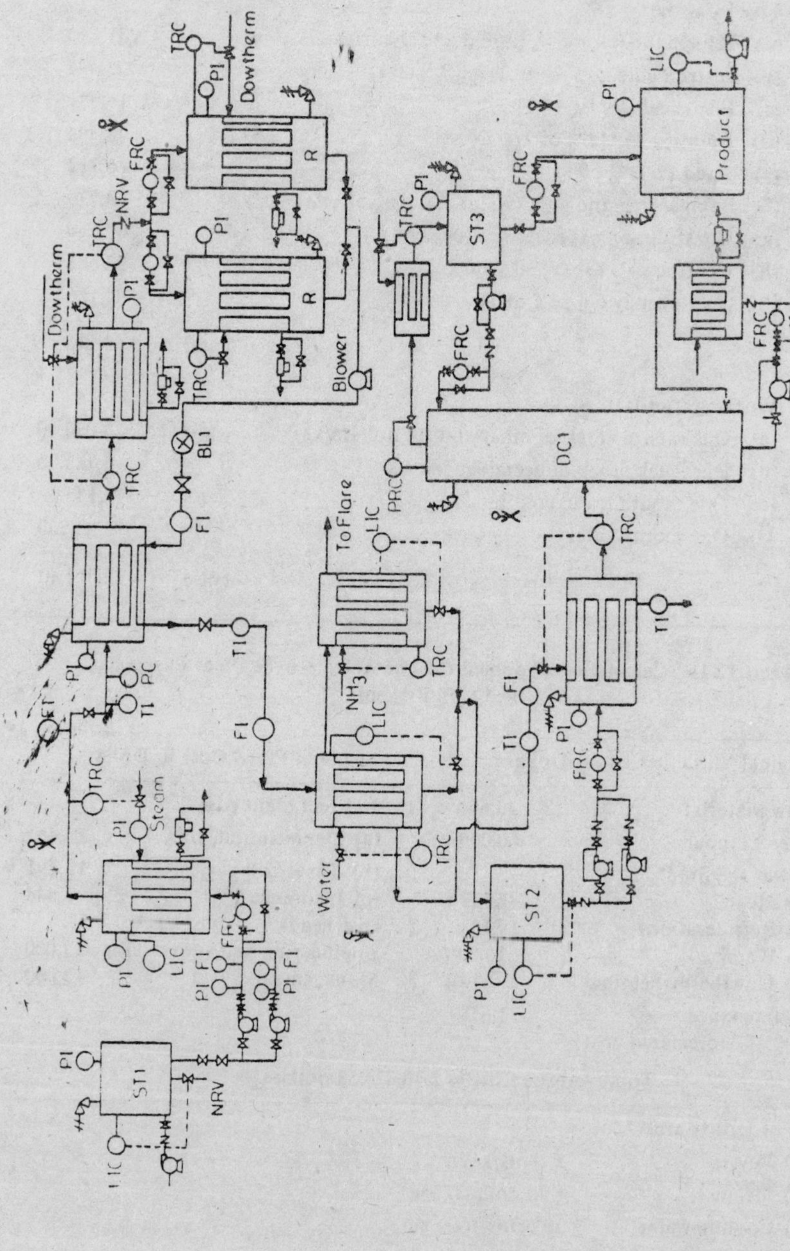

Fig. 10.15 Process flow sheet of the acetaldehyde plant

that the plant cannot function without the heat exchanger HX3, which is also needed during the plant startup. For the purpose of the 'trouble-shooting' analysis, the plant has been divided into four sections:

(1) storage tank and vaporizer,
(2) heat exchangers HX1 and HX3 and reactor,
(3) condensers C1 and C2, and
(4) distillation column,

and the complete analysis is given in Appendix 10.2 and readers must go through it carefully. After this has been carried out completely, it is worthwhile to give attention to problems that can arise during plant startup and shutdown and some of the considerations are given in Table 10.12.

10.11 PLANT DESIGN UNDER THE RISK OF INVESTMENT

Whenever an investment has to be made, different alternatives available to the entrepreneur are evaluated based on the following two factors:[16, 17]

1. Expected rate of return, and
2. the risk of investment.

In practice, the rate of return from an investment is never definite and in view of this, it is possible to associate a probability distribution of return with a given project. The mean of this distribution function is called the expected return and the standard deviation, the risk of the investment.

According to the theory of dynamic equilibrium, if a commodity is manufactured at a production level more than its demand, the market prices would fall causing a reduction in the profit per unit weight of the product produced[18]. Thus the demand of the product represents an extremum on the choice of plant capcity. Not only this, when there are several competitors already existing in the market, a new entrepreneur, at times, faces the problem of 'squeezing out', in which the market price is purposefully reduced to evict him. There are also instances where the plant size is limited by the availability of the raw materials even though there may exist an infinite demand of the product.[11]

Among the factors limiting the plant capacity, which an entrepreneur more commonly faces, is the risk of investment itself. In this section, optimal capacity of a chemical plant has been calculated assuming an exponential risk of investment.[19]

When an entrepreneur is in the process of making up his mind, he is greatly influenced by

1. the rate of return, and
2. the capital layout required for the capacity chosen.

The larger this capital layout, the more hesitant he is to go ahead with the investment because of the risk of the loss of capital. This is mathematically written by a probability, p[19],

$$p = \exp\left(-c^{*}\Sigma C\right) \qquad (10.11.1)$$

Table 10.12 Some of Startup and Shutdown Considerations of Acetaldehyde Plant

Startup Consideration

1. Flush the line with N_2 to remove air.
2. Start the controller system.
3. Check the level in storage tank (ST1) and pump ethanol to get minimum required level.
4. Open the main line valve between pump p_2 and the vaporizer. Simultaneously start the pump p_2 and steam flow to the vaporizer.
5. Check the vapour flow rate out of the vaporizer and as soon as vapours start coming out, open the main line valve and start the dowtherm stream to HX2 and simultaneously start preheating the reactor.
6. Start the blower BL1 when the temperature of the reactor reaches the reaction temperature.
7. Ensure communication between all sets of simultaneous operations.
8. Open the line valve before and after HX1.
9. Start precooling condenser Q by opening NH_3 circuit.
10. Start water flow rate in condenser C1 as soon as FM on C1 inlet shows the flow.
11. Open the liquid outlet valve on C1 and start pump p_3.
12. Open the valve between C1 and C2, start blower BL2.
13. Open the line valve and start pump p_4.
14. Open the line valve on flare and start the continuous ignition system.
15. Start water line on preheater.
16. Open the line valve before DC1 condenser.
17. Start the cooling fluid flow when pressure indicator before the condenser indicates 1.5 atm.
18. Start the reflux line pump and open the valve; initially total reflux.
19. Open the valve before and after reboiler and start the reboiler.
20. Open the line valve and pump between storage tank and reflux splitter on DC after the desired temperature composition in DC is reached.

Shutdown Consideration

1. Announce shutdown throughout the plant.
2. Shutdown all pumps, flows, and auxiliary fluids (like steam, NH_3, dowtherm etc.) and control system, etc. successively from the raw material end.
3. Isolate various sections in the system having same phases by closing appropriate manual valves.
4. For the section which have vapour phase, flush the line with N_2 and send the stream to the flare.
5. For the sections, which have liquid, can be drained by using liquid bleed line and then flush it with N_2.
6. Close the ignition on flare when only N_2 goes through it.

where c^* is a parameter characterizing the personal attitude of the entrepreneur and (ΣC) is the total capital layout. This is assumed to consist of:

1. the total fixed cost of the plant,
2. a few month's salary (say m) of workers, and
3. a few month's raw material requirements.

The plant cost C_p is written as

$$C_p = AQ^x \tag{10.11.2}$$

The labour cost C_L as

$$C_L = \beta Q^y \tag{10.11.3}$$

The raw material cost C_R as

$$C_R = \gamma Q \tag{10.11.4}$$

Therefore the plant layout cost is given by

$$\begin{aligned} C &= AQ^x + m\beta Q^y + m\gamma Q \\ &\equiv (AQ^x + BQ^y + CQ) \end{aligned} \tag{10.11.5}$$

Because an entrepreneur can either make an investment or refrain from it, one has the following two states:

1. To make an investment with probability p and earn the rate of return π given by

$$\pi = \eta Q - \left(\frac{A}{n} Q^x + \beta Q^y + \gamma Q\right) \tag{10.11.6}$$

where n is the number of years of depreciation and η the price of the product; or

2. To refrain from the investment with probability $(1 - p)$ and earn nothing.

The expected return, $E(\pi)$, therefore is given by

$$E(\pi) = \exp\,[-c^*(AQ^x + BQ^y + CQ)][\eta Q - (\alpha Q^x + \beta Q^y + \gamma Q)] \tag{10.11.7}$$

where α is equal to (A/n). The optimal capacity Q_{opt} can be obtained by equating $\dfrac{dE(\pi)}{dQ}$ to zero and solving it by trial and error.

For the acetaldehyde plant, the different factors involved in Eq. 10.11.7 have been calculated in rupees in Table 10.13 and the expected profit is given by

$$\begin{aligned} E(\pi) = {} & \exp\,[-c^*(167830Q^{0.63} + 43050Q^{0.25} + 374Q)] \\ & \times[8143Q - 16783Q^{0.63} - 172200Q^{0.25}]. \end{aligned} \tag{10.11.8}$$

In Fig. 10.16 the value of c^* has been systematically varied and $E(\pi)$ calculated for different values of Q. For a given c^*, the curve first rises

Table 10.13 Calculation of Parameters of the Acetaldehyde Plant

Selling cost of acetaldehyde[11] = Rs. 9640.4 per ton
Cost of ethanol[11] = Rs. 1496.9/ton

$\therefore \quad \delta = \eta - \gamma = 8143$

$x = 0.63$

$y = 0.25$

The fixed cost of the acetaldehyde plant = $167830Q^{0.63}$.

Assuming the plant depreciates in 10 years linearly

$\alpha = 0.1 \times 167830 = 16783$

$A = 167830$

Labour cost of a 180 ton per year plant = Rs. 630741/year

$$\therefore \quad \beta = \frac{630741}{(180)^{0.25}} = 172200$$

Assuming 3 months of raw material stock and three months of salary of workers is always maintained,

$m = 0.25$ years

$\therefore \quad B = m\beta = 43050$

$C = m\gamma = 374$

Therefore the profit function is

$$\pi = [\exp(-c^*(167830Q^{0.63} + 43050Q^{0.25} + 374Q))] \times [8143Q - 16783Q^{0.63} - 172200Q^{0.25}].$$

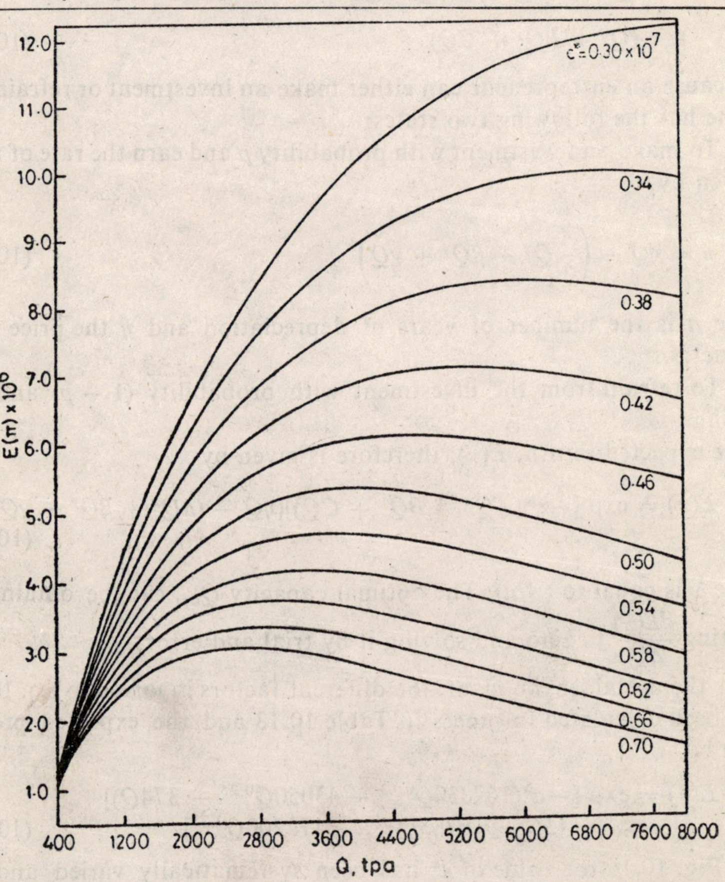

Fig. 10.16 The expected rate of return as a function of Q with c^* as parameter

and then falls for higher values of Q undergoing a maximum. In Fig. 10.17, the optimal plant capacity Q_{opt} has been plotted against c^* and in

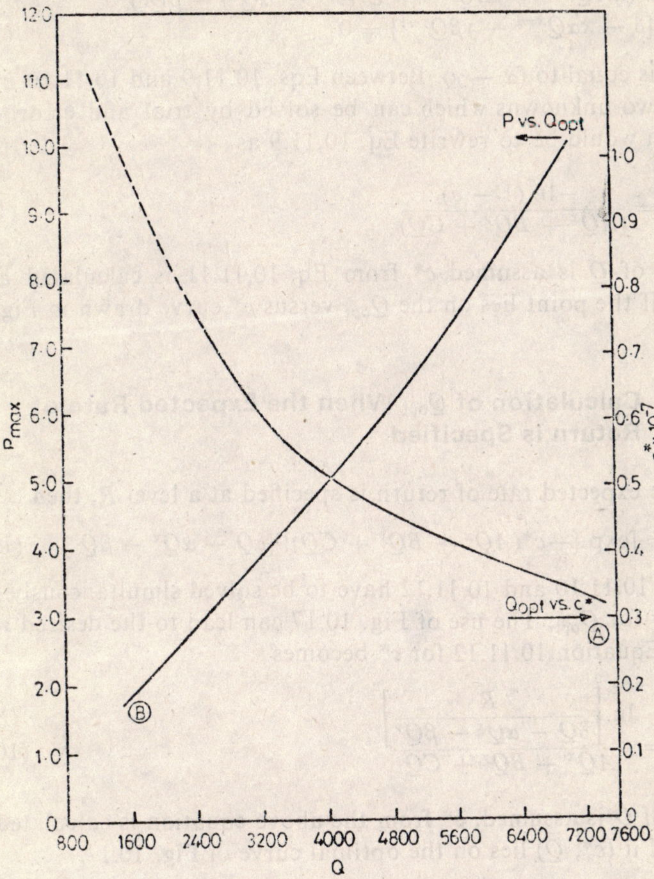

Fig. 10.17 Q_{opt} versus c^*

the same graph the expected optimal return has also been shown. Increasing c^* signifies that the entrepreneur is more and more conservative in his attitude towards investment and in view of this he would tend to settle for lower profits which is precisely what is depicted in Fig. 10.17.

10.11.1 Calculation of Q_{opt} for the Specified Equity

Equity is the fraction of the capital invested by the entrepreneur and the remainder is the borrowed money (preferably from the bank). In case the venture fails, this would represent his risk, *e*. In other words,

$$e = (1 - p) = 1 - e^{-c^*}(AQ^x + BQ^y + CQ) \tag{10.11.9}$$

The optimal capacity Q_{opt} is given by setting $\frac{dE(\pi)}{dQ}$ equal to zero. When this is carried out, one obtains

$$-c^*(xAQ^{x-1} + yBQ^{y-1} + C)(\delta Q - \alpha Q^x) - \beta Q^y) + [\delta - x\alpha Q^{x-1} - y\beta Q^{y-1}] = 0 \qquad (10.11.10)$$

where δ is equal to $(\alpha - \gamma)$. Between Eqs. 10.11.9 and 10.11.10 c^* and Q are the two unknowns which can be solved by trial and error. A better approach would be to rewrite Eq. 10.11.9 as

$$c^* = \frac{-\ln(1 - e)}{(AQ^x + BQ^y + CQ)} \qquad (10.11.11)$$

A value of Q is assumed c^* from Eq. 10.11.11 is calculated and it is checked if the point lies on the Q_{opt} versus c^* curve drawn in Fig. 10.17.

10.11.2 Calculation of Q_{opt} When the Expected Rate of Return is Specified

When the expected rate of return is specified at a level R, then

$$R = [\exp\{-c^*(AQ^x + BQ^y + CQ)\}][\delta Q - \alpha Q^x - \beta Q^y) \qquad (10.11.12)$$

and Eqs. 10.11.10 and 10.11.12 have to be solved simultaneously by trial and error for Q_{opt}. The use of Fig. 10.17 can lead to the desired result as follows. Equation 10.11.12 for c^* becomes

$$c^* = \frac{\ln\left[\dfrac{R}{\delta Q - \alpha Q^x - \beta Q^y}\right]}{AQ^x + BQ^y + CQ} \qquad (10.11.13)$$

A value of Q is assumed, c^* from the above equation is calculated and it is checked if (c^*, Q) lies on the optimal curve of Fig. 10.17.

10.12 CONCLUSIONS

In this chapter, steady state plant design is considered when the flow sheet of the process is given. It is shown how the plant throughput is decided from economic considerations. It was found that there is a minimum capacity, called the break-even point, at which the total cost is equal to the total revenue. A method is suggested on how this can be determined.

The decision on the plant capacity is shown to be a problem of long-range planning and an algorithm has been developed which minimizes the entrepreneur's risk of capital. The optimal capacity can be uniquely determined either when the equity is specified or the level of return is known.

APPENDIX 10.1

Detailed Design of Reactor and Distillation Column for Acetaldehyde Plant

A10.1.1 Reactor

A static bed reactor is designed by cost considerations, along the same lines as presented in reference 10. The Cu on asbestos catalyst is used with particle size −28 to +200 Tyler mesh as in the reference. The weight of catalyst required W, the length L and diameter D of the reactor are found as follows.

According to reference 10, the optimum reaction conditions are

$$\text{Reactor pressure } p = 1 \text{ atm} \tag{A10.1.1a}$$

$$\text{Temperature of the reaction } T = 310°\text{C} \tag{A10.1.1b}$$

$$\text{Conversion } x = 0.94 \tag{A10.1.1c}$$

Kinetic equation for the first order reaction

$$C_2H_5OH \rightarrow CH_3CHO + H_2 \tag{A10.1.1d}$$

is

$$\frac{W}{F} = \frac{-x - 2\ln(1-x)}{k_1 p/RT} \tag{A10.1.2}$$

Values of rate constant and gas density at the temperature are

$$k_1 = 3 \text{ L/g h} \tag{A10.1.3a}$$

and

$$\frac{p}{RT} = 0.96 \text{ g/L} \tag{A10.1.3b}$$

In this case the feed rate, F is 30.98 kg/h and feed composition is 100% ethanol. Therefore the weight of the catalyst W is given by

$$W = 30.98 \times \frac{-0.94 - 2\ln(1-0.94)}{3 \times 0.96} = 50.43 \text{ kg} \tag{A10.1.4}$$

To optimize for D one notes that

$$\text{Power requirement} = \Delta P \times V \tag{A10.1.5}$$

where V is the volumetric flow rate of the feed. From Ergun's equation,

$$\Delta P = \frac{150(1-\epsilon)\mu}{D_p G_0} + 1.75 \frac{G_0^2}{\rho} \cdot \frac{L}{D_p} \frac{(1-\epsilon)}{\epsilon^3} \tag{A10.1.6}$$

where μ is the gaseous viscosity, ϵ the bed porosity, ρ the gas density, G_0 the mass flow per unit area and D_p is the average particle size.

$$V = \frac{F}{\rho} = 30.98 \text{ kg/h} \times \frac{10^6}{0.96} \text{ cc/kg}$$

$$= 32.3 \times 10^6 \text{ cc/h} \tag{A10.1.7a}$$

$$G_0 = \frac{4F}{\pi D^2}, \quad L = \frac{4W}{\rho_B \pi D^2} \tag{A10.1.7b}$$

where ρ_B is the density of bed [= 0.347 g/cm³] found from reference 10. If the cost of power C_p = Rs. 0.25/kWh, $\epsilon = 0.885$, $\mu = 0.0165$ cp at 310°C, one has the following

$$\text{Cost of power} = C_p\left[\frac{150(1-\epsilon)\pi D^2 \mu}{D_p 4F} + 1.75\right] \times \frac{16F^2}{\pi^2 D^4 \rho} \frac{4W}{D_p \rho_B \pi D^2} \frac{(1-\epsilon)}{\epsilon^3} V \tag{A10.1.8a}$$

Substituting all the known values in cgs units, one gets the power cost as

$$\text{Power cost} = C_p\left[\frac{150 \times 0.115 \times 3.142 \times (0.0165 \times 0.01)D^2}{0.03315 \times 4 \times (30980/3600)} + 1.75\right] \times \frac{16 \times (30980/3600)^2}{3.142^2 \times 0.00096} \frac{1}{D^6} \frac{4 \times 50430}{0.03315 \times 0.347} \times \frac{0.115}{0.885^3} \times \frac{32.3 \times 10^6}{3600}$$

$$= C_p[0.0076D^2 + 1.75] \frac{1}{D^6} \times 1.0367 \times 10^{15} \text{ ergs/s}$$

$$= C_p[0.0076D^2 + 1.75] \frac{1}{D^6} \times 28.797 \text{ kWh/s}$$

$$= 0.25 \times 28.797\left[\frac{0.0076D^2 + 1.75}{D^6}\right] \text{Rs./s} \tag{A10.1.8b}$$

Fixed cost of the reactor is given by

$$\text{Cost of reactor} = \frac{DLt\rho' C_r}{m} \tag{A10.1.9}$$

where C_r is the cost of the reactor material and is Rs. 100/kg for stainless steel. Steel density ρ' is 7.75 g/cc. Assuming a linear depreciation, life of the reactor m is taken as 10 years and the thickness t as 1/4′; one has

$$\therefore \quad \text{Reactor cost} = \frac{3.142 \times D \times (20D) \times (0.25 \times 2.54) \times 7.75 \times \text{Rs. } 0.1}{10 \times 365 \times 24 \times 3600 \text{ s}}$$

$$= 1.19305 \times 10^{-7} D^2 \text{ Rs./s}$$

$$L = 20D \text{ (assumed)}$$

The total cost C per unit time (in Rs./s) is obtained by taking $L/D = 20$ above and adding pumping cost to it. Or,

$$C = \frac{0.0547}{D^4} + \frac{12.6}{D^6} + 1.19305 \times 10^{-7} \times D^2$$

To find the optimal dia, one sets $dC/dD = 0$

$$dC/dD = 0 = -4 \times 0.0545/D^5 - 6 \times 12.6/D^7 + 2 \times 1.19305 \times 10^{-7} D$$

Solving by trial and error one gets $D = 11$ cm

$$\therefore \quad L = 20D = 20 \times 11 \text{ cm} = 2.20 \text{ m}$$

Allowing for free space, one takes L as 2.5 m.

$$\therefore \quad \text{Residence time} = \frac{\pi(D^2/4) \times L}{V} = \frac{237.6 \text{ cm}^3}{8.97 \text{ lit/s}} = 0.0265 \text{ s.}$$

Results: Weight of catalyst = 50.43 kg
Length of reactor bed = 2.5 m
Dia. of reactor bed = 11 cm
Residence time = 0.0265 s

A10.1.2 Design of Distillation Column

A10.1.2.1 *Number of Trays*

Temperature at the top of the column = 40°C
Temperature at the bottom = 87.5°C
Total column pressure p_t = 1519 mm Hg

The geometric average of these two temperatures is 59.16°C or 60°C. Therefore calculations have been made for the average column temperature of 60°C. At this temperature, vapour pressures of acetaldehyde and ethanol are:

$p_A = 2811.2$ mm Hg (from Eq. 6.8.15)
$p_E = 351$ (from reference 8)

$$\therefore \quad \text{Relative volatility } \alpha = \frac{p_A}{p_E} = 8.01$$

Figure 10.18 gives the plot of y versus x and the Macabe-Thiele plot to

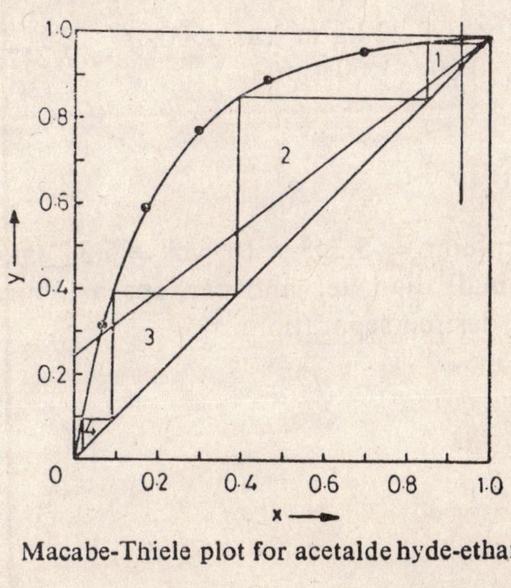

Fig. 10.18 Macabe-Thiele plot for acetaldehyde-ethanol system

obtain the number of equilibrium stages as 4. From reference 15 (Fig. 10.16, p. 151) one obtains

$$\text{Viscosity } \mu = 0.16\ c_p$$

$$\text{Fractional overall tray efficiency} = 0.47$$

$$\therefore \quad \text{Actual number of trays} = \frac{4}{0.47} = 8.5$$

or one would require *nine* actual trays.

A10.1.2.2 *Determination of Column Diameter*

$$\text{Average column temperature} = 60°\text{C}$$

$$\therefore \quad \text{Gas density } \rho_g = \frac{(44 \text{ g/g-mole}) \times (2.00 \text{ atm})}{\left(82.057 \dfrac{\text{atm-cm}^3}{\text{g-mole °K}}\right) \times (333°\text{K})}$$

$$= 3.22 \times 10^3 \text{ g/c.c.}$$

$$\text{Liquid density } \rho_L = 0.74 \text{ g/c.c.}$$

For $L/D = 3$, one finds

$$B = 139 \text{ kg/h} \qquad D = 25 \text{ kg/h}$$

$$L = 75 \text{ kg/h} \qquad L' = 101.39 \text{ kg/h}$$

$$V = 100 \text{ kg/h} \qquad V' = 100 \text{ kg/h}$$

From Chapter 3 for plate spacing of 60 cm

$$\text{Maximum vapour velocity } v_f = 1.603 \text{ m/s}$$

Assuming a safe operating velocity as 60% of this value, one gets the column cross-sectional area A as

$$A = \frac{(25 \text{ kg/h}) \times (1/3.22 \text{ kg/m}^3)(\text{hr}/3600 \text{ s})}{(0.60 \times 1.603 \text{ m/s})} = 2.2423 \times 10^{-3} \text{ m}^2$$

or

$$\frac{\pi d^2}{4} = 2.2423 \times 10^{-3} \text{ m}^2$$

or diameter of the column $= 5.343 \times 10^{-2}$ m $= 5.34$ cm.

This is rather a small diameter and perhaps a laboratory distillation unit would give the desired separation.

APPENDIX 10.2

Trouble-shooting Analysis of Acetaldehyde Plant

A10.2.1 General

For the trouble-shooting analysis the plant is divided into four sections, namely, storage tank ST and the vaporizer; the heat exchangers HX1 and HX2 and the reactor; the two condensers; the distillation column with its condenser and reboiler.

A10.2.2 Storage Tank and Vaporizer Section

A10.2.2.1 *Guide word: None*

Deviation — No flow to the vaporizer.

Possible causes

1. No material in ST.
2. Cavitation due to local heating causing back flow.
3. Flow controller F_1 fails.
4. Temperature controller of vaporizer fails and steam line is fully closed. No material is evaporated and no feed to the reactor section.
5. Valve P_1 fails.
6. Loss of material due to line fracture.

Consequences

1. Loss of feed to the reaction system.
2. Overheating of vaporizer due to lack of material.
3. Spillage of material in adjoining area in case of line fracture.

Action required

1. Ensure a rapid supply of raw material to the reactor section at short notice, if possible.
2. Keep sand bags at places near the pipeline to take care of sporadic fires. Provide proper drainage so that material does not accumulate in case any pipeline fractures. All vertical pipes should be supported, etc.
3. Start the bypass line manually in case of valve blockages.
4. Make sure that liquid-bleed system operates at a pressure higher than the pressure due to liquid level in ST_1.
5. Have flow indicators to detect failure of valves on the line. In case the valve fails, start the manual bypass line immediately.

A10.2.2.2 *Guide word: More of*

(a) Deviation

1. More flow to the vaporizer.
2. More steam flow.

Possible causes

1. Pressure in ST_1 can rise in two ways: first when the level indicator fails and the level in ST_1 rises above the desired level. Secondly if the breathing line on ST_1 is clogged.
2. More steam flow would occur when TC fails.

Consequences

1. Temperature of the outgoing stream from the vaporizer decreases in case of more flow to the vaporizer. High flow to reactor causes a lower residence time in the reactor giving a lower conversion.
2. Temperature of outlet process stream increases due to increase in steam flow which would cause an increase in ethanol vapour generated.

Action required

1. (a) Ensure that the relief line is working.
   (b) Inspect flow controller, TCR and manual bypass line valves at regular intervals.
2. Control the steam flow rate manually through the bypass line in case the automatic controller does not work.

(b) Deviation

More pressure in the vaporizer.

Possible causes

1. Breathing line and bleeding fails.
2. TCR fails.

Consequences

Overheating of product stream occurs in vaporizer. Load on vapour trap of steam side of vaporizer increases.
Change in flow to HX occurs due to change in pressure.

Action required

Provide a safety valve with emission to stack in case of excessive pressure. Cut-off, breathe and bleed temporarily, and operate flow controllers on both the sides of the vaporizer manually.

(c) Deviation

More temperature in the vaporizer.

Possible causes

1. TC's fail.
2. Valve on inlet of the vaporizer clog.
3. Low flow rate from ST_1.

Consequences

Overheating of line from vaporizer to HX occurs.
Pressure rises in vaporizer.

Action required

1. Measure and coordinate the temperature of the stream and flow rate of steam by controlling the latter's flow rate manually by a bypass line.
2. Control and operate manually, the valve of FCV (inlet to vaporizer).
3. Same as (2) and check the level of ethanol in the storage tank.

A10.2.2.3 *Guide word: Less of*

(a) Deviation

Less flow to vaporizer.

Possible causes

1. Leakage or partial clogging of line, because of any impurities (solid) in the feed.
2. Low flow from ST_1.
3. FRC fails.
4. Pump P_2 develops faults.

Consequences

Pressure in vaporizer increases.
There is low flow to HX1.
P_2 operates at low flow rate.
Temperature in vaporizer increases.

Action required

1. Inspect flow meters on the line terminals. Operate manually the bypass line in case of excessive pressure.
2. Communicate with the operator of the storage tank ST1.
3. Measure and control flow rate manually. If the flow to HX1 cannot be corrected, order plant shutdown.

(b) Deviation

Low pressure in vaporizer.

Possible causes

1. Breathing on the shell side of vaporizer fails.
2. Steam production from boiler house at low pressure occurs.

Consequences

There is low vapour production.
Load on liquid trap of vaporizer increases.

Action required

1. Install pressure indicator on vaporizer.
2. Communicate with the boiler house.

(c) Deviation

1. Valves on steam inlet clog.
2. FC fails giving more flow to vaporizer.
3. Steam inlet pressure decreases.

Consequences

Hammering at the outlet of vaporizer occurs

	due to mixed phase. Load on HX1 increases. Efficiency of vaporizer decreases.
Action required	1. Read the temperature of outlet stream with the help of TI control steam flow manually by bypass line. 2, 3. Already taken care of

A10.2.2.4 *Guide word: Part of*

Deviation	Change in composition of the stream due to impurities or auxiliary chemicals.
Possible causes	1. Leakage of steam (tube side) in the vaporizer leading to water as impurity. This can be detected by measuring composition after vaporizer. Composition can be measured by measuring refractive index. 2. Ethanol is not supplied as 100% pure.
Consequences	The stream to the reaction will not be pure ethanol and consequently will lower the final conversion. The load of steam in vaporizer will increase due to leakage on tube side; also in the shell side due to this, steam pressure will decrease, and the steam lost through leakage would show up in reduced ethanol concentration in the outgoing stream.
Action required	1. Once water is detected by composition measurement, check the quality of the raw material. If not found matching, call for plant shutdown to repair the leakage. 2. Continuous testing of the composition of ethanol feed, change the supplier of ethanol if feed is not pure.

A10.2.2.5 *Guide word: More than*

Deviation	Undesirable phase changes in the system; that is, entry of vapour in the pipeline between ST_1 and the vaporizer or output of liquid ethanol from the vaporizer.
Possible causes	1. Any liquid or vapour trap fails. 2. TCR or valve on steam inlet fails. Erosion and hammering in the vaporizer. Load on liquid trap at the outlet of vaporizer increases.

Load on HX1 increases.

Action required
1. Operate manual valves on the traps during the replacement of bad traps.
2. Already taken care of.

A10.2.2.6 *Guide word: Maintenance*

1. Periodic backflushing (cleaning) of the vaporizer.
2. Check the insulation pipelines.
3. Supervision of joints and valves to see if there is any leakage (minor leakages which cannot be detected by flow rates).
4. Check controllers by first isolating them by opening the bypass line. Then pass a known amount of material and measure the output of the controller and compare it with the set point. Do this for a range of set point and input feed rate.

A10.2.3 HX1, HX2 and Reactor Section

A10.2.3.1 *Guide word: None*

Deviation — No flow to the reactor.

Possible causes
1. No material flow from the vaporizer.
2. Valve on pipeline fails shut.
3. Line fracture or major leakage.
4. Pressure controller on the reactor fails, that is, it stops operating while in the closed position. It may stop due to lack of air supply or due to jamming or due to some circuit faults.
5. Release valves on any of the HX′ fail.

Consequences — Reactor tube side pressure decreases. Spillage of material due to line fracture. The loss of material to atmosphere occurs due to cause 5, increasing the danger of fire.
Nitrogen enters into the system due to nitrogen breathing.
Pump P_2 overheats.

Action required
1. Communicate with vaporizer section. Install a flow meter before HX1. Provide one man with instruction to shutdown steam in HX2 in case of leakage and to maintain the desired temperature at the reactor in case

TC fails. He should close down the outlet of reactor to prevent nitrogen from going into the operation.

2. Install an arrangement to stop pump P_2 quickly in case of leakage. Inspect the line to check valves. Flow indicators to be installed at places along the line.
3. Call for plant shut-off when pipelines fracture.
4. Place pressure indicator on reactor and flow indicator on the line and regularly inspect these.
5. Provide arrangement to dispose off the vapour safely to atmosphere. Install a failure alarm on the release valves.

A10.2.3.2 *Guide word: More of*

(a) Deviation — Flow controller on the reactor fails.

Possible cause — More flow to the reactor.

Consequences — Dowtherm pressure on the reactor and HX3 shell side increases as conversion falls. When more flow of material in to the reactor occurs the pressure in the reactor increases and the excess pressure is released through nitrogen breath-out line and hence some material is lost.

Action required

1. Start the bypass line (having the same function as the pressure controller valve, before the reactor) for reducing the flow rate. One man to keep an eye on the flow rate for manually controlling it temporarily.
2. Already taken care of

(b) Deviation — More pressure in the reactor.

Possible causes

1. Breathing fails.
2. Dowtherm pressure increases due to failure of TC.
3. Pressure controller fails.

Consequences — Inlet flow rate decreases because pressure controller after sensing the high pressure in the reactor decreases it. Outlet flow rate will not increase because flow controller is located after the reactor. Temperature of the reactor increases, as a result leading to unwanted side reactions.

Action required

1. Provide a safety valve with emission to stack in case of excessive pressure buildup. Install pressure indicator on line and flow controller on inlet as well as outlet.
2. Install shell side pressure indicator, operate manually the bypass line on the inlet of dowtherm. Bypass line has to be placed on every controller valve so that if the controller or controller valve fails, the operation may still be continued by opening the bypass line and controlling it manually. An arrangement may also exist to remove the controller from the running plant for repairing.

(c) Deviation — More temperature on the reactor.

Possible causes

1. Feed to HX1 at high temperature.
2. Flow controller on outlet of reactor fails, i.e. opens fully.
3. Low flow rate of feed to HX1.
4. TC on HX fails.
5. TC on dowtherm inlet fails.
6. Valves on heating fluid side fails.

Consequences

Conversion is no more optimal since temperature changes.
Pressure in the reactor increases.
Load on condenser increases.
Load on blower increases.

Action required

1. Place temperature indicator before HX1; communicate from HX to vaporizer.

2, 6. Check the flow rate at the outlet. Switch the bypass line for adjusting flow.

3. Check flow rate of inlet to HX1 and communicate with vaporizer.

4, 5. Check temperature on inlet of HX2 and perform manual operation of bypass line to control.

A10.2.3.3 *Guide word: Less of*

(a) Deviation — Low flow rate to the reactor.

Possible causes

1. Leakages and partial clogging of lines before the reactor.
2. Low flow from the vaporizer section.
3. High pressure in HX1 or HX2 leading to loss of material through relief valve.
4. Pressure control on reactor fails.

Consequences — Dowtherm pressure on shell side falls leading to fall in conversion.
Pressure in the reactor fails.
Nitrogen is breathed into the system.
Load on condenser increases.

Action required
1. Install flow indicator before and after every long pipeline section. Measure flow rate at the reactor inlet.
2. Communicate with vaporizer section. Measure flow rate at the inlet of HX1.

3, 4. Check flow rate and temperature of reactor outlet.
Check TC and steam pressure in HX2.
Check flow rate and switch the bypass line in case of problems.

Action required — Place pressure indicator and flow controller on both the sides.

(b) Deviation — Low temperatures in the reactors.

Possible causes
1. Feed to HX1 at low temperature.
2. TC on HX2 fails.
3. TC on dowtherm inlet fails.

Consequences — Non-optimal conversion. Pressure in the reactor falls. Load on the blower increases. Load on the condenser increases.

Action required — Already taken care of in high temperature consideration.

A10.2.3.4 *Guide word: Part of*

Deviation — Change in composition of the stream due to impurities and auxiliary chemicals, the impurities can be in the ethanol itself due to insufficient separation from its secondary products.

Possible causes
1. Composition from vaporiser is changed.
2. Leakage in heating fluids in HX1 or in the reactor.
3. Metallic impurities due to pipe corrosion (ordinarily negligible).
4. Nitrogen breath-in into reactor.
5. Fluctuations in conversion.

Consequences — Low conversion.
Leak of steam or dowtherm into the reactor will adversely affect the catalyst. Reduction in porosity of the catalyst, change of composi-

tion in stream sent to the condenser.

Action required
1. Sampling is required at different places like, before HX1, before HX2, and after the reactor, to detect the location of leakage of auxiliary chemicals.
2. Arrange for plant shutdown if leak detected.

3, 4, 5. No additional action required.

A10.2.3.5 *Guide word: More than*

Deviation
Since 'more than' involves a phase change which cannot occur in this section unless liquid phase comes from the vaporizer, a possibility which has already been considered in the vaporizer section.

A10.2.3.6 *Guide word: Maintenance*

1. Switching of reactors for catalyst regeneration is required. This is carried out after 24 hours of operation; the regeneration takes one hour. This would mean that one should have two reactor system.
2. At times, cleaning or backflushing of heat exchanger is required.
3. Regular pipeline checking for leaks, corrosion and insulation on pipelines.
   See if lagging is intact.
4. Regular checking of measurement instrument and controller.
5. After regular intervals, check release valves and breathing facilities.
6. Change of catalyst after its life time and checking of the reactor.
7. Checking of valves and major joints.

A10.2.4 The Condenser Section

A10.2.4.1 *Guide word: None*

Deviation
No flow in either condenser.
1. No flow from the reactor section.
2. Valves in the path fail.
3. Line fracture.
4. High pressures in C1 or C2 due to failure of breath lines.
5. Pressure controller at C1 output fails.
6. Blower between C1 and C2 stops.

Consequences — Nitrogen is breathed into the condensers. No material would flow to the distillation section. Ammonia pressure in C2 increases. Pumps P_3, P_4 overheat, spillage of cooling fluids.

Action required
1. Communicate with the reactor section. Install flow meter at the inlet.
2. Arrange to stop P_3, P_4. Install flow indicators at places along the line.
3. Monitor leaks through flow indicators.
4. Install pressure indicator on Cl and C2. Provide arrangement to remove the faulty bleed and breath line from the main system while the plant is running.
5. Operate manually the bypass in case of trouble.
6. Check the operation of blower regularly.

A10.2.4.2 *Guide word: More of*

(a) Deviation — More flow in Cl and C2.

Possible causes
1. More flow from the reactor section.
2. Insufficient condensation in Cl and C2 due to failure of TC's.

Consequences — Load on condenser increases.
Pressure in condenser increases.
Material lost in breathing and bleeding.
Increased loss of material to stack.

Action required
1. Communicate with reactor section. Place pressure indicators on condensers.
2. Check the bleed valves and breathers. Check the temperature and flow rates of the cooling fluids. Install an operator at P_5, P_6 and manual bypass at TC's. Arrange also for safe disposal of bleeds.

(b) Deviation — More pressure in Cl and C2.

Possible causes — Breathing and bleeding fails.

Consequences — Poorer condensation.
Liquid level in condenser rises.
More flow of material to C2 and to stack. Load on P_5, P_6 increases.
Inlet flow rate decreases.

Action required
1. Provide a safety valve with emission to stack.
2. Provide arrangement to cut-off the breather and bleed temporarily and continue the operation.

(c) Deviation — More temperature in C1 and C2.

1. TC's fail.
2. Input from reactor section at high temperature.
3. Valves on cooling fluid side fail.

Consequences — Loads on P_5, P_6 increase.
Poor condensation.
Loss of material to stack and to bleeds.
P_3, P_4 overheat due to higher load of flow rate.
Low flow to distillation section.
Low quality of material to distillation section.

Action required

1. Measure and correct the temperature of C1 and C2. Operate manually the bypass line for flow of cooling fluid while TC is repaired.
2. Maintain communication with the reactor section. Arrange safe disposal of bleed from the condensers.

A10.2.4.3 *Guide word: Less of*

(a) Deviation — Less flow to C1 and C2.

Possible causes

1. Leakage and partial clogging of line.
2. Low flow from reactor section.
3. High pressure in C1 and C2 due to failure of bleeding and breathing.
4. High condensation efficiency in C1 and C2 due to failure of TC's.
5. Blower between C1 and C2 fails.

Consequences — Nitrogen is breathed in reducing the vapour composition. Low flow to distillation section. P_3, P_4, P_5, P_6 operate at small flow rates leading to overheating. Composition to distillation section is affected. High flow through flare.

Action required

1. Measure flow rates before C1 and C2.
2. Communicate with reactor and distillation sections.

3, 4, 5. Already taken care of.

(b) Deviation — Low pressure in C1 and C2.

Possible causes — Breathing fails.

Consequences — Poor condensation.
Low quality, low rate of output to the distillation section.
Load on P_5, P_6 increases.
Load on P_3, P_4 decreases.

	Load on blower between C1 and C2 increases. No material flow to flare.
Action required	Install pressure indication on C1, C2. Rest is already taken care of.
(c) Deviation	Low temperature in C1 and C2.
Possible causes	1. Input from reactor at low temperature. 2. TC's fail. 3. Valves on cooling fluid side fail.
Consequences	Load on P_5, P_6 decreases. More condensation. Loss of material through bleeds. Quality to distillation column improves. Flow to C2 and to flare decreases.
Action required	1, 2, 3. Already taken care of.

A10.2.4.4 *Guide word: Part of*

Deviation	Change in composition of streams (due to impurities, auxiliary chemicals and improper condensations).
Possible causes	1. Composition from the reactor is changed. 2. Leaking of cooling fluids. 3. Nitrogen is breathed in. 4. Fluctuations in pressure, temperature, flow rate. Conditions of condensers due to TC failure, etc.
Consequences	Load on condensers changes and their efficiency reduces. Change in composition and temperature of stream to distillation section. Flow rate to distillation column fluctuates. Nitrogen can be breathed in to change the composition further. Material loss from breathe and bleed.
Action required	1. Sample continuously at various points to detect changes in composition. Communicate with reactor section. 2. Provide arrangement to bypass the condensors when there is a leak and have only one condenser operating in case of emergency. 3, 4. No additional action.

A10.2.4.5 *Guide word: More than*

Deviation	Undesirable phase change in the system.

Possible causes	1. Any liquid or vapour trap fails.
Consequences	Bubbling, cavitation in pump P_3, P_4, P_5, P_6. Erosion in blower. Mixed phase stream to distillation section. Liquid reaches flare.
Action required	1. Install an operator at pumps to stop them in case of cavitation. 2. Communicate from pump house with condenser operators. Provide arrangement for repair of the seals and traps in operation or for fast replacement of these; have manual valves on both sides of traps and seals for easy replacement.

A10.2.4.6 *Guide word: Maintenance*

1. Cleaning of condensers periodically.
2. Regular pipeline checking and checking of measurement and controller instruments.
3. Checking up of release valves and breathing facilities at intervals.
4. Periodic maintenance of pumps.
5. Checking of valves and major flanges.

A10.2.5 Distillation Column, Condenser, Reboiler and Preheater

A10.2.5.1 *Guide word: None*

Deviation	No flow to the distillation column, DC.
Possible causes	1. No material flow from the previous section. 2. Valves on the way fail. 3. Line fractures. 4. Flow controller on the line fails. 5. Pressure in DC increases.
Consequences	Temperature inside DC decreases as liquid feed stops; the liquid flow rate in the column will gradually decrease and the vapour flow rate remains the same due to the TCR and FC on the reboiler. This tends to equalise the temperature of the successive plates. No material flow through preheater. Spillage of material due to line fracture. Nitrogen comes in the preheater due to breathing. Poorer separation due to flooding.

	Overchurning of pumps P_3, P_4 resulting in overheating. No products to ST_2.
Action required	1. Communicate with condenser section; install a flow indicator on the line to preheater. 2. Place flow indicators before each equipment to detect leakages. 3. Install flow indicators on the reboiler and have their inspection. Manual operation of the bypass should be possible in case of emergency.

A10.2.5.2 *Guide word: More of*

(a) Deviation	More flow to DC.
Possible causes	1. Flow controller on the line fails open. 2. More flow from the condenser section.
Consequences	Load on preheater reboiler and the condenser increases. Product loss from the bottom. There will be weeping, consequently poorer separation. The average temperature of column decreases.
Action required	1. Provide manually operated flow lines. 2. Maintain communication between condenser and DC.
(b) Deviation	More pressure in DC, condenser, reboiler and preheater.
Possible causes	1. Breathing and bleeding fails. 2. TCR fails on condenser.
Consequences	Vibration in the preheater, condenser and reboiler. Poorer separation. More steam requirement in the reboiler. Product loss through bottom of DC.
Action required	1. Communicate with the preheater and the condenser. 2. Control manually temperature and flow rate of all auxiliary streams. 3. Manipulate vapour reflux manually. 4. Start the standby reflux pump.

A10.2.5.3 *Guide word: Less of*

(a) Deviation	Less flow to DC, preheater, reboiler, condenser.

Possible causes	1. Leakage and partial clogging of line. 2. Less flow from the condenser section. 3. Failure of FRC on line or FC of reflux. 4. TC on reboiler, condenser fail.
Consequences	Less product in ST_2. More pressure in condenser, reboiler. P_3, P_4 operate at low flow rates. Composition in DC is affected. Load on condenser increases and on reboiler decreases.
Action required	1. Measure of flow rate on line before the preheater and before each other equipment. 2. Communicate with condenser section. 3. Already taken care of. 4. Control flow of water in the condenser and steam in the reboiler manually.
(b) Deviation	Less pressure on DC reboiler, condenser and preheater.
Possible causes	1. Breathing fails. 2. TCR of condenser fails.
Consequences	More cooling water required in condenser, i.e. load on condenser increases. Also load in reboiler increases.
Action required	1. Place pressure indicators on all equipment. 2. Control manually water flow rate in condenser.
(c) Deviation	Low temperature in DC section. 1. Liquid reflux controller fails, i.e. opens fully. 2. TC's fail. 3. FC on reboiler outlet fails, i.e. closes partially.
Consequences	Low quality product due to less separation. More load on preheater and reboiler.
Action required	1. Manually operate the liquid reflux. 2, 3. Already taken care of.

A10.2.5.4 *Guide word: Part of*

Deviation	Change in composition of the main stream.
Possible causes	1. Composition controller fails. 2. Leak in auxiliary fluids. 3. Fluctuations in the conditions of operation of equipment.

Consequences — Product quality declines. Material loss in bleed and breathing.

Action required
1. Carry out frequent sampling and testing of the product from the line and maintain communication between DC and condenser section.
2. Provide arrangement to bypass the preheaters or have another standby unit.
3. No additional action.

A10.2.5.5 *Guide word: More than*

Deviation — Undesirable phase changes in the sub-system.

Possible causes
1. Any liquid or vapour trap fails.
2. Conditions of the unit change drastically.

Consequences — Bubbling, cavitation in pumps, poorer separation, acoustic vibration in the preheater, etc.

Action required
1. Place a man at the pump house to stop the pump.
   Communicate with the condenser section and with the pump house.
   Provide arrangement for replacement of traps during operation.
   Have a system to isolate any of the traps in case of replacement.
2. No additional action.

A10.2.5.6 *Guide word: Maintenance*

1. Cleaning of preheater, condenser and reboiler periodically.
2. Regular checking of pipelines, regulator and controllers.
3. Periodic oiling of pumps.
4. Checking the working of valves and joints.
5. Regular checking of the plates and monitoring of leakages of the distillation column.

Acknowledgements

M. K. Agarwal, R. K. Modi and R. M. Khemka of the 1978 graduating batch carried out the calculations of Sections 10.10 and Appendix 10.1 and developed the trouble-shooting analysis of Appendix 10.2; S. Gopala Rao, K. Jhalaria, J. L. Prasad and N. B. Shah developed the procedure of Section 10.11 in a term project as a part of the design course at the Indian Institute of Technology, Kanpur.

REFERENCES

1. Rudd, D. F. and Watson, C. C., *Strategy of Process Engineering*, 1st edn., Wiley, N.Y., 1968.
2. Freund, R. J., *Econometrica*, **24**, 253, 1956.
3. Jelen, F. C., *Cost and Optimization Engineering*, 1st edn, McGraw-Hill, 1970.
4. Peters, M. S. and Timmerhaus, K. D., *Plant Design and Economics for Chemical Engineers*, 2nd edn, McGraw-Hill, N.Y., 1968.
5. Miller, S. A., *Chemical Proc. Engg.*, **49**(3), 75, 1968.
6. Church, J. M. and Joshi, H. K., *Ind. Eng. Chem.*, **43**, 1804, 1951.
7. Franckaerts, J. and Froment, G. F., *Chem. Eng. Sci.*, **19**, 807, 1964.
8. Perry, J. H. (ed.), *Chemical Engineer's Handbook*, 5th edn, McGraw-Hill, N.Y., 1972.
9. Washburn, E. W. (ed.), *International Critical Tables of Numerical Data in Physics, Chemistry and Technology*, Vol. 3, p. 313, McGraw-Hill, N.Y., 1928.
10. Venugopal, B., Kumar, R. and Kuloor, N. R., *Ind. J. Technol.*, **4**(7), 205, 1966.
11. Kharbanda, O. P., *Process Plant and Equipment Cost Estimation*, 1st edn, Sevak Publication, Bombay, 1977.
12. Kern, D. Q., *Process Heat Transfer*, 1st edn, McGraw-Hill, N.Y., 1950.
13. Garcia Borras, T., *Hydrocarbon Process*, 171, January 1977.
14. Gupta, S. K., *Momentum Transfer Operations*, Tata McGraw-Hill, New Delhi, 1978.
15. Treyball, R. E., *Mass Transfer Operations*, McGraw-Hill, N.Y., 1955.
16. Weston, J. F. and Brigham, E. F., *Managerial Finance*, 3rd edn, Holt, Reinhart and Winston, New York, 1969.
17. Haley, C. W. and Schall, L. D., *The Theory of Financial Decision*, 1st edn, McGraw-Hill, N.Y., 1973.
18. Samuelson, P. A., *Economics*, 8th edn, McGraw-Hill, N.Y., 1970.
19. Weisman, J. and Holtzman, A., *I. & E. C. Proc. Des. Dev.*, **11**, 386, 1972.

PROBLEMS

10.1 150 g. of S toothpaste (Family size) in the open market sells for Rs. 7. Since the excise duty on toothpaste is 20% of the ex-factory price and the government gives excise tax rebate to small-scale entrepreneurs, it appears to be an attractive proposition for the latter.

An enquiry from the market has revealed that each empty tube costs Re. 1. Each of the big manufacturers has stockists (who charge about 10% of the retail price) before the product goes to the retailer (who charges 15% of the retail price), and spends about Rs. 8 per kg of the material on advertisement.

The chemical analysis of toothpaste shows that it has following ingredients in it:

	Amount	Cost per kg
Carboxy methyl cellulose (as a binder)	59 gm	Rs. 20
Water	590 gm	
Calcium carbonate (200 mesh)	289 gm	Rs. 10
Detergent (for froth formation on brushing)	59 gm	Rs. 20
Peppermint (for the mint tastes)	3.0 gm	Rs. 70

As you know from your unit operation experiments, CMC dissolves in water with difficulty and takes about one hour with constant stirring. One dissolves detergent, CMC and peppermint before hand and then adds the powder which must be homogenized to obtain the final product.

(1) Suggest a suitable unit operation of homogenizing powder with solution.

(2) How much profit margin exists for large manufacturing units ?

(3) Suggest a suitable flow sheet for a small-scale plant you desire to open up.

(4) Estimate the minimum labour force you would require for your plant after carefully defining the work task.

(5) Calculate the break-even point.

(6) If you sell at the same price as it is available in the market, how much profit would you make if your plant capacity is five times the capacity at the break-even point ?

N.B. (i) Cost of the mixer $= 2400 \times Q^{2/3}$

(ii) Utility charges per month $= 2000Q$

(7) Recalculate the break-even point using nomographs of Chapter 9.

10.2 Dr. Kunzru has been working on the purification of asafoetida (*heeng*). Raw asafoetida consists of mud and impurities and to remove it, one has to dissolve 10% solution by weight in water. The solution is decanted to leave mud behind.

Now it remains to recover *heeng* from this solution. One engineer has suggested that one could boil the water off, but experiments in the laboratory reveal that *heeng* starts colouring beyond 50°C. Hence this method cannot be used at all.

Consequently lowering of vapour pressure by vacuum as well as bubbling of air must be evaluated. From economic considerations, the latter would be preferred as vacuum systems normally are very expensive. However, if one measures the viscosity as a function of *heeng* concentration, air bubbles stop rising and remain trapped in the 90% solution. Hence one decides to remove the bulk of water by bubbling air and then applying vacuum over it to remove the remaining water. Assume that *heeng* is totally inert and cannot be vaporized.

(1) Write down the detailed flow sheet. Write down all process conditions suitable for Kanpur's climatic conditions.

(2) Assuming that raw *heeng* sells for Rs. 60 and Rs. 5/kg are the transportation charges, find the break-even point. A small air compressor costs Rs. 5000 and a vacuum pump Rs. 50,000. Purified *heeng* costs Rs. 100/kg.

(3) Estimate the labour force for a 100 kg per day plant.

(4) What air flow would you choose and how long would it take to dry up ? What vacuum would you apply ?

10.3 Calculate the break-even point of the acetaldehyde plant using Indian cost data of Table 9.16. Why does this value differ from that calculated in the text (if at all) ?

10.4 Modify Eq. 10.9.8 if it takes three years to install the plant. Build in your modified equation the time value of money due to inflation as well as the bank interest rate.

10.5 Calculate the profit per year if the acetaldehyde plant capacity is 100 kg/h.

10.6 How does the break-even point of the acetaldehyde plant calculated in the text change if the depreciation is carried out using the double-declining method in 10 years ?

10.7 For the 100 kg/h plant suggest a plant layout. For doing so, take a graph paper and show all the units. It may be pointed out here that very few works exist on this subject, and in practice it is done by building an actual three-dimensional model (with proper reduction of scale).

10.8 After suggesting a suitable plant layout in P10.7 carry out the detailed labour force analysis (as done in Chapter 9). Define your work task carefully.

10.9 In the text, discounting has been done discretely at the end of every year. Supposing it is done after every nth interval at the interest rate of i per year. Modify the equation for discounting and then develop corresponding results for unacost and capitalized cost.

10.10 Supposing the discounting is done continuously. Derive the result by letting $n \to \infty$ in the result of P10.9. Then determine the expression for unacost.

10.11 There are several cases where one needs to maintain stock so that at any moment one does not run short of the material. As the size of the stock maintained is made larger and larger, the chances of any loss of revenue due to non-availability of material is considerably reduced. However, the storage charges go up correspondingly. Consequently there is an optimal stock level and this is called the inventory problem. The simplest model of this complex situation is the following.

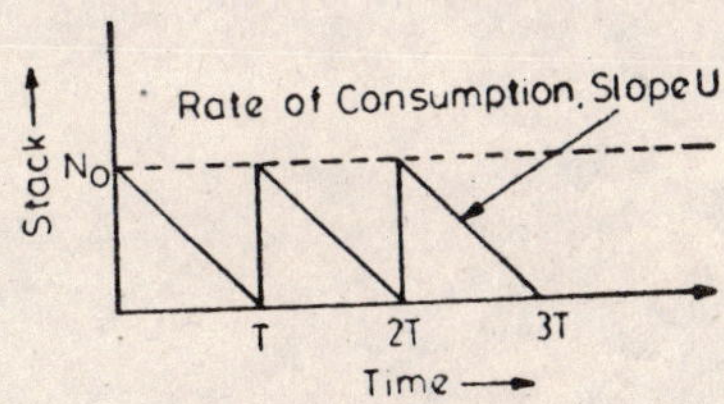

Fig. P10.11

In Fig. P.10.11, N_0 is the optimal stock level, U is the consumption rate (assumed constant). Write an expression for the total cost, C_t per unit time in terms of the following:

C_o Cost of placing and receiving order

C_h Storage or holding cost per unit stock per period.

10.12 In P10.11, the stock at one point goes to zero. Supposing it is decided not to allow the stock below a safe level N_S, what would be the optimal value N_0 in this case ?

10.13 A queuing problem arises whenever there is any kind of servicing involved. A common example would be a service station where cars are waiting to get petrol. If the rates of input of cars are known and time of servicing a single car is known, one can easily find out how many service stations should be installed. The simplest model of this problem can be made after defining the following terms:

T = time, periods

T_a = time between two consecutive arrivals

T_s = time to service one unit

C_w = waiting cost per unit per unit time

C_f = cost of servicing a single unit

If C_t is the total cost per period, write an expression for it in terms of costs defined above. Then find the optimum servicing time for a given T_a.

10.14 Literature consists of sufficient information on manufacturing styrene from ethylbenzene. Do carry out the literature search and then perform the break-even analysis. Subsequently, use the risk analysis discussed to find the optimal capacity for the equity as 25%. Design your plant for this size. Compare the (Q_{OPT}/Q_{BE}) for this with that for the acetaldehyde plant for the same risk factor c^*.

10.15 Carry out the plant design for manufacturing diethyl ether from ethanol after finding out the relevant information from literature. Calculate the plant capacity for 35% equity.

10.16 Carry out the plant design for formaldehyde from methanol. Calculate the plant capacity for 50% equity.

Index